Edited by Peter K. Machamer and Robert G. Turnbull

MOTION AND TIME
SPACE AND MATTER

Interrelations in the History of Philosophy and Science

OHIO STATE UNIVERSITY PRESS

147484

Library of Congress Cataloguing in Publication Data

Main entry under title:
Motion and time, space and matter.

Includes index.
1. Space and time. 2. Science—History. 3. Science—Philosophy. 4. Motion. I. Machamer,
Peter K. II. Turnbull, Robert G.
QC173.59.S65M68   530'.09   75-26517
ISBN 0-8142-0207-1

TO ZENO
who said it couldn't be done

CONTENTS

This volume of essays is the tangible result of a National Endowment for the Humanities curriculum development grant for the academic year 1972-73. The program director was Robert G. Turnbull and the assistant director was Peter Machamer, both of the Department of Philosophy, Ohio State University. The Ohio State University faculty whose essays appear here had primary responsibility for the teaching of a three-quarter course sequence that dealt with the relations between the history of philosophy and the history of science. The other contributors served as consultants for the program. In this capacity they originally visited the O.S.U. campus in pairs to talk to the class on the subject and texts being dealt with at the time of their visit. All the consultants returned to Columbus in June 1973 for a week-long conference. At that conference both they and the relevant Ohio State faculty read versions of the papers that appear in this volume. The success of the June conference was heightened by the attendance of some twenty-five additional scholars, who were specifically invited.

The quality and interest of the papers printed here testify to the success of the conference and to the general curriculum programs of which the conference was part. It is the hope of the editors, shared by most of the contributors, that similar courses stressing the interrelations between philosophy and science in various historical periods will become common in many universities. Likewise it is hoped that historians of philosophy and of science will begin, or will continue, such work of overlapping concern. As the many shared and constantly recurring themes in these papers exhibit, there exists in this area a fascinating subject matter. Academic isolationism of scholars sheltered in different departments can yield, at best, partial insights, at worst, complete misunderstandings.

As mentioned above, many themes keep recurring. In large part the extensive index is meant to serve as a guideline for those wishing to form an overview of the topics here presented. We wish we had the expertise and ability to somehow succinctly cast the issues raised in these papers into an intelligent and intelligible overview. Such is not the case. The alternative is to let the papers speak for themselves and to let the reader forge his own view of the progress, continuity, or revolutions that provide the subject matter for the papers collected here. The subjects of all of the essays should be reasonably clear from their titles. From the title of the volume it should be clear that all the essays deal with inter-face problems in the areas of science and philosophy. Perhaps the general contention emanating from these essays is that nowhere in the history of thought was it the case that science and philosophy were neatly separated and isolated into their own mutually exclusive disciplines. This is not, nor should it be, surprising. From ancient times through modern, the aim of both philosophy and science has been understanding or wisdom. The regrettable fact has been, in recent years, a widening of the gap between academically categorized disciplines, resulting in little interchange between scholars whose specialties overlap. This overlap is the area of concentration in these essays. When Owen's essay picks up the subject of time in Aristotle's *Physics,* he sets a few problems that are treated in the essays of Kretzmann, Koslow, McGuire, Hankins, and Salmon. When Hahm begins to talk of the character of matter in Aristotle and its essential qualities, this again trails through history and is found in the essays on the Atomists, Descartes, Leibniz, and Mach. The influence of mathematics and its relations to metaphysical and epistemological questions occurs in the essays dealing with time, and comes out even more explicitly in the papers by Anderson, Wilson, Hankins, and Hausman. Methodological issues concerning science and knowledge, in general, including the role of mathematics, are treated prominently in the essays of Turnbull, Machamer, Laudan, and Schaffner. Again space plays a prominent role in many of the papers primarily dealing with time and matter and an essential role in the essays by Furley, Grant, Koslow, Anderson, and the relativity theory essays.

The interrelations between scientific and philosophical concepts as explained in these essays are multifarious. Indeed, it is our opinion that in most crucial instances it is impossible to tell where the influence of philosophy on science, or the converse, leaves off and the other discipline begins. In part this is exemplified in the essays through the constantly recurring themes in varying contexts. Over time both philosophy and science underwent many changes, but still the perennial problems arose, in new terms, with new trappings, but still unmistakably the same. The details and complexities of this development are deserving of study. Change is the subject of the studies here and, as a result, the study of the subject of change is the focal point of this collection. The outcome should have bearing upon the subsequent studies in the histories of philosophy and science (insofar as those can be considered distinct), on the philosophy of science conceived historically (and how else could it reasonably be conceived?), on the intellectual history of various periods, and, in general, as adding to the sum of our knowledge and understanding about the ways in which we and our predecessors attempted to understand the world and the men in it.

The editors would like to thank the National Endowment for the Humanities. It was their grant, Curriculum Development grant number 350881, that made possible an exciting and innovative course, and that provided funds to bring together the group of scholars who contributed to this volume. Further, the National Endowment is to be thanked for continuing our grant for another year, so that we may look forward to another course, conference, and volume of essays. The topic for 1973-74 is ''Perception.'' In this regard we must also thank the administration of the Ohio State University for their moral and monetary support, and the chairmen and faculty from other departments who supported our grant application.

We are most grateful to our consultants, who by their work and enthusiasm made the course and conference delightful in all ways. To our wives and those of our colleagues, who helped plan and host the many parties, we are grateful. Thanks also to the graduate students who assisted during the conference. We must pay tribute to the staff of the O.S.U. Philosophy Department: Virginia Foster, Regina Cox, Beatrice Weaver, the work-study students, and especially to our executive secretary, Elizabeth Hellinger. Through the year and during the conference these people were primarily responsible for smooth operation that resulted. Stephen Lunsford receives thanks for his assistance in preparing the index for this volume. Wesley Salmon won the contest for the best dedication, and, thus, took on that responsibility. We thank all who helped to make the year so worthwhile and enjoyable.

Peter Machamer

Robert Turnbull

MOTION AND TIME, SPACE AND MATTER

G. E. L. Owen

Chapter One

ARISTOTLE ON TIME

Aristotle's discussions of time in the *Physics,* principally in IV.10-14, remain a classic source of argument for philosophers and historians of science. His problems are often ours, and his solutions often enlarge our view of the problems. Through Augustine his arguments continued to exercise later philosophers such as Wittgenstein; through other post-Aristotelian schools and commentators they shaped the debates on the subject in pre-Newtonian science. Aristotle's chief concerns are the nature of the present, the reality of time, and the relations between time, space, and motion that he evidently takes to show time both real and measurable.[1] So these are the issues in his argument on which I shall concentrate.

The issues are directly linked, but there is a methodological interest that may also help to draw the discussion together. I can best introduce this by recalling, briefly and with apology, some suggestions made elsewhere.[2] It is a commonplace that Aristotle is alert to philosophically important expressions that have more than one sense or kind of use, and critical of philosophers who innocently build on these without noticing the ambiguities they contain. The fifth book of the *Metaphysics* is given up to exploring thirty such sources of ambiguity. But in a more positive mood, when he is pressing his own analysis of some key-concept, there is often a different impulse in his treatment of such expressions, and it does not readily mesh with the first. He is apt now to start his explanation from some particular, favored use of the expression and deal in various ways with other sorts of use that do not seem to fit the primary case though they bear some relation to it. (This notion of primacy rightly exercised him;[3] we shall have examples to discuss later.) One way of dealing with such recalcitrant uses is to argue them away; another is to extend and often to weaken his account of the basic occasion of use to accommodate them; a third, and perhaps the only one to deserve the name of a method, is to import a device or family of devices, sometimes now discussed under the rubric "focal meaning", which enables him to marry his interest in the paradigm case with his interest in the plurality of an expression's uses by explaining other kinds of use in terms of one that he represents as primary.

The second of these ways does not concern us here: it can be illustrated from many texts such as the analyses of change and finality in the first two books of the *Physics*.[4] The others play substantial parts in Aristotle's analyses of time and motion; and I shall suggest that in fact the first is used where the third would be appropriate, and the third is used where the question of a primary use cannot properly arise at all. But such cavils on method are subordinate to understanding the argument. This is not primarily a study of method.

MOTION AT A MOMENT

There is an important but still debated case where Aristotle seems so preoccupied with paradigm conditions of use that he rejects other uses of an expression—or, in this instance, of a family of expressions, the verbs of motion and rest. I mean his denial that there can be motion or rest at a moment, an unextended point of time.

(Some translators prefer "instant" to "moment"; that does not matter provided it is understood that Aristotle is talking of temporal points with no duration. What matters more is that the same Greek expression is the word for "now" or "the present"; but that connection, and the reason for it, will not concern us yet.)

Aristotle reasonably argues that any movement must be taken to cover some distance in some period of time; for movements differ in speed, and speed is a function of extended times and distances, one measured against the other. And an extended time is one in which different moments can be distinguished and ordered as earlier and later (*Phys.* 222b30–23a15). Not, of course, that he will allow a period to be regarded as merely a class or collection of the moments by which we can mark off subdivisions in it, any more than he will have a line built of points (231b6-10). Start, as Aristotle does, with the thought that points are produced by divisions of a magnitude, and it comes to seem evident

4

that however far such divisions proceed they will at any stage be finite in number and the points produced by them finite and separated by stretches that are the real factors of the magnitude. Still, if motion and velocity require some period of time within which different moments can be distinguished, one conclusion seemed to follow: there can be no talk of something moving, or therefore moving at any velocity, at one moment (234a24–31, recalled at 237a14, 239b1–2, 241a24–26). Nor yet can a body be said to be stationary at such a moment, as Zeno had argued;[5] for if movement entails being at different places at different moments, rest surely entails being at the same place at different moments. If we consider only one moment, neither description applies (239a26-62).

Now the Greeks like ourselves could speak of a body as moving, and moving relatively fast or slowly, just when it reached a winning-post or overtook another body; and this, we may surely object, is to speak of movement as going on at a moment. Aristotle himself rejects Zeno's paradox that a slow runner pursued by a fast one "will never be overtaken when running" (239b15–16, cf. 26–29). Our talk of a body as moving at a moment is common usage, preceding any mathematical theory of limits designed to accommodate it. No doubt it is in an important sense parasitic on descriptions of the body as moving over distances through periods of time, for our assessments of speed begin with these. To consider a body as moving with some velocity at a moment is to consider it as moving at some over-all speed or speeds during periods within which the moment falls.[6] We can allow too that "moving" has a different sense or use in the two contexts; for we can ask how much ground the body moved over in a period, and we cannot ask this of its performance at a moment. But between the two uses there are familiar translation rules, rougher in ordinary discourse, sharpened in post-Aristotelian theories of mechanics. When Aristotle rejects the derivative use of such verbs, he is rightly impressed, but unluckily over-impressed, by the requirements of the paradigm case in which motion takes so much time to cover so much ground. Hintikka has found another instance of the same tendency in Aristotle's unwillingness "to speak of a possibility at an instant".[7] His rejection of such ways of talking of motion and velocity, force and possibility at a moment, seems to have contributed to the final sterility of his mechanics.[8] Yet the author of the *Mechanica,* once ascribed to Aristotle and probably written not

5

much later,[9] can resolve circular motion into two rectilinear components, tangential and centripetal (848b1–49b19), and make the remarkable suggestion that the relation between these components must be instantaneous and not maintained for any time at all; otherwise, so long as it was maintained, the resultant motion would be not circular but in a straight line (848b26–27). This and other constructions in the same author seem to be ruled out for Aristotle.[10]

But this objection to Aristotle itself faces objections. In pressing on him one difference in the senses or uses of verbs of motion, we may have given too little weight to another. For there are two senses in which we might want to deny motion at a moment: we might want to deny that at any moment a body can *move,* carry out a journey from A to B; or we might want to deny that at any moment a body can *be moving* from A to B. In English the senses are divided between the non-continuous (perfective) and continuous (imperfective) forms of the verb; both are covered by the Greek present tense, though in past forms the difference can be marked. Now there is good sense in saying that a body cannot move at a moment, for this is only to say that it takes time to make a journey. It would be a mistake to say that at a moment a body cannot be moving, in transit between A and B. But surely, it may be protested, it is the first point that Aristotle is making. And considerations can be found to strengthen the protest.

First, the phrase I have rendered by "at a moment" is more literally translated "in a moment". That suggests that Aristotle is making the legitimate point: a moment is not the sort of time within which a journey can be carried out. On this view, he has grasped a mistake in Zeno's paradox of the Flying Arrow, the paradox he introduces immediately after denying motion or rest at a moment (*Phys.* 239a20–39b9). In Aristotle's version of the argument—and we have none earlier—Zeno had held that at any moment in its path a flying arrow cannot cover more than its own length and hence (since, presumably, there is no room for movement then) is stationary. And what is true of it at each moment of its flight is true of it throughout, so the arrow is stationary throughout its flight. The second step of the argument, with its inference from the moments to the period, Aristotle attacks by saying that a period is not composed of moments (239b8–9, 30–33);[12] we gave some sense to this earlier, and we shall meet it again. The first step he rejects by arguing that when there is no time for movement there is equally no time to be stationary (239a35–b4). In both replies he can be interpreted as locating Zeno's mistake in treating moments as cripplingly short periods of time, too small to move in and only sufficient for staying in one place. He corrects it, on this reading, by arguing that what cannot move *in* a moment cannot stand still in a moment either. *Move,* carry out a journey; not *be moving,* for the question what if anything can be contained in a moment can only be sharply discussed in terms of the first. Again, he often insists that motion is from one place to another and from one point of time to another (e.g., 224a34–b2, *Eth. Nic.* 1174a29–b5); and this can be taken to show that when he speaks of motion he always has in mind an X that carries out the journey from A to B, not just an X in transit between these points. A little earlier in the *Physics,* for example (237a17–28), he argues that whatever has moved must have taken some time to do so, for it could not have journeyed from A to B in an instant.

Does the issue have a disappointingly verbal air? The brunt of the protest seems to be just this, that Aristotle's rejection of motion at a moment is misrepresented in English by casting it in the continuous forms of the verb; translate it into the non-continuous forms, and it is simply valid. Offer Aristotle a choice such as the past tenses in Greek allow, and (it is implied) he will choose wisely, confining his rejection to the former way of talking. But what has this issue of surface grammar to do with mapping the emergence of an exact science? Well, everything, as the protester recognizes. The question is whether Aristotle does, in theory if not always in practice,[13] discard *all* talk of motion at a moment—including (and, as it will turn out, primarily) such talk as goes better into our continuous tenses. By such a rejection—and it is of course an argued rejection, not a mere confusion drawn from an ambiguous Greek tense—he will equally debar himself from all talk of velocity and rest at a moment; and therewith debar himself from considering, as the author of the *Mechanica* can, the idea of a "force" or "strength" operating on a point whose resultant motion can be characterized only for that instant. So the protest deserves to be met.

First, so far from using the expression "in a moment" to point a logical error, Aristotle takes it as an unexceptionable part of speech. He imports it earlier in his own argument in the *Physics* (234a24, 237a14), and in our context (239a35–b3) he says: "In a moment it (sc. the moving body) is over against some static thing,[14] but it is not stationary itself. . . . What is true of it in a moment is that it is not in motion and is over against something. . . .'' Translators commonly write ''*at* a moment'', knowing that the Greek preposition *en* has other senses than that of containment, and they are right.

Further, on the appeal to Aristotle's general account of movement it is sufficient to recall the theorems about moving bodies that he has defended just before his criticism of Zeno. One is that what moves, or is moving, must already have moved (236b32–37a17). *Moves,* or *is moving*: which? Surely the second, for it would be unreasonable to maintain that whatever makes a journey from A to B must have traveled previously (that way a pointless regress lies), but reasonable to say that something currently engaged in making that journey must have already come part of the way.[15] So too with the logic of ''coming to a halt'', which Aristotle treats in parallel with that of ''moving''. He argues that, given a period in which something is coming to a halt, it must be coming to a halt in any part of that time (238b31–39a10).[16] Read ''come to a halt'' instead of ''be coming to a halt'', and the claim becomes nonsense. So does the connected claim that a thing is moving when it is coming to a halt (238b23–29) if you substitute the non-continuous forms of the verbs. The point that ''the Greek present tense is generally imperfective'' was made (and subsequently forgotten) by T. M. Penner in a recent paper.[17] For verbs of motion, at any rate, it holds good, and it explains much else in Aristotle's account of movement.[18]

In logic and in grammar, then, the defense fails. Aristotle can in theory allow no talk of motion proceeding at a moment. No doubt it might still be protested that the notion of an exact moment having no duration in time is only adumbrated, approached only asymptotically, in the common discourse from which Aristotle professes to set out in shaping the basic concepts of his physics.[19] Modern English can be more precise because it is heir to so much analytic thought on the issues, including Aristotle's. If Aristotle is refining a vaguer notion for technical use, he is surely entitled to stipulate conditions for its use. And if he discounts linguistic clues that lead to motion at a moment in order to safeguard clues that connect motion with periods, his dialectical method allows for that too.[20]

It is no reply to point to the resultant theory of mechanics, cumbrous and impoverished by comparison with the almost Newtonian insights of the *Mechanica*. Quite apart from the question whether we yet have a "theory" on our hands, it was already a major advance in the science to insist that movement and velocity must in any final analysis be treated as functions of distances and periods. Not all advances can be expected at one time, or the papers in this volume would not have been written. But there is another reply to the protest.

Elsewhere in his analyses of time and motion Aristotle several times uses a device, or family of devices, which would have let him exploit the linguistic clues he jettisons, even while recognizing that any talk of motion and speed at some moment in a period must be explained (no doubt roughly at first) in terms of motion and speed over periods. If there are two uses of "moving" here, Aristotle knows ways of analyzing a secondary use in terms of a primary without either conflating them or discounting the first. His stock examples of expressions amenable to such treatment are "medical" and "healthy": it is medical science that is called "medical" in the primary sense, for medical treatment and medical instruments are so called as being a product of that science or adapted to its exercise, and the explanation could not proceed in the opposite direction.[21] These are nursery examples, but (in reply to the protester) when Aristotle puts such patterns of analysis to larger use in his philosophy, he is often trying to put a harder edge on ideas implicit in common speech. It is hard to deny that an entrée could have been found here for motion at a moment.

Yet when he does use such analyses in his treatment of time and motion, we shall find other difficulties arising. Once try to mark off different uses of a key-expression and put them in some order of dependence, and it becomes a question how the primary use is picked out and what sort of primacy is being claimed for it. Let us turn to these troubles.

THE PRIMARY USE OF "NOW"

One example of the kind of analysis just sketched is his review in *Physics* III of uses of "infinite" (which in this context signifies *infinitely divisible,* divisible into parts which are themselves always further divisible). He writes (207b21–25): "The infinite is not one and the same sort of thing in the case of distance and of motion and of time. The term has a primary use and another dependent on it. Motion is called infinite because the distance covered by the motion is called so . . . and the time is so called in virtue of the motion.'' Notice the ordering, on which we shall find Aristotle putting much weight later: first the application to spatial magnitude, then derivatively to the motion that traverses it, then, by a further explanatory step, to the time taken. In his treatment of time in the next book of the *Physics* he repeats the claim (219a10–14, b15–16), substituting ''continuous'' for ''infinite'' (which in effect comes to the same thing), and then he argues from this[22] that the expressions ''before'' and ''after'' are also used primarily of spatial distinctions and only secondarily, ''by analogy'', applied to movement and thence in turn to time (219a14–19).[23] At this stage of the argument I cite these only as illustrations of what Aristotle evidently regards as a valuable (though flexible, and perhaps too hospitable) form of analysis. The results must be questioned later, and that questioning is better approached by considering yet another example of the same technique.

In *Physics* IV.10–14 Aristotle gives much discussion to the character of *now* or *the present*. When he takes himself to have settled its character, he says (222a20–24): ''This is one way we speak of *now*; it is another now we speak of when the time is near to that first now. We say 'He is going to arrive now' because it's today that he's going to arrive, and 'Now he arrives' because it's today he came; but not 'Now the events in the Iliad have taken place' or 'Now the end of the world'—not because the time stretching to these events (sc. from the present) is not continuous, but just because they are not near.'' Not near, that is, to the primary now: but what is the primary use of ''now''?

In the set of puzzles with which Aristotle introduces his discussion of time he argues: How can time exist, except at best in some reduced and obscure sense? For part of it is past and exists no longer, part of it is to come and exists not yet; and between them this pair exhausts the whole of time and any particular stretch of it (217b32–18a3). If we object that this discounts the now, the present time lying between past and future, Aristotle counters with an argument that the now is no part of time; for (1) any part must serve to measure the whole of which it is a part, and (2) time is not composed of nows.[24] But when a thing with parts has no part of it currently in existence, it does not exist itself (218a3–8).

The premises (1) and (2) should show us something of what Aristotle understands by ''now'', but they are surely debatable. How is time not composed of nows? Take all the times that ever have been or are or will be present and that is surely all the time there could be: 1975 is present,

1974 was, 1976 will be, and so on, each year having its turn. So too with premise (1): the present year can surely serve as a measure of time, inasmuch as (e.g.) a decade is just ten times its length. But, other objections apart,[25] it becomes clear that Aristotle will not have 1975, or any other period, as an instance of the now, the present he is trying to explain. For a few lines later (218a18–19) he says: "Let it be taken that one now cannot be next to another, any more than one point can be next to another." So the instances of *now* or *the present* that he is prepared to accept seem plain, and his later discussion bears this out. They answer, we may say (as he could not), to such time-references as "1200 hours 18 May 1974" *provided* this is somewhat artificially understood as fixing a point of time without duration. For on this understanding there is theoretically no *next* time-reference of the same order: however close we call our next reference, provided it has some finite value in the same system some period of time must have elapsed between the two within which a still closer reference could have been called. The spatial analogue is indeed the point.

But why should a point of time seem a satisfactory value of *now* rather than a present period such as 1975? So far from explaining this, our premises (1) and (2) merely presupposed it. But Aristotle's argument is not far to seek. It lies in the comment that past and future time together will be found to exhaust any period whatever (218a1–2). It is a line of reasoning that recurs in the Skeptics and perhaps most famously in the eleventh book of Augustine's *Confessions*. The present century, Augustine observes, is a long time. But is the whole century present? No, for we are living now in one year of it, and the others are past or to come. And so on down, through days and minutes, shedding whatever can be shed into the past and the future until we are left at last with the durationless present moment, which has no parts to shed.

Here is trouble indeed. The present is like the point, and lines are not collections of points.[26] But we are not required to tackle the fairy-tale task of building lines out of points, for there are countless lines in the world that can be drawn and shown and used to measure other lines. When we use the edge of the ruler, it is of no use to know that the twelve-inch mark is to the right of the one-inch mark, or to wonder how many points to the right it is. Distance cannot be measured in points alone. But then how to measure times, when Aristotle's argument leaves us with no lines in time present, only present moments? And apparently

11

all we can know of these is that one is earlier or later than another, which is no more than knowing that the twelve-inch mark is to the right of the one-inch mark. Augustine was so exercised by this that he supposed that what we measure in time is a kind of proxy time-stretch held in our memory. Others such as Marcus Aurelius drew morals: each of us lives only in this instantaneous present. The morals may be excellent, but the argument is insufficient. Aristotle evidently thought his paradoxes could be evaded even though he retained his account of the primary use of "now".[27]

THE RETRENCHABILITY OF "HERE" AND "NOW"

Let us start by considering a way of avoiding the first step into these paradoxes, and then look at Aristotle's way of retrieving the reality and measurability of time. Consider "here" as the spatial counterpart of the temporal demonstrative "now". It is not an exact counterpart, as Dummett and others have pointed out,[28] but for the present argument it is close enough.

When Kingsley Amis entitled a book "I like it here," it was not hard to find what he intended by "here". It became apparent from the disaffection he showed for those living beyond a certain perimeter—in fact, the English coast; and he might well have wanted to narrow that perimeter too. The compass of his "here" was settled by what he took to lie outside. The same holds good of "now". If we are told, once more, that we live now in an Age of Anxiety or an Age of Violence, we shall be clearer on our informant's meaning if we find him, for example, relegating the Boxer Rebellion to the past tense and providing for the future by saying that if certain détentes follow quickly he will retract his observation as spoken too late or too soon. Such dismissals into the past or the future are all we usually require or assume in taking ourselves to understand a claim about the present. Of course there are certain semantic rules that must not be broken except in artificial situations: I must not as a rule relegate the whole year 1975 to the past or the future tense when I am speaking in that year, for instance. But within the application of such rules it remains true that what a speaker counts as present can be seen from what he counts as past or future or both; and the Greeks were as flexible in this as we are. When Plato talks of the order of nature that obtains *now*, he shows the scope of his "now" by contrasting the present with a prehistoric past age in which the order was reversed (*Ptcs* 273e).[29] What a speaker consigns to the past or future—subject to those semantic rules—depends on his immediate purposes and subject-matter (as well as, no doubt, on his training and culture). That is why it can vary, as it familiarly does, from utterance to utterance.

This is troublesome to one like Aristotle, who aims to bring uniformity into the linguistic procedures taken over by a science. But if "here" behaves in this respect like "now", it is tempting to find a counterpart to Aristotle's paradox about "now". Suppose I tell you, "It's all public land here", and you ask, "What do you mean by *here*? Just this field, or more?" Repeating "here", with whatever emphasis and gestures, is useless. What will help will be to specify land and landmarks that lie inside and others that lie outside the borders of whatever my spatial demonstrative was meant to cover. Given an interest in a smaller stretch of land or a different interest in this one (say, the placing of a rare plant), the scope of my "here" could have been narrowed by counting more things into the surroundings, to left or right or beyond. There is no one reason for setting the frontier closer or farther: the demonstrative carries no privileged reference to a measure of ground that could not for another purpose be narrowed. Let us say, for short, that "here" is retrenchable.

But then it might be argued that, understood strictly, "here" cannot be taken to pick out any stretch of ground. For if there is no stretch such that part of it could not equally be counted as *here* and the rest consigned to the surroundings, there is no reason to stop at any of these arbitrary and dwindling frontiers in trying to determine what really answers to "here" on any occasion of using it to make a spatial reference. What really answers to it is of course something within which no relegations to left or right or beyond are possible. In its paradigm use, "here" must on any occasion pick out a point of space. Given that use it can be allowed, derivatively and on sufferance, to indicate any stretch of land containing or (I parody Aristotle's account of "now") near to that point.

As an account of our use of spatial demonstratives this is absurd. In settling the scope of the statement "It's all public land here", we had no need (and probably no time or technique) to come to an agreement on the identifying of some unextended spatial point. To understand "here", as to understand "now", the idea of retrenchability for different purposes is essential, and that of picking out an unextended point is not.[30]

13

This would be a sufficient answer if the pseudo-paradox about "here" were a sufficient parallel to Aristotle's paradox about "now". But there is an asymmetry that seems to tell against it, and no doubt this is why Aristotle does not extend his temporal paradox to space. Past and future, we are inclined to protest, are not comparable to left and right and yonder, just because what *can* be counted on any occasion as past *is* then irretrievably past: it is not up to the speaker to retrieve it by deciding, within certain semantic conventions, what he will then count as past and hence as present. With space it is otherwise: what is counted as lying to the left for one purpose can on the same occasion be included in the central ground for another purpose, for all such options remain open in a space all of whose parts remain accessible. No doubt an observer falling ineluctably like an Epicurean atom in one direction might think of any fixed landmarks he had passed (if there were such landmarks) as irretrievably above or behind. Spatial logic might then match temporal logic. And, conversely, there have been attempts to give sense to the suggestion that progression in time could have more than one direction, like movement in space. But, as things are, the asymmetry seems undeniable so long as we stay within the confines of "here" and "there" and "now" and "then".

There are difficulties in this argument; but I cannot pursue them here beyond asking whether Aristotle is entitled to such a reply after some further arguments we have yet to examine about spatial and temporal order. Those arguments we shall meet when we come to consider how Aristotle tries to vindicate the reality of time in the face of his paradox. But there is another issue to be considered first.

THE NOW AND THE MOMENT

In the opening discussion of motion and rest I made much use of the notion of a moment that occupies no stretch of time. In discussing Aristotle's treatment of "now" I have made more use of it. In the first discussion I warned that the same Greek expression that is translated "moment" in some contexts is the normal Greek for "now" or "the present". Someone impressed by the argument from retrenchability might argue: what Aristotle says of "now" makes poor sense of that expression, but good sense of our talk about moments. Is it not, after all, the idea of a moment that Aristotle in his discussion of time is trying to explain? Perhaps by Aristotle's day the expression has acquired a conventional sense in philosophical contexts that depends on discarding the presentness and stressing the momentariness of a *now*. After all, he refuses to be upset by the paradox he retails. He does not directly confront the argument that, if all present time is momentary, all we have as materials from which to build time is a class of moments that by their nature cannot serve as building material. In *Physics* IV.11 (219a10–b2) he sets up a parallel between the distance covered by a moving body, the movement that it performs, and the time taken by the movement. He remarks there (219a25–30, 220a14–21) that, just as between any two points on the ground there must be a distance that is not a collection of points, so between any two nows some time must elapse that is not composed of nows. Read ''moments'' for ''nows'', and the claim becomes recognizable and valid. In fact, we may feel, it is better served by casting it in those tenseless forms of speech that allow us to treat of moments as standing in a linear order of earlier and later without either explicitly or implicitly importing a reference to the present or its concomitants the future and the past.[31] Science needs the bedrock of the first, not the shifting sands of the second.

The suggestion would be a mistake and an anachronism. Aristotle does not discuss the nature of moments in abstraction form the idea of the present. His paradox about the unreality of time plainly has teeth only if it is understood in his terms, as contrasting a present that exists with the non-existent past and future, not as making the perverse suggestion that there are only moments and not days. Nor, and here I agree with Hintikka, is he near to grasping the idea of tenseless statements; Parmenides and Plato came closer.[32] In his other arguments on time the word for ''now'' brings together what seem to us two distinct concepts, that of the moment and that of the present. When he speaks of the lapse of time as marked by different nows in an order of earlier and later (219a26–30), we think of moments. When he speaks of the now as progressing through time in a way comparable to that of a body progressing through a movement, collecting different descriptions according to the stage it has reached (219b22–33), we think of the present as something continuously overtaking such successive moments and leaving them in the past. When he claims to show how the now is perpetually different yet perpetually the same, since on the one hand there is a succession of nows (219b13–14, cf.219a25–29) yet on the other there is the one progressing now (219b22–28), we cannot think of him as distinguishing the two concepts but rather as conflating them.

How then are the two wedded in the one expression? And how far was the divorce managed by Aristotle's successors?

In reply to the first question we might hazard that, once Aristotle had argued himself into the belief that in its strictest use ''now'' must always denote a present moment, the word became for him, by a natural extension and in default of any technical Greek equivalent, the stock expression for any moment. Such extensions of usage are common enough with him.[33] We might alternatively father Aristotle's ''now'' on Zeno, recalling the account of the Flying Arrow in which the arrow is stationary at any *now* of its flight (239b7, 32). Neither reply is enough. Aristotle's version of the Arrow, even if we could reconstruct the text with certainty, may be a recasting in his own terms of what he takes to be Zeno's point. And there is a fuller and more secure precedent for Aristotle's usage, one in which we can see ''now'' being tailored for its philosophical use, and one which must have shaped the recasting of Zeno's argument if any such recasting took place. I mean an argument in that treasury on which Aristotle drew so extensively for his own *Physics,* the second part of Plato's *Parmenides.*[34] Given the importance of Zeno in that dialogue, I have no doubt that his paradox stands behind the argument, but I shall not debate the extent of the debt.

Plato is coining puzzles from the logic of growing older (*Parm.* 152a-e). Nothing will be lost for our purposes by replacing his special subject ''the one'' by a neutral ''X'', or by omitting arguments which offer to show that if X is becoming older it is also becoming younger.

The reasoning Plato puts into Parmenides' mouth is this. ''Whenever in this process of becoming older X is at the time that is *now,* between *was* and *will be,* then X *is* older. After all, in its journeying from the past to the future it is not going to skip the now. So whenever it coincides with the now, at that time it stops becoming older: at that time it is not becoming, but just is, older. For if it were moving on it could never be caught by the now: what characterizes something moving on is that it is in touch with the future as well as the now, letting go of now and laying hold of the future and carrying out a change between the two. But then, given that nothing in process of change can by-pass the now,[35] it stops becoming on any occasion when it is at the now, and then is whatever it

16

may be becoming.[36] So it is with X: whenever in the process of becoming older it coincides with the now, forthwith it stops becoming: then it *is* older."

Then (152e1 ff.) Plato generalizes this to cover X's whole career. "Moreover the now always accompanies X throughout its existence, for whenever X is it is *now*. So X always *is* older, not just becoming so."

Generally it is poor practice to isolate any argument in the *Parmenides* from the network of paradox that it serves, but this has a special interest for us. It shows "now" being groomed for its Aristotelian part. First Plato maintains that *now* X cannot be becoming so-and-so, and only is so-and-so: if it is getting older, or moving to the left, it cannot be engaged in the business now; now it can only be older or further to the left. And then he argues that this conclusion can be generalized for the whole of X's career: whatever holds good of X must hold good at some time that is then present, a time that is then properly called "now".

The second arm of the argument seems innocuous. It generalizes on the uses of one temporal demonstrative, at the price of one kind of referential opacity; but the possibility of such generalization is already assumed in the first arm. Plato does not go on to say that sooner or later every time becomes a present time, or that a time only exists when it is present. Otherwise he might have been vulnerable not only to some of McTaggart's paradoxes but to a dilemma of Aristotle's (*Phys.* 218a14–18). For when does a moment cease to be present? Not at the time of its own occurrence, nor at any later time; for the first requires it to stop when it is still present, and the second requires it to linger to a later moment and so stretch into a period of time. The dilemma is artificial, depending on treating a moment as something with a career in time and not as itself an element of time. But Plato does not court this difficulty. Nor does he pursue, what this second arm of his argument suggests, the Aristotelian image of the present as progressing through time. He does not exploit the language that presents Aristotle with another paradox (218b13–18). For Plato the nonsense-question, how quickly the present proceeds on its path, does not arise.[37]

It is the first arm of the argument, then, that carries the load. Why cannot we say that X is growing older, or the arrow is flying, now? Because, Plato argues, such descriptions make tacit reference to the future as well as the present. It could not be the case that X is becoming older unless at some later time X will be older than it now is; it could not be the case that the arrow is moving unless it will later be in some place other than where it is now. Otherwise the flight is at an end, the business of growing older is finished. So if we are to say only what is true of the subject now, we have to eliminate this future component from our description. The arrow is now in just the place it occupies, X just is older. The second is Plato's example, and it seems ill-chosen; for if X is now older, it is older than it was, and this implicit reference to the past should be excluded together with references to the future. *Now* X is just wherever in time it happens to be. But Plato uses the Greek idiom ''X is (or becomes) older *than itself*'', and this obscures the point as well as furnishing him with other paradoxes.[38]

What more is there in Plato's argument than we found summarized (surely because it was now too familiar to spell out) in Aristotle's? Nothing, so far as I can see. But here is Aristotle's ''now'' in the making. Its retrenchability is discounted by shedding from its scope whatever can be relegated to the future (and, Aristotle adds, to the past). So ''now'' becomes a paradigm way of referring to a moment, and the way is open for Aristotle to extend it to all moments. But only by considering them all as becoming sooner or later present, in the sequence future-present-past. The word never came to signify the ''moment'' of the translators or of our detensed textbooks in physics. It kept its connotation of time present, the sense on which the arguments of Plato and Aristotle were built.

So the two senses, that of the moment and that of the present, stay wedded in the one expression, and we have to turn to our second question: was the appropriate divorce arranged later in Greek philosophy? I think it was, but the evidence has been sometimes misread.

Consider what Chrysippus the Stoic is reported to have maintained about time present. According to Plutarch (*Comm. Not.* 1081c–82a), his contribution to the topic consisted chiefly in three theses. First, there is no such thing as a shortest time (1081c); secondly, the now is not something indivisible (ibid.); thirdly, any part of present time can equally be taken to be past or future (1081d–82a). Here, surely, is our retrenchable present. Elsewhere in their physics the Stoics were ready enough to admit indivisibles: they insisted, as Aristotle had before them (and to the scandal of Plutarch), that contact between bodies is not the juxtaposition of some smallest possible parts or volumes of those

18

bodies, for there are no such volumes; contact is a function of surfaces without depth or of points without magnitude (1080e–81b). But they were clear that such indivisibles were not the model on which to explain our use of ''now''.

Although I cannot be sure that the Stoics have our point clearly in view, I am sure that they once had less than justice from one of their most persuasive sponsors, Samuel Sambursky. In *The Physical World of the Greeks*[39] he had this to say concerning the third thesis credited to Chrysippus: ''This formulation, with its definition of the present as the centre of a very small, but still finite, portion of time, is clearly an attempt to comprehend the elements of time as finite 'quanta' and not as extensionless points. The present thus becomes, so to speak, an 'atom of time', or, to use the language of the calculus, a differential of time.'' A little later he said: ''So great was the desire of the Stoics to give a clear answer to the paradox of the arrow, that these bitter opponents of the atomic hypothesis and ardent champions of continuum and no compromise had to have recourse to an 'atomic' solution.''

But not only is atomism in general just as repugnant to Stoic physics as Sambursky recognizes, the suggestion that ''now'' is always or primarily used to identify some time-atom is incompatible with the first two of Chrysippus' propositions and, as soon as it is reconsidered, with the third. Sambursky's belief that even for Chrysippus the use of our temporal demonstratives must ultimately reduce to a reference to just one sort of time-element—if not moments, then atomic times or real infinitesimals of time—seems to be one more sign of the spell that Zeno's Arrow or its interpreters cast on philosophy. It is this spell that Chrysippus seems anxious to break.

PRIORITIES AND PARALLELS BETWEEN SPACE AND TIME

We left Aristotle in the grip of his paradoxes. Only the present is real; yet the present is never a stretch of time, and time must consist of successive stretches, not of unextended moments. Moreover, time is a function of change (218b21–19a10), but nothing can be changing at a present moment. How can time be real? Yet later Aristotle claims to have said of time both that it is and what it is (222b27–29). How does he break free?

His reply is that there could not be a present moment *without* time, any

notice at least some mental change in ourselves; and later (223a21–29) he even suggests that there could not be time if there were not a mind capable of counting off and measuring time, even if there were some change going on that would otherwise supply material for the counting and measuring. It is no comfort that he prefaces this last suggestion with an argument that any change entails time, since it can proceed more or less quickly (222b30–23a4). What we need is an argument that any time entails change. For that conclusion his reason seems to be that we do not recognize time if we do not recognize a change we can measure by it. The reason is not enough, and later mechanics dispensed with it.

Still, grant that though time is not change, it can always be plotted against change. Aristotle asks us to think of three lines and evidently draws them on his blackboard (an anachronism: it would have been a whiteboard). One represents the distance covered by a movement, the next represents the movement over that distance, and the third represents the time taken by the movement. Call them "D", "M", and "T", as he does not.

a	b	c	D
a′	b′	c′	M
a″	b″	c″	T

About these lines he argues two things. First, they are parallel in structure, all three being continua having parts that can be correlated, so that each of them can be measured by its neighbor: time by motion and vice versa, motion by distance and vice versa (220b14–32). Moreover, we can take a particular point on M (b′) as representing the moving body at that stage of its movement and correlate it with the appropriate spatial and temporal points (b, b″) on D and T (219b12–20a21).[42]

Secondly (and this is the stronger claim, thought it comes first in his analysis), spatial order is somehow basic to the others: the structure of D prior to that of M, and that in turn prior to the structure of T. "Since the moving object is moving from something to something" (e.g., from a to c in our diagram), "and since every magnitude is a continuum, the movement answers to[43] the magnitude: movement is a continuum because magnitude is a continuum. And time is so because movement is so, for we always take the amount of time that has elapsed to correspond to the amount of the movement. Thus[44] *before* and *after* are found primarily in location, where they depend on position; and since they are found in the magnitude'' (sc. D) "necessarily they are found in movement too, by analogy with those there'' (sc. the before and after in D). "But in turn *before* and *after* are found in time, because the one always answers to the other'' (sc. time to movement).[45] This claim of priority (219a10–19) he recalls later (220b24–28, cf. 219b15–16, 220a9–11). It is the last example we shall meet here of the pattern of analysis we desiderated in his treatment of motion at a moment, and found present but questionable in his account of "now". Here too its use is questionable.

As a preliminary question, what sort of priority or basicness has he in mind? Elsewhere, and notably in *Metaphysics* V.11, he distinguishes some main uses of "prior" and "posterior".[46] It is worth notice that he gives spatial order an earlier place in his account than temporal order (1018b12–19): that suits his analysis in the *Physics,* but it reverses the explanation in the post-predicaments of the *Categories,* if these are his work (14a26–29). But, more important, he distinguishes between what came to be called ontological and epistemological priority. Sometimes, as here, he brings "priority in definition" under the second head (1018b32–33), sometimes he distinguishes the two more sharply (1028a34–b2); once (in our context, 1019a11–14) he implies that the epistemological is a kind of extrapolation from the ontological, the priority "by nature and substance". But the distinction holds firm. If A is prior to B "by nature", then A can exist without B but not B without A. If A is prior "for knowledge", the understanding or explanation of B requires that of A, and not the converse. There are different ways of understanding and explaining, and this priority can accordingly go different ways (1018b30–19a1); what matters for present purposes is that the epistemological priority does not entail the ontological, nor vice versa.[47]

Aristotle might have believed that there could be space but no movement over it; but then it could not be measured (220b28–32), and this he rejects. He might even have believed, for the brief span of one argument we noticed earlier (223a21–29), that there could be movement without time; but this too he consistently rejects elsewhere (e.g. 222b30–23a4, 232b20–23). Ontological priority is not what interests him here. He wants to show that spatial order is conceptually basic to the rest, that by starting from this we can explain the order of movement, and at another step the order of time, without circularity, i.e., without importing into our explanations the things they were meant to explain. The enterprise fails. But enough remains of his parallel between space and time to make an argument for time's reality.

He argues first that, since the distance D is a continuum, so is the movement M over D; and since M is, so is T, for the time taken varies with the extent of the movement (219a12–14). Already we feel a qualm. The parallel between D and T relied, for a first step, on the parallel between D and M: the size and divisions of the movement corresponded directly to the size and divisions of the ground covered. But the time taken does not correspond equally directly to the ground covered, for speed can change. To bring the M/T parallel into line with the D/M parallel, we shall have to say that *at a given speed* the greater distance requires both a greater movement and, in the same ratio, a greater time; but this is to reimport the notion of speed, which employs the idea of time as well as of distance. The qualm can be met by giving up the requirement of a direct ratio between M and T or between D and M, and therewith the simple model of measuring one by its neighbor. Perhaps all we need to retain of the argument is the suggestion that (to put it roughly) further on in space is further on in the movement, and so further on in the time. But this is just the claim that spatial order is basic to the rest, and that claim seems untenable.

Spatial before-and-after is relative to some more or less arbitrarily chosen position on a line (cf. *Met.* 1018b12–14). Thus on D, b is before c relatively to a if and only if ac contains ab but ab does not contain ac. Similarly b is before a relatively to c, because ac contains bc but bc does not contain ac. Now if temporal order is to be explained by the order of motion on such a line as D, evidently the motion must have a direction: for instance, on M, b' must precede c'. But can this direction be derived from the spatial before-and-after we have just defined, without importing just the temporal priority we meant to explain? Evidently not. We might define a direction *abc for D, by saying that ac contains ab but ab does not contain ac; but of course we could on the same terms define the direction *cba. It is of no use to think we have given the movement a temporal direction by saying that it is on a line for which the spatial direction *abc can be defined, for the opposite direction can equally be defined for that line. It is just as useless to suppose we have given it such a direction by saying that the body moving on abc may move over ab without moving over ac but not vice-versa—hoping thereby to indicate that by moving from a toward c it may reach b but not c. For this is compatible with its moving from b to a. And if we try to sharpen the condition by specifying where on the line the movement begins or ends, our explanation of temporal order becomes immediately circular. Spatially, the movement can take either direction, temporally it can take only one. Just this was what prompted the earlier qualm about our appeal to retrenchability.

If Aristotle hopes to show spatial order conceptually prior to temporal order, the attempt fails. But does this spoil the general parallel between space and time by which Aristotle hopes to meet the paradox of time's unreality? Surely not, for the gist of that was that in both cases there cannot be points without the stretches they join and bound. The moment 5:30 and the period from 5:30 to 6:00 are interdependent. Time is left with periods that are, in Aristotle's sense, "measures" of time. What he says of this goes admirably, as we can put it now, into a tenseless language of moments and durations. But it does not meet the paradox that at any real present the measurable stretches of time are behind or in front of us; that, as Marcus Aurelius holds, "each of us lives only this momentary present"—with the possible comfort he derives from this, that dying is no loss since we cannot lose a past or future that is not real now.[48]

1. Since drafting these arguments I have seen a forthcoming paper on some of the same topics by Professor Fred D. Miller, Jr., and felt able accordingly to shorten mine at some points and push it further at others. (Professor Miller's paper has now appeared in *Archiv für Geschichte der Philosophie* 56.2 [1974]: 132-55.)

2. G. E. L. Owen, "The Platonism of Aristotle," *Proc. Brit. Acad.* 51 (1965): 148–49 and nn.

3. E.g., Aristotle, *Metaphysics* V.11, VII.1.

4. Cf. "The Platonism of Aristotle," l.c.; "Aristotle," in C. C. Gillispie, ed., *Dictionary of Scientific Biography*, 1 (New York: Scribner's, 1970): 254–55.

5. This is how Aristotle read Zeno's conclusion, and not merely as holding (as Ross suggests, *Aristotle's Physics* [Oxford: Clarendon Press, 1955], p. 658) that the arrow is not moving (which Aristotle would accept, adding that it is not stationary either): see *Physics* 239b30.

6. "Over-all", d/t, rather than "average", which itself requires analysis in terms of momentary velocities and is therefore doubly general.

7. Jaakko Hintikka, *Time and Necessity* (Oxford: Oxford University Press, 1973), p. 162.

8. Cf. G. E. L. Owen, "Zeno and the Mathematicians," *Proc. Arist. Soc.* N. S. 58 (1957-58): 221–22.

9. Cf. T. L. Heath, *Mathematics in Aristotle* (Oxford: Oxford University Press, 1949), p. 227.

10. Notice, e.g., his refusal to assimilate circular and rectilinear motion in *Physics* VII.4 (248a10–49a25).

11. Cf. n. 5 above.

12. T. M. Penner, "Verbs and the Identity of Actions," in O. P. Wood and G. Pitcher, eds. *Ryle* (New York: Doubleday, 1970), p. 458, contends against some earlier arguments of mine that Aristotle "has no *general* presumption against saying that if X was Φ-ing throughout a period *p*, then at any moment *t* during *p* X was Φ-ing" (or, we may add in view of Penner's argument, against the converse). "It is just that this cannot be said with movement, or, more generally, with *kineseis*" (the latter being Aristotle's word for all changes for which he takes movement in important respects as a model). So Penner holds that Aristotle would readily concede my point that if I am asleep at any and every moment of the afternoon, I am asleep throughout the afternoon, and the converse; but this would not spoil his reply to Zeno, which is that where movement is concerned no such inferences lie between moments and periods, for sleeping is not a movement or *kinesis* but what Aristotle calls an *energeia*. But by the tests proposed by Penner *being at rest* is also an *energeia;* and Aristotle denies that a body can be at rest at a moment.

13. The corollary for rest is ignored at *Physics* 236a17–18.

14. I retain the *menon* of the mss. at 239a35–39bl against Prantl, etc.: what Aristotle says does not entail that the static body is static at that moment and the repeated *men* is not needed.

15. The point is generally well taken by translators and commentators, together with the equally important point that the perfect tense is (of course) perfective, "have moved" and not "have been moving."

16. The argument requires this, and not just that the object must come, or be coming, to rest in any arbitrarily late part of the time in which it is coming to rest. Any such time can itself be divided into others in any of which the coming to rest is proceeding quickly or slowly (238b26–30).

17. Point made, op. cit. (n. 12 above), p. 400; neglected in the argument manufactured for Zeno, p. 459 ("At a moment no distance is traversed", etc.). To speak of "Aristotle's claim that there is moving only if there is a movement" is either misleading, if it is taken to support this rewriting in the non-continuous present, or transparently acceptable if all it says is that to be moving requires that a movement take place from some A to some B—even though to be moving from A to C does not require the movement from A to C to be completed.

18. E.g., (1) the use of imperfect tenses as proxies for the present in *Physics* 231b30–32a2, (2) Aristotle's unreadiness to use the present tense in saying that A *becomes* B at a given moment rather than to say that it then has become or earlier was becoming B (e.g., *Physics* 263b26–64a1).

19. G. E. L. Owen, "Tithenai ta phainomena," *Aristote et les problèmes de méthode* (Louvain and Paris: Nauwelaerts 1961), pp. 83–103.

20. Thus *Nicomachean Ethics,* 1145b2–5: if all the common conceptions on a topic cannot be vindicated, we must defend most and the most commanding.

21. E.g., *Metaphysics,* 1003a33, 1060b36–61a7.

22. I read $d\bar{e}$ = "therefore" with Ross at 219a14: cf. his synopsis of the argument, *Aristotle's Physics,* p. 386.

23. "Analogy" covers but is normally wider than the asymmetrical relation of prior-posterior between different uses of an expression that is at issue here. Cf. G. E. L. Owen, "Logic and Metaphysics in Some Earlier Works of Aristotle," in I. Düring and G. E. L. Owen, eds. *Aristotle and Plato in the Mid-fourth Century,* (Goteborg: 1960), pp. 179–81.

24. On the implications of this cf. n. 12 above.

25. There are older arguments alleging the lack of parallel between measurements of space and time and based on the possibility of transporting a rigid measuring-rod in space but not in time. Aristotle does not discuss this issue, and the rigid transportable measuring-rod has become an anachronism.

26. They can of course be treated as classes of points for many purposes, e.g., for the inferences sketched in n. 12 above. Aristotle's principal objection relies on the Zenonian argument (p. 4

above) that, on any common rule of division, no exhaustive division of a magnitude into points can be completed.

27. There is a question here: given that the paradoxes are not systematically and directly answered in the sequel, how and when were they prefaced to the argument? Compare the question how the paradoxes of *Metaphysics* III were given their present place.

28. M. A. E. Dummett, "A Defense of McTaggart's Proof of the Unreality of Time," *Philosophical Review* 59 (1960): 500.

29. *Politicus* 270d6–e1 does not say or imply that in some sense the order of time was reversed in that other age; even if it did, the point would not be affected.

30. Of course one can imagine, under artifical conditions, a situation in which my utterance of "now" or "here" will pick out a point of time or space for my hearer. Suppose the chronometer set up to start just when I start to say, "Now I see him pass the starting-point": I could not say it without being able to say, "Now he is running fast."

31. *Generalized* references to future, present, and past go without remainder into this form: cf. Nelson Goodman, *The Structure of Appearance* (Cambridge, Mass.: Harvard University Press, 1951), pp. 298–99.

32. Hintikka, *Time and Necessity,* ch. 4; G. E. L. Owen, "Plato and Parmenides on the Timeless Present," *Monist* 50 (1966): 317-40.

33. N. 2 above.

34. For the documentation see n. 19 above.

35. F. M. Cornford, in *Plato and Parmenides* (London: Routledge, 1939), p. 187, translates the "by-pass" (*parelthein,* 152b7) by "it can never *pass beyond* the present", suggesting that Plato means that it cannot break out of the present. What Plato means is that it cannot avoid or side-step the present.

36. Cornford (ibid.) translates here as "whatever it may be that it was becoming", a possible imperfectival past sense of the participle but not relevant to the argument, which is to show that for a chosen X ("older") if *a* is becoming X, then at any given present time *a* just is X.

37. C. Strang ("Plato and the Instant," *Proc. Arist. Soc.* 48 [1974] 63–79) draws other inferences from the text translated, together with an earlier reference to time as moving on (152a3–4) —a reference at once replaced by an innocuous description of the subject as "proceeding temporally" (not, *pace* Mr. Strang, as "keeping pace with time"). His interpretation is not adopted here.

38. Principally those paradoxes that concern its becoming younger as well as older (*Parmenides* 141a–d, 152a–b).

39. Samuel Sambursky, *The Physical World of the Greeks,* (New York: Humanities Press, 1956), p. 151.

40. This seems the most direct interpretation of his argument and appears to agree with Ross's "we must in thought pause over the point, as it were" *(Aristotle's Physics,* p. 602).

41. Cf. nn. 31 and 32 above.

42. At 219b19 I read a comma after *auto,* and *he* for $\bar{e}$ before *stigmē*: Aristotle says "the point [on my diagram] is [to be taken as] either a stone or something of the kind"

43. The verb is *akolouthein;* Hintikka (*Time and Necessity,* pp. 43-47) makes a case for reading it in some contexts as expressing either compatibility or equivalence, but here as usually it does not stand for a symmetrical relation.

44. Cf. n. 22 above.

45. *Physics* 219a19, *thāteroi thāteron:* "the one to the other", not "each to each"; Simplicius and the Loeb version have it right here, the Oxford and Budé, wrong.

46. Cf. Owen. "Logic and Metaphysics," pp. 170-72.

47. Cf. n. 44 above.

48. Marcus Aurelius, *Meditations,* vii, 27, ii, 14; both references I owe to Professor Miller's paper (n. 1 above).

Robert G. Turnbull

Chapter Two

"PHYSICS" I: SENSE UNIVERSALS, PRINCIPLES,
MULTIPLICITY, AND MOTION

Wicksteed remarks in his (and Cornford's) edition of Aristotle's *Physics* that book I "is a very serious piece of work and in its way a model of discussion. A short introductory chapter gives us the key, if we can but find the skill to turn it through the wards."[1] The remark is, I think, well made. Unfortunately, despite the useful work of Wicksteed and many others,[2] we remain in want not only of the skill to *use* that first chapter; we are in want of the key itself. Indeed, as late as 1970, Charlton notes that 184a23–b14 (the bulk of the chapter) "are obscure particularly as we are told elsewhere that the individual is known before the universal (*An.Po.*II.19) and that what we perceive is individual (e.g., *E. N.* VII.117a26)."[3] Since *Physics* I has been the object of multiple, careful, varied, and able scholarly attention, it is likely that the difficulty lies deep and is rather more philosophical than philological in character. And since *Physics* I is intellectually related to several major Aristotelian texts (notably, but not exclusively, *Posterior Analytics* and *Metaphysics*), any attempt at radical reinterpretation must draw upon, explain, and/or explain away those relations. In attempting fairly radical reinterpretation, I shall, unfortunately (because of space limitations), deal rather summarily with the relations. My intention is to remove the interpretative obscurity from chapter 1 and then, to shift back to Wicksteed's metaphor, to turn the key through a few of the wards of book I. It will facilitate explanation and exposition to have the text of chapter 1 directly before us.

> 184a10 Since knowing from experience [$\varepsilon i\delta \acute{\varepsilon} \nu \alpha \iota$] and scientific knowing [$\varepsilon \pi \acute{\iota} \sigma \tau \alpha \sigma \theta \alpha \iota$] agree—with regard to all inquiries involving principles [$\mathring{\alpha} \rho \chi \alpha \acute{\iota}$] or causes [$\alpha \mathring{\iota} \tau \iota \alpha$] or elements

[στοιχεῖα]—in being derived from acquaintance [γνωρίζειν] with these [i.e., principles, causes, or elements] (for we believe that we make each thing known when we become acquainted with the causes—the first
15 ones—and principles—the first ones—and have gotten back to the elements), it is clear that for scientific knowledge [ἐπιστήμη] of nature [φύσις] we must first attempt to draw distinctions concerning principles.

The natural way to proceed is from distinctions which are more intelligible and more lucid to us to those which are naturally [τῇ φύσει] more lucid and intelligible. For those which are intelligible to us and those which are absolutely
20 [ἁπλῶς] intelligible are not the same. We must, therefore, advance along the route from distinctions which are naturally less lucid but more lucid to us to those which are naturally both more lucid and more intelligible.

At first more or less compounded and confused things [τὰ συγκεχυμένα] are clear and lucid to us. Later—from these [i.e., the compounded and confused]—the elements come to be intelligible, and the principles separate them out [διαιροῦσι ταῦτα; alternatively, divide them]. It is therefore necessary to advance from universals [τῶν καθόλου]
25 to their constituents [τὰ καθ᾽ ἕκαστα]. For, so far as sense perception is concerned [κατὰ τὴν αἴσθησιν] the whole [τὸ ὅλον] is more intelligible, and the universal is a certain kind of whole. For the universal encompasses many—so to speak—parts.

184b10 The same sort of move holds also for terms [τὰ ὀνόματα] in relation to the formula [πρὸς τὸν λόγον]. For here also a whole signifies indiscriminately [ἀδιορίστως] ('circle', for example); but its definition separates out the constituents [διαιρεῖ εἰς τὰ καθ᾽ ἕκαστα]. And children at first call all men fathers and all women mothers, but later distinguish each of them.

Anyone who is familiar with some of the English translations of the chapter will recognize that my translation, though more literal than most, gives clear indication of the interpretative line I shall be taking. Even so, let me state its main features rather baldly. (1) Aristotelian universals include many that are very close (in a sense to be explained below in discussing *Posterior Analytics* II.19) to sense perception. (2) These universals are, however familiar and obvious, sufficiently complex to contain and confuse constituents that, in relatively naïve consciousness, are not even noticed. (3) There is some process or set of procedures by which it is possible to move from consciousness by means of confused ''sense universals'' to consciousness by means of the articulated (and, presumably, interrelated) constituents of those universals. (And this is like the move from a term to its articulated definition.) (4) What are obvious and intelligible ''to us'' are the sense universals (or alternatively, what they make known); what are obvious and intelligible ''naturally'' are the articulated constituents (or, alternatively, what they make known). (5) Principles or causes or elements are (or are made known by) some (at least) of the constituents of sense universals. And thus scientific knowledge [ἐπιστήμη] either is or depends upon ''acquaintance'' with those constituents (or what they make known). The terms 'principle', 'cause', and 'element' are used here and throughout book I virtually interchangeably. They are so used that, if a sense universal contains Φ, Ψ, and X as constituents, then each of the constituents is a principle, cause, or element. And ultimate genera of these would be *first* principles, causes, or elements. (For this usage there is clear textual warrant in *Metaphysics* Δ, the so-called ''dictionary''. Cf. *Metaphysics* Δ, the first three chapters.)

Throughout book I, Aristotle is intent on showing exactly what first principles are necessary in order for there to be a science of nature [φύσις], and this requires that principles be articulated in terms by means of which *change* generally—including change of place, growth and diminution, coming-to-be and passing-away, alteration, and so on—can be intelligible. Chapters 2 and 3 are mainly devoted to refutation of Parmenides' (and Melissus') claim that there is but one (first) principle and the consequent claim that change is unintelligible. Chapter 4 notes that the physicists [οἱ φυσικοί] make opposites [τὰ ἐναντία] principles and moves on to the refutation of Anaxagoras, including the latter's claim that the principles are an unlimited number. Chapter 5 ratifies the claim that opposites are principles and explores the idea of opposites which are more primary than others. Chapter 6 argues that there must be another principle, viz., a nature that underlies the contraries [τις ἕτεραν ὑποτίθησι τοῖς ἐναντίοις φύσιν]. Chapter 7 worries the question whether one may need only the "underlying nature" and one of the opposites and agrees that, in a sense, this may be so, but insists that, in another sense, the principles must be three in number. Chapter 8 argues that earlier philosophers, particularly the Eleatics, misled by the dictum that nothing comes to be from nothing, concluded erroneously that nothing comes to be at all—failing to distinguish between accidental or incidental [κατὰ συμβεβηκός] and essential or *per se* [καθ᾽ αὐτό] non-being [τὸ μὴ ὄν], the former being construed as lack or shortage [στέρησις]. Chapter 9 deplores the identification of matter [ὕλη] and shortage (or lack) and insists that matter alone (not shortage) can, as subject, long for and desire [ἐφίεσθαι καὶ ὀρέγεσθαι] form (as opposite of shortage). Book I ends by leaving the task of determining whether the principle of form [ἡ κατὰ τὸ εἶδος ἀρχή] is one or many and of what sort or sorts it may be to first philosophy and announcing that in what follows only physical and perishable forms will be spoken of.

In chapters 2 through 9 Aristotle operates at a high level of abstraction, fully justifying a first chapter that explains briefly the relation of sense universals to their (ultimate) constituents. Indeed, chapter 2 abruptly introduces the major terms of Plato's *Parmenides,* that most abstract and austere of Plato's writings, viz. (the) *being* [τὸ ὄν] and (the) *one* [τὸ ἕν]. Chapter 1 provides a key to "turn through the wards" of the later chapters in (at least) the clear sense that it tells us how Aristotle is going to use the terms 'principle', 'cause', and 'element' in an attempt to exhibit the logical (and ontological) conditions for the intelligibility of a science of nature. And it provides an equally clear clue to the sort of refutation Aristotle will use in succeeding chapters in dealing with philosophical opponents. Enough, however, of hints and promises. It is time to turn to justification of my reading of chapter 1 and, after that, to more detail from the remaining chapters.

The—alas—widely accepted view (noted by Charlton) that Aristotle in other texts holds that "the individual is known before the universal" and "that what we perceive is individual" makes talk of "sense universals" puzzling if not downright dubious. I should like, however, to show, by summary but somewhat detailed treatment of the *Posterior*

Analytics chapter that Charlton cites (II.19), that one crucial text demands the interpretation which I have given to chapter 1 of Book I of *Physics*. First, a bit of stage-setting.

Aristotelian universals are patently "in the soul" [τῇ ψυχῇ] and are not, *as such,* in the world, though they are *of* and make known various features of the world. General terms, which are emphatically conventional, get their "meanings" by association with universals in the soul, which are emphatically *not* conventional (cf. *On Interpretation,* 16a, 4–8). When a general term is *used,* it is, of course, *not mentioned* (though it *may,* in a different context, be mentioned). One may, as well, speak of universals as being *used,* at least in the sense that there are occasions when one is aware of a kind or of something's being of a kind. Thus one may perceive Callias and, if he (the perceiver) has the universal *man,* also be aware of Callias as being a man or, in perceiving Callias, be aware of *man.* (Cf. *Posterior Analytics,* 100b, 1–2: "For, though the individual [τὸ καθ' ἕκαστον] is perceived, the perception is of-the-universal,[4] for example, of-man, but not of-Callias-man.")

The realism of G. E. Moore, Bertrand Russell, and others (in the first quarter of this century) has so influenced Anglo-American philosophy that many who find it difficult to conceive of awareness as having a structure are likely to dismiss any attempt so to conceive of awareness as creeping Hegelianism. Fortunately, the fear of Hegel has diminished in recent years with the realization that non-paradoxical accounts of meaning require that semantic rules be construed as intralinguistic (or intraconceptual) and that languages (or conceptual systems) be taken as more or less adequate "pictures" of the world. (Where "picture adequacy" is *not* an invitation to more or less naive comparison of conceptual with non-conceptual "peeks" at the world.) With this realization it is possible to make sense of Aristotle's account of universals in *Posterior Analytics* II.19.[5]

Aristotle's task in that nineteenth chapter is twofold: (1) to show how universals (or, if you please, concepts) can be acquired without resorting to Plato's doctrine of Recollection and (2) to round out his account of demonstration by explaining our knowledge of "unmiddled first principles" [τὰς πρώτας ἀρχὰς τὰς ἀμέσους] that are presupposed by the "middled" products of demonstration. (Where a "middled" proposition is one whose terms are shown to be connected by means of a third or "middle" term.)

He accomplishes the first part of the task in the following way. In order to acquire at least our initial universals (concepts), we must be capable of (a) sense-perception [αἴσθησις], (b) memory [μνήμη], (c) integrating "repeated memory of the same" [μνήμη πολλάκις τοῦ αὐτοῦ], the ability so to integrate being what Aristotle calls 'empeiria' [ἐμπειρία, usually translated as 'experience', a better translation, perhaps, being 'concept']. Empeiria in actual operation is identified with "the universal fully come to rest in the soul" (100a, 5–6). As Aristotle actually puts the above in 100a, 3–9, sense-perception gives rise to memory, and repeated memory of the same gives rise to empeiria. Thus, for the actual coming to be of a universal (at least the first ones), there must be sense-perception of something, memory of it, and then "repeated memory of the same" to trigger off empeiria.

In On Memory and Recollection, Aristotle speaks of a "modification" in soul and body which modification, in beings capable of memory, persists. (He writes in somewhat similar vein in Posterior Analytics 100a, 2–3.)

450a26 It might be asked how one can remember something when only the modification is present [ποτὲ τοῦ μὲν πάθους παρόντος], the thing being absent. Obviously one must think of what comes to be through sense-perception in the soul and in the part of the body which has it as modification
30 which is a sort of picture, the lasting state of which we call memory [οὗ ψαμὲν τὴν ἕξιν μνήμην εἶναι].

Memory, so thought of, is a sort of picturing ability, acquired as a result of sense perception and more or less enduring. So understood, memory can be one or the other of two different processes (at a given time): first, in the absence of the (previously) perceived object which brought the particular memory into being, to imagine or picture the absent object, as imagining or picturing a certain (absent) person standing in one's living room; second, in the presence of the perceived object (at a later time) or in the presence of an object of the same kind, to fit or map a memory image or "picture" on the presently perceived object and thus recognize it (as the "same" or of the "same kind").

We may therefore well suppose that Aristotle thinks that universals or concepts belong to a species of memories, namely, those which, while remaining single, may be fitted to, or mapped upon, an indefinite number of numerically distinct perceived (or remembered) objects. (Cf., *Posterior Analytics,* 100a, 5–9: "From *empeiria,* that is, from the universal fully come to rest in the soul, from the one over against the many which is one and the same in all of them. . . .") *Empeiria* as a *faculty* just seems to be the ability to have such memories, and it seems to contrast with what might be called "simple" memory, which would be the ability to remember or imagine individual objects; but the matter is difficult to decide, as the several controversies over the *sensus communis* make clear. What *is* obvious from this chapter of *Posterior Analytics* is the truth of the cliché that all Aristotelian concepts derive from sense perception. What I wish to get to, however, is the claim that, on the evidence provided by that chapter, it is obvious that concepts immediately related to sense are compounded and confused.

Consider Jones's original concept of a horse. He got it, first, by seeing a horse, then, with memory aroused, seeing the same horse or other horses, and finally, with *empeiria* aroused, recognizing this or that horse as falling under a single concept or universal, i.e., as a horse, not as Dobbin, Citation, or Secretariat. But the original Jonesian horse universal may be confused in several ways. Two are obvious. Jones may confuse horses with cows or deer, and he may not recognize some horses (say, horses of a different color) as such. Related to, but distinct from, these confusions is another—and this is directly connected with the first chapter of *Physics.* Jones's original horse concept or universal is inevitably confused in the sense that the "parts" of the original universal are not articulated or separated out. As the *Physics* chapter notes, the articulation of such parts is like the articulation of parts in the definition of a term. Thus, in accord with the standard Aristotelian definition, there would have to be a genus universal and a differentia universal (or universals) as parts of the species universal, *horse.* Indeed, Jones is likely to have the sense universal, *horse,* without even being able to recognize *animals* as such, though, in the process of articulating his *horse* concept, he would have to be able to do so. In that process, Jones would, in Aristotle's view of things (in chapter 1 of *Physics*) be moving from something more intelligible *to him* (the sense universal or sense whole) to something "*naturally*" more intelligible (the part, in particular, the genus universal). Before attending to what that process might be, it is worth noting that, by the same logic, the universal *animal* has parts, including a genus part, and the demand of genus part clarity cannot be met until a universal that has no genus part is reached. In the example at hand (*horse*), the universal that has no genus part would be *ousia* [οὐσία—if you please, "substance"]. In *Posterior Analytics* II.19 Aristotle speaks of a series of "halts" that is not ended "until the universals which are without parts [τὰ ἀμερῆ] take their stand—for example, from such and such animal up to animal, and then from animal in the same way" (100b, 2–3).

I noted in the above paragraph that the first two confusions were related to, but distinct from, the third confusion. It is fairly obvious that if Jones, by use of his sense concept *horse*, really does confuse horses and cows or deer, his concept or universal is badly in need of repair. As I read the last chapter of *Posterior Analytics*, however, if Jones continues to observe the animals around him, and his abilities do not fail, he will acquire separate concepts or universals for deer and cows, partially at least from the bad mapping fits of his confused concept *horse*. And he will, of course, revise and improve that concept. Assuming a relatively complex Jonesian environment, it would seem as well that he will acquire at least a rudimentary concept or universal *animal,* for there will be memory fits or picturings for horses, cows, and deer alike that simply do not fit or picture stones, trees, and so on. The scheme that Aristotle sketches in *Posterior Analytics* does not, of course, suggest any seriate application of the powers of sense, memory, and *empeiria* after initial acquisition of sense universals of various kinds. But it is one that encourages a reader to think that, after acquisition of initial sense concepts or universals, there is a good deal of testing of initial (and partially corrected) sense universals by observation—a back-and-forth process that will yield not merely improved species universals but, as well, a variety of species-genera hierarchies. The process is not unlike (and is, of course, historically related to) the process that Leibniz and other seventeenth-century rationalists called 'analysis'. Aristotle puts the matter in a famous image in *Posterior Analytics:*

100a10　These abilities [ἕξεις—i.e., universals or concepts] are not innate and fully developed, nor do they come to be from other and higher abilities to apprehend. Rather, they come to be from sense perception [ἀπὸ αἰσθήσεως]. It is like a rout in battle where, when one man halts, another does, and

35

then another, until the original position is restored. The
14 soul, being what it is, is capable of undergoing this process
[δύνασθαι πάσχειν τοῦτο].

The picture, of which the rout in battle is an analogue, is of incredible
confusion in aboriginal sense perception, the acquisition of first one,
then another, then another, sense universal, and finally, the articulation
of species-genera hierarchies of universals, so that, with full and articulated conceptualization, the scene around to conceptually informed
perception is as ordered and intelligible as an organized and completely
disciplined army.

There are, of course, universals that make known other objects of
sense perception than *ousiai* or substances.[6] In a well-known passage in
book II of *On the Soul,* Aristotle distinguishes between three objects of
sense perception: (1) "that which is peculiar to each sense"; (2) "that
which is common to all [the senses]"; and (3) "the incidental [or
accidental, κατὰ συμβεβηκός] object of sense" (418a). Presumably
the perceptual basis for the distinction between *ousia* and accident
categories is found in the distinction between the third and the first two
objects of sense. Colors, sounds, and the like, as well as motions,
numbers, shapes, and so on (i.e., both peculiar and common objects of
sense) are always *of* something or other and thus (in the *Categories* sense
of the term) *accidents* [συμβεβηκότα] *of* things (*ousiai* or substances).
This sense is, of course, the one Aristotle invokes in making the *On the
Soul* distinction. The "incidental" or "accidental" object of sense is
just that because a peculiar or common sensible, in being "in" or "of"
it, is an accident of it. It is, I think, fairly obvious that an account of the
acquisition of accident universals can be given that is exactly parallel to
that we looked at (in the light of the last chapter of *Posterior Analytics*)
for Jones's acquisition of the *ousia* universal, *horse*—an account starting with sense-perception, moving then to memory and *empeiria.* And,
as in the case of Jones's original *horse* universal we sketched a hierarchy
of genera parts up to *ousia,* we can think of similar sketches from
original accident universals through the genera for peculiar and common
sense universals to universals with no parts, i.e., to the several accident
categories: quality, quantity, and so on.

Now, of course, the differentiae parts of *ousia* universals will be
accident universals, though such parts will not be accidental to the *ousia*

universals in question. Thus, if *A* is a sense universal, and it is determined after the back-and-forth process described above that it has the *genus* part, Φ, and the differentiae parts, α and β, then if *A* holds for anything, α and β essentially [καθ αὐτό] hold for that thing, and, of course, Φ essentially holds for it. If one perceives anything by means of the articulated universal *A* (say, this *A*), thus having what Aristotle speaks of as an "of the universal perception" [ἡ αἴσθησις τοῦ καθόλου], *Posterior Analytics* 100b1), he will perceive it as being "by virtue of itself" [καθ' αυτό] and thus *necessarily* α, β, and Φ. If *A* is unarticulated, he will not perceive it as necessarily α, β', and Φ, though he may find *A* and its use familiar and obvious—in much the way that the concept of circle may be familiar and obvious to a person who has no formal training in geometry. If *A* does not contain, say, γ as a part, then, if something that is *A* (or that is perceived as *A*) happens to be γ, it is so accidentally (κατὰ συμβεβηκός). Since all of Aristotelian "science" [ἐπιστήμη] is a matter of what obtains *necessarily,* and the whole-part articulations that are paradigms of necessity are made possible by definitions, it is obvious why the entire Aristotelian tradition sets so much store by definitions.

Though it would seem that, given articulation of a universal—say, *A*—the recognition that whatever has it has some or all of its parts—say Φ, α, or β--is immediate or unmiddled, special attention is to be given to the unmiddled recognition of genera parts of species-wholes. For a series of such unmiddled recognitions would, of course, culminate in the discernment of universals that have no parts, and these would be *first* principles. This point is made by Aristotle in *Posterior Analytics* immediately after the famous image of the rout in battle and the reassembling army.

100a15   If what was just said was expressed unclearly, let us state it again. When one of many undiscriminateds [ἀδιαφόρων] halts, the first universal is in the soul; for, though the individual is perceived, the perception is of-the-universal,

100b   for example, of-man, but not of-Callias-man. Again a halt is made among these, until the universals which are partless [ἀμερῆ] take their stand—for example, from such and such animal up to animal, and then from animal in the same way. And so it is clear that it must be by assemblage [induction, ἐπαγωγῇ] that we apprehend the firsts [τὰ πρῶτα]. For even perceiving produces the universal in this manner.

The translation of ἐπαγωγή as 'assemblage' comports with my earlier treatment of sense universals as assembling many individuals under, or by means of, a common picturing or mapping ability.

The above quotation is followed immediately by Aristotle's making a sharp distinction between "scientific knowledge" [ἐπιστήμη] and "intuitive knowledge" [νοῦς], the former being, as we noted earlier, knowledge of "middleds", the latter being immediate. After noting that "a principle of scientific knowledge [ἐπιστήμη] is not itself scientific knowledge", he concludes the *Posterior Analytics* with the following:

100b14   If, therefore, we have no other true kind [of knowledge] besides scientific knowledge ͺἐπιστήμη] [and intuitive knowledge], intuitive knowledge [νοῦς] must be a principle [ἀρχή] of scientific knowledge. And as, on the one hand, it [intuitive knowledge] must be [stand to scientific knowledge as] the principle of the principle [ἡ ἀρχὴ τῆς ἀρχῆς], so, on the other hand, the whole [of scientific knowledge] stands likewise to the whole [body of] fact [τὸ πᾶν πρᾶγμα].

Thus, in anachronistic echo of Locke's contention that every step in a demonstration must be intuitive, Aristotle treats "middled" knowledge as derivative from intuitive knowledge and all "fact", *qua* demonstrated, as derivative from scientific knowledge.

Though I wrote, in the last paragraph but one, of unmiddled recognitions which may go up the species-genus hierarchy (in any category) and culminate in universals that have no parts, there is a sense in which the discernment that any species has a genus part, where that genus, in turn, has a genus part, is middled. If x is A and A has the genus part B, then x is B. And if B has the genus part C, and so on until we reach a genus that has no parts, the final discernment that something has the genus which has no parts would be *strictly* unmiddled. It may be, as Aristotle suggests in the first chapter of *Physics*, that the goal of inquiry is "acquaintance [γνωρίζειν] with" principles, i.e., universals that have

38

no parts, and that intuitive knowledge simply is such acquaintance. Even if this were the case, however, there would have to be some sort of immediate apprehension that such parts which have no parts are parts of this or that whole, at whatever level in the species-genus hierarchy.

This line of thought which takes genera (relatively and absolutely) as principles is subjected to scrutiny in various parts of *Metaphysics*, notably in *B*, chapter 3. Aristotle concludes that chapter in the following way:

999a17 For the principle or the cause [αἰτίαν] must be beyond the things [τὰ πράγματα] of which it is the principle, and it must be capable of being separated from them. But why should anyone suppose any such thing to be beyond the individual except that it is predicated universally and of all?

20 But, if this is the reason offered, it must be set down that the more universal things may be, the more they are principles [τὰ μᾶλλον καθόλου μᾶλλον θετέον ἀρχάς]. So that the first genera would be principles.

It may be worth noting as well that the lexicon of *Metaphysics* Δ lists genera as a meaning of both 'causes' and 'elements' (cf. especially 1013b-1014a26 and 1014b9-16). In this sense, e.g., 2:1 and number are causes of the octave. Summa genera are thought especially to be elements because there is no definition of them (i.e., they have, in the usage of this paper, no parts). I think that it is in this general way that Aristotle uses the terms 'principle', 'cause', and 'element' in the first chapter of *Physics* where the terms are used as though they meant the same. This usage continues through book 1.

Before attempting to use this line of interpretation of the first chapter of *Physics* I as a key to the later chapters, I should like to advert to a further matter that Aristotle most pregnantly discusses in *Metaphysics* Γ, viz., the matter of what I shall call (for want of a better expression) the "super-genera" in particular, *being* [ὄν], *one* [ἕν], *same* [αὐτό], and *different* [ἕτερον]. That being and one may be appropriately called genera is implied by 1003b33–34, where Aristotle remarks that "there are exactly as many species [εἴδη] of being as there are of one". In many other places, including the context of the above remark and in *Physics* I

39

(185b6-8), Aristotle claims that *being* and *one* are "said in many ways" or, to put it misleadingly, have many senses. This is usually qualified by the insistence that the terms are not, strictly speaking, ambiguous, for they are "said", or their senses are, *pros hen* ($\pi\rho\grave{o}\varsigma$ $\check{\varepsilon}\nu$), i.e., related to, or dependent upon, some one primary "saying" or sense, namely, that of *ousia* or substance "saying". At *Metaphysics* 1004a20-24 he says the same sort of thing concerning *difference* or *different:*

> For contrariety is a sort of differentia and differentia is a sort of difference. So that, since one is said in many ways, these also will be said in many ways, yet it is by means of a single [science] that all are discerned.

The difficult issues in what Aristotle calls "first philosophy" concerning the nature and character of super-genera are not, fortunately, our immediate concern. It helps, however, in understanding the first chapter of *Physics* to see how far from the confused universals of sense perception Aristotle is prepared to go in the search for principles, elements, and/or causes. And the evidence is that he is prepared to entertain principles in first philosophy (the super-genera) that are principles, as it were, for the principles of the so-called special sciences, the latter being the sort of principles we have been discussing in connection with the *Posterior Analytics* chapter.

The notion of super-genera combined with the notion that the related terms are "said in many ways" suggests that such super-genera stand to the common or garden variety genera of the special sciences as the rules (or, if you please, the truths) of logic stand to the substantive premises. And this would be in keeping with Aristotle's view that first philosophy is presupposed by any and all of the special sciences. Indeed, he insists, in *Metaphysics* Γ (1005b1–5), that persons who pursue one or other of the special sciences ought beforehand to have had training in first philosophy or logic (analytics, $\tau\tilde{\omega}\nu$ $\dot{\alpha}\nu\alpha\lambda\upsilon\tau\iota\kappa\tilde{\omega}\nu$):

> Physics is also wisdom of sorts, but it is not first wisdom [$\sigma o\varphi\acute{\iota}\alpha$ $\pi\rho\acute{\omega}\tau\eta$]. And the efforts on the part of some of those persons who say what in the way of truth necessarily follows from what are due to lack of instruction [or training] in analytics. For they must have previous knowledge of these [i.e., analytics] before coming to [special study or a special science] and not be searching them out [i.e., analytics or its principles] while listening [to special study or special science lectures].

In Aristotle's *Metaphysics* discourses concerning first philosophy and what I have called super-genera, he seems to be concerned with how it is possible, without paradox or contradiction, to say or think intelligibly strings or thoughts with the forms '_____ is -----', '_____ is the same as -----', and (perhaps among others) '_____ is different from -----', without regard to the particular "non-logical" terms or *incomplexa* that fill the blanks (or, at any rate, without that being the focus of concern). At least part of the point of the insistence that *being* is "said in many ways" is that, for example (and contra Parmenides), the same thing can "be" a man and also "be" white, though *being a man* and *being white* are different (and different in a different way from the way in which *being a man* and *being a horse* are different). Indeed, Aristotle makes precisely this point against Parmenides in chapter 3 of *Physics* I (186ba25 ff.). But now I am anticipating a bit the "turning of the key through the wards."

That the many ways in which super-genera terms are "said" are all *pros hen* or related to some "primary saying" reflects Aristotle's sense that the fundamental things in the world are primary *ousiai* (or, if you please, "substances"), for the *hen* that the others are *pros* is *ousia* "saying". Again I anticipate, for, in chapter 2 of *Physics* I, Aristotle insists (while toying with the notion that Parmenides and Melissus might think that the "All" is a single quality or quantity) that "everything is said with regard to substrate [predicated of, καθ' ὑποκειμένου λέγεται] of *ousia*" (185a32). It is tempting indeed to think that, in moving from sense universals to principles, Aristotle moves up the species-genera scale emphasizing the unity of the several *ousia* species and genera and treating the many differentiae universals as ancillary to the successively more general *ousia* universals. "What is an *A*?" "An *A* is a *B* which α's and β's." '*B*' is a generic *ousia* term, and 'α' and 'β' are accident terms. This is the basic Aristotelian question and answer, whether α and β are proper differentiae or merely purported to be so. The 'is' of 'is α' (or 'is β') or 'α's' (or 'β's') is therefore strictly dependent for its logical intelligibility upon the basic intelligibility of the *ousia* 'is', and that dependence is a consequence of the dependence of differentiae upon genera in definitions. This, in turn, reflects the dependence of accident universals upon *ousia* universals "in the soul", whether in sophisticated consciousness or in the confused and compounded universals of sense. Finally, *this*, in its turn, reflects the

41

difference between the peculiar (or, if you please, special) and common sensibles, on the one hand, and the incidental objects of sense, on the other hand. The super-genus *being* or *is,* so to speak, is the principle of this structuring and is brought to explicit recognition by instruction in, or reflection upon, the sort of matters that Aristotle calls "analytics".

Though the "basic" principles are *ousia* genera, there are, as we have seen, accident genera, and, as such, they are principles, causes, or elements. We can, moreover, go through what I called above "the basic Aristotelian question and answer" where it goes as follows. "What is an α ?" "An α is a φ which β's and γ's". In this, 'α', 'β', and 'γ' are, of course, accident terms, and 'φ' is a generic accident term. Concomitantly, it is possible not only to speak of *the* (or *this*) *A* (an *ousia* or substance) but also of *the* (or *this*) α (an accident). (And this possibility obtains for terms at every level in the genera-species hierarchy, whether of *ousiai* or accidents.) Thus there are *ones* in every category, or, alternatively, *one* is "said in many ways." And once again the "*pros hen*" points to *ousia*. For, whenever it is said that the (or this) α β's (or is β), where both are accident terms, it is always intelligible to ask, "*What (ousia* or substance) α's and β's (or is α and is β)?*" On the other hand, it is unintelligible to ask, "*What B's or A's (or is B or A)?*", where '*B*' and '*A*' are *ousia* terms. The dependence of accident *ones* upon *ousia ones* thus parallels the dependence of accident *beings* (or, if the barbarism be permitted, *ises*) upon *ousia beings* (or *ises*).

The difference between the super-genera, *one* and *being,* can be gotten at linguistically. *One* seems to be the principle of subject terms as such, and *being* the principle of predicate terms as such. Both reflect the structure of Aristotelian definitions, with the *one* reflecting the genus and *being* reflecting the differentiae. And accidental ($\kappa\alpha\tau\grave{\alpha}$ $\sigma\upsilon\mu\beta\epsilon\beta\eta\kappa\acute{o}\varsigma$) predication, as contrasted with essential or *per se* ($\kappa\alpha\theta$' $\alpha\grave{\upsilon}\tau\acute{o}$) predication, seems to be a weakened form of this structuring. I cannot avoid some adversion at this point to Plato's *Parmenides,* for its intense and puzzling preoccupation with "logical" matters concerning *one* and *being* (as well as *same* and *different*) seems clearly a source, if not *the* source, of Aristotle's preoccupation with the same matters in *Metaphysics* and, of course, *Physics* I. Early in the process of "drawing consequences" for the so-called second hypothesis in *Parmenides,* Plato notes (1) that the hypothesis is that *one is,* with "'is' signifying something other than 'one'" (142 B 4), (2) that the one and the being are "parts of the one-being" and that neither "can forsake the other" (142 E 1), (3) that "being [οὐσία] is distributed through all the many beings [πάντα πολλὰ ὄντα] and is wanting in none of the beings" (144 B 1–2), and (4) "[Being] is not distributed into more [parts] than the one, but into an equal number, as it seems, to the one. For the being [τὸ ὄν] is not absent from the one, nor the one from the being, but both are equally present in everything always" (144 E 1–3).

What I should like to claim for Plato as well as Aristotle—though the above quotations from *Parmenides* are, of course, not conclusive—is that *one* and *being* are, respectively, principles of subjects as such and predicates as such. The use of the terms for them—as super-genera—is in articulation of logical form, or, if you please, *logischer Raum*. And both of them are surely pervasive, for it is obviously impossible to say anything about anything without presupposing that about which it is said (the one) and what is said about it (the being). The Platonic conception of the equal distribution of *one* and *being* seems indeed to require a world of "facts" (Plato's term is 'one being', ἒν ὄν) the inseparable components of which are "ones" and "beings". What Aristotle seems to have in mind in *Metaphysics* Γ are super-genera which are principles and causes in the sense that the articulation of their several species is presupposed by any of the sciences in which, in Quine's phrase, descriptive terms occur "non-vacuously". We are, in all this, a very long way from the naïve and compounded universals of sense, but Aristotle himself takes us there as early as chapter 2 of *Physics* I.

A word about *same* and *different* before turning to the other chapters of *Physics* I. (As we shall see, this has a bearing on the explanation of "opposites" as principles in later chapters of the book.) For our rather cursory treatment, *same* presents no great problems, for it seems clearly the super-generic principle for what is commonly referred to as the "is of identity" and thus, after the fashion of Plato's much-discussed *Sophist* account,[7] is pervasive. It will be recalled (from *Metaphysics* 1004a20–34) that Aristotle took contrariety [ἡ ἐναντιότης] as a "sort of" [τις] differentia or variance [διαφορά] and variance as a sort of difference [ἑτερότης] and claimed that all three, though "said in many ways", are, in so doing, "discerned by means of a single science"— first philosophy. A bit later, at 1005a6–8, he suggests that their "many sayings" are, like those of *one*, *pros ti*, i.e., related to a "primary" saying.

CHAPTERS 2 AND 3 OF "PHYSICS" I

Aristotle starts chapter 2 by distinguishing between those inquiring whether there is one principle or many to be the logical or ultimate subject of propositions about the world and those inquiring whether there is one or many "beings" [ὄντα] to be the logical or ultimate predicate. He concludes (at 184b25–6) that the latter as well as the former are inquiring after principles or elements. He does not press the distinction, for he is really after what he takes to be the contention of Parmenides and Melissus that there is but one ultimate subject and that it is unchangeable [ἀκίνητον]. Only later, at 186a30–32, does he note that Parmenides "had not yet seen" the distinction between a subject and what "holds for" [ὑπάρχει] a subject.

Aristotle notes almost at the outset that the investigation of the question whether "being is one and unchangeable is not a question for investigation in physics" (184b27–185a1). Even so, after making the remark, he continues in the interest of making some general points to discuss Parmenides and Melissus. He says, first, that, if the so-called monists are serious about their "one" claim, then they cannot be said to have a principle at all, for a principle must be *of* something. In the light of our earlier discussion, this amounts to saying that, on the supposition of one utterly simple being, there can be no relation between what has parts and the principle (which has no parts) either relatively or absolutely. Commencing at 185b20 he toys with the ideas (a) that the monist might be claiming that *is* is said in only one way, so that there would be a species-genera hierarchy in only one of the categories, as *ousia* or quality, and (b) that, even more constricting, the "all" would be one man or one white or one hot. These obviously will not do at all. But he goes on to say that, if there should be *ousia,* quantity, *and* quality, and so on, whether these are "apart from" each other or not, then "the beings will be many". This is not, of course, another way of saying that there will be many things in the "ordinary" way, but rather a conclusion that *being* will be said in many ways and thus the principles will be many. The qualification 'apart from' is simply the claim that this holds whether or not Aristotle's own doctrine of *pros hen* holds. He does, however, invoke that doctrine in next claiming that "none of the others [i.e., quality, quantity, and so on] is separable from *ousia*" (185a31–33). (And this latter is not to be read as claiming or implying, as some translations suggest, that, while beings in other categories cannot "exist" alone, *ousiai* can.)

There follows an attack on Melissus' doctrine that the one is *apeiron*, unlimited or infinite. Its interpretation is rather obvious. But it does serve to introduce Aristotle's remark that "*one* is itself said in as many ways as *being*" (which we have noted earlier). He follows it by suggesting that Melissus might be thinking (albeit incoherently) that there is an indefinite or infinite multiplicity all of which are "one in definition or formula". Aristotle points out the absurdity of this possibility by drawing the Heraclitean conclusion that, if this were so, then everything would be equally a horse, a man, a quality, good, and so on—with no distinctions at all.

Chapter 2 concludes with Aristotle scorning an attempt—attributed to Lycophron—to do away with *is* by the expedient of substituting, e.g., *walks* for *is walking,* and so on. He simply points out that one can be many, provided that we recognize that *is* can be said in many ways and refrain from contrary or contradictory "sayings" within one category, as one and the same man may be both white and cultivated [μουσικός].

Early in chapter 3 a puzzling argument is attributed to Melissus. It is: "If whatever comes-to-be has an *arche* [αρχή] , it follows that what does not come-to-be does not" (186a7–9). Whether '*arche*' is read as 'beginning' or as 'principle', the argument is an obvious howler. Read as 'beginning', though the argument remains a howler, there seems no great harm (as Charlton, among others, notes).[8] Read as 'principle', we can construe the argument as denying that there can be any principle for coming-to-be on the ground that, if there were, then there would be no principle for what does not come-to-be, and that this is absurd. Unfortunately for this reading and for the notion that the argument is a howler, Melissus, *Frag. 2,* seems to argue that, though what comes-to-be has a beginning (or, possibly, principle), what does not come-to-be does not have a beginning (or principle) and is thus unlimited. And this, though it might be false, does not involve any obvious logical mistake.

Immediately after citing this puzzling argument, Aristotle takes Melissus as failing to recognize that "instantaneous" change from this to that is as much a coming-to-be as change that is divisible into temporally successive parts. And he insists that a spatially extended thing, though very large, can change from this to this without there being successive changes of its spatial parts. In this same vein, he asks why, even granting that the all or whole is one, the whole, as such cannot move in place. And why, granting the same, it cannot as such undergo or suffer qualitative change (ἀλλοίωσις). These arguments seem clearly based on the idea that something can be simply and solely one and without parts but also be many in the sense that it may at least successively admit of different predicates and are thus exploitations of the distinction between *one* and *being* which we discussed earlier. Aristotle goes on to point out that (186a23–33), still consonant with something's being simply one, if we (or, in the case at hand, Parmenides) will grant that *being* is said in many ways, there is nothing to prevent a single thing's simultaneously admitting different predicates. And, if the thesis is that *being* is said in only one way, there is nothing to prevent any number of things from *being,* say, white. It is at this point that Aristotle accuses Parmenides of not arriving at or recognizing the distinction between *being* (white or whatever) and whatever it is that is (white or whatever).

The remainder of chapter 3 is devoted to exposing paradoxes which result from failure to recognize this distinction. Since he fails to do so, Parmenides "must therefore assume not only that 'being' signifies one (of which it is predicated), but also [that it signifies] that very being or that very one [ὅπερ ὄν καὶ ὅπερ ἕν]" (186a33–34). With that assumption, *being* cannot be said with regard to substrate (or, if you please, predicated) of anything else [καθ᾽ ὑποκειμένου τινὸς λέγεται]. For then it would have to be said of what strictly and simply is not, for by hypothesis, *being* signifies and can be predicated only of itself, and nothing else *is* at all. And with that assumption, nothing can be predicated of *being.* For, not being "that very being", what is predicated must not *be* at all.

Again, with the assumption, *being* can have no magnitude. For a magnitude has parts that are different from one another, and thus *being* would not be "that very being" or simply and solely being, but rather *beings.*

Nor, on the assumption, can being have parts in the way a universal may have parts. Suppose, Aristotle says, that *man* is *that very being.* But *man* has the parts, *animal* and *biped.* If each of these is *that very being* (*man*), we have made no progress at all. If neither of them is, then they must either be predicated of (literally, they will be accidents or "happenings to" [συμβεβηκότα ἔσται]) either *man* or some other subject [ἄλλῳ τινὶ ὑποκειμένῳ]. But both of these are impossible.

The first is impossible. For something is said *symbebekos* or predicated of a subject in such a way that it may hold of the subject or not hold of it [ὑπάρχειν καὶ μὴ ὑπάρχειν], in which case we are back to the difference problem. Or something is said *symbebekos* or predicated of a subject where what is predicated enters into the definition of the subject (as animal and *biped* enter into *man*). But this cannot hold or not hold on pain of admitting the possibility of men who are neither bipeds nor animals. Since *biped* and *animal* are parts of the definition (here λόγος, perhaps better rendered as 'formula') of *man,* and *man* is not part of the definition of either of them, their predication once again gives us the difference problem.

The second is impossible. If *animal* and *biped* hold for something else that is other than *man,* then (since they form the definition of *man*) *man* must hold for it also. But then *man* could not be *that very being.* Aristotle ends this argument by noting that, failing these possibilities and continuing the assumption, the "all" would have to consist of or be from "undivideds" [ἐξ ἀδιαιρέτων].

And he concludes the chapter with a plea for the recognition that being is said in many ways and that, consistently with denying that contradictories can be true together and that whatever *is* is *that very being,* we can admit what is not (but not, of course, what is not purely and simply) and a plurality.

CHAPTERS 4 AND 5 OF "PHYSICS" I

Aristotle starts discussion of the "physicists" by dividing them into two kinds. First, there are those who make their *one* an underlying body (whether air, water, fire, or some other) that can *be* or be characterized by opposites [ἐναντία] and thus admit of change. Though Plato is listed with them, Aristotle notes that Plato's underlying stuff, "the great and small", contains the opposites, and his *form* [εἶδος] supplies the principle of the *one.* Second, there are the others, notably Anaxagoras, who have a *one* (stuff) that contains the opposites, the latter being separated out [ἐκκρίνεσθαι] in change. Though Empedocles is also counted with these, Aristotle devotes almost all of chapter 3 to an attack on

Anaxagoras. I shall ignore most of the attack, for it turns largely on fairly obvious difficulties in connection with Anaxagoras' infinite or unlimited quantity of "seeds". At 188a6–9 Aristotle gets to what, for our purposes, is the crucial matter.

> It is correctly said [by Anaxagoras] that the separating out is never completed, but not said knowingly. For undergoings [or affections, τὰ πάθη] are not separable. If colors and states [αἱ ἕξεις] are mixed together and then separated out, there will be a certain white or a certain healthy which is nothing other than just that and is not [predicable of, holding for] any subject [or substrate, καθ᾽ ὑποκείμενου].

The interesting problem with Anaxagoras is thus—in Aristotle's view—his failure to grasp the difference between *being φ* and the *thing* (or, if you please, substrate) *which is φ*, in a word, the need for both the principles of the *one* and *being*.

Chapter 5 commences with the astonishing claim that "everyone makes the opposites principles" including Parmenides, who is alleged to "make hot and cold principles". The latter is a bit less disturbing when it is recognized that Parmenides, for whatever obscure reason, thought his *being* to be made out of fire and earth. More important is what he says almost immediately afterward concerning Democritus, whose atoms have "position, shape, arrangement", namely, that each of these is a genus of opposites. The first has species like above and below, the second like straight and curved. He gives no species for arrangement. He goes on to say:

> 188a26   And so that all in some way [πως] make the opposites principles is obvious. And this is reasonable [εὐλόγως]. For it is necessary that the principles not be from one another or from others and that all be from them. And these conditions hold for the primary opposites—because of their being primary [πρῶτα] they are not from others; because of
> 30   their being opposites, they are not from one another. But we must see how this follows also from the way we speak [ἀλλὰ δεῖ τοῦτο καὶ ἐπὶ τοῦ λόγου σκέψασθαι πῶς συμβαίνει].

It is important to notice that the above claims that opposites cannot *be* from one another. The *be* is in sharp contrast to what we shall shortly attend to, namely, that opposites *come-to-be* from one another. (εἶναι is in marked contrast to γίγνεσθαι.) They are not from one another in the sense that an opposite is a principle (or genus) of its opposite. And, of course, if an opposite is primary, there is no principle or genus above it. Part of Aristotle's complaint against a number of his predecessors is that, though they recognize opposites as principles, they have chosen subordinate ones. The last sentence of the quotation has been the subject of some dispute,[9] a dispute that turns on the meaning of ἐπὶ τοῦ λόγου. Though this is not the appropriate place for full discussion of the dispute, it does seem clear to me that Aristotle does indeed turn immediately from making the comment to an exhibition of what has recently come to be called the "logical grammar" of our (or, if you please, Greek speakers') talk of coming-to-be this or that.

The next section is devoted to pointing out that it is incoherent to say that something comes-to-be φ "from" ψ, unless ψ is appropriately opposed to φ. For example, one does not come-to-be pale from knowing music (or, if you please, being cultured). One comes-to-be pale from, say, being dark. Aristotle completely generalizes this claim about coming-to-be: Whatever comes-to-be comes to be from its opposite; whatever passes away passes into its opposite. There is some qualification, however, for there are, in many cases at least, opposites that are "between" extreme opposites, as most of the colors are between black (or the dark) and white (or the light). And Aristotle does not fail to point out that we do not have terms for a number of opposites. This is especially the case for "arrangements" of various kinds. A house is such an arrangement; but we have no name for the lack of arrangement which is that in a pile of bricks from which the house comes-to-be. Presupposed in Aristotle's discussion but not explicitly formulated is the idea that an opposite cannot be a principle for its appropriate opposite, since opposites are species under a common genus and thus one cannot be the genus or principle of the other.

He concludes the chapter by complaining, as I noted earlier, that some of his predecessors have chosen subordinate and not primary opposites as principles. In this connection he accuses them of "laying down" their opposites "without reason" or "without an account" [ἄνευ λόγου]. Because of this they have no clear understanding of which opposites are prior and which are posterior. There results the anomaly that a pair of

49

opposites may be genuinely different from each other, but generically the same. Two colors, for example, are genuinely opposites, but they are both, generically, colors and in this way the same. Clearly Aristotle, with it in view to uncover the first principles of any change whatsoever, is in search of opposites for which there is no such anomaly.

The chapter ends with a passage that is prima facie incompatible with the interpretation I have given chapter 1.

> 189a3   In this way, therefore, they speak in the same way and differently and better and worse. And some [lay down] those which are more knowable [γνωριώτερα] by means of reason
> 5   [or by an account, κατὰ τὸν λόγον], others, [those which are more knowable] by perception [κατὰ τὴν αἴσθησιν]. For the universal is knowable by reason, but the constituent [in most translations, individual or particular, τὸ καθ' ἕκαστον] by perception. For reason [λόγος] is of the universal, but perception is of the according to part [τοῦ κατὰ μέρος]. For example, the great and the small by reason, and the porous and the compact by perception.

I should *like* to read this in the following way. Since Aristotle has just been concerning himself with the distinction between primary and subordinate opposites, both taken as universals, with the former taken as genera and the latter as species, so also in this passage. And the great and small stand to porous and compact as genus (at some remove) to species (or, if you please, genera to species). It is only by reflection or reasoning (logos) that one arrives at the more universal or generic principles. Put slightly differently, genera are not perceived at all in the sense that, for example, one never perceives something that is just an animal and fails to be some sort of animal. So too with small and porous or great and dense, color and black, plane figure and triangle (or triangle and equilateral triangle), and so on. This requires reading τὸ καθ' ἕκαστον as something like 'lowest level species' or, as terms were used earlier in this paper, as 'sense universal'. There is some textual warrant for reading this expression in some such manner. For example, at 331a20–24 of *On Generation and Corruption* the contrast between καθόλου and καθ' ἕκαστον appears to be between earth, air, fire, and water on the one hand and their sensible features (hot, cold, moist, and dry) on the

50

other. And the contrast in the present passage (next to the last sentence) between καθόλου and κατὰ μέρος does not seem to be one between universal and particular, though it parallels the earlier distinction. Indeed, there seems at this point some echo of the chapter 1 distinction between what is intelligible naturally and what is intelligible to us, the latter being sensible. In the early sections of this paper, however, I treated the sense universal as the *whole* and generic removes from it as *parts*. I do not know how to resolve this conflict of usage between chapter 1 and chapter 5 except to note that there are other Aristotelian usages of the terms in question that are naturally read in one or the other of these two ways and to note as well that there is nothing strange about Aristotle's thinking in one context of "parts" of universals "intensionally" as genera and in another "extensionally" as classes whose members are sub-groups of the membership of larger classes.

CHAPTERS 6 THROUGH 9 OF "PHYSICS" I

Chapter 6 raises the question: How many principles are needed for making change intelligible? There must be more than one, for we need "opposites", and this requires no less than two. And there cannot be an unlimited number of principles without sacrificing intelligibility. After noting that some opposites are prior to others, and others are derived from others [γίνεται ἕτερα ἐξ ἄλλων] and giving sweet-and-bitter and white-and-black as derivative examples, he says (at 189a20) that "principles must always stand fast" [τὰς δ' ἀρχὰς ἀεὶ δεῖ μένειν]. I understand this "stand fast" or "remain" to mean that principles must be a pair of *basic* or, if you please, generic opposites which are present in every pair of species opposites in the way that *animal* is present in *dog, horse,* etc. But this theme is by now very familiar.

But there would seem to have to be a "third nature" to "underlie" [ὑποτίθημι] the opposites, and this would seem to have to be an *ousia*. And this underlying *ousia* cannot, of course, be derived from either or both of the opposites which are "said of" it, the opposites not being themselves *ousiai*. Furthermore *ousia* has no opposite. Aristotle even suggests that it is an "ancient opinion" that "the one and excess and defect are principles of 'the beings'" [τὸ ἕν καὶ ὑπεροχὴ καὶ ἔλλειψις ἀρχαὶ τῶν ὄντων εἰσί] at 189b12–13. And he obviously approves the opinion. And in this, of course, 'the one' seems to indicate

147484

51

generically a or the proper holder of the subject place, namely, *ousia* (or, if you please, *ousia* proper). He concludes the chapter by declaring it a problem whether the elements [στοιχεῖα] are two or three in number. Here obviously 'element' is used as a synonym of 'principle'.

Chapter 7 begins with the declared intention to attend to "all (or the whole of) coming-to-be" [πάσης γενέσεως]. To do this, he says, it is "natural" to proceed from "the common [features]" [τὰ κοινά] to "the peculiar [features] concerning each" [τὰ περὶ ἕκαστον ἴδια]. I mention this latter in part because I wish to note another use of ἕκαστον where it does not mean 'individual' or 'particular' unless 'individual or particular *case*'.

Aristotle proceeds to note that we may speak of one thing's coming-to-be "from" another either by using incomposite or simple terms, on the one hand, or at least one composite term, on the other. Thus, if β and γ are "opposites", and 'A' is an *ousia* term, we can say 'The A comes-to-be γ', 'The β comes-to-be γ', or 'the β A comes-to-be (a) γ A'. The terms alone, of course, are or get at incomposites or simples. Joined, they are or get at composites. A, β, or βA Aristotle calls a "coming-to-be thing", γ or γA he calls "what comes-to-be". As noted in earlier chapters, γ comes-to-be "out of" or "from" β, but it can hardly come-to-be "out of" or "from" A. (As, in Aristotle's example, musical comes-to-be from unmusical, but not from man, though a man may come-to-be musical.)

He notes as well that in the coming-to-be, A remains, but β does not. Furthermore, though βA may be "one in number", it is not one "in form" [εἴδει], for the "formula" or "definition" of A and that of β are different. If, therefore, there is something that remains through the coming-to-be, and we think of the coming-to-be thing as composite, the principles are two. This possibility—that the principles are two—is given considerably greater plausibility if one considers cases of coming-to-be that are on the surface at least rather different from the schematic example above, for example, a statue's coming-to-be from bronze (or *this* bronze). Here, however, the bronze is the subject, and the opposites are shapes. Or the coming-to-be of an animal from the sperm. Here (and in a few other cases) the composite character of the coming-to-be thing is not so clear. At 190b28, Aristotle introduces the term 'shortage' [στέρησις] into the discussion. The suggestion is that, in the cases where we are tempted to take the principles as two, we can think of an underlying substrate [matter or ὕλη, as subject] and shortage (i.e., shortage of form or some such) as making the composite coming-to-be thing and of what comes to be as the form or accident (the "shortage" of which disappears in the coming-to-be).

Aristotle ends the chapter expressing a slight misgiving concerning "whether the form [τὸ εἶδος] or the subject [τὸ ὑποκείμενον] is *ousia*" (191a20–21). What, indeed, are we to say of "this (bit of) matter" which, from shortage, comes-to-be, say a man?

Aristotle begins chapter 8 expressing the hope that the distinctions made in the previous chapter will show a way out of the following familiar dilemma of his Eleatic predecessors. There can be no coming-to-be, say, φ. For it must be *either* from or out of what *is* φ (but that is φ already) *or* from or out of what *is not* φ (but for this there would have to be something underlying [which would presumably have to be φ]). Since these are both impossible, coming-to-be φ is impossible. And at 191a33–34, he writes, "And thus, carrying on through to the next conclusion, they say that many are not but only being itself [καὶ οὕτω δὴ τὸ ἐφεξῆς συμβαῖνον αὔξοντες οὐδ᾽ εἶναι πολλὰ φάσιν ἀλλὰ μόνον αὐτὸ τὸ ὄν]". To see the relevance of this, I think that one must suppose that, should, per impossibile, there be a coming-to-be φ out of what is not φ, but rather, say, ψ, then there must *be* both φ and ψ (and there would then be many). But there can be no such coming-to-be, so there is the presumption that "many are not." I might note the similarity of this line of thought to that expressed in the Zenonian argument stated early in Plato's *Parmenides* (127E, 1 ff.). The argument is: If the beings (those which are) are many, then they must be likes and unlikes; but they cannot be likes and unlikes; so the beings cannot be many. Here, unlike the above, the syllogism is valid. It does not seem to stretch the sense of the *Parmenides* argument to find coming-to-be φ from or out of being ψ as requiring the same "being" to be both like and unlike and thus warranting the monistic conclusion.

Aristotle goes on in his own way to agree with the coming-to-be dilemma of the Eleatics but insists that the statement of it is ambiguous [τοῦτο διχῶς λέγεται (191b2)], and exhibits that ambiguity with the use of the distinctions made in chapter 7. If the substrate or *ousia* is, say, A, then there can be a coming-to-be of φ out of what *is*, that is what *is* A (not, of course, what *is* φ). Again, if the substrate is A, there can be a coming-to-be of φ out of what *is not* in that what is not φ, namely ψ, belongs to the composite ψA and is thus an "accident" of A. If, therefore, one considers ψ and φ just by themselves or in themselves (καθ' αὐτό), the Eleatic dilemma holds. But if one notes the distinction between *just by themselves* and *accidentally* [κατὰ συμβεβηκός] and thus recognizes the need for a substrate or *ousia,* the dilemma fails. And this, of course, is tantamount to recognizing the need of both a substrate and a pair of opposites as principles or elements.

In chapter 9 Aristotle is at pains to distinguish his "triad" of "matter" (the substrate or subject), "shortage" (one of the opposites), and "form" (the other of the opposites) from Plato's apparent triad of "great and small" and "form" [ἰδέα]. His claim is that Plato's "great and small" fuses or confuses the roles of subject (which remains and thus "is" through a "change") and shortage [στέρησις] (which does not remain and constitutes the "non-being" from which the "being" of the form comes). In making the claim, Aristotle is quite explicit that by 'matter' he does *not* mean sheer potentiality or "prime matter". "For I mean by matter the primary subject in each case" [λέγω γὰρ ὕλη τὸ πρῶτον ὑποκείμενον ἑκάστω]. And this is always something definite—whether a piece of leather (in the coming-to-be of a purse), a pile of bricks (in the coming-to-be of a house), a horse (in the coming-to-be running of that horse), or whatever.

And this matter, always being something definite, is capable (in the "nature" cases) of "desiring" [ἐφίεσθαι—the metaphor being explicitly sexual] and "longing for" [ὀρέγεσθαι—192a19] the form. The "great and small", identified with the form's opposite or "non-being", can hardly do so on pain of its own destruction. Nor can the form, for it *is* already, as it were. This, of course, is more evidence that Aristotle is taking *matter* as definite, i.e., as having its own nature and hence capable of efficient causality. I think that Charlton is quite right in noting allusions to Plato's *Timaeus* in the chapter.[10] Aristotle, at 192a13–14, seems to concede Plato a persisting entity which is a joint cause with the form [συναιτία τῇ μορφῇ] in coming-to-be and is, as it were, a "mother" [ὥσπερ μήτηρ— as in the *Timaeus* metaphor of the "receptacle" as a mother]. So Aristotle's complaint with Plato's great and small is that, as subject, it has no nature and hence no desire and that, as opposite (or shortage), it cannot persist. He concludes by eschewing investigation of "(the) principle so far as it concerns form" [περὶ δὲ τῆς κατὰ τὸ εἶδος ἀρχῆς] ("whether the principle is one or many and what it or they are''), leaving it to first philosophy and another occasion. The concern of physics, as he puts it, is rather with ''natural and perishable forms''.

Book I of *Physics* ends with Aristotle proclaiming that he has determined in the book ''that there are principles, what they are, and how many there are''. It will be recalled that chapter 1 began with the claim that we can have ''scientific knowledge'' of nature only when we have pushed the inquiry back to the first principles and have gotten ''acquainted'' with each of them. It seems only appropriate to end by claiming success after unearthing and arguing for there being the three principles: subject or substrate, shortage and form (or, if you please, a pair of opposites).

1. Philip H. Wicksteed and Francis M. Cornford, *Aristotle: The Physics,* Loeb Classical Library, vol. 1 (Cambridge, Mass., and London: Harvard University Press, 1929), p. 4.

2. For example: W. D. Ross, *Aristotle's PHYSICS,* (Oxford: Oxford University Press, 1936); W. Wieland, *Die aristotelische Physik,* (Leiden: 1959); W. Charlton, *Aristotle's PHYSICS I, II,* (Oxford: Oxford University Press, 1970) (hereafter, ''Charlton, PHYSICS''); G. E. L. Owen, ''Logic and Metaphysics in Some Earlier Works of Aristotle,'' in I. Düring and G. E. L. Owen, eds., *Aristotle and Plato in Mid-Fourth Century,* (Goteborg: 1960), pp. 163–90.

3. Charlton, *PHYSICS*, p. 52.

4. The Greek text reads as follows: καὶ γὰρ αἰσθάνεται μὲν τὸ καθ' ἕκαστον, ἡ δ' αἴσθησις τοῦ καθόλου ἐστίν, οἱον ἀνθρώπου, αλλ' οὐ Καλλίου ἀνθρώπου. I understand the contrast in the sentence to be between what is perceived (the individual) and a characteristic or feature of the perception (viz., its being an ''of-the-universal'' perception). I do *not* understand it to be a contrast between two different items which are perceived.

5. What follows on the last chapter of POSTERIOR ANALYTICS is presented in much greater detail in my article, ''POSTERIOR ANALYTICS II, Chapt. XIX: Sense Perception, Universals, and Intuitive Knowledge'', *Proceedings of the Ohio Philosophical Association*(Akron: 1970).

6. There is, of course, some discomfort in translating οὐσία by 'substance', if only because of possible confusion with ὑποκείμενον. In what follows, I shall stay almost exclusively with the term 'ousia'.

7. I have in mind, of course, the *Sophist* notion that *Being, Same,* and *Different* pervade (at least) all of the forms; for every form *is, is the same* (as itself), and *is different* (from every other). I think that Plato's inclusion of *Motion* and *Rest* among the ''greatest kinds'' in *Sophist* is due to his thinking of *Motion* as ''always in a different'' and rest as ''always in the same''. Cf. *Parmenides* 146A, 2–5.

8. Charlton, PHYSICS, p. 59.

9. Charlton, PHYSICS, pp. 65–66.

10. Charlton, PHYSICS, pp. 82–83.

David E. Hahm

Chapter Three

WEIGHT AND LIGHTNESS IN ARISTOTLE
AND HIS PREDECESSORS

Early Western philosophy and science has always seen a close connection between weight and the movement of elemental substances. Aristotle's formulation of this connection in the fourth century B.C. was, no doubt, the most influential, eventually coming to dominate Western thinking on the subject and serving as the starting point for subsequent formulations. Aristotle himself was fully aware of the originality of his theory. In his treatise "On Heavy and Light" he complained, "Everyone uses the powers [of heavy and light], but none save a few have defined them" (*Cael.* IV.1.308a3–4).[1] And even these, he maintained, have spoken only of the relatively heavy and light; the absolutely heavy and light have never before been discussed (*Cael.* IV.1.308a9–13). To bear out his claim of originality and to demonstrate the superiority of his own theory Aristotle then presented and criticized the views of his predecessors. Though, in retrospect, Aristotle's originality on this subject seems unquestionable, the precise nature of his achievement does not seem to be fully revealed in his own discussion. To really understand Aristotle's relation to his predecessors and his own contribution to the topic of weight and the movement of elemental substances, we must make a thorough, independent comparison of Aristotle's view with those of his predecessors.[2]

Aristotle begins his primary discussion of weight and lightness in *On the Heavens IV* by defining his subject and distinguishing two senses of the terms "heavy" and "light". "Heavy" and "light", he maintains, are applied "with reference to the capacity to move naturally in some direction" (*Cael.* IV.1.307b31–32). The actualization of the capacity or potential, Aristotle claims, has no specific name, except perhaps "inclination" [$\dot{\rho}o\pi\acute{\eta}$]. Becoming more precise, Aristotle points out that the terms "heavy" and "light" may be used in a strict, absolute sense or in a relative sense, as we say of two heavy things that one is lighter and the other heavier. By definition "absolutely light" is "that which moves up and toward the extremity of the cosmos", and "absolutely heavy" is "that which moves down and toward the middle". In contrast, "relatively light" or "lighter" is "that one of two bodies of equal size, each possessing weight, which moves down by nature less swiftly than the other" (*Cael.* IV.1.308a29–33). The remainder of the book is devoted to an explanation of the causes of the phenomena denoted by the terms "heavy" and "light" and to a discussion of the specific differences and properties of heavy and light substances.[3] But before Aristotle expounds his own theory in the third chapter, he makes it clear that he is most proud, not of his causal explanation of the movement of heavy and light, but of his discovery of heavy and light in the absolute sense. For it is not only for minimizing the topic of weight and lightness that he castigates his predecessors, but for failing even to distinguish heavy and light in the absolute sense and for mistaking an explanation of relative weight and lightness for an explanation of absolute weight and lightness (*Cael.* IV.1.308a9–13; IV.2.308a34–b2).

This assessment of his predecessors' failure is underscored in his review of their opinions. In chapter 2 Aristotle first reviews the theory of Plato's *Timaeus* that the heavier is made up of a greater number of parts, which Aristotle takes to be the triangles of which Plato's elementary bodies were composed. He refutes this theory by pointing out that it does not fit the facts of absolute weight and lightness. For example, it cannot explain why a larger quantity of fire should move up more quickly than a smaller quantity. On Plato's theory the larger quantity would have more parts and would therefore be heavier. Again, Plato's theory would permit one to imagine a quantity of air of more parts and therefore heavier than some given quantity of water. Yet observation shows that any quantity of air whatsoever will rise in water. In fact, the larger the quantity of air, the more it will rise in water (*Cael.* IV.2.308b3–28). Hence Plato's theory is not acceptable.

The second theory he discusses and refutes is that of the atomists, who maintain that some bodies are lighter than others because they contain more void. Aristotle first improves the theory by adding that the lighter is what contains not only more void but less solid.[4] Then, he dismisses it because it fails to account for the fact that fire will always be lighter than

57

earth, regardless of the quantity. For the atomists would have to concede that there is a quantity of fire so large it contains more solid than a given quantity of earth and that this quantity of fire is heavier than the quantity of earth. Similarly, one can imagine a quantity of earth so large it contains more void than a given quantity of fire, and hence one would have to admit that it is lighter than fire and so moves up. After enumerating a number of such absurd consequences, Aristotle suggests the atomists might want to resort to the ratio of solid to void as an explanation for heaviness and lightness. But then, he points out triumphantly, they will not be able to explain why a larger quantity of the same substance moves faster (*Cael*. IV.2.308b28–309b15). In the last analysis, the atomists have failed to account for absolute weight and lightness.

The third theory discussed, but not identified, by Aristotle is one that differentiates heavier from lighter by greatness and smallness. This view, too, embodies a basic flaw. In making everything quantitative variations of one nature or substance (*physis*), it precludes any absolutely light substance or upward movement except by "being left behind or being squeezed out". Secondly it implies that many small bodies may be heavier than a few large ones, and consequently a large quantity of air or fire will be heavier than a small amount of water or earth (*Cael*. IV.2.309b29–310a13). Thus this theory, like the former, fails to define and account for what Aristotle calls "the absolutely light", i.e., the tendency of fire in any amount whatsoever to rise to the top of everything. More generally, we may say that what Aristotle misses in all his predecessors' theories of heavy and light is his own theory of absolute weight and lightness and its corollaries.[5]

After his review of his predecessors Aristotle is satisfied that he has made a momentous, unquestionable discovery. But to the historian his discovery appears to be rather a fertile, new point of view, the essence of which is revealed only by a direct comparison of Aristotle's theory with those of his predecessors.

The evidence for the atomists' ideas on weight and lightness is not completely adequate, but it would seem that Democritus did not assign weight to the atoms as an inherent property.[6] Instead, he maintained that atoms are endowed with size, shape, and an eternal random motion in the void. Very likely Democritus did not assign any particular cause to the eternal random motion that makes the atoms collide with and jostle one another eternally. In the course of the constant jostling the atoms tend to form a whirl or vortex, and this vortex tends to make atoms of similar shape and size congregate. Gradually the larger atoms move toward the center and the smaller ones toward the edge of the vortex, and so the universe is formed. As the mass of atoms in the center of the vortex grows denser, the atoms strike each other and press against each other to such a degree that the smaller and smoother atoms are forcibly squeezed out. Compound bodies behave in accord with their constituent atoms; and so a dense body composed of tightly interlocked atoms will tend to move to the center of the vortex, whereas a body made up of atoms interspersed with much void will be forced out toward the periphery. In this way individual atoms and groups of atoms in the cosmos will move either down toward the center or up toward the periphery. So far no mention has been made of weight. This is because heaviness and lightness seem to be for the fifth-century atomists, not inherent properties of matter, but acquired by atoms and compounds in the vortex. What is more, if we may believe Theophrastus, Democritus ranked heavy and light among the objects of sense; and though heavy and light are rooted in the size and quantity of the atoms, an observer is required to perceive and distinguish them.[7]

Though Democritus' views are known only imperfectly because of the absence of original sources, Plato clearly considered heavy and light to be sense-impressions produced by certain objects. For he treats them in the section of the *Timaeus* which deals with "affections of the whole body", like hot and cold, hard and soft, rough and smooth, and pleasure and pain. He approaches the definition of heavy and light by picturing a man dragging things into an alien element, as earth may be dragged from its own region near the center into the air, or fire may be dragged from the periphery into the air. Such bodies, he claims, will resist the forcible removal from their main masses and will tend to rejoin their own kind, since "like moves to like". If two pieces of the same substance are so removed from their own kind by dragging them off in the two pans of a balance, the larger will resist more, whereas the smaller will follow the constraint more readily and swiftly. The large mass, Plato maintains, is called "heavy and downward moving" [βαρὺ καὶ κάτω φερόμενον] and the small mass "light and upward moving" [ἐλαφρὸν καὶ ἄνω]. The terms "light", "heavy", "up", and "down", are used relatively to the observer, so that to a man standing on the interior surface of the sphere of fire a large mass of fire weighed in a balance will be heavy and a small mass light. Plato concludes that "the one thing to be realized in all cases is that it is the travelling of each thing to its kindred which makes the moving object 'heavy' and the place to which it moves 'down' and that contrary names are given to their contraries" (*Tim.* 62c–63c). Plato clearly treats weight and lightness, not as inherent properties of matter, but as sense-impressions produced in an observer by the motion of an element toward its own kind. The cause of the motion and thus of weight and lightness is the tendency of like to move toward like, a tendency that is ultimately caused by the shaking motion of the receptacle.[8]

Against this background the nature of Aristotle's innovation becomes clearer. He has obviously abandoned the notion that heavy and light are sense-impressions and has made them inherent properties of matter.[9] This change is not a new physical theory but a redefinition of heavy and light in terms of the natural movements of elements, rather than in terms of the effects of such movements upon an observer. Aristotle's physical theory of natural places and motions was already present in embryonic form in Plato's account of weight.[10] So, too, was the assumed fact that fire always moves toward the periphery of the cosmos. Plato had even explicitly equated "heavy" with "downward moving" and "light" with "upward moving". However, he had denied that the cosmic movements of the elements as such constitute the heavy and light. Asserting that there are no absolute directions "up" and "down" in the cosmos, Plato had insisted that up and down are relative to the observer.[11] Aristotle, in contrast, has removed the observer from the picture. He has defended absolute directions in the cosmos against Plato by defining "down" as "toward the middle" and "up" as "toward the periphery" (*Cael.* IV.1.308a17–29); and he has postulated an absolutely heavy and an absolutely light by defining "heavy" as "that which moves down by nature" and "light" as "that which moves up by nature". This may have been only a redefinition of terms, but it was a redefinition that proved to be of the utmost consequence for the subsequent history of mechanics, since it both presupposed an explanation for freely falling bodies and entailed the corollary that "more moves faster" or, in other words, that velocity is proportional to weight.[12]

This assessment of Aristotle's historical position is satisfactory as far as it goes, but there is also some evidence that the historical situation was more complex. In setting out his theory of weight and lightness Aristotle

60

carefully distinguished "absolutely heavy and light" from "relatively heavy and light" [τὸ πρὸς ἕτερον βαρὺ καὶ κοῦφον], which he defined and then quickly dismissed from his attention (*Cael.* IV.1.308a8–13, 29–33). This is unfortunate because not only is the concept interesting in itself but it seems to bear a close relation to Aristotle's notion of absolutely heavy and light and to its corollary that "more moves faster". To find out how it is related to his concept of absolute weight and lightness and to the ideas of his predecessors, we must examine it carefully on the basis of the few incidental references we possess.

In distinguishing the notion of relatively heavy and light, Aristotle claims to follow common usage and quotes as evidence an example: "For we say that of things which possess weight one is lighter and the other heavier, as, for instance, bronze is heavier than wood" (*Cael.* IV.1.308a8–9). This introduction to the subject indicates that "relatively heavy and light" come into consideration when we wish to compare things which possess weight in the absolute sense, i.e., when we are speaking of things whose natural movement is down. He does not suggest that there is a counterpart among things that are absolutely light. Further, he shows that the comparative form of the adjective ("heavier", "lighter") is equivalent to "relatively heavy or light". Finally, his examples (bronze and wood) suggest he is thinking of comparisons, not of total weight, but of specific weight or weight per unit of volume; for not every volume of bronze would be heavier than every volume of wood.[13] His actual definition of "relatively heavy and light" comes at the end of the chapter: "Relatively light or lighter is that one of two bodies of equal size, each possessing weight, which moves down by nature less swiftly than the other".[14] The text, as it stands, does not define the relatively heavy, but it is not hard to infer that the "relatively heavy" or "heavier" of two objects of the same size is the one that falls more quickly. Thus, to use Aristotle's own example, if a piece of bronze and a piece of wood of the same size were allowed to fall from the same height, the bronze, being the heavier, would reach the ground first, whereas the lighter wood would take longer.

On the surface, the definition seems clear enough, but as soon as one begins to probe and bring it into line with other aspects of Aristotle's theory of heavy and light, several problems emerge. First of all, it seems clear from the definition itself that Aristotle means to apply the notion of relative weight and lightness only to comparisons of objects having weight in an absolute sense (i.e., tending to move down). This was suggested by Aristotle's statements introducing the concept of relative weight and lightness, and now it is not only mentioned again in the definition ("of two objects having weight"), but the definition is in terms of the velocity of *downward* motion. However, a few chapters later (*Cael*. IV.4–5) Aristotle explains that only the element earth is absolutely heavy and fire absolutely light. Air and water are heavier than fire, but lighter than earth, because they descend in fire and rise in earth. With respect to each other they are absolutely heavy and light, since air rises above water and water sinks below air (*Cael*. IV.4.311a15–29). This might suggest that air and fire should be called relatively heavy or light, because their weight or lightness is relative to the (unnatural) position from which they begin movement to their own place.[15] A closer look, however, shows that Aristotle never calls water and air relatively heavy or light.[16] What he says is that, whereas earth or fire are absolutely heavy or light in that they fall below or rise above everything, air and water are heavy and light in another sense [ἄλλως], in that they possess *both* properties [οἶς ἀμφοτέρα ὑπάρχει]. Thus they sink below some things and rise above others (*Cael*. IV.4.311a22–24). Air and water, then, are not *relatively* heavy and light but *both* heavy and light. Their weight and lightness are really forms of absolute weight and lightness.[17] Thus we might say air and water possess a combination of weight and lightness, whereas earth and fire possess only one, either absolute weight or absolute lightness. The behavior of air and water, then, has nothing to do with the relative weight and lightness which we are investigating. We may go even further and assert that absolute weight and lightness are essentially designations of the *direction* of natural movement, whereas relative weight and lightness are designations of the *speed* of natural motion.

This brings us to a second problem. Aristotle's definition of lighter and heavier in terms of the speed of falling objects suggests that relative weight is a quantitative measure. Yet Aristotelian commentators from antiquity on have maintained that Aristotle considered heavy and light to be qualities, not quantities.[18] As evidence they may cite the omission of weight and lightness from the discussion of quantity in *Categories VI*. 4b20–6a35. Furthermore, Aristotle lists heavy and light as a kind of condensation and rarefaction in the context of a discussion of *alloiōsis*, change of quality (*Phys.* VIII.7.260b1–13). Finally, heavy and light are classed with hot and cold, wet and dry, and hard and soft, as "opposites according to touch" (*Gen. Corr.* II.2.329b18–20); and Aristotle definitely maintained there is no true opposite in the category of quantity (*Cat.* VI.5b11–13). Aristotle's apparent uncertainty whether to consider heavy and light qualities or quantities is easily resolved if we keep in mind that heavy and light have at least two senses—an absolute sense referring to the direction of natural motion and a relative or comparative sense referring to the degree (specifically, the velocity) of the motion. To maintain that heavy and light are qualities is not to deny that in another sense they may be quantities. Aristotle himself asserted in the *Categories* that qualities may admit of degrees (*Cat.* VIII.10b26–29). So, too, heavy and light admit of degrees, as Aristotle clearly suggests, saying that every heavy thing may also be heavier (*Cael.* III.1.299a31–b6). In fact, on at least one occasion Aristotle explicitly lists weight along with length, breadth, depth, and speed as quantities and says weight may mean "either that which has any degree at all of inclination or that which possesses an excess [$ὑπεροχή$] of inclination; for even the lighter has some weight."[19]

Though it seems clear that relative or comparative weight is a quantifiable entity, we are still faced with the problem of deciding what that entity is. Logically there are two possibilities, total weight and specific weight or weight per unit of volume. Strictly speaking, the definition allows either interpretation, for in specifying that the objects compared be "of equal size" [$τὸν ὄγκον ἴσον$], Aristotle makes the lighter object less both in total and in specific weight. However, we have already noted that Aristotle's first mention of the subject of relative weight and the examples chosen to illustrate it, wood and bronze, suggest he is thinking of specific weight. Moreover, the stipulation that the objects to be compared be of equal size would be superfluous and even misleading if Aristotle were trying to define total weight. Thus, it is hard to escape the conclusion that Aristotle is here attempting to define specific weight or weight per unit of volume. But then we must relate this definition to Aristotle's principle that the speed of descent or ascent of a naturally moving body is proportional to its quantity and hence to its total weight.[20]

Logically Aristotle's definition of relative weight and the principle that velocity is proportional to total weight are perfectly compatible. In the definition of relative weight Aristotle keeps the volume constant and asserts that the specific weight is proportional to the velocity. In the refutation of his predecessors he keeps the specific weight constant by comparing two quantities of the same element and asserts that the velocity is proportional to the volume. This procedure of comparing one variable while holding all other variables constant is characteristic of Greek science; and elsewhere Aristotle uses it to make comparisons between the force, the quantity, the distance, and the time of moving bodies (*Phys*. VII.5.249b27–250b7). Its obvious disadvantage is that no comparisons can be made when two factors vary, and so Aristotle could make no statements about the relative velocity of two bodies whose size and specific weight are both different.

For purposes of a definition of relative weight Aristotle does not have to work out the connection between total weight, specific weight, and rate of fall; but some statements in his later discussion of the "differences and properties" of heavy and light suggest he had thought about the connection (*Cael*. IV.4.311a29–b13). There he says that the heaviness or lightness of compound bodies depends on the nature and amount of the elements which constitute them. This presumably means that whatever element predominates in a compound will determine the absolute weight or lightness of a compound and whether it will move up or down. The exact proportions will determine the degree of weight or lightness. So Aristotle points out in air a *talent* of wood will be heavier than a *mina* of lead, but in water it will be lighter.[21] The paradoxical behavior of these compounds is explainable because earth and whatever consists mainly of earth is heavy everywhere, water everywhere except in earth, air everywhere except in water and earth; and all except fire have weight in their own place.[22] Thus something composed mainly of earth, like lead, moves down in both air and water. But something like wood, which consists predominantly of air, may be lighter in water, where its predominant element rises; but in air it may be heavier than a piece of lead, if its total volume is large enough. Similarly, Aristotle says oil floats on water, because "the air in it moves it upward and raises it to the surface, and is the cause of its lightness" (*Gen. An*. II.2.735b24–27, Cf. *Meteor*, IV.7.383b20–26).

The theory that the speed and direction of the natural movement of a compound body is a resultant of the speed and direction of the natural movements of the elements of that body allows us to relate Aristotle's principles of elemental motion to his concept of specific weight. At the level of the elements the nature of the element combined with its location at the start of its movement determines the direction of its movement; the nature of the element and its total amount determine the speed of that movement. To illustrate how this can explain Aristotle's statements, let us imagine two substances (call them "lead" and "wood") made of air and earth. Let the "lead" consist of two parts earth and one part air, and let the "wood" consist of two parts air and one part earth. Then assign the elements velocities in air and water to reflect the direction of their natural motions and the proportionality of their downward tendencies (i.e., earth is heavier than water, and water is heavier than air):

	Earth	Air
In air	$+10$	$+1$
In water	$+ 5$	-5

The downward movement of a compound will then be the total of the movements of its elements. Specifically, the resultant velocities of "lead" and "wood" in air and water will be:

	"Lead"			"Wood"			"Wood" × 105 (=1 Talent)		
In air	$2 \times$	$10 =$	20	$1 \times$	$10 =$	10	$105 \times$	$10 =$	$1,050$
	$1 \times$	$1 =$	$\underline{1}$	$2 \times$	$1 =$	$\underline{2}$	$210 \times$	$1 =$	$\underline{210}$
			21			12			$1,260$
In water	$2 \times$	$5 =$	10	$1 \times$	$5 =$	5	$105 \times$	$5 =$	525
	$1 \times$	$-5 =$	$\underline{-5}$	$2 \times$	$-5 =$	$\underline{-10}$	$210 \times$	$-5 =$	$\underline{-1,050}$
			5			-5			-525

These results now conform perfectly to Aristotle's observations. When the quantities are equal, lead is heavier than wood, because its descent in air is faster ($21 > 12$). A talent of wood with a downward inclination 60 times that of a *mina* of lead (1 *talent* = 60 *minas*) would be much heavier than the lead in air ($1,260 > 21$); but in water it would move up (represented by a negative velocity, -525) and would thus be lighter. Hence it is quite possible to relate Aristotle's theory of elemental motion and his principle that "more moves faster" to his concept of relative weight.

Besides the formal definition of relatively heavy and light and the suggestion that differences in the relative weight of compounds can be explained by the proportion of their constituent elements, there are a few more statements that shed light on Aristotle's concept of relative weight. In his refutation of Plato's theory that the elementary bodies are made up of triangles, Aristotle uses as a premise the contention that the heavy is something dense [πυκνόν] and the light is something rare [μανόν]. He then asserts that "dense differs from rare in that there is more in equal bulk" [τῷ ἐν ἴσῳ ὄγκῳ πλεῖον ἐνυπάρχειν] (Cael. III.1.299b7–9). The equation of heavy with dense and the definition of density as "more in equal volume" occur in a conditional clause, and so Aristotle does not here commit himself to their validity. It is always possible that they are Platonic principles, brought in to refute Plato on his own premises.[23] But Aristotle himself subscribes to a similar view in the Physics: "The dense is heavy and the rare light, for two characteristics apiece accompany the dense and the rare, inasmuch as the heavy and the hard seem to be dense things, and their opposites, the light and the soft, seem to be rare things" (Phys. IV.9.217b11–18). Furthermore, he considers condensation and rarefaction to be the source [ἀρχή] of all changes in qualities [παθήματα], and says, "Heavy and light, soft and hard, hot and cold are condensations and rarefactions" [πυκνότητες καὶ ἀραιότητές τινες] (Phys. VIII.7.260b7–10). However, he is forced to admit that the heavy and the hard do not coincide in lead and in iron (Phys. IV.9.217b19–20); for lead is the heavier of the two, whereas iron is the harder. Obviously Aristotle is aware that substances vary in weight per unit of volume, and he uses the terms "heavy" and "light" for this property. Moreover, he hopes to explain specific weight in terms of density, though he is troubled by the fact that hardness, which also seems to him to be connected with density, is not necessarily proportional to weight. His own explanation for differences in density and changes in density is that the same matter is capable of different degrees of magnitude and a given amount of matter may be either large or small or change from one size [ὄγκος] to another (Phys. IV.9.217a21–b12). Thus not only does Aristotle in the Physics show an awareness of the notions of specific weight and of density, but he attempts to explain these properties of matter in terms of the amount of matter in comparison to its volume.

Finally, there is one hint that Aristotle may even have been aware of attempts to make precise quantitative determinations of specific weight. In his disproof of the existence of void in the *Physics* he says: "For just as a cube placed in water displaces [ἐκστήσεται] as much water as the volume of the cube, so it does also in air" (*Phys.* IV.8.216a28–29). He also refers to "the volume [ὄγκος] of the wooden cube" (*Phys.* IV.8.216b6) and discusses the relation of this volume to the region of void occupied by the cube if it were moving through a void. The argument seems to turn on the displacement of the medium through which a falling body moves and deals with the necessity for postulating an "empty" space for a body to fill. The precise measure of the volume of the body is not essential to the argument nor is either the specification of the material of the cube or the notion of placing a cube in water. In fact, Aristotle cites the displacement of a volume of water equal to the cube as an analogy to the idea he needs for his argument, not as an integral part of it. This suggests that he was citing a fact that was well known to his readers and one that might be part of a body of knowledge concerning the specific weight of various materials (like wood) and the behavior of solids in water.

These tantalizing hints are all we have of Aristotle's notion of relative weight and lightness. The fact is that the whole idea is quite inconspicuous in Aristotle's works. Even though he wrote a whole treatise on heavy and light and their causes (*Cael.* IV) and so had ample opportunity to discuss the relatively heavy and light, all he provides is a brief definition to distinguish them from the absolutely heavy and light. Nor is there anywhere the characteristic discussion of predecessors on the subject. When Aristotle does discuss predecessors on heavy and light, it is to criticize them for expecting their theories of the relatively heavy and light to account for the absolutely heavy and light; he does not test their theories for validity as theories of relative weight. It appears, then, that Aristotle was not really interested in explaining relative weight, but merely in distinguishing it from absolute weight and lightness to eliminate a source of confusion in his central concern.

Though Aristotle does not seem to have considered relative weight and lightness central to his physics and cosmology and so does not develop the idea fully, he gives enough indication of his view to allow us to see how original he was. Only two pre-Aristotelian attempts to account for differences in "weight" are known with any certainty—that of Democritus and that of Plato. Of Democritus' views little is known apart from what Aristotle and Theophrastus tell us. But it is enough to give us an idea of Democritus' conception of what Aristotle calls relative weight. There are really two aspects to the question as we look backward from Aristotle. One question is how one determines which of two bodies is heavier and the other is how one accounts for the difference in terms of one's basic assumptions. Determining which of two bodies is heavier has one obvious simple solution: place the two objects on the pans of a balance and observe which pan sinks. But this procedure reveals only the total weight of the object. Democritus was fully aware that some substances are heavier than others even when both are of the same size. Aristotle says Democritus accounted for the heaviness and lightness of compound bodies by the presence of void in their constitution (*Cael.* IV.2.308b30–309a11). The more void, the lighter the body will be (Theophrastus, *De Sensu* 61). But according to Aristotle, Democritus said an atom's weight is proportional to its size;[24] and according to Theophrastus, he said the fine [λεπτόν] is light (*De Sensu* 61). Theophrastus sees this as an alternative explanation for lightness, found in a different context; and perhaps it was. But more likely Democritus used it as an alternative description of the same theory. If Democritus saw weight as proportional to the total amount of solid matter in an object, then of two objects that have the same *number* of atoms, the one with the smaller and finer atoms will be lighter. Theophrastus also records Democritus' distinction between weight and hardness: "Hard refers to [literally: "is"] the dense, and soft to the rare; and the degrees of these qualities are proportional. But the position and grouping of the void spaces in the hard and the soft and the heavy and the light differ somewhat. Thus iron is harder, but lead is heavier. For iron has an uneven composition and contains void in many places and in large units [κατὰ μεγάλα], but in some places iron is condensed; as a result it has more void over-all. Lead, on the other hand, has less void, but is of even and uniform composition. Hence it is heavier than iron, but softer" (*De Sensu* 62). Clearly Democritus was trying to relate both weight and hardness to density. The obvious fact that there is not perfect correspondence between weight and hardness in nature requires some special pleading. Democritus' solution was to retain the direct correspondence between weight and the amount of void (per unit of volume), and to save the superior hardness of a less dense, lighter substance by attributing the hardness to the fact that the atoms are distributed in denser clusters and therefore offer more resistance to intrusions from outside.

How Democritus would have decided which of two substances is heavier in specific weight, we can only guess; but is it absurd to suggest on the basis of the correlation with density and hardness that the decision was made by "feel"? Democritus considered the sensible qualities to be alterations in the observer, not properties of the objects, and Theophrastus saw a bald contradiction between this notion and Democritus' founding of heavy and light and dense and rare on the size of the atoms.[25] But if Democritus did regard heavy and light as sense-impressions, it seems likely that he would make the observer's experience the ultimate judge.

Plato's views on the subject of relative weight can be known with somewhat more certainty because they are preserved in his own words. It appears that Plato agreed in many respects with Democritus, if we may judge from his discussion of melting and solidification and of the properties of different metals (*Tim.* 58d–59d). Most revealing is his description of three metals:

> Of all the varieties of water which we called fusible, one, the densest, consisting of the finest and most uniform particles, . . . endowed with a shining, bright color, namely gold, is compacted, being filtered through rock. The scion of gold, which is hardest on account of its density, and is dark-colored, is called adamant [tempered iron?].[26] Another substance [copper] has parts nearly like those of gold, but of more than one grade. In density it is in one way more dense and contains a small, fine portion of earth, so that it is harder; but on the other hand it is lighter by having large gaps within it. (*Tim.* 54a8–c1)

Like Democritus, Plato wishes to correlate density, hardness, and heaviness. But Plato, too, is bothered by the fact that copper is both harder and lighter than gold. So like Democritus he attributes the lightness of the harder metal to the large gaps within it—a solution that is possible only because Plato's elements come in geometrically shaped particles. The hardness he explains, unlike Democritus, by an admixture of earth, thereby saving the direct correspondence between hardness and density. Weight, as in Democritus, is correlated with the quantity of matter, be it the number of triangles in a particle of fire (*Tim.* 56a6–b1), the larger grade of particles of water in metals (*Tim.* 58d8–e2), or the higher proportion of solid to void in gold than in copper (*Tim.* 59b8–c1).

In these passages Plato gives no clue how one might measure which material is heavier. It could be simple "feel" in the hand. But when he proceeds to discuss the affections of the body and turns his attention to heavy and light, he makes it clear that the balance is used to aid the observer in deciding which of two things is heavier. Here, however, he is not looking any longer at specific weight. Instead he says, if a man detaches two portions of earth from its kindred and drags the two portions into the dissimilar air by means of a balance, "the smaller will follow what forces it upward more easily than the larger" (*Tim.* 63c6–d2). Conversely, the larger will resist removal from its kindred more. Thus the larger is called "heavy" and "downward-moving" and the smaller is called "light" and "upward-moving" (*Tim.* 63b-d). Moreover, since "up" and "down" are defined with reference to the position of the observer (*Tim.* 62c–63a), the terms "heavy" and "light" are applicable only with reference to an observer (*Tim.* 63d–e).

The comparison here is between different quantities of the same material, and there is a direct correlation between weight and quantity. The weight being measured, therefore, is total weight. Moreover, heavy and light are defined in terms of the direction of movement on the pans of a balance; and this direction, in turn, is in terms of the position of the observer holding the balance. Thus the observer with his instrument is the *sine qua non* for deciding which of two things is heavier. For us it is easy to imagine how one might go from this to a determination of the relative specific weight of any two or more substances. The simple procedure of weighing on a balance equal volumes of two substances would give the correct answer. But the simplicity of the procedure is no justification for assuming that Plato (or Democritus) had done so or thought of doing so. We must simply suspend judgment on how Plato determined specific weight.[27]

Against this background Aristotle's debts and originality stand out in bold relief. In general, Aristotle subscribed to the principle defended by Democritus and Plato that specific weight and hardness bear some relation to density; and he also admitted, as did his predecessors, that in some cases (e.g., lead and iron) superior heaviness does not accompany superior hardness (*Phys*. IV.9.217b11–20). But this is as far as the similarities go; and even though we have no extended discussion of the subject in Aristotle, we can detect a new point of view. First of all, one senses in all Aristotle's statements dealing with specific weight a quantitative precision that was absent from Democritus and Plato. For example, in his refutation of Plato's views on weight Aristotle presents Plato in more general and more quantitative language than the *Timaeus* uses, and Aristotle fills it with specific examples (lead, bronze, wood) not found in Plato (*Cael*. IV.2.308b3–28). He also says, "Some bodies are obviously smaller in volume, but heavier" (*Cael*. IV.2.308b32–33). Neither Plato nor Democritus (as far as we know) made explicit the comparison between weight and volume. Then, too, the definition of the dense as "more in equal volume" (*Cael*. III.1.299b7–9) and the statement that a cube displaces an equal volume of water (*Phys*. IV.8.216a28–29) suggest a more quantitative approach. But the most novel aspect of Aristotle's idea of relative weight is his definition of the lighter of two bodies of equal size as the one whose rate of descent is less. Plato's balance is gone as a measure of weight, and instead we find the velocity of free fall used as a measure.[28] Plato had made some tentative suggestions of a connection between weight and movement. But he associated lightness with mobility (*Tim*. 55e7–56b2, 58d5–e2). From this starting point one ought to conclude that the lighter would move faster. Obviously, Aristotle has come a long way from Plato.

Aristotle's new conception of relative weight and its measurement is in complete agreement with his innovation in the conception of weight and lightness in the absolute sense as inherent potentialities for movement. Aristotle's innovation in absolute weight and lightness was to remove them from the realm of sense-impressions and to make them basic properties of the elements. It is now entirely understandable that he could not accept any notion of specific weight that relied on the observer for its definition or measurement. Accordingly, he has pulled the balance out from under Plato's moving objects and has redefined the magnitude of weight and lightness in a way that does not depend on an observer. As absolute weight and lightness had been redefined in terms of the directions of natural motions in the cosmos, so now relative weight and lightness are defined in terms of the velocity of this natural motion. Relative weight becomes an inherent property of the elements

71

and does not depend on an observer. The real surprise in all this is not that Aristotle adopted this new view but that he did nothing to explain and publicize so fundamental and far-reaching an innovation.

However, before we accuse Aristotle of any uncharacteristic humility, we had better be sure this view was indeed original with him. In the second chapter of *On the Heavens IV* Aristotle rejects the views of three opponents, accusing them all of trying to make a theory of relative weight and lightness serve where only a theory of absolute weight and lightness will do. The first opponent, Plato, receives nothing but criticism. The atomists, though earlier in time, are discussed in the second place and receive a slight compliment for being more modern (*Cael.* IV.2.308b30–32). The third opponent receives the briefest treatment, but the most praise: "The view that distinguishes heavier and lighter by greatness and smallness is less [the manuscripts read "more"[29]] like fiction than the former; and because it can differentiate the four elements individually, it is more firmly established against the previous objections. But because it imagines that the things which differ in magnitude have one single substance [$\varphi\acute{\upsilon}\sigma\iota\varsigma$], it is in the same position as those who assume a single matter [$\ddot{\upsilon}\lambda\eta$]; there is nothing absolutely light and upward moving, except in being left behind [$\dot{\upsilon}\sigma\tau\epsilon\rho\acute{\iota}\zeta o\nu$] or being squeezed out [$\dot{\epsilon}\kappa\vartheta\lambda\iota\beta\acute{o}\mu\epsilon\nu o\nu$]; and many small bodies will be heavier than a few large ones. If so, a large quantity of air or of fire will be heavier than a small quantity of water or earth. But that is impossible" (*Cael.* IV.2.310a3–14).

This third view, to my knowledge, has never been specifically identified.[30] There are sufficient statements in Plato (*Tim.* 63a) and in Aristotle's and Theophrastus' accounts of Democritus (Arist. *Gen. Corr.* I.8.326a9–10; Theophr. *De Sensu* 61–63, 68, 71) to refer this summary to Plato or Democritus if necessary.[31] Simplicius thought it referred to a view which made relative weight and lightness depend on density (*Cael.* 692.25–28 [Heiberg]). This, of course, is quite likely. But who held this view? Again, both Democritus and Plato come to mind. But the third view cannot be that of Democritus or Plato, or else Aristotle would not have discussed it separately. Moreover, it seems to embody two additional features: (1) it permits the distinction of four elements, which the atomists' theory failed to do;[32] and (2) it explains, or at least could potentially explain, upward motion only by "being left behind" or by "being squeezed out"; and so it cannot apply to Plato.

Even if the view cannot be identified, it must not be dismissed; for it seems to be regarded as independent of Plato and Democritus and substantially superior to their views. One thing Aristotle approves in the theory is its ability to differentiate four elements, and very likely the differentiation was quantitative. This would accord well with Aristotle's own theory of four elements and his own view that different degrees of density and relative weight characterize the four elements (*Phys.* IV.9.217a21–b20; *Cael.* IV.2.309b5–8). But the most tantalizing clue to the nature of this third theory is Aristotle's statement that it could potentially explain upward motion by "being left behind" [$\dot{v}\sigma\tau\epsilon\rho\dot{\iota}\zeta o\nu$], or by "being squeezed out" [$\dot{\epsilon}\kappa\vartheta\lambda\iota\beta\acute{o}\mu\epsilon\nu o\nu$]. "Being left behind" suggests that falling bodies fall at different rates of speed and that the lighter body falls less quickly. This is precisely the notion Aristotle expresses in his definition of relative weight and lightness. There is, of course, no proof that Aristotle owes any direct debt to this third view, but it looks as though he knows of a third explanation of relative weight that involves different degrees of speed. Since he never cites or evaluates any rivals for his own theory of relative weight, we simply cannot determine his estimate of this third theory's validity; all we know is that it fails to satisfy his criteria for a theory of *absolute* weight. His deferential treatment of it suggests it may have contained much to his liking. It is unfortunate that the over-all scheme of his science did not require further discussion of the subject.

In sum, there appears to be substantial and consistent evidence that behind Aristotle's notion of relative weight and lightness lies a body of knowledge of a quantitative nature, dealing with specific weight and defining it in terms of the rate of speed of falling bodies. Furthermore, this body of knowledge may have included some knowledge of the behavior and weight of bodies immersed in water.[33]

This, indeed, is a surprising conclusion, since knowledge of this sort has not yet been firmly documented before the third century B.C. and is generally associated with the name of Archimedes. But before doubting it, we must look at one neglected text which may hold the key to Aristotle's relation to his predecessors. An Arabic list of Euclid's works includes among the treatises regarded as genuine "The Book of the Heavy and Light." This work, no longer extant in Greek, was translated from Arabic into Latin and was published with the Latin translation of Euclid in the sixteenth century.[34] The authenticity of the work has been questioned because there is no other evidence that Euclid wrote on mechanics and because the work contains a notion of specific weight so clear that it ought to be later than Archimedes.[35] However, neither of these objections is sufficient to condemn the testimony of the manuscripts and the Arabic catalog, and so the Euclidean authorship of the work may be provisionally accepted.[36]

An examination of the postulates of the treatise shows immediately that its author and Aristotle share many assumptions: ''(1) Bodies equal in magnitude are those which fill equal places. (2) And those which fill unequal places are said to be of different magnitude. (3) And those which are called large [grandia] among bodies are called capacious [ampla] among places. (4) Bodies equal in force[37] are those whose motions are over equal places in equal times, in the same air or the same water. (5) And those which traverse equal places in different times are said to be different in force [fortitudo]. (6) And that which is greater in force is less in time. (7) Bodies of the same kind are those whose force is equal when they are of equal magnitude. (8) When bodies are equal in magnitude but different in force with respect to the same air or water, they are different in kind. (9) And the denser is more powerful.'' These postulates set out in precise proportions the relations between magnitude, the force of weight, velocity, and density—the very relations we have found mentioned in Aristotle. Here in the Euclidean treatise ''On the Heavy and Light'' we find that the force of weight is directly proportional to the speed of descent, so that the heavier body falls faster. We also find that there is a correlation between weight and density. The denser the body, the heavier it is, provided it has the same volume. Moreover, these postulates allow us to infer that velocity will be proportional to total weight if the substances are of the same kind, and that the velocity will be proportional to the specific weight if they are of the same volume. In addition, the medium through which the object moves is always kept the same, presumably because the medium will affect the velocity, as Aristotle also realized (Phys. IV.8.215a25–216a21). So, also, determinations of density are made ''in the same air or the same water'', presumably because the weight is known to be affected by its environment. This suggests some knowledge of determinations of density, presumably by a process like the one mentioned in Aristotle, namely, submerging cubes in water. Finally, like Aristotle's third opponent in On the Heavens IV, chapter 2, weight is proportional to magnitude, since all relations between force, volume, and density are proportional;[38] and at least two of Aristotle's four elements are explicitly granted recognition.[39]

The similarity between Aristotle's assumptions, his third opponent, and the Euclidean treatise "On the Heavy and Light" suggests there has to be some connection. Traditionally it has been assumed that the Euclidean treatise incorporates Aristotle's mechanics and gives it a mathematical formulation. So whether by Euclid, living shortly after Aristotle, or by a later author, it is dependent on Aristotle.[40] This assumption must now be questioned. Aristotle's ideas on specific weight and its definition and measurement are so incidental to his exposition of his theory of weight and lightness that it seems hard to believe they could have had an extensive influence on Euclid or any other mathematician. Either we will have to assume that Aristotle expressed his ideas on this subject at greater length on other, unknown occasions and so influenced the mathematicians; or we will have to admit that there was a body of mathematical knowledge similar, if not identical, to that of the treatise attributed to Euclid, and that this was the basis both for Aristotle's own statements on the subject and for the Euclidean treatise. By far the simplest hypothesis, I believe, is that Euclid did write a treatise "On the Heavy and Light" from which the extant Latin version ultimately derives and that in it Euclid made use of the work of earlier mathematicians on the subject, as he did also in his *Elements*. These earlier mathematicians of the mid-fourth century, then, were known to Aristotle, were taken along with Plato and the atomists as a third rival to his own view, and provided him with his own definition of relative weight and lightness and with the principle that the velocity of a falling body is proportional to its weight.[41]

If this hypothesis is correct and Aristotle's ideas on relative weight and lightness are in any way dependent on some earlier mathematical source, Aristotle's role in the history of the concepts of weight and the movement of elemental substances takes on a new light. When Aristotle changed the definition of heavy and light in the absolute sense from an effect on an observer to an inherent potential for natural motion, he made a substantial break with his predecessors, Democritus and Plato. But though his concept of the relatively heavy and light as degrees of this natural motion was found to be consistent with, and at first sight apparently deduced from, his concept of absolutely heavy and light, it

75

now appears to have been adapted from some earlier mathematical treatment of the subject.[42]

This suggests that Aristotle drew inspiration from two distinct sources, the well-known physical tradition of seeking the causes of phenomena, and in addition a mathematical tradition of laying out the proportional relationships which the phenomena manifest. Since Aristotle's own primary interest was the quest for causes, this aspect of the topic was most fully and explicitly developed and showed the greatest originality. In contrast, the other side of Aristotle's debt almost passes unnoticed, as he tacitly defers to the authority of the mathematicians for the mathematical relationships.[43] Yet, in the last analysis, his theory of absolute and relative weight and lightness must be regarded as the product of an interaction between contemporary mathematics and a quest for the causes of natural phenomena; and some of the credit for the resulting theory must go to the anonymous fourth-century mathematicians who were beginning to develop the notion of specific weight. In fact, Aristotle may never have come to the view that weight and lightness must be tied directly to the movement of the elements if some mathematician had not already been analyzing weight in terms of the behavior and movement of falling bodies.

This interpretation in no way diminishes Aristotle's achievement in laying the foundations of the normative theory of the next two millenia. For what Aristotle did was to draw out some of the underlying assumptions of the mathematicians and elevate them to cosmological principles in harmony with the rest of his physics and cosmology. The result was an unprecedented definition of weight and lightness and one that integrated weight and lightness into the Aristotelian theory of elemental motion.

Furthermore, at the heart of Aristotle's theory was the simple but provocative notion that the balance and its behavior may serve as a model for the behavior of naturally moving bodies. This idea was not unprecedented. Plato had used the balance to measure the strength of natural motion, and the mathematical tradition saw natural motion as the appropriate phenomenon by which to define the property measured by the balance. In fact, popular Greek vocabulary already saw a connection between falling bodies and the fall of the heavier pan of a balance.[44] What Aristotle did that none before him had done was to use the recent progress in the science of weights, based on the new mathematical theory of proportion, as an aid in explaining the behavior of all naturally moving bodies in the cosmos.[45]

It may have been the choice of this model for explaining falling bodies that led Aristotle and whatever predecessors he had to the fateful error of thinking the rate of descent is directly proportional to the weight and inversely proportional to the resistance of the medium.[46] Historians have long pondered how Aristotle could have made an error so easily refutable, not only by simple experiment, but by any chance observation of the fact that large bodies seem to fall at approximately the same rate as smaller ones of the same shape and material, and certainly not proportionately faster. However, if Aristotle and his predecessors were looking only at the model and extrapolating to falling bodies, the error is understandable. For if a very small weight is added to one side of a balance that has been in equilibrium, the heavier pan will sink slowly. If the added weight is larger, the pan will sink much faster. Similarly, if the two pans of the balance are viewed as analogous to the falling body and the resisting medium, the inverse proportionality of speed and resistance makes sense. For when a given weight is added to one pan of a balance in equilibrium, that pan will descend the more quickly, the less weight there is in the other pan. This might suggest that the velocity of a moving body is inversely proportional to the resistance to be overcome.

But before condemning Aristotle and any predecessor who may have subscribed to these principles, we must consider the positive effects of the use of the balance and its behavior as a model for the behavior of naturally moving bodies. The foremost of these was to bring into focus an important new scientific problem, namely the speed and cause of free fall. For this Aristotle himself seems to have been primarily responsible. Even if the association of weight and elemental movement was assumed before Aristotle, it was Aristotle's elevation of this assumption to a major physical principle and his search for its causes that first opened the door to investigation of the connections between weight, natural places and movements, and the speed of falling bodies. More specifically, it seems to have been because Aristotle brought the mathematical theory of proportion as developed in the emerging science of weights into contact with the cosmological problem of falling bodies and the movement of the elements, that the velocity and acceleration of falling bodies emerged as a subject of scientific inquiry.

These conclusions must, of course, remain hypothetical since the works of some of Aristotle's predecessors are no longer extant. Nevertheless, it seems safe to say that although Aristotle played a fundamental role in establishing a close connection between weight and the movement of elemental substances, he was not totally without predecessors, and his own highly original theory must have arisen from a complex interplay of previous philosophical and scientific ideas.

1. The treatise "On Heavy and Light" stands as book IV of *On the Heavens*. Hereafter, Aristotle's works will be cited by abbreviated Latin titles, e.g., *Cael.* = *De Caelo (On the Heavens)*.

2. F. Solmsen, *Aristotle's System of the Physical World: A Comparison with His Predecessors,* Cornell Studies in Classical Philology 33 (Ithaca:Cornell University Press, 1960), has done this for Aristotle's physical philosophy as a whole. His work is the starting point for the present study.

3. His program is laid out in *Cael.* IV.1.307b28–30 and IV.3.310a16–20.

4. At this point the received text contains a digression that says Anaxagoras and Empedocles had no explanation of light and heavy, and others who denied the existence of void had no explanation for absolute weight and lightness and no clear explanation of how some bodies can be smaller in size, but heavier (*Cael.* IV.2.309a19–27). This looks like a later interpolation, either by Aristotle himself or by an editor.

5. The most important corollary is that more of any element moves faster in its natural direction than less and must, therefore, be called either heavier or lighter (as the case may be) than less.

6. Aetius I.3.18; I.12.6; Cicero, *De Fato* 20.46 = H. Diels and W. Kranz, *Die Fragmente der Vorsokratiker,* 6th ed. (Berlin, 1952) vol. 2, p. 96, number 68 A 47. This is a disputed point. D. Furley, "Aristotle and the Atomists on Motion in a Void" (in this volume), tries to revive the old theory that Democritus' atoms do possess weight. He is right in pointing out the difficulty involved in explaining the perpendicular fall of bodies toward the surface of a flat earth solely by means of vertical motion. But this difficulty suggests at most that the atoms possess some kind of natural tendency to fall vertically, not that Democritus called this tendency "heaviness". There is no evidence that Democritus anticipated Aristotle in thinking of heaviness as a tendency to move in some direction. Since Furley's argument is not conclusive, I follow the currently favored reconstruction. For fuller discussion and citation of evidence and bibliography see W. K. C. Guthrie, *History of Greek Philosophy,* 2 (Cambridge: At the University Press, 1965): 396–413.

7. *De Sensu* 71, see 68(=Diels-Kranz[above, note 6] vol. 2, pp. 119–20, number 68 A 135.71, cf. 68). Theophrastus says Democritus distinguished heavy and light by size [$\mu\acute{\varepsilon}\gamma\varepsilon\theta os$], so that the larger the atom, the heavier it is. In the case of compounds the more void a body contains, the lighter it is; and the less void, the heavier (*De Sensu* 61, 68). In this Theophrastus finds the contradiction he so desires in his opponents and consequently contends that Democritus gave heavy and light some substantive being [$o\dot{v}\sigma\acute{\iota}\alpha$], and yet believed them to be relative to the observer [$\pi\rho\grave{o}s\ \dot{\eta}\mu\tilde{\alpha}s$] (*De Sensu* 71, cf. 68).

8. *Tim.* 52d–53a. See H. Cherniss, *Aristotle's Criticism of Plato and the Academy* (Baltimore: Johns Hopkins University Press, 1944) p. 449, note 393, and Solmsen, *Aristotle's System of the Physical World,* p. 282.

9. For this analysis of Aristotle's originality with respect to his predecessors I am indebted to Solmsen, *Aristotle's System of the Physical World,* pp. 275–80.

10. Cf. ibid., pp. 95–102, pp. 279–80, and W. K. C. Guthrie, *Aristotle: On the Heavens,* Loeb

Classical Library (London and Cambridge, Mass.: William Heinemann and Harvard University Press, 1939), pp. xvii–xx.

11. This may be the reason that Aristotle in refuting Plato's theory ignored Plato's primary discussion of heavy and light (*Tim.* 62a–63e). He may have seen no connection between Plato's comments on the movement of the elements in *Tim.* 63a–e and weight and lightness as he defined these terms. So he simply disregarded the passage except to refute its relative definition of up and down. Instead, he concentrated his attack on Plato's remark that fire is lightest and quickest because it consists of the fewest parts and accordingly attributed to Plato the theory that the relatively heavy or heavier is that which has more of the same parts (*Cael.* IV.2.308b3–28 seems to refer primarily to *Tim.* 56b1–2, but cf. also 58d8–e2, 59a8–c3 [esp. b8–c3]). Though Aristotle does not seem to make use of *Tim.* 62c–63e, we cannot say he was unaware of it. Theophrastus, *De Sensu* 88, definitely refers to *Tim.* 62c–63e in defining heavy and light in terms of the difficulty of dragging pieces of elementary matter into an alien place; yet it, too, states Plato's definition of heavy as that which has more of the same kind of [ὁμογενεῖς] parts. So, too, Aristotle's ὁμοειδῶν and ὁμογενῆ (*Cael.* IV.2.308b8, 22) might not exclude the theory of *Tim.* 62c–63e. Whatever Aristotle's reasons may have been for disregarding Plato's theory of weight and lightness, the result is that Aristotle's criticism ultimately fails to touch Plato's theory (cf. Cherniss, *Aristotle's Criticism of Plato and the Academy,* pp. 161–65).

12. Aristotle was fond of this corollary and used it with devastating force against Plato and the atomists (*Cael.* IV.2.308b18–21, 27–28; 309b12–15).

13. Strictly speaking, we should call this a comparison of densities. But "density" must be reserved for παχύτης, which is a distinct, though related, concept in antiquity. Hence I shall use "specific weight" to denote "weight per unit of volume" in contrast to "total weight". It should be noted that this "specific weight" is not the same as the modern term "specific gravity", which involves a comparison with a standard (like water).

14. *Cael.* IV.1.308a31–33: πρὸς ἄλλο δὲ κοῦφον καὶ κουφότερον, οὗ [Stocks, Allan, Guthrie: MMS ὅ] δυοῖν ἐχόντων βάρος καὶ τὸν ὄγκον ἴσον κάτω φέρεται θάτερον φύσει θᾶττον. There is a slight problem with the text. The emendation of Stocks, accepted by Allan and Guthrie, is simplest, but produces an awkward sentence. Longo, following the paraphrase of Simplicius, proposes an emendation which would bring in a definition of the relatively heavy. But this emendation gives the relatively light an upward motion (again following Simplicius) and so runs counter to everything we know about Aristotle's concept of relative weight and lightness. The text would conform best with Aristotle's theory if the ὅ of the manuscripts were retained and the latter part were emended as follows: τὸν ὄγκον ἴσον κάτω φέρεται [θατέρου] φύσει [βραδύτερον · πρὸς ἄλλο δὲ βαρὺ καὶ βαρύτερον ὅ κάτω φέρεται θατέρου φύσει] θᾶττον. I have translated the sentence to give the sense required by the context.

15. This is the conclusion that seems to be drawn by W. D. Ross, *Aristotle* (London: Methuen, 1949) p. 99. G. A. Seeck, *Über die Elemente in der Kosmologie des Aristotles: Untersuchungen zu "De Generatione et Corruptione" und "De Caelo,"* Zetemata 34 (Munich: C. H. Beck, 1964): 110–12, too, considers air to possess relative weight and lightness; but, in keeping with this hypothesis that Aristotle's treatises contain incompatible theories of elements, he carefully distinguishes this sense of relative weight from that of chapter one.

16. There may be one exception, *Cael.* IV.5.312b16–19, but it is not clear how "the relatively heavy" is to be understood in this difficult passage.

17. Even when speaking specifically of their weight and lightness "relative to each other" [πρὸς ἑαυτά], Aristotle does not hesitate to say one is "absolutely heavy" and the other "absolutely light" (*Cael.* IV.4.311a27–28).

18. For the view of the commentators of late antiquity see S. Sambursky, *The Physical World of Late Antiquity* (London: Routledge & Kegan Paul, 1962), pp. 82–85.

19. *Metaph.* X.1.1052b16–31. Modern writers are sometimes misled by Simplicius' contention that Archytas made the inclination of weight one of the three species of quantity, along with extension and number (Simplic. *Cat.* 128.18; 269.30[Kalbfleisch]). For example, Sambursky, *The*

Physical World of Late Antiquity, p. 83, calls Aristotle's omission of weight from the types of quantities in the *Categories* "a retrograde step" compared with Archytas. What Sambursky does not realize is that Simplicius is not quoting from a lost work of the fourth-century Pythagorean, but from a Hellenistic forgery entitled "On Universal Words" or "Categories", now edited by H. Thesleff, *The Pythagorean Texts of the Hellenistic Period,* Acta Academiae Aboensis, Ser. A, 3 (Abo, Finland: Abo Akademia, 1968): 25.1–2. Ironically, the work is deeply influenced by Aristotle, and this particular idea may well go back to *Metaph.* X.1.1052b16–31. L. Elders, *Aristotle's Cosmology* (Assen, Netherlands: Van Gorcum & Co., 1966), p. 331, also takes Simplicius at face value.

20. *Cael.* III.5.304b17–18; IV.2.308b18–19, 309b12–15; cf. III.2.301a22–b17; IV.2.308b27–28; *Phys.* IV.8.216a13–21. *Cael.* IV.2.309b11–15 rules out the possibility that the specific weight *per se* may affect the velocity.

21. The *talent* (which is also a Greek word for "balance") and the *mina* are Greek standards of weight. 60 *minas* = 1 *talent.*

22. Aristotle's proof that air has weight in its own place is the curious "observation" that an inflated bladder weighs more than an empty one. For later repetitions of the experiment with different results see Simplicius, *Cael.* 710.24–711.4 (Heiberg). See W. D. Ross, *Aristotle's Physics* (Oxford: Clarendon Press, 1936) pp. 26–27.

23. See below, pp. 69–70.

24. *Gen. Corr.* I.8.326a9–10; cf. Theophrastus, *De Sensu* 68, 71. Aristotle's statement and similar statements in Simplicius suggest that Democritus' atoms are endowed with weight. But weight is an ambiguous term. Though it seems probable that Democritus denied that atoms have weight in the sense of an internal mover (like Epicurus), he may well have attributed to them weight in some other sense. If he regarded "weight" as the effect of the movement of atoms or groups of atoms in the whirl that drives larger atoms toward the center (see above, page 59), the "weight" of an individual atom will be the force with which an atom moves toward the center of the whirl and so will be proportional to the size of the atom.

25. W. K. C. Guthrie, *History of Greek Philosophy.* 2:438–40. See also Theophrastus, *De Sensu* 68, 71.

26. The identification of *adamas* is disputed. Though the word often means "diamond", it is hard to see how Plato can here intend the diamond, since the diamond is not metallic (see A. E. Taylor, *A Commentary on Plato's Timaeus* [Oxford: Clarendon Press, 1928] p. 416). However, the word also seems to have been a poetic term for steel or tempered iron, and it may be this that Plato has in mind, possibly confusing this with the diamond, which was associated with gold (see F. M. Cornford, *Plato's Cosmology* [London: Routledge & Kegan Paul, 1937; repr., New York: Bobbs-Merrill Co., 1957] pp. 251–52.

27. We might notice that Plato does not arrange his three metals in order of descending specific gravity (gold, 19.3; steel, 7.8; copper, 8.9). But he may have been trying to do so, since gold is first. Without the use of measuring and weighing devices the difference between steel and copper would be unnoticeable.

28. A balance may lie behind the measurement of total weight in *Cael.* I.6.273a21–b27, where Aristotle is attempting to refute the existence of an infinite body by showing that its weight cannot be finite (but then he goes on to tackle the other horn of the dilemma, namely, that the infinite body has infinite weight, on the basis of the proportionality between weight and speed of movement). The only incontrovertible appearance of the balance in Aristotle is *Gen. An.* IV.9.777a28–31, and here the movement of the balance is an analogy for the fetal position of animals, not a means of comparing weights. In *Cael.* II.14.297a27–b14, in a proof of the sphericity of the earth, the different degrees of weight result in a sort of shoving contest.

29. Alexander of Aphrodisias (third century A.D.) already pointed out that Aristotle's refutation of this theory would also apply to the atomists (Simplic. *Cael.* 693.25–32 [Heiberg]).

30. H. Cherniss, *Aristotle's Criticism of Presocratic Philosophy* (Baltimore: Johns Hopkins

University Press, 1935), p. 213, assumes it refers to the atomists, but most editors and commentators do not attempt to identify it.

31. L. Elders, *Aristotle's Cosmology*, pp. 339–40, has suggested reading "less" rather than "more". His emendation can be defended on the grounds that the second half of the sentence, joined by □, is complimentary, and the view in general receives so little criticism. A further argument might be that Aristotle seems to present the three views in order of increasing approval, and so it would be appropriate to call this view *"less* like fiction than the former views".

32. Aristotle had voiced this objection against the atomists only a few lines earlier (*Cael.* IV.2.309b34–310a3).

33. Cf. the cube that displaces an equal volume of water (*Phys.* IV.8.216a27–29). The *ekthlipsis* of *Cael.* IV.2.310a10 could come from a discussion of the behavior of wood (cf. the wooden cube of *Phys.* IV.8.216b6) in water; for Archimedes uses *ekthlipsis* in a similar way in *On Floating Bodies,* Postulate and Proposition 5 (= Archimedes, *Opera Omnia,* ed. J. Heiberg, [Leipzig: B. G. Teubner, 1913], pp. 318, 328–30).

34. A critical edition of the Latin text with English translation can be found in E. A. Moody and M. Clagett, *The Medieval Science of Weights* (Madison: University of Wisconsin Press, 1952), pp. 26–30.

35. Cf., e.g., G. Sarton, *Introduction to the History of Science* (Baltimore: Williams & Wilkins Co., 1927), p. 156; and T. L. Heath, *The Thirteen Books of Euclid's Elements*[2] (Cambridge, 1908; repr. New York: Dover, 1956), 1:18, who follows J. L. Heiberg, *Litterargeschichtliche Studien Uber Euklid* (Leipzig: B. G. Teubner, 1882), pp. 9–10.

36. The argument that there is no other evidence that Euclid wrote on mechanics is an argument from silence. The argument that the work contains too clear a notion of specific weight to be prior to Archimedes simply begs the question. Since Archimedes' priority in recognizing the concept of specific weight cannot be established independently, there are no grounds for maintaining no earlier writer could conceive of specific weight. Cf. Moody, *The Medieval Science of Weights,* p. 23. (This argument was later quietly abandoned by T. L. Heath in *A History of Greek Mathematics* [Oxford: Clarendon Press, 1921] 1:445–46). Furthermore, to deny Euclidean authorship entails rejection of the judgment of the Arabic scholars who conscientiously tried to identify pseudepigrapha in the Euclidean corpus and who found nothing objectionable about "The Book of the Heavy and Light". Though their judgment is in no way sacrosanct, it may be significant that none of the other five books accepted as genuine in the Arabic list (*Elements, Phaenomena, Data, Division, Canon*) is now considered certainly spurious, and only one is questioned (*Canon,* the identity of which is disputed). For the full list see Heath, *Euclid's Elements,* p. 17.

37. The Latin word is *virtus,* which P. Duhem, *Les origines de la statique* (Paris: A Hermann, 1905), 1:69, and Heath, *A History of Greek Mathematics,* 1:445, take to be a translation of δύναμις and ἰσχύς of *Physics* VII.5. But the "power" in *Physics* VII.5 is not the power which moves freely falling bodies. *Virtus* may well represent δύναμις and ἰσχύς, but in the sense these words have in, e.g., *Cael.* IV.6.313b19; *Phys.* IV.8.216a19. Or else *virtus* is a translation of the tendency to move down, as in *Cael.* IV.1.307b31–33.

38. Proposition 3 shows that weight is proportional to magnitude if the bodies are of the same kind. But weight might also be considered proportional to magnitude when the volumes are equal and the densities vary, if "denser" is interpreted as "more in an equal volume" (cf. *Cael.* III.1.299b7–9).

39. Presumably the elements could be differentiated by density.

40. A dependence on Aristotle is stated or implied by Duhem, *Les Origines de la statique,* 1:68–69; Heath, *A History of Greek Mathematics,* 1:446; Moody and Clagett, *The Medieval Science of Weights,* p. 23; and M. Clagett, *The Science of Mechanics in the Middle Ages* (Madison: University of Wisconsin Press, 1959), p. 430. Moody goes so far as to admit the possibility of Euclidean authorship.

41. The date and identity of the originator of these views cannot be known. If *On the Heavens* is

an early work, written either during Aristotle's stay at the Academy or shortly thereafter (cf. I. Düring, *Aristotles; Darstellung und Interpretation seines Denkens* [Heidelberg: Carl Winter, 1966], p. 50, and W. Jaeger, *Aristotle: Fundamentals of the History of His Development*,[2] trans. R. Robinson [Oxford: Clarendon Press, 1948], pp. 293–308), these views must have originated around the middle of the fourth century B.C. or earlier. The only mathematician of this time who is said to have written on mechanics is Archytas of Tarentum (Diogenes Laertius 8.83). However, we know nothing specific about his works on mechanics, if indeed he wrote any; and it is not impossible that other mathematicians were interested in the subject, particularly those in and around the Academy.

42. It is possible that Aristotle also derived his notion that velocity is inversely proportional to the density of the medium from the same source.

43. On another occasion of borrowing mathematical facts Aristotle explicitly acknowledges his debt to the mathematicians and disclaims personal authority and originality (*Metaph.* XII.8.1073b10–17).

44. Aristotle's standard word for the descent of falling bodies is $\dot{\rho}o\pi\acute{\eta}$ or $\dot{\rho}\acute{\epsilon}\pi\epsilon\iota\nu$ (cf. H. Bonitz, *Index Aristotelicus*[2] [Graz: Akademische Druck, 1955] 665b34–45, 668b30–46), words that from Homer to Aristotle denoted the sinking of the heavier pan of the balance (cf. H. G. Liddell and R. Scott, *A Greek-English Lexicon*[9] [Oxford: Clarendon Press, 1940], s.v. $\dot{\rho}\acute{\epsilon}\pi\omega$, $\dot{\rho}o\pi\acute{\eta}$).

45. On the new theory of proportion, devised by Eudoxus and represented by Euclid, *Elem.* V, see Heath, *A History of Greek Mathematics,* 1:325–27.

46. If, as I have suggested, the basic idea was already stated by Aristotle's mathematical source, the error will have to be attributed also to his source.

David J. Furley

Chapter Four

ARISTOTLE AND THE ATOMISTS ON MOTION IN A VOID

I. INTRODUCTION

This paper is part of an attempt to study the controversy between the Greek Atomists and their opponents. The Atomists are Leucippus and Democritus (their individual contributions cannot be separately identified from the available evidence), Epicurus and his Greek followers, and Lucretius. Among their opponents I include Plato, Aristotle and his Peripatetic followers, and the Stoics.

The controversy was systematic and fundamental. The two groups took opposed positions on all these problems: Is the universe infinite or finite? Are there many worlds or only one? Is our world a mortal compound, having a beginning and an end in time, or is it everlasting? Have living species or natural kinds in general developed out of simpler states of matter, or are they eternal? Is matter atomic or continuous?[1] Is there void or not? Is all change reducible to the rearrangement of changeless elements or not? Are we to seek only for mechanistic explanations or for teleological ones?

The task is to analyze and evaluate the arguments set out by each side in support of its position, in the hope of contributing to an understanding of the post-Classical history of the controversy. But there is a snag. In all the ancient polemical texts that bear on these questions, only Aristotle habitually identifies his opponents and systematically criticizes their arguments. At the opposite extreme, Epicurus, it appears, took a perverse delight in presenting his philosophy as something new, without antecedents. Lucretius was more generous, at least to those whom he could regard as the grandfathers of atomism, such as Empedocles; but he too preferred to identify his opponents as *quidam* or *stolidi*. So the interpreter of the Epicureans has to work hard to identify the targets of their polemics—and he can make mistakes in this work. If he looks everywhere for anti-Aristotelian polemic, like Zeal-of-the-Land Busy looking for Enormity at Bartholomew Fair, he may find it where it does not exist; for no one quite knows how much of Aristotle Epicurus had at his finger tips. On the other hand, it may be argued that this does not matter very much in the long run. What we have to do is to put ourselves in the position of someone like Cicero or Atticus, who could read both the Atomists and their opponents and assess the claims of each on their argumentative merits. For such an assessment, if a particular Epicurean argument *can* be turned against Aristotle, it may not matter much whether it was framed with that intention or not.

Within this general field, this paper takes as its topic part of the controversy about motion. The Atomists held that atoms move in the void, and indeed that this is the only kind of change that there is. Their opponents denied the existence of void space (except that the Stoics allowed it outside the cosmos), and held that all locomotion is a matter of swimming through a medium. The first idea, of motion in the void, is the main subject of this paper; the alternative will feature only in passing.

II.   DEMOCRITUS' THEORY OF MOTION

In the present context, the creation of the notion of void need not detain us long. Melissus, supporting Parmenides, argued that since the void is nothing, and a thing moving needs the void to move into, there can be no motion.[2] The Atomists turned the argument around: if there were no void, there would be no motion; but there is motion; therefore, there is void.[3] The void is thus introduced into the theory in order to allow for the motion of atoms: it is a necessary condition of motion. Aristotle sometimes writes as though he were criticizing a theory in which void is a sufficient condition of motion also;[4] but we shall see later exactly what the target of his criticism is.

The sources establish clearly enough certain properties of Democritean atoms. Their constituent material is qualitatively uniform: they differ from each other only in shape and size. Since the material is uniform, greater size in an atom entails greater weight—but how Democritus thought of weight is problematic, as we shall see. Atoms are not liable to any kind of change, except change of place or position. So it seems clear *a priori* that the only possible form of interaction between atoms is collision, and the only differences between interactions are those brought about by differences in the shape, size, and weight of the colliding atoms, and differences in the direction and speed of their motion before colliding. *A priori,* there appears to be no room in the theory for any kind of attraction or action at a distance.

What the atomic theory is required to explain is the whole of the natural world and our knowledge of it; and one of the problems to which it is applied is how this or any other cosmos is formed in the first place. We would expect *a priori,* before looking directly at the relevant evidence, that the Atomists would attempt to form a theory of motion that would account not only for the observed changes in the natural world but also, without any additional *ad hoc* assumptions, for the formation of the world itself.

The evidence confirms our expectations, but with a good deal of ambiguity; the source materials are severely at odds with each other about some of the details. The thoughts on the subject that follow are by no means comprehensive, and focus as narrowly as possible on the question that most irritated Aristotle—the question of natural motion, to use Aristotle's own phrase.

Ancient Atomism was a steady-state theory, not a big-bang theory. The matter that composes the universe is constant in quantity and quality and is always distributed through void space in more or less the same density. The motion of atoms had no beginning; no atom ever moved for the first time. A cosmos begins to take shape when a number of atoms at random, by some unexplained mechanism, happen to distinguish themselves from all the others by joining together in a vortex (*dine*), the first result of which is that the component atoms begin to be sorted like to like.[5]

Some historians have apparently thought of this sorting process as presupposing a form of attraction between atoms of similar size and shape. There is some evidence that suggests this. Democritus describes examples of the sorting process: birds of a feather flock together, pebbles of similar size and shape are grouped together by the action of waves on a beach, grains are sorted by size and shape by the motion of a sieve, "as if the likeness in these had some [force] of attracting things."[6] It is likely, however, that the examples (including the birds) are meant to illustrate that there are instances of natural sorting without the action of a discriminating mind, and that the last clause, which is almost certainly added by someone other than Democritus in any case, is meant in a strictly figurative sense—it is *as if* there were an attraction.[7] There are also statements about Democritus that say he held that only like can act on like;[8] but these are either parts of his argument for the uniform quality of all atoms, or they are about the qualities of compounds. The evidence, in fact, does not support the view that there is any irreducible law in Democritus' dynamics that gives special properties to atoms of similar size and shape. The sorting of like with like, so far as we can judge, happens because similar atoms are similarly affected by the same causes.

The cosmic vortex was evidently conceived on models familiar in ordinary life—whirlpools and eddies in water, tornadoes or twisters of one sort or another in air. The essential element in these models is that objects caught in them are mechanically sorted, to some extent, into kinds. The actual dynamics of vortices are complex:[9] the one extended text that describes the early Atomists' use of this model, obscure though it is, makes it clear that in their theory the vortex has the effect of bringing larger and heavier objects to the center, where they tend to lose their rotatory motion, and sending smaller and lighter objects away from the center.[10]

Now, where does the concept of weight fit into this theory? The evidence is contradictory: "Democritus said [that the elements have] two properties, size and shape, but Epicurus added a third to these—namely weight."[11] "Democritus says the primary bodies do not have weight, but move in the infinite through collisions with each other."[12] Against this, we have the testimony of Aristotle: "Democritus says each of the indivisible [bodies] is heavier the bigger it is";[13] and of Theophrastus: "Democritus distinguishes heavy and light by size: for he says that if each one [i.e., uncompounded body] when separated according to shape differs in weight, it differs in size."[14]

Faced with this contradiction, modern interpreters have propounded a clever solution. So long as atoms are not involved in a cosmic vortex, they are weightless, and collision is the only factor that explains their motion.[15] The vortex, however, drives *larger* atoms to the center, and this tendency to move toward the center is what "weight" means.[16] So this cake can be had and eaten.

But in spite of the general agreement about this solution, it seems to me very dubious. It breaks down on an ambiguity that has plagued both ancient and modern commentators—the ambiguity of "center". Let us agree that the vortex produced a tendency in some bodies to move toward the center; this sounds just like what "weight" means in the familiar Aristotelian cosmology: a tendency to move toward the center *point* of the cosmic *sphere*. But this cannot be what Democritus meant, for two reasons. First, a vortex turns about an axis; the center of a vortex is a line, not a point, and although it may account for the motion of bodies towards the central axis, it does not yield an explanation of why bodies should congregate at the midpoint of the central axis, at least unless some extra assumptions are made. Second, the Aristotelian view of weight as a tendency to move toward the midpoint of the cosmic sphere entails that the earth itself is spherical, as he argues himself.[17] But Democritus believed the earth to be flat and shaped like a drum.[18] The vortex that originally shaped the cosmos still continues, and is seen in the movements of the heavenly bodies. It is true that the axis of the continuing vortex no longer coincides with the vertical, because a tilt is supposed to have entered the system (to account for the rising and setting of some heavenly bodies). But even with its axis tilted in relation to the earth's surface, the vortex seems totally inadequate to explain weight— that is to say, to explain why a piece of rock dropped from a height in Abdera falls on a line perpendicular to a stationary, flat earth.

If the vertical fall of heavy bodies on the earth's surface cannot be explained by the vortex, and cannot be explained by the attraction of like to like, then we seem to be forced back on the interpretation that weight, meaning a tendency to fall vertically, is a primary, irreducible property of the atoms.[19] The direction that we call vertical will be a datum of the system: the earth must take the shape it has *because* atoms fall in that direction. In this respect, although not in others, Democritus' theory of motion will be the same as the Epicurean theory.[20]

But what then are we to make of Aristotle's repeated complaints that Democritus did not define *natural* motion and had no theory of absolute weight?[21] The question cannot be answered until we have examined Aristotle's criticisms of motion in the void.

III. ARISTOTLE'S CRITICISM OF THE CONCEPT OF VOID

Aristotle criticized the Atomists for their theory of motion directly on two grounds: that they could explain neither the speed nor the direction of motion of bodies moving in the void. These two lines of criticism will be explored in sections 4 and 5. First we must review briefly the grounds on which Aristotle argued that there can be no such thing as void space.

He regards the question as being dependent on the concept of *place*. A void, if there could be such a thing, would be an empty place. In fact, his own idea of what place is makes it impossible that there could be an empty one.

We need the concept of place, he says, for two reasons: (1) we want to talk about displacement (*antimetastasis*): this pot is full of water, but when I pour the water out, it becomes full of air. Air has taken the *place* of water; (2) we want to talk about natural place, i.e., the destination, the resting place, of bodies with respect to their natural motion. A useful concept of place will be one that helps us to do these two things without confusion.[22]

Working mainly from the first criterion, Aristotle argues that place must be the inner surface of the containing body: the place of the water in the pot is the inner surface of the walls of the pot.

He considers and rejects another possible definition of place: according to this, place would not be the surface of the containing body but the interval (*diastema*) defined by that surface.[23] If we analyze a case of displacement into its component parts using the Aristotelian definition, we have three items: the body that first occupies the place, the body that displaces it, and the container. If we analyze it using the alternative definition of place as interval, we have four items: the first body, the second body, the container, and the place. Now Aristotle intends to go on, as he does in book IV.7, to show that we could have an empty place only if we adopt this alternative definition of place as interval; and so his refutation of the concept of void is just his argument against this alternative definition of place. How, then, does he argue against it?

His first move amounts to saying that we do not need this fourth item in the analysis of displacement: "It is thought that there is an interval as something distinct from the changing body. But this is not so: one or other of the changing bodies capable of being in contact replaces the outgoing body."[24] Simplicius fills out the details in his commentary: think of a wine skin full of wine; the place of the wine is the inner surface of the wine skin. Empty the wine out, and either air replaces it, or the skin simply collapses, in which case there is no place left to think about.[25] He quotes Galen: "But let us suppose that when the water is emptied out of the pot, no other body flows in: then the interval between the inner surfaces remains as a distinct entity (*kechorismenon*)."[26] But this, says Simplicius, is just an *alogos hypothesis*. If there is nothing to take the place of the departing water, then it will not leave—as is proved by clepsydras.

In this argument, then, Aristotle observes that we do not need a place that is not the place of either the first body or the second body in order to account for displacement, because in our experience of displacements there never is such a distinct entity. And so he concludes that place is not the interval between the inner surfaces of the container, but is just those surfaces themselves. Then he goes on to say that since void means an empty place, there cannot be a void at all, because there cannot be an empty place, by his definition.

So what looks at first sight like an *a priori* argument based on an analysis of place turns out to be an *a posteriori* argument that begs the question raised by the Atomists' theory. If this were the only argument of Aristotle's against the interval concept of place, his opponents had only to point this out. But in fact he had more arguments against it.[27] They are, however, very obscure, at least in the received text, and it would be impossible to attempt an analysis of them here without taking too much space. They appear to depend on certain difficulties that arise in the interval theory of place when you consider either the parts of the contained substance taken as divisible *ad' infinitum,* or the parts of something moving in a complex of moved containers. Perhaps all we need say about them in the present context is this: insofar as they depend on infinite divisibility, they would not worry the Atomists, who denied infinite divisibility; insofar as they depend on puzzles about moved containers, Aristotle's own theory was no better off. In any case they apparently failed to convince even his own supporters, such as Strato and Simplicius.[28]

On this score, then, the Atomists had little to trouble them.

IV.   ARGUMENTS ABOUT SPEED OF MOTION IN A VOID

Aristotle's first objection under this heading[29] grows straightfor-wardly out of his assumption that the time taken by a given object to traverse a stretch of the continuum under constant conditions varies in proportion to the resistance offered by the medium: the thinner, the shorter.

A body A moves through a thick medium B in a relatively long time C, and through a thin medium D of the same length in time E; and the times C and E are to each other as the media B and D are to each other in thickness.

$$C : E :: B : D$$

Now suppose, as the Atomists do, that there exists a void stretch Z of the same length as B and D, and that the body A would take some time to traverse it, say, H. Then there is a ratio between this time and time E, and by the proportion already established we can show that there is a very thin medium (not void) L that bears the same relation to medium D that H bears to E.

$$E : H :: D : L$$

And according to our assumptions, body A will get through medium L in time H. But our statement of the Atomists' postulate held that body A gets through a *void* stretch in that time. It cannot take the same time to get through a void and a medium, however thin; and hence the Atomists' postulate is wrong.

This argument proceeds by taking the nature of the moving body as a constant and varying the thickness of the medium. Aristotle goes on to generate an argument of the same kind by assuming a constant medium and varying the nature of the moving body.[30] Bodies that have a greater force (*rhope*) of weight or lightness get through the same distance in a shorter time (sc. than bodies that are less heavy or less light). But this is because the bodies that have more *rhope* cleave the medium more quickly. In a void all bodies would move with the same speed, which is impossible.[31]

The Epicurean Atomists accepted that in a void there can be no explanation of why things move at different speeds, and asserted that all atoms moving in the void move at the same speed.[32] That assumption was also necessary if they were to meet another of Aristotle's arguments, about the motion of indivisible magnitudes; but we will leave that aspect aside now.[33]

They refused, rightly, to accept Aristotle's proportion sums, which entailed that a medium of zero resistance takes no time to traverse, and instead they said that the speed of atoms through the void is a natural limit. It is not infinite, but simply faster than any other speed.[34]

Atoms are never stationary, and never slow down, even when they are involved in compounds; but they do change direction when they collide. So the speed of motion of a compound may vary from nil, when all the component atoms are simply colliding with each other within the same volume, to atomic speed, when all the component atoms are moving in the same direction and not colliding with anything.[35] At the level of compounds, therefore, the Epicureans would agree with Aristotle that a thing goes faster if the resistance is less or its weight is more; but as the medium gets thinner, the speed approaches the fixed atomic speed, not infinity.

I have not observed any *arguments* in Epicurean texts that directly attack Aristotle's proposition that speed varies inversely with the thickness of the medium. The Epicurean position appears to be to show that the phenomena *can* be explained consistently on the assumption that speed has a natural limit. They did, however, directly attack the Aristotelian theory of motion as swimming through a medium;[36] but an examination of that argument would be out of place here.

V.   ARGUMENTS ABOUT THE DIRECTION OF MOTION IN A VOID

To understand Aristotle's criticisms of the Atomists under this heading, it is necessary to recall some points about his own theory of natural motion.

He marks out certain instances of motion as natural—the motion of earth and water toward the center of the universe and of fire and air away from the center, and the motions of living beings—and he defines nature as an inherent source of motion and rest.[37] This is by no means the end of the matter, however. It would be, if he thought of nature as some kind of immanent force or power. The author of the pseudo-Aristotelian *De mundo* seems to have had some such idea, in that he ascribes the operations of nature to the power (*dynamis*) of a cosmic deity who presides over the cosmos using his *dynamis* as an executive civil service, just as the king of Persia presides over his empire from his palace in Ecbatana.[38] Presumably no further explanation is needed then; heavy things fall, light things rise, because they are interfused with this *dynamis*. We can describe in more detail, but no more causal explanation is called for. Something similar is true of Stoic cosmology, in which all things are permeated with a divine *pneuma,* an agent that distributes active properties throughout the cosmos. In both cases we can add teleological explanations for this and that feature: the divine *dynamis* or *pneuma* acts in such and such a way, rather than somehow else, because the first is better. But nothing is lacking from the causal explanation.

It is reasonably certain, however, that Aristotle did not content himself with a concept of nature of this kind. In his most explicit discussion of the problem[39] he distinguishes between living creatures, which can correctly be said to move naturally and to move themselves, and the simple bodies, which move naturally but do not move themselves. "These are what might raise the problem: what are they moved by? I mean, the light and the heavy. They are moved to places opposite to their own by force, but to their own place (the light up, the heavy down) by nature. And yet it is not clear even so by what they are moved, as it is when they are moved contrary to nature. For it is impossible that they are moved by themselves, since that is a property of life. and peculiar to living things, and they would in that case also be able to stop themselves."[40]

Are they moved by nothing? That would contradict the proposition that stands at the beginning of *Physics* VII: "everything that is moved is moved by something."

In the face of this, many commentators have taken the view that the mover that causes the natural motion of the simple bodies is their natural place, which exercises an attraction on them, as the Unmoved Mover does on the sphere of the stars. But this is contradicted by his statement that place is not a cause in any of the four categories of cause,[41] and it appears to be inconsistent with his analysis of natural motion in *Physics* VIII.4.

That analysis appears to evade the problem, rather than to solve it.[42] A body that is actually heavy is potentially light; the change from heavy to light is caused, as such changes are in Aristotle's theory, by the action of something that is already actually light. Once it has become light, it will move to its natural place *unless prevented*. That is to say, no further direct cause is to be sought for its motion.

It is important to notice, however, that even if we reject natural place as a *cause* of natural motion, it is still a necessary condition. Aristotle gives a causal account of how a body changes from heavy to light or vice versa, but he *defines* ''heavy'' and ''light'' in terms of natural place: ''The question is asked, 'why do light things and heavy things move to their own place?' The reason is that their nature is to go somewhere: that is what it is to be light and heavy—the former is defined by 'up' and the latter by 'down'.''[34]

We must now recall that place for Aristotle is not a part of geometrical space defined by some system of coordinates but the inner surface of the containing body. So we can define the place of the whole homogeneous body of cosmic fire, for example, as the inner surface of the aetherial sphere of the moon on one side and the outer surface of the sphere of air on the other side. An individual portion of fire, however, has a place of its own only if it is surrounded by something non-homogeneous with itself. We can say, then, that it is only if a portion of one of the simple bodies is surrounded by a different body that its lightness or heaviness is manifested in motion. To be light is to be at rest in the natural place of the light body and to move upward, if not prevented, from any other place.

Let us compare the naturally falling body with a projectile. The thrower imposes a motion on the projectile, the projectile moves the air, and the air keeps the projectile going.[44] Aristotle once remarks that natural motion can be *assisted* in the same way: for example, if you throw a stone downward, the air ''blows helpfully on'' the natural motion.[45] Now, the fact that it is just in this situation, in which a heavy body is *thrown* downward, that air is mentioned as contributing to the motion shows that in the free fall of a stone air does not function in the same way. The medium is not a sustaining cause of natural motion in the way that it is a sustaining cause of forced motion. It is nevertheless a necessary part of the explanation of natural motion, in that natural motion involves place in its essence, and there is no place if there is no containing body.[46]

With all this behind us, let us come back to Democritus' theory of motion and Aristotle's criticisms of it. In *Physics* IV.8, before the criticisms on the subject of speed of motion that we have examined in section IV, Aristotle has several criticisms that bear on the problem of direction of motion. He first argues that the void cannot explain any motion, and then goes on to argue that, so far from explaining motion, it actually makes motion impossible. The point of the first argument is repeated in the second; so we can look directly at that.

"If they say that void is necessary if there is to be motion, examination shows the opposite to be the case: if there is void, then nothing can move at all."[47] First, just as some say the earth is at rest because it is in equipoise, so things must be at rest in the void, because there is no difference (sc. of place) in the void. Second, the idea of forced motion logically presupposes that of natural motion: but there can be no natural motion in the void and in the infinite, since there is no up, down, or center in the infinite, and no difference between up and down in the void, which is equivalent to "nothing".[48] Furthermore, there can be no motion of a projectile in a void. A projectile continues to move after losing contact with the thrower either by antiperistasis[49] or because the air that is pushed pushes with a stronger motion than the projectile's natural motion.[50] This cannot happen in a void. Moreover, no one could explain why a thing moved in a void should ever stop. "For why should it stop *here* rather than *here*? So it must either rest or continue to move *ad infinitum,* unless something more powerful impedes it."[51] Again, the idea is that a thing moves into the void because it yields; but in a void everything is all alike, so that it will move in every direction.

The sentences translated verbatim above sound like an early statement of a principle of inertia.[52] But that is only one point in Aristotle's argument. What he is saying in the whole passage is that in a void there is no reason why a moving object should stop *here* rather than *here*, but also there is no reason why it should move in this direction rather than that, or why it should go on moving at all rather than come to a baffled halt.

Void cannot be a place, in Aristotle's view, and consequently has no power (*dynamis*), as place has.[53] We ought to try to see what he means without any presuppositions about inertia at all. He thinks that even the natural fall of a stone depends on its being all the time in a place. Suppose now that a stone is falling high up in the air, and between the inner surface of the sphere of air and the outer surface of the sphere of water and earth there is (*per impossibile*) a void interval.[54] It is consistent with Aristotle's doctrine, I think, that the stone should not carry over the interval in a straight line with some kind of inertial motion; its motion should rather become completely random.

If Aristotle had thought of nature as an immanent force of some kind, we should expect him to explain why this force could not account for a falling stone's continuing motion through a void. If he had thought of the natural place of an object as exercizing an attraction on the object, we should expect him to explain why that attraction could not be exercized through a void interval. If he had thought of a projectile as having in it some kind of impetus imparted by the thrower, we should have expected an explanation of why the impetus could not carry over a void interval. In fact, however, his criticisms of the void under our present heading can be reduced to two: forced motion needs a mover always in contact with the thing that is moved, and obviously the void cannot provide one; natural motion depends on the properties of weight and lightness, they depend on place, and the void is no place.

Whatever can be said about motion in the void can also be said about rest, if we use Aristotle's premises. Forced rest, as he calls it,[55] needs something to explain why the thing that is at rest does not move with its natural motion, and void cannot provide that. Natural rest occurs only in natural place; and there is no place in a void.

Is Aristotle's criticism of Democritus consistent with the explanation of the Atomists' theory of motion that has been put forward in section II of this paper?

"When Leucippus and Democritus say that the primary bodies always move in the void and the infinite, they ought to say with what motion, and what is their natural motion. For if the elements move each other by force, still each of them must have some natural motion too, which the forced motion is contrasted with; and the primary movement cannot move by force, but only by nature; otherwise there will be an infinite regress, if there is to be no first thing that imparts motion naturally but always a prior thing itself moved by force."[56] Could Aristotle have written this if he had been aware that Democritus believed atoms to have a natural tendency to fall downward? We shall have to assume that the burden of his objection is that the Atomists did not distinguish motion brought about by weight in the void from motion brought about by collisions. They said that both kinds of motion had been going on, naturally, from infinite time. There was no state of the atoms when no collision took place, nor was there ever a time when any atoms collided for the first time. From all eternity, atoms moved in the void in a jostling crowd.[57]

"They say that the nature of the atoms is single—as if each were a separate piece of (say) gold. But in that case, as we say, they must have a single motion; for a single clod and the whole earth go to the same place, and so do the whole of fire and a spark.''[58] Would Aristotle say this if he knew that all atoms have a tendency to fall downward because of their weight? In the context it appears that his meaning may be just that the Atomists, lacking the idea of absolute lightness, have no good explanation of upward motion. Some atoms must move upward, in their cosmology; but if there are only differences between more and less heavy things, then there can be no explanation, according to Aristotle's view, of any natural upward motion. He concludes: ''Necessarily then neither do all things have weight, nor do they all have lightness, but some have one, some the other.'' We might find some confirmation of this interpretation in a casual remark he makes elsewhere, that earlier philosophers did not explain what is heavy and what is light, but only what is relatively heavy and relatively light *among things that have weight*.[59]

Is Aristotle's complaint that the Atomists assign no cause for motion[60] compatible with his knowing that they said atoms have a tendency to fall downward due to their weight? Both passages where this complaint is made occur in the context of a discussion of *first* causes of motion. It is quite clear that, whatever account we may give of Aristotle's own theory of weight and lightness, these two properties do not satisfy his criteria for a first cause: his own theory of the natural motions of the simple bodies does not preclude the need for an unmoved prime mover, and it seems reasonable enough therefore that he should not say anything about the Atomists' notion of weight here.

Epicurus defended and revised the Atomist theory of motion by introducing something like the concept of a vector.[61] In the infinite, obviously, there is no ''highest'' or ''lowest'' point. But he rejected Aristotle's assumption that two extremes are needed to define a movement and claimed that it is enough to specify a point, a direction, and a sense. Take the point on the earth's surface ''wherever we stand'', the direction as vertical, and the sense either up or down. In both senses, the vertical line can be prolonged to infinity, and ''down'' is the direction of motion of atoms due to their weight.

Insofar as it was acceptable, the introduction of this concept into Epicurean physics served to disarm all of Aristotle's objections to motion in a void that were based on his notion of place. A body moving through the void is no longer a body fatally deprived of a place and therefore deprived of direction and sense. There is no reason to think it will not continue to move in the same direction through the void until something causes a change, since the void offers no resistance.

Epicurus counterattacked the Aristotelian theory by pointing out that one and the same line prolonged in the same direction is both up and down in his system: if you produce the vertical line downward until it passes through the center of the Aristotelian cosmos, downward changes to upward at that point. Why should the geometrical center of the cosmic sphere have such an effect? Perhaps it was partly to answer this counterattack that the Stoics modified Aristotle's theory, substituting the center of the *earth* for the center of the cosmos as the focus of natural motions.[62]

I have been attempting to revive an old view of Democritus' theory of motion according to which the differences between Democritus and Epicurus are smaller than most interpreters allow them to be. These differences, however, still exist. We have already noted in section IV that Epicurus was persuaded by Aristotle's arguments that there can be no explanation for different speeds of atoms in the void. He may also have been influenced by Aristotle's complaints that the early Atomists failed to distinguish the natural motion of atoms from motions caused by collisions. He was left with the proposition that the natural motion of all atoms is to fall downward in parallel straight lines at constant speed, and hence the explanation of the occurrence of collisions— that faster atoms catch up with slower ones—that was available to Democritus was closed to him. To deal with this problem, he introduced the notorious swerve.[63]

This Epicurean theory has been much criticized since antiquity on two grounds. The first is that the swerve is a completely arbitrary assumption, "a piece of childish fiction", as Cicero called it.[64] It *is* arbitrary, and can be defended only on the ground that it is the most economical hypothesis that will save the appearances. The second line of criticism is that it is impossible to specify an "up" or "down" in an infinite void. This is, however, no more difficult, so far as I can see, than the idea of the existence of infinitely numerous atoms having shape, size, and solidity. These properties of the atoms are inferred from sense experiences; and it is an argument of just the same kind that established the downward direction of motion of the atoms.

There is a feature of the Atomists' theory of motion that is much more vulnerable to criticism, and it is surprising that so little attention is drawn to it. It is simply that the theory depends on the proposition that the earth is flat. The direction up-down is a datum of the system. We know what the direction is by observation—that is, by taking the vertical from "wherever we stand", as Epicurus says. All perpendiculars to the earth's surface must be parallel, if the theory is to work. If these perpendiculars are not parallel, the theory is faced with one of three equally fatal consequences: either the downward fall of the atoms focuses on a single point, as in Aristotle's cosmology, and in that case *our* cosmos gets back the uniquely privileged position that the Atomists wanted to deny it; or there is no single focus but a random assortment of directions, and in that case the theory is left with no coherent explanation of weight; or the motion of atoms in the cosmos must be different from that of atoms at large, and the theory cannot explain either that difference itself or our knowledge of the different conditions outside our cosmos.

So the plausibility of Atomism as an alternative to Aristotelianism hinges in part on this question: was it reasonable to believe that the earth is flat? What arguments were used to defend the sphericity of the earth *before* Epicurus, and was it reasonable to reject them? But that is the subject of another paper.

1. Individuals who in general qualify as opponents of the Atomists sometimes differ from the others on particular points, as Plato does on this one.

2. Melissus B 7.7.

3. Leucippus A 7 = Aristotle, *De generatione et corruptione,* I.8.323a23 ff.

4. E.g., *Physics* IV.8.214b14 ff.

5. H. Diels and W. Kranz, *Die Fragmente der Vorsokratiker,* 6th ed. (Berlin: Weidmann, 1951), 67 A 1 (hereafter cited as DK).

6. Ibid., 68 B 164.

7. This is the view, for example, of Kirk, in G. S. Kirk and J. E. Raven, *The Presocratic Philosophers* (Cambridge: At the University Press, 1960) (hereafter cited as KR); and Carl W. Mueller, *Gleiches zu Gleichem* (Wiesbaden: Harrasonritz, 1965).

8. DK, 68 A 38 (Simplicius); A 63 (Aristotle); A 135 (Theophrastus).

9. I am indebted to John Ferguson, "Dinos," *Phronesis* 16 (1971): 97–115; and Steven S. Tigner, "Empedocles' Twirled Ladle and the Vortex-Supported Earth," *Isis* 65 (1974): 433–47.

10. DK, 67 A 1.

11. Aetius I.3.18 in DK, 68 A 47.

12. Aetius I.12.6, ibid.

13. *De gen. et corr.* I.8.326a9. This translation is disputed by Harold Cherniss, *Aristotle's Criticism of Presocratic Philosophy* (Baltimore: Johns Hopkins University Press, 1935; repr. New

DAVID J. FURLEY

York: Octagon Books, 1964), p. 97 n. 412, but comparison with *De caelo* IV.2.308b9 seems to me to prove it correct.

14. Theophrastus, *De sensibus* 61, according to the text and interpretation of John B. McDiarmid, "Theophrastus *De sensibus* 61–62: Democritus' Theory of Weight," *Classical Philology* 55 (1960): 28–30.

15. In some versions of this interpretation, atoms are allowed to have weight at all times, but weight is construed as a power of resistance, not as a tendency to move in any direction.

16. J. Burnet, *Early Greek Philosophy,* 4th ed. (London: A. C. Black, 1945), pp. 343–45; C. Bailey, *The Greek Atomists and Epicurus* (Oxford: Oxford University Press, 1928; repr. New York: Oxford University Press, 1964), p. 83; Kirk in KR, p. 415 f.; W. K. C. Guthrie, *A History of Greek Philosophy* (Cambridge: At the University Press, 1962 ff.), 2:400–404; V. E. Alfieri, *Atomos Idea: L'Origine del Concetto dell'Atomo nel Pensiero Greco* (Florence: Le Monnier, 1953), pp. 88 ff.

17. *De Caelo* II.14.

18. DK, 59 A 88.

19. We had better call it a property of the atoms, not of the infinite void. By calling the void "what is not", the Atomists must have meant to deny it all positive properties.

20. This is a return to the interpretation of E. Zeller, *Die Philosophie der Griechen,* 6th ed. (Leipzig: Reisland, 1920), 1:1076 ff. I hope to defend it with more documentation elsewhere.

21. *De caelo* III.2.300b11, I.7.274b30.

22. *Physics* IV.1.

23. Ibid., IV.4.211b15–29.

24. Ibid., IV.4.211b18–19.

25. Simplicius, *Physics,* 573, 2–27.

26. The reference is not known.

27. *Physics* IV.4.211b20–29.

28. Strato, fr. 55 (Wehrli), with H. B. Gottschalk's Appendix II, in his *Strato of Lampsacus: Some Texts* (Leeds: Philosophical and Literary Society, 1965), p. 169; Simplicius, *Physics* 577, 24 ff.

29. *Physics* IV.8.215a24–216a11.

30. Ibid., IV.8.216a11–21.

31. Aristotle is thinking about his own theory of natural motions, of course, in which lightness is not defined as the absence of weight, but as a tendency to move upward.

32. Epicurus, *Ep.* I.61–62; Lucretius II.225–39.

33. *Physics* VI.1–2. I have discussed this at length in my *Two Studies in the Greek Atomists* (Princeton, N.J.: Princeton University Press, 1967), pp. 111–29.

34. "As fast as thought": Epicurus, *Ep.* I.61. "Faster than lightning"; Lucretius VI. 325–47.

35. Epicurus, *Ep.* I.62. For more defense of this interpretation, see my *Two Studies,* pp. 121–25.

36. Lucretius I.370 ff.

37. *Physics* II.1.192b21.

38. *De mundo* 6.397b9 ff.

39. *Physics* VIII.4.

40. Ibid., 255a1–7.

41. Ibid., IV.1.209a20. This statement occurs in the context of a possible *aporia.* Aristotle ought, I think, to qualify it; see below, n. 46.

42. Ibid., VIII.4.255a24 ff.

43. Ibid., 255b13–16.

99

44. Ibid., VIII.10.267a2 ff.

45. *De caelo* III.2.301b29. The verb is *synepourizo*. Simplicius has one of his engagingly erratic moments: he notes that it probably comes from *ourios,* which is used of favorable winds, but adds that perhaps the metaphor comes from lions, "which are said to goad themselves into motion by lashing themselves with their own tails (*oura*)" (597, 1–5).

46. Hence I think Aristotle should modify his statement in *Physics* IV.1.209a20 (above, note 41). He says there that place is not a cause in any of the four categories. It is, however, a factor in the formal cause, the *logos* or *eidos,* of the primary bodies.

47. *Physics* IV.8.214b28 ff.

48. Note that the infinity of the Atomists' void is mentioned here. Guthrie, criticizing the view of Democritus' theory that I am defending, says (II:401): "Neither here [sc. *De caelo* 275b29 ff.] nor anywhere else does [Aristotle] criticize the atoms for the absurdity of moving downwards in a *apeiron.*" Not so.

49. Perhaps Plato's explanation, *Timaeus* 59a.

50. Aristotle's explanation, *Physics* 266b27–267a12.

51. Ibid., 215a20–22.

52. See Edward Grant on this, "Motion in the Void and the Principle of Inertia in the Middle Ages," *Isis* 55 (1964): 265–92.

53. "Void is that in which nothing is heavy or light" (*Physics* IV.1.214a2).

54. Buridan's argument; see Grant, "Motion in the Void," p. 278.

55. *De caelo* III.2.300a29.

56. Ibid., 300b9–16.

57. Cf. the fragment of Aristotle's book *On Democritus* (= DK, 68 a 37), where the verb *stasiazein* ("to quarrel with each other") is applied to atoms in the void.

58. *De caelo* I.7.275b32.

59. Ibid., IV.1.308 a11.

60. *Metaphysics* I.4.985b19; *Physics* VIII.1.252a32.

61. *Ep.* I.60. I am indebted to David Konstan's discussion in "Epicurus on 'Up' and 'Down' (*Letter to Herodotus* 60)," *Phronesis* 17 (1972): 269–78.

62. The two centers coincide, in Stoic cosmology, but it is by virtue of its being the center of the earth that this point is the focus of natural motions.

63. Lucretius II.216–50.

64. Cicero, *De finibus* I.19.

Chapter Five

INCIPIT/DESINIT

Beginning no later than the middle of the thirteenth century, medieval logicians regularly included the verbs 'begins' (*incipit*) and 'ceases' (*desinit*)[1]—along with quantifiers, truth-functional connectives, and modal auxiliaries—among words and phrases whose logical and semantic properties constituted part of the subject matter of logic. This is an odd fact. The verbs 'begins' and 'ceases' seem not to exercise any standard logical function, and the medieval discussion of them introduces considerations of time and change that are typical of physics or of metaphysics rather than of logic.[2]

In this paper I am going to try to explain the logicians' concern with 'begins' and 'ceases' and the peculiar combination of physical and logical issues that characterizes their discussion of those verbs. I shall investigate some of the problems they were dealing with and their solutions to them, particularly as they relate to the general topic of this volume.

There is a great deal of relevant medieval literature, and almost all of it is difficult to get at. My results in this paper are based on such texts as I happen to have had time and opportunity to study.[3] I believe that the texts I used are representative of the literature generally, but they are only a scattered sampling. There is much more to be done.[4]

I

Since the general topic of this volume is the treatment of the notions of space, matter, and motion in the history of philosophy and science, I want to leave the medieval logicians and the verbs 'begins' and 'ceases' aside temporarily in order to show how the nature of beginning and ceasing constitutes a problem for the treatment of those notions.[5] And since Professor Grant and I have taken on the specific job of discussing medieval reactions to Aristotle's *Physics,* I shall sketch a part of the problem as Aristotle raised and attempted to solve it in the *Physics,* especially in book VI, chapter 5. (I shall be turning to the *Physics* more than once in this discussion. My aim in doing so now is simply to introduce the physical side of the problem in its historical setting, not to provide a close reading of Aristotle's text.)

Aristotle maintained that time is continuous in the way a geometrical line is continuous (232b24 ff.). Temporal instants, somewhat like geometrical points, are not real but theoretical entities (222a10 ff.). They are like geometrical points in that time is not *made up of* instants although it is infinitely *divisible into* instants; between any two instants there are infinitely many instants (218a18 ff.). We may, then, describe the Aristotelian problem of beginning and ceasing very broadly as the problem of assigning temporal limits to a process of change measured against a continuum. The acquisition of any designated changed state takes place instantaneously (235b31 ff.), and so each temporal limit must be identified with some instant. Aristotle recognized that this problem of beginning and ceasing varies from one sort of change to another[6] (236b15 ff.). In order to introduce the problem I shall consider a case of simple local motion.

Imagine a ball resting on a platform at the top of a ramp, then rolling down the ramp, and finally coming to rest against a barrier at the bottom of the ramp. On the face of it this may seem to be a case in which assigning temporal limits to the motion is childishly simple. Is it not obvious that the first instant of the ball's motion constitutes one of the temporal limits and the last instant of its motion the other? But as Aristotle viewed it, against the background of his theory of time, this apparently obvious account is absurd. There can be neither a first instant nor a last instant of motion.

I can simplify the problem without distorting it by considering it as it arises in connection with just the first instant; exactly analogous considerations apply to the last instant. Suppose that t_1 is the first instant of the ball's motion. Because motion takes time and because Aristotle took instants to be indivisible, he denied that motion can take place at an instant (234a25 ff.). But we may suppose that t_1 is the first instant of motion in the sense that it is the first instant at which it is true to say of the ball that it is in motion. Now suppose that t_0 is an instant at which it is true to say of the ball that it has not yet moved. On Aristotle's continuum theory of time, t_0 cannot be the immediate predecessor of t_1, but we may consider it as arbitrarily near t_1, recognizing that there are nevertheless infinitely many instants—$t_{.5}$, $t_{.25}$, $t_{.125}$, . . .—between t_0 and t_1. If we rule out the possibility of teleportation, then it is true to say of the ball at a given instant t_n that it is in motion if and only if it is at t_n in a position different from its position at any arbitrarily near preceding or succeeding instant. It follows that for any instant t_n at which it is true to say that the ball is in motion there is an earlier instant t_m such that it was true to say at t_m that the ball is in motion. "With this shown," as Aristotle says, "it is evident that every thing which is in motion must have moved before" (236b34). Therefore, the notion of a first (or a last) instant of motion is an absurdity.

And yet, of course, we must be able to designate the beginning of the ball's motion. Since Aristotle denied the possibility of an intrinsic limit to mark the beginning of the motion, he had to be able to assign an extrinsic limit. If there cannot be a first instant at which it is true to say that the ball is in motion, there must be a last instant at which it is true to say that the ball has not yet moved. That last instant of not moving, then, and not the inconceivable first instant of moving, must mark the beginning, constitute the extrinsic limit, of the ball's motion.

(Of course this is an incomplete account of Aristotle's treatment of the problem of temporal limits in the *Physics,* and it is only a sketch even of the material it deals with. Serious difficulties are discernible even in this sketch, and we shall be considering some of them.)

Insofar as the medievals were presented with a *physical* problem regarding beginning and ceasing, that was it. Its basic ingredients are the theory of time as a continuum, the notion of indivisible instants, the doctrine of the instantaneous acquisition of the changed state, intrinsic or extrinsic temporal limits (first and last instants), and different analyses of change associated with different subjects of change. We shall see what some medieval logicians made of these ingredients.

II

Before considering the work of the logicians directly, I want to describe the historical development of their concern with beginning and ceasing.

As far as I know, the only other person who has studied the medieval treatment of beginning and ceasing in any detail is Curtis Wilson, in the second chapter of his book *William Heytesbury: Medieval Logic and the Rise of Mathematical Physics,* first published in 1956. I have benefited greatly from his pioneering work. More source material has become available since Wilson wrote, and as a result I can move into my own discussion of the medievals by offering a correction of Wilson's historical hypothesis. The evidence for the correction will emerge in the course of my discussion.

Wilson's hypothesis has three parts: (1) the medieval discussion of beginning and ceasing stems from Aristotle's *Physics,* especially from parts of books VI and VIII (p. 29); (2) it develops in two "phases", the physical and the logical (pp. 31–32); and (3) "it is from the discussion in the realm of physics that the discussion in the realm of logic arises" (p. 32). It certainly is essential to distinguish the physical and the logical strands in the medieval discussions of beginning and ceasing, and so I would accept part (2) of Wilson's hypothesis if it were not for its suggestion, borne out in part (3), that the physical and logical strands are historical stages—"phases"—in the development of the discussions. And part (3) of the hypothesis, which must have seemed obvious at the time Wilson wrote, is almost certainly false. As I see it, the truth is something like this. The discussion in the realm of logic, stimulated by considerations that had nothing to do with Aristotle's *Physics,* was already under way when the rediscovery of Aristotle's discussions of the temporal limits of change began to provide medieval logicians with interesting new problems and with criteria for evaluating their analyses. And if I am right about this, then part (1) of Wilson's hypothesis is true only as regards the physical strand of the discussion, not as regards the discussion generally.[7]

My own historical hypothesis is that the earliest medieval literature of the sort Wilson and I are interested in occurs in the twelfth-century treatises on fallacy, which were stimulated by the rediscovery of Aristotle's *De sophisticis elenchis*.[8] As we shall see, the authors of those treatises recognized that 'begins' and 'ceases', along with several other words, could give rise to ambiguities regarding their scope, that they involved covert reference to the past or the future despite their occurrence in the present tense, and that they involved covert negation despite their affirmative form. Thus, to begin with, the logicians were concerned not with the nature of beginning and ceasing viewed against a certain theory of time but with the analysis of sentences that have 'begins' or 'ceases' as their main verb,[9] especially when such sentences occurred as premises in inferences. And this logical analysis seems to have begun before Aristotle's *Physics,* with its discussion of temporal limits, was generally available again in western Europe.[10] When the Aristotelian physical problems did become known, they were associated with the sort of purely logical analysis that had already begun for other reasons. It seems likely that 'begins' and 'ceases' would have received very little attention as medieval logic developed, and certainly nothing like the detailed attention they did receive, if their role as fallacy-generators had not been expanded and enhanced by connecting them with Aristotelian physics just as the *logica moderna* was coming into its own, in the early thirteenth century.[11]

III

By presenting samples of the earliest treatments of 'begins' and 'ceases' I know of, I can provide some idea of the character of the logical strand when it evidently first appeared, soon after the middle of the twelfth century.[12] In the texts of the late twelfth and early thirteenth centuries edited and discussed by L. M. de Rijk in the two volumes of his *Logica Modernorum,* the logicians were preoccupied generally with the difficulties that the interpretation of ordinary speech put in the way of standard acceptable patterns of inference. The recently recovered Aristotelian treatment of fallacies provided the categories into which these difficulties at first were sorted, sometimes pretty haphazardly. Regardless of the labels pinned on them by the authors of these early texts, for our purposes the fallacies may be conveniently divided into four groups.

Fallacies in the first and least interesting group depend very little or not at all on distinguishing characteristics of these verbs. For instance, in the *Fallacie Parvipontane* (from the latter half of the twelfth century) we find this example of the fallacy of division (*LM* 1.581):

> Those two men cease to be.
> If anyone ceases to be, he dies.
> Therefore those two men die.

The fallacy, of course, consists in the fact that the pair of men may cease to be in virtue of the death of only one. If any special characteristic of 'cease' affects this case at all, it is that a pair of men *considered as a pair* may cease to be, while it is only *considered as two men* that a pair of men may die. But that characteristic is one that 'cease to be' shares with countless other predicates.[13]

Each of the other three groups of fallacies does depend on one of the distinguishing characteristics of 'begins' or 'ceases' that came to play an essential role in the fully developed analyses we encounter beginning around the middle of the thirteenth century.

Thus fallacies in the second group depend on the fact that in using these verbs one makes a covert reference to times other than the time indicated by the tense of the verb. For instance, again in *Fallacie Parvipontane,* we find these examples (*LM* 1.563–64):

> (A) Every man is white.
>
> No man begins to be white.
>
> Therefore every man was white.

> (B) Every man is white.
>
> No man ceases to be white.
>
> Therefore every man will be white.

Later medieval logicians would have provided analyses of the second premises bringing out the veiled tenses; but this anonymous twelfth-century writer rejects both inferences as they stand, claiming an equivocation on 'man' in the first premises (where it is related to [*habet se ad*] present men) and in the conclusions (where it is related in [A] to past men and in [B] to future men).[14]

I divide these fallacies into the third and fourth groups on the basis of considerations that may amount simply to two aspects of one distinguishing characteristic of 'begins' and 'ceases', but at any rate they were not always recognized as such.

Fallacies in the third group depend on the fact that these verbs involve covert negations. It is primarily for that reason that logicians of the twelfth and thirteenth centuries often group them among syncategorematic words, linked especially with other covertly negating words—the exclusives (such as 'only' and 'alone') and the exceptives (such as 'but' and 'except'). In the *Dialectica Monacensis,* dating from some time in or near the decade 1160–70, 'begins' and 'ceases' are described as syncategorematic words "containing the force of negation" and associated with the exclusives (*LM* 2(2).590). It is that characteristic of them which is supposed to account for the flaw in this inference, described as an instance of the fallacy of composition (*LM* 2(2).591):

> Socrates begins to be white.
>
> Therefore he begins to be colored.

If Socrates now begins to be white, then immediately before now he was *not* white—thus the negation. And, of course, having been not white is compatible with having been some other color—thus the fallacy. The analysis developed by later logicians explicitly brought out the negation as well as the implicit temporal reference.[15]

The fourth group of these fallacies is best exemplified in the *Tractatus Anagnini,* dating from the early thirteenth century, when the doctrine of the supposition of terms had already emerged. For fallacies in this group depend on the effect 'begins' and 'ceases' have on the supposition of terms that occur in incipit- and desinit-sentences. Later logicians sometimes attributed this sort of effect to the implicitly negating function of the verbs.[16] Thus the *Tractatus Anagnini* groups these verbs together with the exclusives and exceptives as affecting univerally quantified expressions in such a way as to block universal instantiation. We may not infer 'Almost every man is running; therefore almost Socrates is running'; 'Every man but Socrates is running; therefore Plato but Socrates is running'; or 'Socrates ceased to see every man; therefore Socrates ceased to see this man' (*LM* 2(2).292).[17] And only in the case involving 'ceases' does the syncategorematic word produce genuinely altered supposition as a result of which the inference is invalidated. In the first two cases the result is merely incoherence. Thus the distinguishing logico-semantic characteristics of 'begins' and 'ceases' might be thought to have an even higher degree of practical importance and theoretical interest for logicians than do the characteristics of the exclusives and exceptives.

This representative sample of the earliest medieval treatments of 'begins' and 'ceases' should be enough to show two things about the development of the logical strand after Abelard and before William of

107

Sherwood (d. 1266/71) and Peter of Spain (d. 1277). First, 'begins' and 'ceases' came to be recognized not merely as troublesome but as covertly complex in ways that link them with the exclusives and exceptives among syncategorematic words. Secondly, this development shows no signs of having grown out of, or been influenced by, the discussion of temporal limits in Aristotle's *Physics*.[18] The treatment of 'begins' and 'ceases' in these early texts is therefore purely logical—not by choice[19] but because no one had yet made the connection between the logical puzzles and the physical theory. I cannot tell who made the connection or say precisely when it was made, but I can offer some observations that may narrow the gap within which the missing link must be located.

The early texts included in De Rijk's *Logica Modernorum* date from the hundred years between Peter Abelard and William of Sherwood—approximately 1130 to 1230. It is so difficult to identify logicians of that period and to connect books with their authors that it might be called the Anonymous Century in the history of logic. During the Anonymous Century there was some development in the treatment of 'begins' and 'ceases'. From being casually noted as the source of variously classified fallacies they became the focal point of sophismata—logically puzzling sentences[20]—designed to expose their distinguishing characteristics.[21] (Many such sophismata were produced during the four hundred years in which logicians discussed 'begins' and 'ceases',[22] and the treatment of sophismata often constituted the main body of a logician's discussion of those verbs.)[23]

'Begins' and 'ceases' are identified as syncategorematic words in one very interesting but isolated passage in the early texts.[24] But when we encounter them after the Anonymous Century, in the works of the terminist logicians William of Sherwood and Peter of Spain,[25] they have already become established as standard items in the inventory of syncategorematic words, on which both William and Peter wrote treatises.[26] The discussion of 'begins' and 'ceases' in those two treatises is enormously more detailed and sophisticated than the best of the corresponding discussions we have from the Anonymous Century. No doubt there are (or were) treatises that will help to bridge this gap[27]—between any two medieval logicians there is another medieval logician—but in our present state of ignorance we can only speculate on the reasons for the rapid and extensive improvement. It does seem clear, however, both from the nature of the improvement and from historical considerations, that the single most effective reason is likely to have been the fact that William and Peter knew Aristotle's *Physics*[28] and had linked the already developing logical analysis of 'begins' and 'ceases' with the Aristotelian discussion of the temporal limits of change. I see no evidence that either of them actually forged the link, but if they were heirs of a tradition of taking a hybrid physical and logical approach to the problems of beginning and ceasing, it cannot have been a long tradition by the time they wrote their treatises. It did become a long tradition, however, characterizing very many and probably most logicians' treatises on 'begins' and 'ceases' for at least three hundred years afterwards.[29]

Peter of Spain's discussion of 'begins' and 'ceases' in his *Tractatus syncategorematum* is a good example of the hybrid approach.[30] After making a few remarks about the signification and the logical interrelation of the two verbs, he presents an account of permanent and successive things or states and of first and last instants, the medieval logicians' standard counterparts of basic ingredients in Aristotle's theory of change. From this account he then derives certain rules for the exposition of incipit- and desinit-sentences.[31] A few general remarks on time and tense follow the rules, and then he uses the rules to resolve the difficulties arising in five sophismata, introducing supplementary rules at more or less appropriate points among the discussions of the sophismata.

With minor variations and often a good deal of elaboration this seems to have been the pattern of the hybrid approach. The physical strand appears in the presentation of the permanent/successive distinction and the assignment of limiting instants, and sometimes also in statements on the nature of time and the instantaneous acquisition of a changed state. The logical strand appears in the author's adoption of one or another pattern of exposition for incipit- and desinit-sentences and his consequent treatment of sophismata. The two strands are interwoven in the more or less explicit derivation of the pattern of exposition from the presentation of the permanent/successive distinction and the assignment of limiting instants.

I want to lay out the physical strand, drawing mainly on Peter of Spain, and then to consider some problems it presents.

IV

The basic permanent/successive distinction is that those things or states are *permanent* the parts of which occur at one and the same time, whereas those things or states are *successive* the parts of which occur at different times.[32] Standard examples of permanent things or states are a man and being white; standard examples of successive things or states are a motion and becoming whiter.

A more elaborate, and in some respects more satisfactory, version of the distinction can be constructed to take account of details that can be discerned in Peter's discussion and that sometimes occur in other discussions as well.[33]

    I. Things or states that occur *at an instant only.*
       (Instantaneous—e.g., an instant, the acquisition of a changed state)

   II. Things or states that occur *either at an instant or during a period of time,*
      (Permanent—e.g., a man, a man's full height)
      IIa. *and acquire their being (naturally) at an instant (All-or-nothing permanent*—e.g., a man)
      IIb. *and acquire their being (naturally) during a period of time (Little-by-little permanent*—e.g., a man's full height)[34]

  III. Things or states that occur *during a period of time only.* *(Successive*—e.g., a motion, becoming taller)[35]

Limiting instants are assigned to things or states depending on their classification within this schema. Intrinsic limits are first and last instants of being, which we may designate t_{Ba} and t_{B2}; extrinsic limits are last and first instants of not being—t_{NB2} and t_{NBa}. Peter's assignment of limiting instants may be summarized in this table.[36]

	A-O-N permanent	L-B-L permanent and successive
t_{NB2}	NO	YES
t_{Ba}	YES	NO
t_{B2}	NO	NO
t_{NBa}	YES	YES

110

For purposes of raising some questions I want to consider, we may limit our attention to the basic A-O-N permanent/successive distinction. Using a horizontal time line, standard representations of intrinsic and extrinsic limits, and two standard examples, we may present Peter's view in the diagram below.[37]

not being a man	being a man	not being a man
not being in motion	being in motion	not being in motion

Peter's standard expositions of incipit- and desinit-sentences are based on his presentation of the physical strand.[38] In each exposition the first sentence is the incipit- or desinit-sentence to be expounded and the second and third sentences are the first and second exponents.

(1) A man begins to be: (1E1) A man now is, and (1E2) Immediately before this he was not.

(2) A man ceases to be: (2E1) A man now is not, and (2E2) Immediately before this he was.

(3) A motion begins to be: (3E1) A motion now is not, and (3E2) Immediately after this it will be.

(4) A motion ceases to be: (4E1) A motion now is not, and (4E2) Immediately before this it was.

V

What justification, if any, can be offered for Peter's assignment of limiting instants? For some features of the assignment the justification seems obvious.

For instance, a t_{Ba} is assigned to permanent things or states. Regardless of the grounds for that assignment, it is clear that if we assign a t_{Ba} we cannot also assign a t_{NBz}; for between t_{Ba} and any arbitrarily near preceding instant there will be another instant. Therefore, in general, the assignment of a t_{Ba} rules out the assignment of a t_{NBz}, and conversely; and the assignment of a t_{Bz} rules out the assignment of a t_{NBa}, and conversely.

As we saw in the introductory discussion of Aristotle's doctrine of temporal limits, it also seems easy to provide grounds for the assignment of extrinsic limits to motion in particular or to successive things or states in general. If it is true to say of x at t that x is in motion, then x occupies a position at t different from the position it occupied at any arbitrarily near preceding instant and likewise different from the position it will occupy at any arbitrarily near succeeding instant. Therefore there is no instant such that it is the first (t_{Ba}) or the last (t_{B2}) at which it is true to say that x is in motion.

We may then, at least temporarily, concentrate on the assignment of limiting instants to permanent things or states. I can sharpen the issue by pointing out that the *Tractatus exponibilium* formerly attributed to Peter of Spain offers a different exposition of a desinit-sentence whose subject is a permanent thing.[39]

> (5) Socrates ceases to be a man: (5E1) Socrates now is a man, and
> (5E2) Immediately after this he will not be a man.

The author of *Tractatus exponibilium* has nothing to say about first and last instants, but this exposition shows that he holds that the instant of ceasing to be a man and beginning not to be a man is t_{B2}, not t_{NBa}; or that he takes the instant of Socrates' death to be the intrinsic, not the extrinsic, limit of Socrates' being a man (or of the being of the man Socrates).

I want to try to adjudicate this disagreement between Peter and pseudo-Peter. In doing so I think I can uncover what are likely to have been the grounds for Peter's assignment of temporal limits. I believe that one of the conditions of Peter's assignment is the assumption that there is no foreknowledge—at least none on the part of anyone engaged in the assignment of temporal limits. In any case my adjudication is based on the assumption that there is no foreknowledge.

Since the disagreement occurs in connection with the cessation of a permanent thing or state, we may begin with that. There is an instant at which x ceases to be—x's desinit-instant. That much is taken for granted. The problem is to identify x's desinit-instant either as x's t_{B2} or as x's t_{NBa}. (Of course there can and must be only one of those two.) In order to identify x's desinit-instant as t_{B2}, one must be able to identify x's t_{B2}. At t_{B2} x still is; but as long as x still is, no instant can be *identified* as t_{B2}. x's desinit-instant can be identified as soon as x no longer is—that is, as soon as x's t_{NBa} occurs—but no sooner. But x's t_{NBa} *can* be identified *at* t_{NBa}. And once t_{NBa} has been identified, there can be no t_{Bz} to be identified. Therefore x's desinit-instant cannot be identified as x's t_{Bz} and must be identified as x's t_{NBa}. Notice that nothing in this statement of the problem or its solution depends on knowing whether x is permanent or successive. On this basis, therefore, Peter's uniform exposition of desinit-sentences[40] is justified, and pseudo-Peter's special exposition of desinit-sentences whose subjects are permanent things or states is not justified.[41]

How does this approach apply to the problem of beginning? Again, there is an instant at which x begins to be—x's incipit-instant—and again the problem is to identify x's incipit-instant, either as x's t_{NBz} or as x's t_{Ba}. In order to identify x's incipit-instant as t_{NBz}, one must be able to identify x's t_{NBz}. At t_{NBz} x still is not; and as long as x is not, no instant can be identified as x's t_{NBz}. (Moreover, as long as x is not, it is not even possible to identify anything as x.) x's incipit-instant can be identified as soon as x is, but no sooner. In the case of A-O-N permanent things or states, those that acquire their being at an instant, x is as soon as x's t_{Ba} occurs; and x's t_{Ba} *can* be identified *at* t_{Ba}. And once t_{Ba} has been identified, there can be no t_{NBz} to be identified. Therefore if x is an A-O-N permanent thing or state, x's incipit-instant cannot be identified as x's t_{NBz} and must be identified as x's t_{Ba}.

In the case of successive things or states, on the other hand, x's t_{Ba} cannot be identified at all. Therefore if x is a successive thing or state, its incipit-instant must be identified as its t_{NBz}. But in this case the incipit-instant can be identified only retrospectively. And that means that it will not do in general to identify x's incipit-instant with the instant at which it is true to say that x begins, unless one is willing to designate that instant vacuously. For, while in the case of a successive x, x's t_{NBz} is the instant at which it is true to say that x begins, it is also an instant that cannot be identified as t_{NBz} when it occurs. Therefore it is impossible (except by mere accident) that anyone say truly at t_{NBz} that x begins, where x is successive.

It is on grounds such as these that I think Peter's assignment of limiting instants is likely to have been based. Several questions are raised and left unanswered by this account—for instance, the question of whether it is reasonable to permit retrospective identification in just one case. One of those questions seems more important than the others, however, especially because it applies to Aristotle's treatment of temporal limits as well as to Peter's.

VI

Following Peter's assignment of limiting instants, no thing or state has a t_{B_2} with the single exception (which Peter leaves undiscussed) of the prior not being of a successive thing or state—for example, not yet being in motion. But not yet being in motion is evidently being at rest, and being at rest is itself a successive state. Since being at rest is a successive state, it has no t_{B_2}. But the t_{NB_2} of a state of motion, its incipit-instant, seems to coincide with the t_{B_2} of the preceding state of rest. How, if at all, can this inconsistency be avoided?[42]

It is not clear to me that any medieval logician perceived this difficulty. Aristotle may have recognized it and devised a way around it, or it may be that his theory of change developed in such a way that he never encountered the problem. In either case his theory seems to offer a means of avoiding an inconsistency in the account of the change from one successive state to another. I shall call this part of his theory the neutral-instant analysis.[43]

Suppose we go back to considering the example of the ball and the ramp. First the ball is at rest on a platform at the top of the ramp, then it rolls down the ramp, and finally it comes to rest against a barrier at the bottom. Consider just the ball's changing from being at rest to being in motion. An essential part of Aristotle's theory of change is the doctrine of the instantaneous acquisition of a changed state.[44] "It is clear, then, that that which has changed, when it has changed first, exists in that to which it has changed. Now that in which the changed thing first has changed must be indivisible . . . " (235b31 ff.)—that is, the acquisition of a changed state must occur at an indivisible instant. But no motion or rest can occur at an instant. "Nothing, then, can be in motion at an instant . . . nothing can be at rest at an instant either" (232a32 and 34). Even more pertinently, "there is no first [time] in which a thing in motion is moving. . . . Nor again can there be a first time in which that which is resting has rested" (239a1 and 12). Therefore neither motion nor rest can be a changed state; no change can be a change to motion

(224b15) or to rest. And yet, of course, the ball is first at rest and then in motion.

What Aristotle seems concerned to avoid is the analysis of that transition as a change of rest to motion.[45] But we are still faced with the problems of accounting for the relationship between the two states and of providing extrinsic limits for each. It is those problems which give rise to the neutral-instant analysis.

We may think of the state of rest and the state of motion as consecutive states (227a9 ff.), divided by the instant t_C. Let β be the ball, A its position on the platform, and Z its position against the barrier. Then the figure below depicts the neutral-instant analysis. The strictures imposed by the doctrine of the instantaneous acquisition of a changed state are observed in this analysis. There is no change of rest to motion or of motion to rest. Nor is there a change of rest to the neutral instant t_C or a change of the neutral instant to motion. The instant t_C is simply the extrinsic limit of the state of rest as well as the extrinsic limit of the state of motion. But how can it be both at once? On Aristotle's view the instant t_C belongs to both the intervals it divides (234a35–b6). Considered as belonging to the interval during which the ball is in its initial state of rest, t_C is the t_{NBz} of the interval during which the ball is in motion; considered as belonging to the interval of motion, t_C is the t_{NBa} of the interval of rest: "the instant must be the same extremity of both times" (234a5).

The ball, β, is described as being at A at the instant t_C. There is no

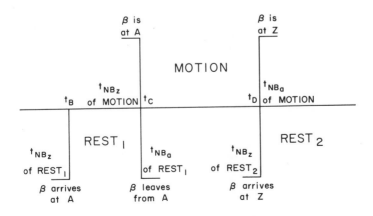

difficulty in describing a subject as being at a position at an instant, but that description applies to β equally correctly at any instant in the interval (t_B, t_C). What distinguishes t_C from any such instant? For any instant t_R during the interval, the positions of the ball at any arbitrarily near preceding or succeeding instant will be the same as its position at t_R. For any instant t_M during the interval (t_C, t_D), the positions of the ball at any arbitrarily near preceding or succeeding instant will be different from each other and from the ball's position at t_M. For t_C, however, the position of the ball at any arbitrarily near preceding instant will be the same as its position, A, at t_C, and its position at any arbitrarily near succeeding instant will be different from A. The neutral instant, t_C, is therefore uniquely determinable.[46]

The ball, β, is also described as leaving from A at t_C, and it may look as if leaving is itself a successive state and thus incapable of occurring at an instant. I think this appearance can be dispelled by considering the verb 'leave' (or 'arrive') in this context as being closely analogous to the verb 'begin' (or 'cease') in the appropriate respect. In one ordinary use of 'begin' and of 'leave' we speak of a thing's beginning to exhibit some property or leaving some location as if beginning and leaving were processes. But in the context of the Aristotelian and medieval discussions, we have to consider beginning as instantaneous. It is this consideration that is at the heart of the doctrine of the instantaneous acquisition of a changed state. The use of the participial construction—'x is beginning to be Φ'—suggests a process where there need be none; and certainly such a process is ruled out in this context. Very similar observations apply to leaving and to 'leave'. 'β is leaving from A' might describe a process. But we may adopt a strict interpretation of 'β leaves from A'—an interpretation included in the range of ordinary uses of 'leave'—according to which the sentence describes an instant at which β is still at A. And we can distinguish that instant, t_C, from any preceding instant at which β is at A just as we did in connection with t_C considered as the extrinsic limit of the interval of motion.

The neutral-instant analysis seems to offer the most effective way of dealing with the transition from one successive state to another, both for Aristotle and the medievals.[47] Many things Aristotle says call for this analysis, and some things he says confirm it;[48] but, as far as I know, he never clearly presents it in detail.

116

VII

Although we have not considered nearly all the interesting or difficult features of the physical strand, I want to conclude by presenting one example of the logical strand in a late development.

Once the hybrid approach to beginning and ceasing had become established, around the middle of the thirteenth century, a logician who took a purely logical approach must have done so as the result of a deliberate choice, motivated perhaps by a logical purism. As a consequence of his choice, either he would have to narrow his treatment of the topics his predecessors had dealt with, or he would have to try to provide some purely logical devices correlative with the physical ingredients in the hybrid approach. Several later logicians did adopt a purely logical approach.

The approach taken by William Ockham (d. 1349) is perhaps idiosyncratic, but it may stand as the type of the purely logical approach which narrows the treatment of 'begins' and 'ceases', rejecting or ignoring the physical strand and most of the problems and technicalities it gives rise to in the hybrid approach. Ockham's discussion of 'begins' and 'ceases' is markedly different from any other I have seen. He might be described very broadly as rejecting the detailed technical analysis of his predecessors in favor of an ordinary language approach. For this reason (and because his writings are less inaccessible than those of most medieval logicians) it does not seem profitable to examine his sort of purely logical approach in this article. [49]

The other sort of purely logical approach may be exemplified in an excerpt from the *Logica* of Johannes Venator Anglicus (d. 1427?). [50] Unlike Ockham, Venator does not explicitly reject any features of the physical strand, but neither does he make any use of them. He begins abruptly with the exposition of incipit- and desinit-sentences. There is no preamble on the reason for expounding sentences generally or for considering such sentences as these in particular, no references to kinds of things or states or to first and last instants. Indeed, in his rather long chapter on 'begins' and 'ceases' [51] there are no references to the *Physics* or to Aristotle at all. Moreover, he casually introduces logical examples in which one instant figures as the immediate successor of another. In all these respects he differs from many and probably most of his predecessors and successors. The level of his logical sophistication strongly suggests that his indifference to the physical strand is studied, not naive.

There are many difficulties even in the short excerpt I have provided from his treatise. I want just to call attention to the nature of the pattern of exposition he adopts and then to go on to look at his treatment of one rather complicated sophisma.

Venator rejects the two ordinary patterns of exposition—labeled [E1] and [E2] in the excerpt—and then offers [E3], the pattern he prefers. Since his rejection of [E1] and [E2] is founded on his arguing that we cannot always discriminate *prima facie* between exponible sentences suited to the one or the other of the standard patterns, [E3] is, understandably, a disjunction of [E1] and [E2].[52] Venator's purely logical approach seems typified by [E3], which gives the appearance of doing away with the need for the permanent/successive distinction and for worrying over the assignment of first and last instants. "But," as Hieronymus of St. Mark put it about a hundred years later, "how does one know which part of the disjunction is true [in a particular case]?"[53] Hieronymus took the hybrid approach, and the answer he provides is in terms of first and last instants. It is hard to see why such a question is not legitimate or how it could be answered without recourse to the physical strand. Venator does not raise the question, and from a purely logical point of view he is not obliged to do so. One or the other disjunct is true if and only if the exponible sentence in question is true, and from a purely logical point of view nothing more need be said.[54]

The sophisma I want to present occurs in connection with the rule labeled [C6] in the excerpt. I have included [C5] as well because [C6] is obviously intended to contrast with [C5]. The contrast focuses on *verifiability for an instant* in [C5] and *verifiability for a finite time* in [C6]. If the analogy is not pushed very hard, it may help at first to think of this distinction between truth-conditions as a purely logical analogue for the permanent/successive distinction.[55]

The contrast between [C5] and [C6] might be put in this way. Broadly speaking, in terms of the [E2] exposition of incipit-sentences, a present-tense incipit-sentence follows from a conjunctive sentence made up of (1) an affirmative future-tense singular sentence and of (2) the negative present-tense singular sentence corresponding to it. But if (1) is verified for an instant and is verifiable only for an instant, then it is not a present-tense incipit-sentence that follows but rather the disjunction of a present-tense and a future-tense incipit-sentence. And if (1) requires not an instant but a finite time for its truth, then no incipit-sentence at all follows from the conjunctive sentence.

The discussion of [C6] occupies half of Venator's entire discussion of 'begins' and 'ceases'; we can consider only the opening moves. [C6A1] and [C6A2] purport to be the first two (of eighteen) affirmative arguments associated with [C6], but they are more conveniently considered as a unit.

The following argument, presented in two slightly different versions, is supposed to be invalid.

(1) Socrates will be as white as Plato will be.

(2) Socrates is not as white as Plato will be.

Therefore,

(3) Socrates begins or will begin to be as white as Plato will be.

And yet, of course, its form (on one level of analysis) is quite the same as the form of the argument presented as valid in [C5A].

(1) This is not white.

(2) This will be white.

Therefore,

(3) This begins or will begin to be white.

What makes the [C5] argument valid and the [C6] argument invalid? The conclusion of the [C6] argument does not follow because "the future-tense proposition (1) 'This (indicating Socrates) will be as white as Plato will be' does not require an instant for which it will be verified". This explanation generates two further questions: (A) How can we tell that (1) requires not an instant but a finite time for which it will be verified? (B) How does that fact, if it is one, block the inference?

Venator offers answers to these questions in the passages immediately following the explanation, but they are difficult to appreciate in the absence of the illustration provided in his discussion of the

119

sophisma in [C6A2], where we are offered a counter-instance as a proof of the invalidity of the [C6] argument.

The "Proof" provides us with a case that is supposed to verify

(1) Socrates will be as white as Plato will be.

(2) Socrates is not as white as Plato will be.

while it falsifies

(3) Socrates begins or will begin to be as white as Plato will be.

It is easiest to begin with the verification of (2). At t_0 (now) "Socrates is white in degree 4 and Plato in degree 8", and during the ensuing hour both Socrates and Plato get whiter. Socrates now is exactly half as white as Plato; therefore Socrates is not as white as Plato is. But Plato will be whiter than Plato is; therefore Socrates is not as white as Plato will be. Thus premise (2) is verified on the hypothesis.

The verification of premise (1) is not so easy. We will do best to begin with this explanation: "Socrates will be as white as Plato will be, because however intense a whiteness Plato will have, Socrates will have one as intense". We may therefore take Venator to be reading (1) in this way:[56]

(1a) However intense a whiteness Plato will have, Socrates will have one as intense.

In quantificational analysis (1a) becomes

(1b) For every future instant t, if Plato has some degree of whiteness at t, then there is some future instant t' such that at t' Socrates is as white as Plato is at t.

The hypothesis does verify (1b). Let t_0 be the present instant, and let the ensuing hour be divided into sixty equal parts marked off by designated instants, so that t_{60} is t_0 + one hour. Suppose that at t_0 Socrates has degree 4 of whiteness and Plato has degree 8 of whiteness. Suppose also that Socrates' whiteness intensifies at the constant rate of 8 degrees per hour and Plato's whiteness intensifies at the constant rate of 4 degrees per hour. Suppose also that at t_{60} both Plato and Socrates cease to exist.[57] Thus in the time interval (t_0, t_{60}) Socrates passes through all the degrees of whiteness in the interval (4, 12), and Plato passes through all the degrees of whiteness in the interval (8, 12). Now let t_m be some instant after t_0 and before t_{60}. At t_m Plato has some degree of whiteness in the interval (8, 12), some degree d such that $8 < d < 12$. Since Socrates passes through all the degrees in the interval (4, 12) during the time interval (t_0, t_{60}), there will be some instant after t_0—i.e., some future instant—t_n such that at t_n Socrates will have degree d of whiteness. Therefore: (1a) However intense a whiteness Plato will have, Socrates will have one as intense. Therefore: (1) Socrates will be as white as Plato will be.

But on the hypothesis the instant t_n at which Socrates will have degree d of whiteness will always occur after the instant t_m at which Plato will have degree d. Thus there is no instant at which it is true to say: (3) Socrates begins or will begin to be as white as Plato will be.

Venator's invalidation of the [C6] argument may have been intended as a rejection of one of the theses regarding beginning and ceasing maintained by William Heytesbury (d. after 1371): "If something will be of a certain sort and is not of that sort now, then it begins or will begin to be of that sort".[58]

Tense logic seems to be the nearest twentieth-century equivalent to the logical strand of the medieval discussion of 'begins' and 'ceases', and Venator's invalidation of the [C6] argument might also be viewed as a rejection of what seems to be an intuitively acceptable thesis of tense logic: If x will be A, then there will be a time at which x is A. (Let 'x' be a place-holder for proper names.) Where A is the simple predicate 'white', this thesis holds good, as [C5] shows; but where A is the predicate 'as white as y will be', the thesis collapses, as [C6] shows.[59]

Curtis Wilson suggests some mathematical parallels to certain elements of the logicians' discussions,[60] but it remains to be seen whether these discussions had any actual influence on the development of mathematics or physics during or after the Middle Ages.

Appendix A

PETER OF SPAIN (d. 1277)
Tractatus syncategorematum (excerpt)

[This translation is based on the text of the 1489 printed edition, which I have emended in a few places. It comprises Peter's entire discussion of 'begins' and 'ceases'.]

[THE VERBS 'BEGINS' AND 'CEASES']

Having discussed words associated with conditional propositions, we must now discuss the verbs 'begins' and 'ceases'. They are called syncategorematic words [for two reasons]—because [1] an affirmation and a negation are understood in the exposition of them, and because [2] their signification is altered by the things they are connected with. They are expounded in one way or another depending on whether they are connected with permanent or successive things, for their signification depends on those things.

Such verbs as 'begins' and 'ceases' either signify the beginnings or ceasings of mutable things or they signify the being or not being of things at their initial or final limits. For 'begins' signifies beginning, which is the start of a thing's being or not being; and so it denotes being together with a limit. And 'ceases' signifies ceasing which is the end of a thing's being or not being.

Moreover, they follow from each other. For whatever begins to be ceases not to be, and whatever ceases to be begins not to be. Therefore both occur at one and the same indivisible [instant]. For example, when someone begins to be a man, he ceases not to be a man; and when someone ceases not to be a man, he begins to be a man.

Since the signification of these verbs varies with different sorts of things, in order to consider their signification in detail, one must first consider things. Some things are permanent, others successive. Those are called permanent the whole being of which occurs at one and the same time, and not one part after another—for example, a man, a piece of wood, a stone. But those are called successive the being of which is a whole not at one and the same time but in accordance with before and after, in such a way that one part succeeds another—for example, a motion or change, an action, a period of time.

There are two differentiae of these things. The first is that the being of permanent things is a whole at one and the same time while the being of successive things is not a whole at one and the same time but only in the succession of the parts. The second differentia is that the parts of permanent things can occur at one and the same time—for example, the part[s] of a man or of a stone—while the parts of successive things cannot occur at one and the same time. Two parts of one and the same motion, for example, or two parts of a period of time cannot occur at one and the same time; instead, one occurs after another successively. It is impossible that more than one time occur at one and the same time. The third differentia is that permanent things are naturally prior and successive things naturally posterior, since permanent things are the cause of successive things. The fourth differentia is that permanent things are intrinsically limited while successive things are not limited intrinsically but are limited by permanent things—for example, a change is limited by a quantity or a quality.

Here is another point. Some permanent things are obtained at an indivisible [instant]—light in the air, for instance—while others are obtained over a period of time through an increase in intensity or an increase in extension—for example, those things the being of which is divisible into degrees of intensity, such as whiteness or blackness, or into degrees of extension, such as length or width.

Thus as regards permanent things the being of which is obtained at an indivisible [instant], one can give the first instant of their being and of their not being afterwards, but one cannot give the last instant of their not being beforehand nor of their being. For example, one gives the first instant of being and does not give the last instant of not being or of being of a man, a horse, a length of two feet, or a length of three feet. But as regards successive things—such as a motion or change, or a period of time—and those permanent things the being of which is not obtained at an instant, one gives neither the first nor the last instant of their being but gives the last instant of their not being beforehand and the first instant of their not being afterwards. For while they have their being in a certain time, they have not being at the limit of that time; and just as there is no interval between a time and its limit, so there is none between their being and their not being.

On the basis of these considerations it is clear that permanent things simply have being at their outset and not being at their limit, since one could give a first instant of being as regards them and not a last. Successive things, on the other hand, have neither a first nor a last of their own kind—for the outset or the end of a period of time is not a time but an instant, and the limit of motion or change is not motion or change but a changed state—and so they have being neither at their outset nor at their end.

Because the verb 'begins' indicates the outset of a thing's being it has a different signification with permanent things and with successive things. For this reason certain rules are laid down.

Rule One. When the verb 'begins' occurs with permanent things the being of which is obtained at an indivisible [instant], it indicates an assertion of the present and a negation of the past. For example, when one says 'Plato begins to be a man'—that is, 'Plato now is a man, and [immediately] before he was not a man'.

Rule Two. When the verb 'begins' occurs with successive things or with permanent things the being of which is not obtained at an instant, it indicates a negation of the present and an assertion of the future, because successive things do not have being at their outset. For example, when one says 'The motion

123

begins', the sense is 'The motion now is not, and immediately after this it will be'. Or 'Socrates begins to be white', for something is said to be white only in virtue of an excess of white over black, which it obtains through a change. And its obtaining of this excess occurs not through an indivisible [instant] but through something that is divisible and not infinite. Therefore one cannot give the first instant at which something can be called white, unless one is speaking of the finished state of being white, which is obtained at the limit of the change. Thus the sense of the proposition is 'Socrates now is not white, and immediately after this he will be white'. Similarly in this case 'Socrates begins to be equal to Plato' the sense is 'Socrates now is not equal to Plato, and immediately after this he will be equal to Plato'. But being equal is something that gets established at an indivisible [instant]; therefore when 'begins' occurs together with the name 'equal', it must be expounded in the same way as when it occurs together with other permanent things the being of which is obtained at an instant.

Rule Three. When being is added to the verb 'ceases', it indicates a negation of the present and an assertion of the past, no matter what sort of thing it occurs together with. The reason for this is that one cannot give the last instant of a thing's being whether the thing is permanent or successive. For example, when one says 'Plato ceases to be a man', the sense is 'Plato now is not a man, and immediately before this he was a man (or Plato at the last instant before now was [a man])'. And when one says 'The motion ceases to be', the sense is 'The motion now is not, and immediately before this it was'.

[Rule Four.] Similarly, when the not being of things that are simply permanent is added to the verb 'ceases', it indicates a negation of the present and an assertion of the past. For example, when one says 'The matter of the water ceases not to be air', the sense is 'The matter of the water now is not under the not being of air, and immediately before this it was under that not being'.

[Rule Five.] But when the not being of successive things is added to 'ceases', it indicates an assertion of the present and a negation of the future, since one does give the last instant of not being of successive things. For example, when one says 'Plato ceases [not] to be in motion', the sense is 'Plato now is not in motion, and immediately after this he will not have not being in motion'.

On the basis of what has been said so far it is clear that verbs of this kind—that is, 'begins' and 'ceases'—convey a notion of different times by reason of what they signify. They do not signify those different times equally from the first, however, but first one and then another. The reason for this is that they invariably indicate an affirmation of one time and a negation of another, while they consignify a thing's being together with its limit. They always indicate the time regarding which they indicate an affirmation before they indicate the time whose negation they indicate. This is because the affirmation pertains to the thing's being but the negation to its not being.

Suppose someone says '[These verbs] are said to be of different times, since they always imply different times'.

In reply it must be said that 'time' can be taken in two ways. In one way strictly, in the sense in which it is the measure of motion or change. In this way it is implied by the significata of those verbs, but the verb is not said to be of a certain time in that sense of 'time'. In the other way it is taken as indicating a mode of being which is considered under a time. In this sense it is a mode of signifying, and it is in this sense that the verb is said to be of a certain time. Thus as regards that mode of signifying, the verb 'begins' is only of the present time, although it implies different times as regards its significatum.

In connection with what has been said so far a question arises regarding the sophisma 'Socrates ceases to be the whitest of men' (supposing that Socrates has always been the whitest of men and now one whiter than he is born). Proof: Socrates now is not the whitest of men, and immediately before this he was the whitest of men; therefore the original proposition is true. Counterargument: Socrates ceases to be the whitest of men; therefore either of the men who are or of the men who are not or of the men some of whom are and some of whom are not. Each of these alternatives is false. Solution: The original proposition is true, and the proof holds good. But the disproof commits the fallacy of accident, for the ceasing (or the 'cease[s] to be the whitest') in the original proposition signifies in respect of men absolutely in general and not in respect of the men who are or the men who are not in particular. Therefore it is not necessarily the case that if one thing applies correctly to another in relation to something that is superior, then it will on that account apply correctly to it in relation to its inferiors. On the contrary, in inferences of that sort one commits the fallacy of accident.

A question arises regarding the sophisma 'Socrates ceases to know whatever he knew' (supposing that Socrates always knows three stateables—named A, B, and C—and that D is a fourth stateable, the knowledge of which he now loses). Proof: Socrates now does not know whatever he knew, and immediately before this he did know whatever he knew; therefore Socrates ceases to know, and so on. Counterargument: Socrates ceases to know whatever he knew, but Socrates knew A; therefore he ceases to know A [which is false]. Solution: The original proposition is false, as the disproof makes clear. In reply to the proof it must be said that the major premise—'Socrates now does not know whatever he knew'—is ambiguous. For the negation can precede the distribution, and in that case it is true—supposing that Socrates does not know D; but on that interpretation the conclusion does not follow. On the other interpretation the negation can follow the distribution, and the sense is 'Whatever Socrates knew he now does not know', which is false; but the conclusion follows.

The following rule is worth noting in this connection. Whenever 'begins' or 'ceases' is connected with a group of things or a plural noun, they must be expounded by removing the whole group or each member of the group. For example, the sense of 'Socrates ceases to know whatever he knew' is 'Whatever Socrates knew he now does not know, and immediately before he did know'. Other cases are to be handled in a similar way.

A question arises regarding the sophisma 'Socrates ceases to know that he ceases to know nothing' (supposing that Socrates knows three stateables—A,

B, and C—necessarily, so that he never forgets them, and that along with them he did know the stateable that he ceases to know nothing, which he now forgets). Proof: Socrates now does not know that he ceases to know nothing, and immediately before he did know; therefore Socrates ceases to know that he ceases to know nothing. Counterargument: Socrates ceases to know that he ceases to know nothing; therefore he did know that he ceases to know nothing. But whatever he knows is necessary; therefore that Socrates ceases to know nothing is necessary. Therefore he ceases to know nothing. And consequently he does not cease to know the stateable that he ceases to know nothing. Solution: The original proposition is true, and the proof holds good. But the disproof commits the fallacy of reasoning from something taken in a certain respect to the same thing taken absolutely, when it draws the inference 'Socrates ceases to know nothing; therefore he does not cease to know the stateable that he ceases to know nothing'. For his knowing that stateable or any other does not occur unless he knows the other three stateables. But to know that he knows those three stateables is only to know reflexively and not to increase his knowledge, just as when someone sees something colored and sees that he sees something colored he does not see more than when he only sees something colored. Therefore in inferring 'He ceases to know nothing; therefore he does not cease to know that he ceases to know nothing', the argument proceeds from something taken in a certain respect to the same thing taken absolutely.

There are certain rules regarding the supposition of terms in connection with the verbs we have been discussing.

Rule One. Common terms following the verbs 'begins' and 'ceases' have a pair of suppositions. This is because those verbs assert both an affirmation and a negation. For in the affirmation the terms have determinate and simple supposition, but in the negation they have confused supposition. For example, when one says 'Socrates begins to be a man', the sense is 'Now he is a man, and before he was not a man'. In the first part 'man' has simple supposition; in the second, distributive confused supposition.

Rule Two. An argument from an inferior to its superior or a superior to its inferior in the predicate does not hold good in connection with the verbs 'begins' and 'ceases'. This is because of the affirmation and the negation which they imply in their exposition. For it is thanks to the affirmation that the argument from the superior to its inferior does not hold good, but thanks to the negation that the converse argument does not hold good. For example, this does not follow: 'Socrates begins to be two feet tall; therefore he begins to be of some size'; nor does the converse follow: 'Socrates begins to be of some size; therefore he begins to be three feet tall'. Rather, in each case the fallacy of the consequent is committed, as has been said.

The phrase 'in the predicate' used in the rule is important, for such verbs do not affect supposition in the subject. An argument from an inferior to its superior proceeding affirmatively or, conversely, negatively can therefore hold good. For example, 'A man begins to move; therefore an animal begins to move', or 'No animal begins to run; therefore no man'.

In connection with what has been said so far a question arises regarding the sophisma 'Socrates ceases to be not ceasing to be' [*Socrates desinit esse non desinendo esse*] (supposing that Socrates is at the instant of death). Proof: Socrates is not not ceasing to be, and immediately before he was not ceasing to be; therefore he ceases to be not ceasing to be. Counterargument: Socrates ceases to be not ceasing to be; therefore Socrates ceases to be *while* he does not cease to be, or *if* he does not cease to be, or *because* he does not cease to be. For a gerund ending in '-*do*' has to be resolved into 'while', 'if', or 'because', just like an ablative absolute. But each of these is false; therefore so is the sophisma. Solution: The original proposition is ambiguous in that the determining expression 'not ceasing to be' can determine either the verb 'ceases' or the verb 'to be'. If it determines 'ceases', the proposition is false because opposites are asserted at the same time about the same thing—the opposites ceasing and the removal of ceasing. In this case the disproof holds good. But if it determines the verb 'to be', the proposition is true because its sense is that Socrates ceases to be not ceasing to be, or ceases to have unfailing being. Thus although he does not cease to be, he ceases to be absolutely of such a sort—that is, unfailing or without ceasing—because for the future he has not that sort of being but failing being. On the first of these two interpretations the original proposition occurs in the compounded sense and is false; on the second, it occurs in the divided sense and is true. Thus there is a fallacy of composition in this case.

Here is another rule. Whenever the verbs 'begins' and 'ceases' are added to a term that contains a suppositum [reading '*suppositum*' for '*suppositionem*'] and an accident or a substance and an accidental form, the resultant expression is ambiguous in that the beginning or ceasing can be taken to occur either in respect of the subject or in respect of the accident. For example, 'Plato begins to be white' can be understood either in the sense that he begins to be simply or in the sense that he begins to be white, which is accidental.

A question arises regarding the sophisma 'Plato begins to be one or the other of them' (supposing that Plato was before [reading '*prius*' for '*post*'] and that Socrates begins to be for the first time). Proof: Plato is one or the other of them, and [immediately] before this he was not one or the other of them; therefore the sophisma is true. Counterargument: Plato begins to be one or the other of them; therefore he begins to be Plato or Socrates, since no one is one or the other of them except for Plato and Socrates. But he does not begin to be Plato, because he was Plato already; and he does not begin to be Socrates, because they are distinct things. Therefore the original proposition is false. Solution: The original proposition is ambiguous in that the term 'one or the other' contains two things—the suppositum in which the accident is and the accident itself, which is otherness. Therefore the beginning can be denied. That which is can be one or the other either as regards the suppositum [reading '*suppositum*' for '*suppositionem*'] or as regards the accident. If as regards the suppositum [reading '*suppositum*' for '*suppositionem*'], the original proposition is false and the disproof holds good. For the sense of 'Plato begins to be one or the other of them' is 'Plato begins to be Plato or Socrates' in that case. If as regards the accident, the original proposition is true, as is proved. And in that case the sense of 'Plato begins to be one or the other of them' is 'Plato begins to be under the otherness of them', for although Plato's being does not begin simply, he does begin to be under otherness. On this interpretation there is a fallacy of accident in the disproof; on the first interpretation the disproof holds good.

In general, therefore, regarding every distinction that is valuable for a solution of sophismata, the sophisma is always true in one sense and false in the other. In the sense in which it is acknowledged to be true the proof is true and the disproof is fallacious, and vice versa.

Appendix B

JOHANNES VENATOR (d. 1427?)

Logica, Tractatus III, Capitulum 6 (excerpt)
[This translation is based on Francesco del Punta's transcription of the manuscript. I have emended the text.]

PROPOSITIONS EXPONIBLE IN RESPECT OF
THE VERBS 'BEGINS' AND 'CEASES'

A proposition exponible in respect of the verb 'begins' is usually expounded in one of two ways. [E1] It is expounded in one way by positing the present—that is, by means of a present-tense affirmative—and removing the past—that is, by means of a past-tense negative. [E2] Alternatively it is expounded in another way by removing the present—that is, by means of a present-tense negative—and positing the future—that is, by means of a future-tense affirmative. An example of [E1]. 'This begins to be' is expounded as follows: 'This now is, and this immediately before the present instant was not'. An example of [E2]. 'This begins to be white' is expounded as follows: 'This now is not white, and this immediately after the present instant will be white'.

[E1N] One argues against these expositions in the following way. I ask whether or not the copulative expounding 'This begins to be' is interchangeable with what is expounded. If not, then the copulative does not expound it fully. If it is claimed that it is interchangeable, then one argues to the contrary as follows. From 'This begins to be' it does not follow 'therefore this is, and this immediately before the present instant was not'; consequently this copulative, which is called the exponent, is not interchangeable with 'This begins to be', which is called the exposite. *Proof of the antecedent.* Suppose that this now is not and immediately after the present instant it will be. In that case the antecedent is apparent and the consequent vanishes. (Let the 'this' in both cases indicate instant B.)

[E2N] There is a similar proof regarding the second example, for this does not follow: 'This begins to be white; therefore this now is not white, and immediately after the present instant it will be white'. Consequently, the consequent of this consequence is not interchangeable with its antecedent. *Proof of the assumption.* Suppose that this begins to be white by means of positing the present and removing the past. Then the whole thing is obvious.

[E3] Therefore 'begins' is expounded by means of a disjunctive taken disjunctively, both the principal parts of which are copulatives taken copulatively. An example of [E3]. 'This begins to be' is expounded as follows: 'This is, and immediately before the present instant this was not; or this is not, and immediately after the present instant this will be'. Therefore, and so on. Any other [such proposition is to be expounded] in just the same way. . . .

[C5] A present-tense or a future-tense incipit-proposition follows from a copulative proposition made up of an affirmative [future-tense] singular proposition which is verified for an instant and can be verified only for an instant and of the negative present-tense [singular] proposition corresponding to it.

[C5A] Once [C5] has been understood, this follows: 'This is not white, and this will be white; therefore this begins to be white or this will begin to be white'; 'This will be in Rome, and this is not in Rome; therefore this begins or will begin to be in Rome'. Thus the future-tense proposition 'This will be white' is verifiable for an instant, and as long as it signifies primarily that this will be white it cannot be true except for an instant. Hence this follows: 'This will be white; therefore this will be white at some instant'.

[C6] No incipit-proposition follows from a copulative proposition made up of an affirmative [future-tense] singular proposition which requires not an instant but a finite time for its truth and of the negative [present-tense] singular proposition corresponding to it.

[C6A1] For example, in the opinion of many men this does not follow: 'This will be as white as Plato will be, and this is not as white as Plato will be; therefore it begins to be or will begin to be as white as Plato will be'. The cause [of its not following] is that the future-tense proposition 'This (indicating Socrates) will be as white as Plato will be' does not require an instant for which it will be verified, for 'This will be as white as Plato will be [; therefore there is some instant at which this will be as white as Plato will be]' does not follow. This is the case generally as regards all comparisons of this sort, in which both that which is compared and that in respect of which the comparison is made are in the future, [not] limited to a certain instant, restricted to a certain degree, or signified by means of a certain differentia of this sort.

[C6A2] Thus this does not follow: 'Socrates will be as white as Plato will be, and Socrates is not as white as Plato will be; therefore Socrates begins or will begin to be as white as Plato will be'. *Proof.* Suppose that Socrates is white in degree 4 only and Plato in degree 8, and that Socrates intensifies [with respect to whiteness] twice as fast as Plato during this whole hour, with the result that at the end of the hour they cease to be together with that which is at the end the least degree which both of them will not have. In that case this is obvious:

129

'Socrates will be as white as Plato will be, because however intense a whiteness Plato will have, Socrates will have one as intense (other things being equal [and] the contrary having been ruled out, as I am assuming); therefore Socrates will be as white as Plato'. The consequence is obvious, and on the basis of the supposition the antecedent is obvious as well. The following argument is to show the falsity of the consequent of the main consequence: 'If Socrates begins or will begin to be as white as Plato will be, and Socrates does not already begin to be as white as Plato will be, then Socrates will begin to be as white as Plato will be'. But I can prove that this is not the case. For if it is the case, it is so either at the last instant of that hour, or earlier, or later. Not at the last instant or later, for then it will not be. And not earlier, for at each instant earlier than the last instant Socrates will not be as white as Plato will be. . . .

I am grateful to Gareth Matthews, Gabriel Nuchelmans, and Richard Sorabji for their comments on an earlier draft of this article. I received a great deal of valuable help in working over this material from John Longeway (particularly on William of Sherwood), Stanley Martens (particularly on Johannes Venator), Eileen Serene (particularly on Walter Burley), and Eleonore Stump (particularly on Aristotle).

1. I translate *desinere* as 'cease' rather than as 'stop', which would be less stilted, because 'cease', like *desinere* and unlike 'stop', regularly takes an infinitive complement and because there is no semantic distinction between the infinitive complement and the participial complement in the case of 'cease' as there is in the case of 'stop'. 'He ceased to breathe' = 'He ceased breathing' = 'He stopped breathing' ≠ 'He stopped to breathe'.

2. Thus William Kneale remarks, "The fact that medieval logicians found it worthwhile to write separate treatises about such [syncategorematic] words shows that they appreciated their importance for formal logic. But it is probably a mistake to suppose that these signs were universally recognized as formal in a very strict sense. For the words *incipit* and *desinit* (meaning 'begins' and 'stops') were sometimes included among *dictiones syncategorematicae*, although they are concerned with temporal distinctions" (*Development of Logic*, pp. 233–34). (I have commented on this remark in *William of Sherwood's Treatise on Syncategorematic Words* [hereafter cited as *WSTSW*], p. 106 n. 1.) From a twentieth-century point of view, medieval logic consisted largely in semantic theory and linguistic analysis. The fact that certain topics are not "formal in a very strict sense" is no barrier at all to their being considered purely logical by medieval logicians.

3. Here is a roughly chronological list of the texts I have made more or less use of. (1) *Summa sophisticorum elencorum* (ca. 1155–60); (2) *Ars Meliduna* (ca. 1160); (3) *Dialectica Monacensis* (1160–70); (4) *Fallacie Vindoboneses* (latter half of twelfth century); (5) *Fallacie Parvipontane* (latter half of twelfth century); (6) *Tractatus de univocatione Monacensis* (third quarter of twelfth century); (7) *Tractatus Anagnini* (early thirteenth century); (8) Richard Fishacre [?] (d. 1248): *Sophismata;* (9) William of Sherwood (d. 1266/71): *Syncategoremata;* (10) Peter of Spain (d. 1277): *Tractatus syncategorematum;* (11) William Ockham (d. 1349): *Summa logicae;* (12) Walter Burley (d. after 1349?): *De puritate artis logicae;* (13) Richard Kilmington (d. 1361/62): *Sophismata;* (14) Johannes Venator Anglicus (d. 1427?): *Logica;* (15) Paul of Venice (d. 1428/29): *Logica parva;* (16) *Tractatus exponibilium* [attributed to Peter of Spain]; (17) Commentary on (16) by fifteenth-century Cologne Thomists; (18) The Hagenau Treatise (1495); (19) Hieronymus of St. Mark: *Compendium praeclarum* (1507); (20) George Lokert: *Tractatus exponibilium* (1522). Texts (1)–(7) have all been published ([2] only in part) in L. M. de Rijk's *Logica Modernorum*. Texts (8), (13), and (14) exist only in manuscript. I am indebted to Francesco del Punta for making available to me microfilms of (8) and (13), for first calling my attention to (14), and for allowing me to use his transcription of the relevant portions of (14). Texts (9), (11), and (12) are available in recent editions; (10) and (15)–(20) are available in fifteenth- and sixteenth-century editions. Texts (9) and (10) are available in English translation. I have given text (18) the designation "The Hagenau Treatise" after its place of publication; it is mistakenly atrributed to Marsilius of Inghen in a recent photographic reproduction of the 1495 printing. I am indebted to E. J. Ashworth for making available to me her photocopies of (19) and (20); I first learned of these texts in her article "The Doctrine of Exponibilia in the Fifteenth and Sixteenth Centuries" (forthcoming in *Vivarium*).

4. There are many more logical texts on 'begins' and 'ceases', some of which I know of but have not yet seen. And there are whole ranges of literature that I have not yet investigated—the treatises *De instanti, De primo et ultimo, De maximo et minimo,* some examples of which are discussed by Curtis Wilson in his book *William Heytesbury;* and especially the medieval commentaries on Aristotle's *Physics*. I have looked at some of these commentaries in a cursory way without finding anything that seemed especially relevant to my purposes, but I have barely scratched the surface.

5. As far as I can see, the nature of beginning and ceasing constitutes no sort of problem for the notion of matter and not much of one for the notion of space. But it does, of course, have a great deal to do with the notion of motion, especially in the broad Aristotelian sense of 'motion'.

6. See p. 110 below for the roughly correlative medieval distinctions.

7. A different and much sketchier historical hypothesis was offered in 1945 by J. P. Mullally in his introduction to *The Summulae logicales of Peter of Spain*, where he says, "The exposition of 'beginning' and 'ending' in the *Summulae* is based to some extent upon Aristotle". (Mullally is

131

writing about the *Tractatus exponibilium,* which is neither part of the *Summulae* nor by Peter; see notes 26 and 39 below.) He suggests *Metaphysics,* book V, chapters 1 and 17, and *Physics,* book IV, chapter 11, 219a 22–23; 219b 11 (pp. lxxv–lxxvi). But the passages he refers to are of too general a nature to have had much influence, if any, on the discussion as it developed among medieval logicians.

8. According to De Rijk, *De sophisticis elenchis* became known in Paris around 1130 (*Logica Modernorum [LM]* 1.24).

9. I shall call them 'incipit-sentences' or 'desinit-sentences'.

10. "Although the text of the *Physics* had been available in Latin translation perhaps since 1150 [we have one fragment that may be that old] and certainly since 1187 (the year of Gerard of Cremona's death), there is no evidence that it had yet been lectured on in the schools or was known to more than a few unusually inquisitive men during the first quarter of the thirteenth century, although the *De Anima, Sophistici elenchi,* and *Posterior Analytics* were certainly being lectured on in the Arts faculties of Oxford and Paris during this same period. Grosseteste's *Commentary on the Physics* then stands as one of the earliest (in fact, the earliest of which I know) attempts in Latin Europe to provide a systematic investigation of the text of Aristotle's *Physics* and a discussion of the problems raised by the work" (Richard C. Dales, in the introduction to his edition of Grosseteste's *Commentary,* p. ix). The *Commentary* dates from around 1230.

11. The treatments of 'begins' and 'ceases' by William of Sherwood and Peter of Spain are the earliest I know of that show unmistakable signs of the influence of the *Physics.* See note 28 below.

12. I have not found any relevant discussion of 'begins' and 'ceases' in Peter Abelard (d. 1142) or in earlier writers.

13. Consider, for example, 'weigh more than they did' contrasted with 'have put on weight'. Other examples that fall more or less clearly into this group may be found in *Ars Meliduna, LM* 2(1).331; *Dialectica Monacensis, LM* 2(2).573; *Fallacie Parvipontane, LM* 1.577; *Tractatus de univocatione Monacensis, LM* 2(2).348; *Tractatus Anagnini, LM* 2(2).256, 268–69.

14. Other examples that fall more or less clearly into this group may be found in *Fallacie Vindobonenses, LM* 1.517; *Fallacie Parvipontane, LM* 1.562; *Tractatus de univocatione Monacensis, LM* 2(2).344; *Tractatus Anagnini, LM* 2(2).265.

15. There is another example in *Dialectica Monacensis* (*LM* 2(2).590) that falls more or less clearly into this group.

16. For example, William of Sherwood (*WSTSW,* p. 111) and Peter of Spain (in the extract from *Tractatus syncategorematum* appended to this article (Appendix A), p. 126).

17. Other examples that fall more or less clearly into this group may be found in *Ars Meliduna, LM* 2(1).341; *Tractatus Anagnini, LM* 2(2).291–93.

18. There are, however, some uncertain signs that some of the authors of the texts De Rijk has published may have had access to the *Physics.* For example, in *Summa sophisticorum elencorum* Aristotle is quoted as having said "Si fit, non est, et si est, non fit", a remark De Rijk quite plausibly associates with *Physics* 237b9 ff. (*LM* 1.369). *Dialectica Monacensis* closely paraphrases the definition of time from *Physics* 219b1—"Tempus est mensura motus primi mobilis secundum prius et posterius"—without attributing it to Aristotle (*LM* 2(2).518). The same treatise contains an uncertain allusion to nature as the "causa efficiens" of motion and rest, which is at least suggestive of the doctrine in *Physics* 192b21–22, as De Rijk points out (*LM* 2(2).541). See also *LM* 2(1).410 and the sources mentioned in note 6 on that page.

19. As in some fourteenth- and fifteenth-century texts; see pp. 117–18 below.

20. More nearly like 'The King of France is bald' and 'The Morning Star is the Evening Star' than like fallacies.

21. At least one subsequently much-discussed sophisma—'Socrates ceases to be not ceasing to be' [Sor desinit esse non desinendo esse]—makes its appearance already in the early texts (*Dialectica Monacensis, LM* 2(2).573). (See Peter of Spain's discussion of it in the portion of *Tractatus syncategorematum* appended to this paper, p. 127 below.) The *Sophismata* contained in the Bodleian Library Ms. Digby 24 and attributed by De Rijk to Richard Fishacre contains twenty-one sophismata involving 'incipit' or 'desinit', most or all of which are discussed also by later logicians. See De Rijk's list of these sophismata, *LM* 2(1).62–71, especially sophismata (177)–(199).

22. In an initial and somewhat haphazard survey I have collected 246 sophismata involving 'incipit' or 'desinit' or both, many of them discussed by more than one logician.

23. Most notably in the *Sophismata* of Richard Kilmington, whose entire discussion of 'incipit' and 'desinit' consists in a discussion of sophismata. Of his forty-nine sophismata twenty-four have to do with 'incipit' or 'desinit' explicitly, and many of the others are intended to illuminate those twenty-four. See Wilson's list of these sophismata, *Heytesbury*, pp. 163–68. (I have undertaken an edition of Kilmington's *Sophismata*.)

24. *Dialectica Monacensis, LM* 2(2).590.

25. I have found nothing on 'incipit' and 'desinit' in Lambert of Auxerre, the terminist logician who has traditionally been associated with William and Peter.

26. 'Incipit' and 'desinit' are discussed not only in Peter's *Tractatus syncategorematum* but also in the *Tractatus exponibilium* that has sometimes been attributed to him—recently by I. M. Bocheński, J. P. Mullally, and Curtis Wilson. De Rijk offers conclusive evidence from the manuscript tradition against the genuineness of the latter treatise (in the introduction to his edition of Peter of Spain's *Tractatus, called afterwards Summule logicales,* pp. LIV-LV). Internal considerations weigh against the attribution as well. For one thing, the technique of exposition in *Tractatus exponiblium* is already stereotyped in a way quite out of keeping with the few stray hints of exposition to be found in the genuine writings of William and of Peter. For a more weighty consideration, see note 39 below.

27. William, unlike Peter, refers (five times) in his chapter on 'incipit' and 'desinit' to predecessors, who have not yet been identified.

28. I have found no explicit reference to the *Physics* in William's writings and none in Peter's discussion of 'incipit' and 'desinit'. Peter does refer to the *Physics* explicitly once in his *Tractatus* (ed. De Rijk. p. 230) and once in his *Tractatus syncategorematum* (tr. Mullally, p. 105). De Rijk has identified two references to Aristotle in the *Tractatus* as allusions to the *Physics* (pp. 28, 102). In this connection it is interesting to note that Robert Grosseteste certainly knew the *Physics* well and played an important part in its dissemination (see note 10 above) and that William, because of his positions at Oxford and Lincoln, was closely associated with Grosseteste, his older contemporary. Richard Fishacre, too, is likely to have known Grosseteste. See p. 9, n. 28, in my *William of Sherwood's Introduction to Logic*.

29. The sixteenth-century treatises by Hieronymus of Saint Mark and George Lokert, for example, are squarely in the tradition of the hybrid approach.

30. My translation of Peter's discussion is appended to this article. Cf. J. P. Mullally's translation of the same material on pp. 58–65 of his translation of the *Tractatus syncategorematum*. My translation of William's discussion appears on pp. 106–16 of *WSTSW*.

31. Exposition was one of three main varieties of linguistic analysis developed by medieval logicians. Even more clearly than the other two, exposition has its roots in the concern of the terminist logicians with syncategorematic words or, more fundamentally, in the recognition that some words or constructions ("mediate terms", as they came to be called) mask logical operations, such as negation. In the writings of William and Peter this development is still in its early stages. In order to bring sentences containing mediate terms into forms suitable for syllogistic reasoning or for verification, later logicians developed detailed methods of analysis under three general heads, depending on whether the mediate terms were (a) resoluble, (b) exponible, or (c) functional. The method for dealing with (a) was aimed primarily at providing straightforwardly verifiable sentences, suggestive of twentieth-century "protocol sentences"; thus 'A man is running' was resolved into 'This is running' and 'This is a man'. The method for dealing with (b) shared the aim of resolution but was more obviously oriented toward unmasking logical operations; thus 'Only Socrates is running' was expounded into 'Socrates is running' and 'Nothing other than Socrates is running'. The mediate terms grouped under (c) were primarily the modal operators and the verbs of propositional attitude—words which affected some complete embedded sentence; thus 'You know that a man is running' was analyzed so as to avoid the problematic expression 'that a man is running' (*hominem currere*): "You know this proposition: 'A man is running' (*homo currit*)". See Jan Pinborg, *Logik und Semantik im Mittelalter*, pp. 103–11.

32. The distinction as presented here is close to the version given by William, *WSTSW*, p. 109. The distinction stems from Aristotle's *Categories,* chapter 6, especially 5a15–37. Even a portion of

the terminology can be found there, since the Latin word *permanens* was used by Boethius to translate Aristotle's ὑπομένον: "none of the parts of a time endures, and how could what is not enduring (ὑπομένον) have any position? Rather might you say that they have a certain *order* (τάξιν = *ordinem*) in that one part of a time is before and another after" (5a27–30; Ackrill's translation). The terminology is not developed further in Boethius' commentary on this passage (Migne, *P.L.* 64.207–9). But Abelard's *Glossae in Categorias* does present a further development: "Continuum etiam duobus modis accipitur, scilicet in designatione rerum et vocum. Acceptum in designatione rerum sic diffinitur: continuum est mensura adiacens substantiis secundum quam ipsae substantiae dicuntur continuae *succedenter* aut *permanenter*, succedenter ut *tempus*, permanenter ut *linea*. Acceptum in designatione vocum sic diffinitur: continuum est vox significans mensuram adiacentem substantiis secundum quam ipsae substantiae dicuntur continuae succedenter aut permanenter" (ed. Dal Pra, pp. 62.38–63.5; some emphasis added).

33. Instantaneous things or states seem not to have been clearly recognized by Peter, but they are more likely to appear in the discussions of later logicians—e.g., in Walter Burley's—than are Peter's little-by-little permanent things or states.

34. The designations 'all-or-nothing' (A-O-N) and 'little-by-little' (L-B-L) are mine, not Peter's.

35. The parts or degrees of L-B-L permanent things or states accumulate; the parts or degrees of successive things or states succeed one another.

36. Aristotle recognizes "three [kinds of] motions, namely, those of quality, those of quantity, and those with respect to place" (*Physics* 225b8–9). Is it possible that this tripartite division is the prototype of Peter's division of subjects of beginning and ceasing into A-O-N permanent, L-B-L permanent, and successive things or states, the order of the three kinds of subjects paralleling that of their counterparts in Aristotle's division? An instantaneous thing or state, although not discussed by Peter, can also be accommodated in the table. It has—or, more accurately, is—a t_{Ba} and a t_{B2}.

37. This diagram presents the view of Aristotle as well. It is not the unanimous view of medieval logicians, as we shall see, but it is perhaps the majority view.

38. Although I use Peter's expositions, I have modified his examples in some cases.

39. This egregious difference of *Tractatus exponibilium* from *Tractatus syncategorematum* is another consideration in favor of rejecting the attribution of the former treatise to Peter.

40. In (2) and (4) on p. 111 above. Peter's uniform exposition may have been designed to conform with *Physics* 263b13–26, where Aristotle argues that the instant dividing two intervals of time "belongs to the earlier as well as to the later time, and it is the same and numerically one, but it is not one in formula, for it is the end of the earlier but the beginning of the later time. *But it belongs always to the thing in its later affection*". See also the remainder of this passage, in which Aristotle discusses the example of a thing's ceasing to be white.

41. I do not know of anything medieval that is precisely like this approach, but a near approximation may be found in the notion of verification as employed by Johannes Venator. See pp. 118–20 below for an introduction to this notion.

42. Notice that the inconsistency might also have been generated regarding the desinit-instant of the motion in virtue of there being no t_{Ba} for the succeeding state of rest.

43. I am indebted to Eleonore Stump for correcting, reorganizing, and clarifying my initial attempt at expounding the neutral-instant analysis.

44. The acquisition of a changed state is the *mutatum esse*—the having changed—frequently mentioned by the logicians and discussed by commentators on the *Physics*.

45. He is, however, apparently prepared to consider it as a change of the ball from one state to another (225b10–34). Perhaps the way in which to view the situation is this. A change must have a subject, and a change itself is not a subject (225b21). Thus there is no change from the successive state of resting to the successive state of moving. But if those states are considered as accidents of the ball, then the ball may be described as changing from the *permanent* state of *being* at rest to the *permanent* state of *being* in motion. Something like this seems to be suggested in the last sentence of the cited passage: "But this [change of a subject from one kind of change to another] will be by accident; for example, in changing from remembering [a successive state] to forgetting [another successive state], it is that to which these belong [e.g., a man] that changes now toward knowledge [a permanent state] and now toward ignorance [another permanent state]".

46. I am grateful to John Longeway for help with this account of the identifiability of the neutral instant.

47. I have not found the neutral-instant analysis in any medieval logician. But because I have not found any of them to be clearly aware of the problem, that is not surprising.

48. Particularly things he says in 225a1ff. and 228a21ff. G. E. L. Owen has suggested to me that the argument and example in 264a10–21 may constitute evidence against Aristotle's acceptance of the neutral-instant analysis. Cf. Plato's suggestion of something very like the neutral-instant analysis in *Parmenides* 156c–157a. This passage must be taken into account in any assessment of Aristotle's position. Lisa Jardine pointed out to me a discussion of the instant of change in Aulus Gellius (*Attic Nights*, bk. 7, chap. 13), where Plato's *Parmenides* is cited (cf. bk. 6, chap. 21).

49. Ockham discusses 'incipit' and 'desinit' in *Summa logicae* (written about 1324), Pars prima, Capitulum 75, and Pars secunda, Capitulum 19.

50. The excerpt appended to this article (Appendix B) is my translation of the text as transcribed from the sole surviving manuscript (for this portion of the *Logica*) by Francesco del Punta and emended at certain points by me.

51. Only about one-sixth of the chapter is included in the excerpt in Appendix B.

52. The roots of [E3], if not the whole of it, can be found in Walter Burley's discussion (*De puritate artis logicae, Tractatus longior;* Pars tertia, Capitulum 4). Instead of offering a single disjunctive exposition, Burley simply says that any incipit-sentence (or desinit-sentence) can be expounded in either of two ways. And he makes a good deal of use of this doctrine in his discussion. Paul of Venice, contemporary with Venator, puts it as Burley does (*Logica parva,* Tractatus IV, Capitulum 14). The Hagenau Treatise (unpaginated) attributes the very same [E3] exposition to Marsilius of Inghen (d. 1396?).

53. Hieronymus is referring to his own version of the disjunctive exposition, not to Venator's. (The 1507 printing of his book is unpaginated.)

54. In a portion of the text not included in the excerpt Venator does, however, provide three "conclusions" and two deliberately and instructively mistaken expositions that might be said to do by example what he does not do directly in his discussion.

55. For medieval logicians considerations of truth-conditions count as purely logical. See note 2 above.

56. I am indebted to Stanley Martens for correcting, reorganizing, and clarifying my initial attempt at expounding this portion of the proof.

57. Although Venator does not and would not say so, t_{60} must be considered to be the *extrinsic* limit of their being, the t_{NBa}, not the t_{B2}, for each of them.

58. Quoted by Wilson (in Latin and in a different English translation), *Heytesbury,* p. 44.

59. I owe this observation to discussions with Richard Boyd and Robert Stalnaker.

60. See, e.g., *Heytesbury,* p. 31.

SELECT BIBLIOGRAPHY

Ackrill, J. L. *Aristotle's Categories and De Interpretatione. Translated with Notes* (Oxford: Clarendon Press, 1963).

Anonymous. *Commentum emendatum et correctum in primum et quartum tractatus Petri Hyspani et super tractatibus Marsilii de suppositionibus, ampliationibus, appellationibus, et consequentiis* ["The Hagenau Treatise"] (Hagenau, 1945; repr. Unveränderter Nachdruck; Frankfurt: Minerva, 1967).

———. *Tractatus exponibilium* [attributed to Peter of Spain] in *Copulata omnium tractatuum Petri hispani etiam sincathegreumatum et parvorum logicalium cum textu secundum doctrinam divi*

Thome Aquinatis iuxta processum magistrorum Colonie in bursa Montis regentium (Cologne: Henricus Quentell, 1489).

―――. Commentary on *Tractatus exponibilium* in edition cited immediately above.

Apostle, Hippocrates G. *Aristotle's Physics. Translated with Commentaries and Glossary* (Bloomington: Indiana University Press, 1969).

Ashworth, E. J. "The Doctrine of Exponibilia in the Fifteenth and Sixteenth Centuries," *Vivarium* 11 (1973): 137–67.

Boehner, Philotheus, ed. *Walter Burleigh: De puritate artis logicae tractatus longior,* Franciscan Institute Publications Text Series No. 9 (St. Bonaventure, N.Y.: The Franciscan Institute, 1955).

―――, ed. *William Ockham: Summa logicae,* Franciscan Institute Publications Text Series No. 2 (St. Bonaventure, N.Y.: The Franciscan Institute, [Pars prima] 1955; [Pars secunda et Tertiae prima] 1962).

Dales, Richard C, ed. *Roberti Grosseteste . . . Commentarius in VIII libros Physicorum Aristotelis* (Boulder: University of Colorado Press, 1963).

Dal Pra, Mario, ed. *Pietro Abelardo: Scritti di logica,* 2nd edition (Florence: La Nuova Italia Editrice, 1969; 1st edition, 1954).

Fishacre, Richard. *Sophismata* (Oxford: Bodleian Library Ms. Digby 24).

Hieronymus of St. Mark. *Compendium praeclarum, quod parva logica seu summulae dicitur* (Coloniae: 1507).

Kilmington, Richard. *Sophismata* (Vatican City: MS Vat. Lat. 3088 and other MSS).

Kneale, William, and Martha Kneale. *The Development of Logic* (Oxford: Clarendon Press, 1962).

Kretzmann, Norman. *William of Sherwood's Introduction to Logic,* Translated with an Introduction and Notes (Minneapolis: University of Minnesota Press, 1966).

―――. *William of Sherwood's Treatise on Syncategorematic Words,* Translated with an Introduction and Notes (Minneapolis: University of Minnesota Press, 1968).

Lokert, George. *Tractatus exponibilium multo aliis lucidior* (Parrhisius, [1522]).

Mullally, J. P. *The Summulae logicales of Peter of Spain,* Partially edited and translated with an Introduction, The University of Notre Dame Publications in Mediaeval Studies, VIII (Notre Dame, Ind.: University of Notre Dame Press, 1945).

―――. *Peter of Spain: Tractatus syncategorematum. And Selected Anonymous Treatises,* Translated and with an Introduction by Mullally and Roland Houde (Milwaukee: Marquette University Press, 1964).

Paul of Venice. Logica ["Logica parva"] (Venice: 1472; repr. Unveränderter Nachdruck; Hildesheim: Olms, 1970).

Pinborg, Jan. *Logik und Semantik im Mittelalter. Ein Überblick,* Mit einem Nachwort von Helmut Kohlenberger (Stuttgart-Bad Cannstatt: Friedrich Fromann-Günther Holzboog, 1972).

de Rijk, L. M. *Logica Modernorum: A Contribution to the History of Early Terminist Logic* (Assen: Van Gorcum & Co., [Vol. I], 1962; [Vol. II, Parts 1 and 2], 1967).

―――. *Peter of Spain (Petrus Hispanus Portugalensis): Tractatus, Called Afterwards Summule logicales.* First critical edition from the manuscripts, with an Introduction (Assen: Van Gorcum & Co., 1972).

Venator, Johannes. *Logica* (Vatican City: MS. Vat. Lat. 2130).

Wilson, Curtis. *William Heytesbury. Medieval Logic and the Rise of Mathematical Physics,* The University of Wisconsin Publications in Medieval Science, 3 (Madison: University of Wisconsin Press, 1956 [2nd printing 1960]).

Chapter Six

PLACE AND SPACE IN MEDIEVAL PHYSICAL THOUGHT

As in so many of the fundamental topics of medieval science, it was Aristotle who furnished the basic doctrine and context for medieval discussions of place and space. Almost as important were the lengthy commentaries on Aristotle's texts supplied by Averroes, Islam's greatest Aristotelian commentator. Not only did Averroes provide detailed explanations of Aristotle's alleged meanings and intent, but he also occasionally transmitted opinions of late Greek commentators such as Simplicius, Philoponus, Themistius, and Alexander of Aphrodisias, as well as reactions of such Arabic commentators as al-Farabi, Avicenna, and Avempace. Made available to the Latin West in the late twelfth and early thirteenth centuries, this body of literature served as a repository of issues and opinions on place and space that was destined to generate nearly four centuries of discussion and debate. Centered largely on Aristotle's discussion of place in the fourth book of the *Physics,* Latin scholastics, for the most part, adopted the essential characteristics of place as described there. They conceived the place of a body as the innermost, immobile surface of the containing body in direct contact with the contained body. The containing surface was held to be exactly equal to the body it contained and distinct and separable from it.

An examination of the numerous extant *Questiones* on the *Physics* of Aristotle reveals that among the most frequently discussed problems concerning place were the following: (1) whether place is the ultimate or innermost surface of the container; (2) whether place is immobile; (3) whether place is equal to the body in that place; and (4) whether the world is in a place. Included in the question on whether place is the innermost surface of the container, or taken up in separate questions, were discussions on three conceptions of place that Aristotle had rejected, namely that place is form, or matter, or the interval or dimension between the sides of a container. The last opinion, which proclaims that place is a separate three-dimensional space, was of great significance in the subsequent development of the history of spatial concepts. Ideas akin to it gained acceptance in the sixteenth century and prepared the way for ultimate acceptance of an infinite three-dimensional void space in the Scientific Revolution of the seventeenth century. The medieval reaction to place as a tridimensional space[1] will be one of the primary concerns of this paper.

Throughout the Middle Ages, the idea of place as a three-dimensional space was characterized as the "common" or "vulgar" opinion.[2] Ironically, this characterization was invoked in the Middle Ages in order to save Aristotle from the charge that he himself subscribed to the rejected opinion. For it was generally assumed that in the *Categories*,[3] or *Praedicamenta*, as it was known in the Middle Ages, Aristotle had assigned the properties of tridimensionality and divisibility to place and body. As Pseudo-Siger interpreted Aristotle, "as are the dimensions of a body so are the dimensions of its place and conversely; and that the parts of a body are joined (*copulantur*) to the parts of its place."[4] Thus, no exegesis, however subtle, could explicate the passage in the *Categories* as in any manner suggesting place as a two-dimensional surface, the position Aristotle actually adopted in the *Physics*. The dilemma was resolved by appeal to Averroes, who had declared that in the *Categories* Aristotle frequently presented common, or vulgar, opinions, whereas he sought the determination of truth in other parts of philosophy.[5] It followed from this that Aristotle's genuine opinion on place was to be found in the *Physics* and not in the *Categories*.

The Latin Middle Ages did, however, learn that at least one Aristotelian commentator had assumed that place was a three-dimensional space. In his commentary on Aristotle's doctrine of place in the *Physics*, Averroes had declared that John Philoponus, or John the Grammarian, as he was called, had rejected the containing surface as the place of a body and had opted for place as a void dimension that was, however, inseparable from the bodies that came to occupy it.[6] Averroes notes further that in this Philoponus disagreed with those who argued that void dimensions could exist independently of the bodies that occupied them.

Philoponus' bold departure found virtually no support prior to the sixteenth century. Except for Hasdai Crescas in the late fourteenth century, who, in a Hebrew treatise, *Or Adonai (Light of the Lord)*, assumed the existence of three dimensional void places,[7] the conception of place as three dimensional was repudiated in favor of Aristotle's two dimensional containing surface. The latter formed too integral a part of Aristotle's hierarchically structured universe. The doctrine of natural place, on which Aristotelian physics and cosmology depended, was intimately linked to it. Despite its many difficulties, which were frequently noted and discussed, there was little inclination to abandon it. To the contrary, a vigorous defense of it is manifested by the powerful arguments formulated against its major rival in hopes of reducing the concept of three dimensional place to total absurdity. The sorts of arguments that were developed are succinctly summarized by Franciscus Pitigiani, a seventeenth-century Scotistic commentator. Proceeding by a series of dichotomies, Pitigiani declares that

> either this space (*spatium*) is something in itself, or it is nothing; if nothing, therefore it is not a place. If it is something, therefore it is a substance or an accident; if a substance, it is, therefore, corporeal or incorporeal; if incorporeal, it will not be extended or have depth (*profunda*) with the located body because this is a function [or characteristic] of a corporeal thing. If it is corporeal, therefore it has quantity and a penetration of bodies will result. If, [however,] it is an accident, it is either a quantity or a quality; if it is a quantity there will be a penetration of bodies in all dimensions; if it is a quality, it will be an accident without a subject and will not be extended to permit a body to be extended in it.[8]

Included in this passage are the most basic counter-arguments against the concept of place as an interval or dimension between the surfaces of a container. Foremost among these was the interpenetration of bodies, which depended on the conviction that if place is a void interval, it must be assumed to possess the dimensions of width, length, and depth and therefore be equivalent to a three dimensional body. It followed inevitably, then, that occupation of such a place by a body would involve the interpenetration of the three dimensions of the body with the three dimensions of the immobile separate void place.[9] It was the seeming inextricable connection between dimensionality and corporeality that made the concept of place as an interval unacceptable. One minor attempt to break this connection only seemed to confound the issue further. According to Pseudo-Siger of Brabant, Philoponus called the dimensions of place nothing in themselves so that no actual interpenetration occurred. Our author was quick to note, however, that if "such dimensions are assumed to be a place *and* since such dimensions are nothing, it follows that place is nothing, which is impossible.''[10] Not until the sixteenth century, perhaps first by Francesco Patrizi, was the nexus between dimensionality and corporeality broken by assigning tridimensionality to place essentially and to body only accidentally. For Patrizi, place is prior to, and wholly separate from, body, whose primary property is resistance.[11]

The assumption of place as tridimensional space posed other serious problems, as Aristotle himself had indicated.[12] Not only would an object have an infinite number of separate places, but each part of the object would actually occupy an infinity of partially overlapping places. For, as Averroes explained, if water enters a vessel, its three dimensions will be located within the three dimensions of the empty vessel or place.[13] But just as the whole quantity of water occupies a place, so will each of its potentially infinite parts, thus resulting in an infinity of places. Moreover, the place of each part of the water is nested in an infinity of other places that surround it. By virtue of its infinite divisibility, any part of the water can be said to occupy simultaneously an infinite number of partially overlapping three-dimensional places,[14] a consequence avoided in Aristotle's concept of place, where the water can only occupy a single place, namely, the innermost surface of the containing vessel in contact with the water.

A further consequence of the preceding argument arises when the vessel is moved. If place is a three-dimensional space between the surfaces of a vessel, the motion of the vessel, according to Thomas Aquinas, would transport not only the contained body, but also the place of that body.[15] As the vessel moves to new places, the original three-dimensional place of the body will also occupy new three dimensional places. Indeed, it seems that the contained body will come to occupy many places simultaneously, for the motion of the vessel to each successive place signifies that the previously added place is now in a place, and so on. This absurdity was avoided by Aristotle, who insisted that water in a transported vessel is only moved *per accidens* and therefore does not change place.

Whatever the merit of such arguments, the consequence of an infinity of simultaneous places served as a powerful deterrent against the concept of place as three dimensional space. Occasionally, the superfluous character of such a concept was also stressed. John Buridan, one of the

140

most famous of fourteenth-century scholastics, argued that we could easily imagine the existence of a separate tridimensional space capable of receiving bodies equal to itself in every dimension.[16] This, after all, was the common sense view of place. We could even imagine that prior to the creation of the world, a separate, tridimensional homogeneous space existed equal to the world and in which the world was subsequently created. But since every material body from the smallest to the largest, namely the world itself, has its own dimensions, it does not require another. Indeed, if a body had a separate tridimensional place, that place would require a place, and so on ad infinitum.[17] A separate preexistent space for the world is also rejected, since it presupposes that God requires a space in which to create the world. But God has no need of such a space, since He can create corporeal magnitudes directly without need of spatial dimensions in which to place them.

Like the preceding arguments, there were others also rooted in the Aristotelian tradition. For example, it is frequently mentioned that in tridimensional, continuous, and homogeneous world space, every part of space is identical so that local motion as described by Aristotle would be impossible, since no good reasons could be offered as to why bodies should move in one direction rather than another.[18] Indeed, natural places, and therefore natural motion, would be impossible under such conditions.[19] These Aristotelian arguments may, for convenience, be classified as "secular", or better yet "natural", by contrast with other arguments developed in a theological context and tradition that emphasized supernatural actions, and that we must now discuss.

The origin of this tradition lies in events of the thirteenth century that centered around the role of Aristotelian natural philosophy and its relation to Christian theology. More concretely, it was a struggle between theologians and masters of arts at the University of Paris over the manner in which God's relationship to the world and its physical operations was to be interpreted and understood. The bitter controversy culminated in a theological condemnation of 219 articles by the bishop of Paris in 1277.[20] A diverse collection of propositions that were neither to be held or defended under pain of excommunication, the Condemnation of 1277 was directed primarily at Aristotelians and alleged Averroists at the University of Paris. It was designed to subvert the Aristotelian philosophical determinism that had reached alarming proportions among the masters of arts, or natural philosophers, of the University of

141

Paris. The theologians who drew up the condemnation sought to achieve this purpose by emphasizing the absolute power of God (*potentia absoluta Dei*) to do whatever He pleased short of a logical contradiction. Appeals to God's absolute power were made in order to justify formulation of a host of thought experiments that were, in one way or another, contrary to Aristotelian physics and cosmology. As a result, it became commonplace, especially in the fourteenth century, to propose hypothetical situations in which God was imagined to produce naturally impossible phenomena the consequences and implications of which were then developed and discussed. Some of these, in which God creates a separate vacuum inside or outside the world, were relevant to the doctrine of place.

The impact of these developments may be seen, first, on the subject-accident relationship so fundamental to Aristotelian thought. We saw earlier that Franciscus Pitigiani used this fundamental relationship to reject the claim that place might be a tridimensional space. If place is tridimensional, it must be corporeal and, therefore, either a quantity or a quality. It cannot be a quantity since two tridimensional entities, namely place and the body it contains, would interpenetrate; nor can it be a quality, for then place would be an accident without a subject—that is, its three dimensions, which are its properties and its only defining characteristics, would lack any subject in which to inhere, and thereby violate the universally accepted Aristotelian subject-accident relationship. In the ordinary understanding of natural phenomena in the Middle Ages, that is, "naturally speaking" (*naturaliter loquendo*), this was universally accepted. But "supernaturally speaking" matters were quite otherwise.

As a consequence of three articles in the Condemnation of 1277,[21] it became an excommunicable offense to deny God's power to create an accident without a substance or to deny the possibility of the separate existence of a quantity or dimension on grounds that such an entity would be a substance. It was probably with these articles in mind that John Buridan, who had denied the *natural* possibility of a three-dimensional place or vacuum[22] for the kinds of reasons already mentioned, concedes that God could not only produce an accident without a subject and separate accidents from their subjects, but He could also conserve these accidents independently of anything else.[23] Consequently, if He wished, God could surely create an absolute dimension wholly separate and distinct from any substance and accidents (presumably accidents other than dimension). Moreover, Buridan also allowed that by His absolute power God could create many bodies in the same place simultaneously. From this it followed that God could surely produce an absolute separate dimension, or space, that was capable of receiving natural bodies despite the interpenetration of space and body.

Although such concessions to divine power became commonplace in the fourteenth century, they did not alter the fact that as far as the material plenum of our natural world was concerned, Aristotle's doctrine of place and his total rejection of void space were accepted without serious challenge. It was with respect to the realm beyond the cosmos, a realm whose very existence was denied by Aristotle, that new developments emerged in the fourteenth century.

Despite mention of changeless, perfect, and eternal beings beyond the last sphere of the universe, Aristotle had emphatically denied existence of bodies, places, time, or void beyond that same sphere.[24] The question of extramundane existence was, however, raised in antiquity, especially by the Stoics, one of whose arguments reached the Latin West in William of Moerbeke's 1271 translation from Greek into Latin of Simplicius' Commentary on De caelo.[25] If someone, standing motionless at the extremity of the world, extended his hand beyond, the hand would either be received into a vacuum or meet a material obstacle preventing its extension. Should the latter occur, the hand would then be extended from the extremity of that material obstacle, a process that could be repeated until no obstacle was met and the hand was actually in a vacuum. Since the Stoics, like Aristotle, assumed a finite universe, material impediments to the hand's extension must eventually cease at which point a vacuum is reached.

The role of this well-known, and frequently cited, Stoic argument in stimulating discussion about extramundane spaces and vacua is unclear.[26] It is virtually certain, however, that the responses to the seemingly irrepressible question about what might lie beyond the world were fundamentally shaped by theological considerations peculiar to medieval Latin Christendom. In the long aftermath of the Condemnation of 1277, with its emphasis on God's absolute power, it must have seemed incongruous to some, at least, that the presence of an infinitely powerful God should extend no further than the finite bounds of the world that He Himself created. Was God to be confined to a finite cosmos simply because Aristotle had denied extracosmic existence? Indeed, article 49 of the Condemnation would in and of itself have proved sufficient to raise the issue directly. For it declared, in effect, that it was an excommunicable offense to hold that God could not move the world with a rectilinear motion because a vacuum would remain.[27] It was an obvious implication of this article that if God did move the world in a straight line, not only would the place left behind be void, but so would the successive places prior to their occupation by the world. Theological considerations were probably reinforced by a strong intuitive sense that, contrary to Aristotle, something must lie beyond the world, with some kind of space the most plausible candidate. As Nicole Oresme expressed it, "Human understanding consents naturally that beyond the heavens and world, which is not infinite, there is some space, whatever it may be; and one could not readily conceive the contrary."[28] Considerations of this kind were probably instrumental in leading theologians such as Thomas Bradwardine, Nicole Oresme, John de Ripa, and others, to assume the existence of an infinite void space beyond the world, a space in which God Himself was necessarily omnipresent.

It was God's omnipresence that sharply contrasted medieval conceptions of an extracosmic infinite void from that of their Stoic predecessors. Theologians had long held that God was omnipresent within the world, from which it seemed to follow that if something lay beyond the world, God would also be omnipresent there. Thus was a link forged between extracosmic void space and God's immensity (*immensitas*), a term that not only signified that God was unlimited and uncircumscribable, but also had intensive and extensive connotations. With God's immensity associated with extracosmic space, it was almost inevitable that this space would be conceived as infinite in extent. In Oresme's words, it is "infinite and indivisible, and is the immensity of God, and is God Himself, just as the duration of God, called eternity, is infinite and indivisible, and is God Himself,"[29] a description with which Bradwardine seems to have concurred when he insisted that "the whole of an infinite magnitude and imaginary extension, and any part of it, coexist fully and simultaneously, for which reason He can be called immense, since He is unmeasured. . . ."[30]

God's immensity and infinite void space seem, in some sense, coextensive. It is almost as if infinite void space were an attribute of God, a position that Henry More and Samuel Clarke would later adopt.[31] It was no doubt because of this that Bradwardine considered infinite space eternal and uncreated. In these properties, Bradwardine saw the answer "to the old question of the gentiles and heretics—'where is your God? and where was God before the [creation of the] world?' "[32] He was, and is now, in His own infinite void space, the very space in which He eventually created the world, freely choosing a place from an infinity of possibilities. Now the place that received the world, and consequently all of the infinite possible places that might also have received it, can have "no positive nature, for otherwise there would be some positive nature which is not God, nor from God. . . . Such a nature would be coeternal with God, something no Christian can accept."[33] Since the totality of infinite possible places that might have received the world makes up the whole of infinite void space, it follows that this space has no positive attributes or properties of its own. It is the place of God, who is not confined to the place where He made the world, for "it is more perfect to be everywhere in some place, and simultaneously in many places, than in a unique place only."[34] Bradwardine concludes that "God is, therefore, necessarily, eternally, infinitely everywhere in an imaginary infinite place, and is truly omnipresent."[35]

In this intimate association of God and space lurks an almost insoluble dilemma. For though Bradwardine proclaims that "a void can exist without body but in no manner can it exist without God,"[36] he also declares that God "is infinitely extended without extension and dimension."[37] Oresme poses the problem even more starkly when he says that "by His infinite magnitude, called immensity (without quantity and absolutely indivisible)," God "is necessarily wholly in every extension or space or place which is or could be imagined."[38] In what conceivable sense could a dimensionless, non-extended God be omnipresent in an infinite void space? Or is void space itself dimensionless by virtue of its eternal association with a dimensionless God? But if the infinite void is conceived as truly dimensional, and terms like "space," "place," and "extension" are regularly applied to it, how can God be everywhere present in it? Was Nicolas Malebranche expressing the medieval view when, in the *Dialogues on Metaphysics and Religion,* his spokesman, Theodore, declares that "God, then, is extended, no less than bodies are; since God possesses all absolute realities or all perfections. But God is not extended in the way in which bodies are. . . ."[39] For, as Theodore

145

subsequently explains, "The immensity of God is His substance itself spread out everywhere, filling all places without local extension, and this I submit is quite incomprehensible. . . . Assuredly, Theotimus, if you judge of the immensity of God by means of the idea of extension, you are giving God a corporeal extension."[40]

Or perhaps what Bradwardine and Oresme had in mind was made explicit by the sixteenth-century Jesuit commentator on Aristotle at the University of Coimbra who, in responding to the question "What is imaginary space?", argued that true space inside and outside the world is not a dimensional entity.[41] Its imaginary character derives not from fictions or mental conceptions, or from its incomprehensibility, but rather because we can conceive dimensions in this space as if they corresponded to the real and positive dimensions of bodies.[42] The correspondence between space and dimensional body follows from the fact that God can create bodies outside the world that are received into this imaginary space. But that imaginary space is not itself a "real and positive being since besides God, no such thing could exist from eternity." Despite its lack of reality, however, imaginary space is not a pure nothing. It is an existent entity of some kind, if only by virtue of the fact that "God is actually in this imaginary space, not as in some real being but through his immensity, which, because the whole universality of the world cannot accommodate it, must of necessity also exist in infinite spaces beyond the sky." Indeed, God cannot be excluded from any place, true or imaginary.

Some two centuries earlier, around 1350, after Bradwardine and before Oresme, John de Ripa also considered the relationship between positive and imaginary space. In what seems thus far the lengthiest extant discussion on extracosmic space, John elaborated his views in Book I, Distinction 37, of his *Sentence Commentary,*[43] which considered the problem of "The Mode of Existence [or Inherence] of the Divine Essence in all Creatures." Unlike Bradwardine and Oresme, who largely ignored the problem, de Ripa argued against those who judged imaginary void space to be nothing, no more than a fiction of the mind. In their view, only positive and real things could exist. But if only real or positive things existed, none of which can lie beyond the world, how could God move the world rectilinearly? Without some kind of place or places to receive the world, God would be unable to move it rectilinearly, which constitutes a restriction on His absolute power and implies acceptance of an article at Paris in 1277.[44]

Since a vacuum was not considered a positive place, rejection of all but positive places had also prompted some theologians to deny John's claim that "God is really present in an infinite imaginary vacuum beyond the world."[45] Once again, John resorts to the absolute power of God. For if the theologians are correct, not only would God be absent from the vacuum but so also would His creations. As a consequence, God could move neither angel nor world in a vacuum, which violates the very same article mentioned above.[46] Moreover, if God can only be present in real, positive places, it would follow that He is now present in the world but was nowhere present prior to its creation, an opinion John deems absurd,[47] as did Bradwardine and probably Oresme. It is essential to assume the existence of imaginary as well as real and positive spaces. For without imaginary space, any angel that God might have created before the creation of the world would have been created nowhere and, therefore, could not have been separated from anything by any distance.[48]

For, contrary to the opinion of most of his contemporaries, John believed that distances were as measurable between places in imaginary space as they were in a plenum with its positive places. To demonstrate this, he derives a series of absurdities involving positive places in a plenum.[49] He first assumes that he is in a place surrounded by air. If the air moves and varies around his body, as it will, the immediate container, or place, surrounding his body is continually altered. Since his place continually changes, he must be in motion even while he is at rest. But if, contrarily, the place surrounding him remained constant as he approached some fixed terminus, he would undergo no local motion, since he has not changed place. Even more relevant are two other arguments. On the assumption that distances are measurable only by means of intervening bodies, then, if two bodies, A and B, are separated by any whatever distance, the corruption or destruction of the intervening bodies would cause A and B to come into contact even though they remain motionless in their original places. Finally, if the bodies intervening between A and B were destroyed part by part, A would draw closer to B with the destruction of each successive part, although neither A nor B are moved.

The inference seems plain enough. If the matter between two bodies or places is destroyed, which would leave an imaginary space, the bodies and distances would remain unaltered. For if the corporeal dimensions intervening between two bodies or places were destroyed, the distance separating them in the imaginary space that now intervenes would surely be identical with the measurement obtainable from the material plenum that previously intervened. It follows that the distance between two angels or spiritual creatures in imaginary space is as measurable as a distance between two bodies in positive places in a plenum.

What then is imaginary space or place? Though it is not real or positive, it is obviously something. Indeed, it is the space of spiritual creatures, such as angels or intellectually simple substances, who occupy limited places within it, places whose "extent" is always proportional to the intensive perfection of the particular spiritual being.[50] Moreover, it seems that an imaginary place has the possibility of becoming a positive space when a body is created or placed in it. Although John does not expressly declare himself, it would appear that the creation of the world exemplifies the transformation of an imaginary to a positive place. The distinction between imaginary and positive place seems to depend on the absence or presence of body. But if an imaginary place is convertible to a positive place by the advent of a body, it would appear that John de Ripa did not subscribe to Aristotle's conception of place as the container of a surface. Since imaginary space is devoid of containing surfaces, the appearance of a body there, which transforms it to a positive place, implies that its place is conceived as three- rather than two-dimensional. And if John adopted Peter Lombard's definition of place, it would indeed be three-dimensional. Writing before the works of Aristotle were available in the Latin West, Peter Lombard, in Book I, Distinction 37, the very section on which John de Ripa was commenting, declared that "a place in a space is what is occupied by the length, width, and depth of a body."[51] Although it was not uncommon for theologians to adopt Peter's definition, they did not, to my knowledge, expressly challenge Aristotle's radically different description. Without a body to occupy it, however, what is the nature of an imaginary place or space? Is it dimensional or non-dimensional?

Unfortunately, de Ripa left this question unresolved and ambiguous. He occasionally indicates that infinite imaginary space is only a possible entity, one of many that might have been realized by virtue of God's absolute creative power.[52] As a possible entity its precise character may therefore have been left undefined. In one sense, it might have been conceived as dimensionless, since it was the abode of spiritual substances themselves dimensionless. This is perhaps reinforced by the necessity to contrast imaginary place with positive place, the latter associated with body. However, none of this would have precluded John de Ripa from assuming imaginary place as three-dimensional. Indeed, in one place he speaks of a three-dimensional volumetric space.[53] If we conceive of a pure volume of one cubic foot, which John describes as an imaginary place, God could replicate this space an infinite number of times to produce an imaginary infinite space of three dimensions, a space characterized as "possible." Is this the true character of imaginary space? Or is it merely another illustration of the way in which God could employ His absolute power? There is nothing in John's discussion that allows for a determination of this important problem.

On one significant theme, John differed radically from Bradwardine and Oresme. Despite agreement that God is omnipresent in infinite imaginary space, Bradwardine and Oresme equated God's immensity with infinite imaginary space. God surrounds and limits all things because infinite space is His immensity, which nothing can exceed. Nor was it created by God, but rather exists co-eternally with Him as a seeming property or attribute. It was this move that enabled Bradwardine to retain the concept of infinite imaginary space, while also denying that God could create actual infinites. For if God could create actual infinites, it would be possible for Him to create another God, or Gods,[54] with powers equal to His own, a potential consequence repugnant to many scholastics.

In dramatic contrast, John de Ripa dissociated infinite void space from God's immensity. Although John also conceived of infinite space as eternal, he describes it as flowing from God's presence as from a cause and therefore totally dependent on Him, thus suggesting a kind of creation for it.[55] Not only is infinite void in some sense created by God, but it is immensely exceeded by Him. In John's words: "The infinity of a whole possible vacuum or imaginary place is immensely exceeded by the real and present divine immensity."[56] God may be described as "an intelligible infinite sphere"[57] who surrounds and circumscribes every imaginary place. In effect, John distinguished two infinites, the ordinary kind, an example of which is imaginary void space, and a single superinfinite that is God's immensity.

The distinction between infinites was an outgrowth of John's conviction that the hierarchy of creatable spiritual or intellectual substances, with its appropriately proportioned degrees of intensive perfection and places, had no theoretical finite limit, as was widely believed, but could go all the way to actual infinity.[58] To reject this claim would constitute a denial of God's absolute power.[59] It seemed illogical to de Ripa that the great chain of creatable substances should culminate in a finite entity, as if God were incapable of proceeding to infinity. He concluded that not all infinites are equal, a rather unusual opinion in the Middle Ages.[60] We do justice to God's great and incomprehensible power only when we understand that not only can He create all manner of possible infinites, but that He immeasurably exceeds and circumscribes them all.[61] It was probably the same theological tradition that led John de Ripa and others to speak of the creation of possible and actual infinites that prompted John Buridan to proclaim God's power to create an immobile, infinite three-dimensional space in which an infinite body might be moved, and to accomplish this despite the interpenetration of two corporeal dimensions, body and space.[62]

In the course of the fourteenth and fifteenth centuries, the actual existence of extracosmic space was accepted by relatively few. Perhaps John Buridan represented a common attitude when he cautioned that "an infinite space existing supernaturally beyond the heavens or outside this world ought not to be assumed because we ought not to posit things that are not apparent to us by sense, experience, natural reason, or by the authority of Sacred Scripture. But in none of these ways does it appear to us that there is an infinite space beyond the world. Nevertheless, it must be conceded that beyond this world God could create a corporeal space and any whatever corporeal substances it pleases Him to create. But we ought not to assume that this is so [just] because of this."[63] But if few accepted its real existence, many discussed the hypothetical consequences that would follow upon God's creation of bodies or worlds beyond our cosmos. And despite the fact that Aristotle could offer little or no guidance in coping with problems that arose in a context whose very existence he had denied, it was often in terms of Aristotelian science that these problems were discussed.

150

A major concern centered on the possibility of measuring distances in extramundance void space. It seemed to most scholastics that the measurement of a distance between two distinctly separate bodies required the existence of an intervening dimension, which, in the material plenum of the Aristotelian world, meant an intervening body. As Buridan expressed it, "The space [or distance] between me and you is nothing but the magnitude of the intervening air or of another natural body, if one should intervene."[64] It would follow, then, that wherever God may have created bodies in a vacuum beyond the world, they could not be separated by any measurable space or distance. But if bodies are not separated by distance, they must be in contact. According to Marsilius of Inghen,[65] God could cause a stone to separate from the convex surface of the last celestial sphere only if He also created a body or bodies between the celestial surface and the stone. Otherwise, says Marsilius, a void space would intervene without the capacity to function as a corporeal interval of separation. In another example, Marsilius argues that if God created three spherical worlds in mutual contact, no measurable distances could be said to intervene in the vacua that lie between their convex surfaces. In such a configuration of worlds, distances could only be measured curvilinearly between the points of any particular surface, since these would be separated by the continuous matter of the surface.

The measurement of distances in void space would, however, be possible if the void was conceived three-dimensionally. This was the view adopted by Henry of Ghent (d. 1293), who distinguished two ways of measuring distances.[66] In one way, distance is measured *per se,* which mode is applicable only to the plenum of our world, since it requires the intervention of corporeal, material dimensions; in the second way, measurement may be *per accidens,* since, despite a lack of positive distance, a measurement can be made by virtue of positive dimensions existing in something—say, a three-dimensional vacuum—that lies between two distinct and separate bodies. Although a vacuum is nothing in itself, it qualifies as something by virtue of its dimensions. Henry therefore argues that if God created a body beyond the world, or even if He created another world, that is not in contact with the outermost celestial sphere of our world, we may infer the existence of an intervening three-dimensional vacuum. Should God decide to

151

create a body in a non-dimensional "pure nothing" (*purum nihil*) beyond the world, that body could not be separated from the world and would, of necessity, be in contact with it.[67]

Aristotle's doctrine of place was, of course, inapplicable to the bodies that God might create in extracosmic void space. If God chose to move such bodies, or even to move the entire world, a possibility conceded by everyone, it could not be from place to place, since places as containing surfaces cannot exist in a void. But if such bodies are not in places, Buridan, for example, was prepared to argue that they are in space, though not the space of an infinite void, since God may not have created such a space. The space Buridan has in mind is the dimension of the body itself. Employing a version of the old Stoic argument derived from Simplicius, Buridan imagines a man placed at the last sphere, who raises his arm. In the absence of a separate extracosmic space and assuming no resistance to the movement of the arm, Buridan declares that "before you raise your arm outside this [last sphere] nothing would be there; but after your arm has been raised, a space would be there, namely the dimension of your arm."[68] It would appear that Buridan has applied the criteria for measurement in a plenum to the situation just described. Between any two material extremities of the arm, matter intervenes, and thus the conditions for spatial separation or distance measurement are met. While the arm lacks a place, since it is not contained by any other body, it is in a space equal to its own dimensions. Like Descartes, Buridan equates matter and extension or dimension, the latter equivalent to space.[69]

On the assumption that, in the infinite space beyond our world, God could create many other worlds identical to ours, Nicole Oresme was led to challenge Aristotle's concept of natural place.[70] What would the behavior of formally identical elements in the various worlds be like? Would Aristotle's claim of a single natural place for earth, namely, the center of our world, apply to the earth in other worlds? That is, would the earth, or earthy substances, of these other worlds tend naturally toward the center of our world? Oresme emphatically rejects this possibility by formulating a different concept of "heavy body" and the direction "down". A body is heavy when it is surrounded by light bodies. In this circumstance, the heavy body is said to be "down" and the surrounding light bodies "up". Heaviness was thus divorced from a motionless, spatial natural place. Earth is heavy and down because it is surrounded

152

by the lighter elements water, air, or fire. From this, it was an easy inference that the earth of each world would remain at the center of its own world. For the parts of earth in each world tend to locate themselves in the middle of light bodies, which is to say they tend downward toward the center of their own world. Thus was the Aristotelian concept of a single universal center repudiated. Despite the enormous potential destructive power of Oresme's concept of place, it had no subsequent impact because he terminates his discussion with a categoric denial of the existence of other worlds.[71] He had only sought to demonstrate that, contrary to Aristotle, the physical phenomena associated with the existence of other worlds implied no contradictions.

Infinite void space was utilized in yet another way. It formed the backdrop for a hypothetical absolute motion. If God chose to move our sole existing world rectilinearly through infinite space, as was possible by article 49 of the Condemnation of 1277, Oresme was prepared to argue that such a motion would be absolute since there was no other body outside to which its motion would be related.[72] It is of some interest that this same illustration was invoked by Samuel Clarke against Leibniz.[73] In Clarke's view, if God moved the world "with any swiftness whatsoever," Leibniz would have to deny movement since it would fail to meet his relational criteria for motion. Despite the assumption of motion, then, Leibniz was locked into a position that compelled him to deny its occurrence.

The divinization of space and its extension to infinity was the work of a few theologians who conceived the absolute power of God in accordance with the Condemnation of 1277. Many proposals alien to, and potentially subversive of, Aristotelian physics and cosmology were conceived and debated in the framework of infinite space or in spaces within our world that God had made void by annihilating the matter therein. While some undoubtedly reveled in the glories of God's absolute power, others, who were seriously committed to Aristotelian physics and cosmology, must have found it a heavy burden to bear. As natural philosophers, they were compelled to cope with a physics of the supernatural imposed upon them by theologians. Absurdities from an Aristotelian standpoint had to be discussed with the utmost seriousness, including themes such as the creation of accidents without substances; worlds, bodies, and vacua beyond our cosmos; and the creation of vacua within the world by the annihilation of vast extents of matter. Among those discussed here, I would judge that John Buridan, Albert of Saxony, and Marsilius of Inghen found little joy in the forced discussion of possibilities they could only deem absurd.

Perhaps it was an urge to retaliate in kind that produced an occasional illustration that might well have caused even the minds of medieval theologians to boggle. It was possibly in this spirit that Albert of Saxony[74] declared that just as God could annihilate everything within the concave surface of the world and yet prevent that surface from collapsing and coming together in the vacuum thus created, so also could God place a body as large as the world inside a millet seed. And He could achieve this without any condensation, rarefaction, or penetration of bodies! For it is in this manner that the body of Christ is lodged in the host. God could create in that millet seed a space of a hundred leagues, or a thousand, or however many are imaginable, so that any man who existed in that millet seed could traverse all those many leagues simply by walking from one extremity of the millet seed to the other! Could Walt Whitman, in his wildest flights of fancy and speculation, have had anything like this in mind when he penned the line "Every cubic inch of space is a miracle"?[75]

Despite the inapplicability of the Aristotelian doctrine of place to extracosmic void space, the opinion that place is a two-dimensional container was never seriously challenged during the Middle Ages. Not until dramatic new intellectual currents took root and developed in Italy in the sixteenth century did Aristotle's concept of place suffer its most serious challenge and ultimate destruction. A growing impatience with University scholasticism combined with the introduction of Greek and Latin texts unknown, or only partially known, to the Middle Ages effectively generated an intellectual climate that would eventually erode the foundations of Aristotelian science. Noteworthy among the newly edited or translated treatises were the works of Plato, the numerous treatises of the Hermetic corpus, Aristotelian commentaries of the late Greek commentators, especially the *Physics* commentary of John Philoponus, the *Pneumatica* of Hero of Alexandria, and the *De rerum natura of* Lucretius with its description of Epicurean atomism. Along with Hebrew treatises such as the Cabbalah and the *Physics* commentary of Hasdai Crescas, a significant quantity of anti-Aristotelian literature was now available.

With Italy as the locus of this new and vigorous anti-Aristotelianism, it is not surprising that Italian authors led the onslaught against Aristotle and his followers. A major target was the Aristotelian doctrine of place. Of the numerous authors who contributed to the assault, two will be considered here: Franciscus Toletus (1532–93), a Jesuit scholastic sympathetic to Aristotle but destined to extend the doctrine of place in directions that undermined it, and Francesco Patrizi (1529–97), a staunch Neoplatonist and vigorous anti-Aristotelian.

No better testimony of Franciscus Toletus' scholastic ties could be had than the context in which his discussion of place appears: a commentary on Aristotle's *Physics* with a series of *Quaestiones* interspersed to elucidate special problems within each topic.[76] Following upon a lengthy consideration of place, Toletus declares that "up to this point, we have defended Aristotle's opinion on place as much as we could because it seemed to suffer from fewer and lesser difficulties than certain other opinions."[77] If Aristotle's opinion was the least objectionable, however, it was far from adequate. The difficulty with all the rival theories, in Toletus' view, was that each sought to assign a variety of properties to a single conception of place that could not bear the burden.[78] For place is a much broader conception than that put forward by Aristotle or his rivals. Fortunately, a recent opinion, whose origin is not otherwise described, succeeds in reconciling all the difficulties and properties.[79] It is this opinion that Toletus adopts and explains in the final question on place.[80]

True place is twofold, external and internal. Toletus explains that "external [place] is what surrounds the located itself, namely the containing body or its ultimate surface, of which Aristotle has spoken. Internal [place] is the place of the thing [or body, namely] the space itself which the thing occupies within itself in accordance with its corpulence."[81]

External place may be subdivided into three parts: (1) it may be immediate when the concave surface of the container is in direct contact with the contained; (2) it may bear a more remote relation to the contained, as when "something is in a room, the room and its walls are its place; or the house within which it is contained; or (3) there is a most remote place, which Aristotle calls the heaven of all things because all things are contained by it."[82] Finally, Toletus distinguishes between partial and total external places.[83] A partial place is an incomplete instance of immediate place, as when a man stands on the earth out in the open. He is immediately surrounded by air on all sides except the bottoms of his feet, which are in direct contact with the earth.[84] A total place is exemplified by a fish submerged in water.

Thus far, we have little more than an elaboration of the Aristotelian concept of place. It is in Toletus' concept of internal place that radically non-Aristotelian elements appear which, when added to the old will, he believes, enable him to reconcile the many problems and difficulties associated with place. Internal space is the extended quantity of a body's matter. "Now this internal [space] of every body cannot be denied," Toletus insists, "because they infer each other as a mutual consequence. For if there is a body, there is a space; and if there is a true space, there is a body in it."[85] Internal space is strikingly similar, if not virtually identical, to the earlier Cartesian-like view formulated by John Buridan for bodies located beyond the world, which we described earlier. Indeed, Toletus presents an illustration that is essentially the same as Buridan's when he declares that "if a man were placed outside the heavens [or world], he could truly be said to be in a place, namely in that space that is proper to his body; and if he were placed in a void medium, a true space would be there within his body, which his body properly occupies, since what surrounds [the space] within his body would only be an imaginary space."[86]

Although at one point Toletus declares that it is of little significance whether this internal space is really (*realiter*) and truly distinct from quantity or only formally (*formaliter*), or perhaps by reason alone, he concurs with John Philoponus that it is only a proper accident inhering in bodies rather than an actually separable space. Toletus concludes that only formally or by reason can we distinguish the internal space of a body from its quantity.[87]

If true space or place is the internal space of particular bodies, what then is imaginary space? In an unusually clear discussion of this elusive topic, Toletus distinguishes two kinds of imaginary space. It is either a mere figment of the mind, or it is something real arrived at by a process of abstraction.

Toletus explains the first way as follows: "Imaginary space, indeed, is that imaginary space beyond the heaven [or world] which anyone can imagine there. And if there were a vacuum here in the world, it would be a place, namely an imaginary space. Moreover, a surface could also be imagined void."[88] It is obvious and noteworthy that Toletus conceives

156

of extracosmic space and void space in the world as wholly fictitious and non-existent entities.

In the second way, imaginary space is arrived at by abstraction. For "by abstracting from this or that space of particular bodies, [we arrive at the] space common to the whole world in which there are only bodies; by abstracting, I say, from this or that body; and this consideration is not a fiction."[89] Indeed, the imaginary space of the whole world that is abstracted from the particular spaces of bodies is no more a fiction than the abstract quantities considered by the mathematician. When a mathematician considers quantity *in abstracto,* he does not conceive it as the property of particular bodies, but as common to them all indifferently. Such an abstract quantity, moreover, is not a fiction, since it is derived from the quantities of actual bodies. In the same way, Toletus holds that the common world space abstracted from particular bodies is not the space of any particular body but the space of all of them.[90] And although it is not a true or real space, it is no fiction either, for by analogy with the abstract quantities of mathematics, abstracted common world space, which is immobile, is derived directly from the real spaces of particular bodies. It therefore differs from any vacua i ide or outside the world, which are not abstracted from anything, but, rather, are conceived by the mind as mere fictional entities.

But how, Toletus asks, can we, who judge only by our senses, abstract a common space? This is possible because a given space seems to remain equal even though different bodies occupy it. Although the space is occupied by different bodies, where the space of each body is numerically distinct (presumably because each body has its own internal space), yet these "spaces" are identical in all other respects, or, as Toletus expresses it, they are identical in species and equality.[91] Moreover, the accident of distance is always constant between two fixed points, say between the sides of a vessel or any two points on the circumference of the world. Although different bodies may successively occupy the vessel and bodies in the world may constantly alter their positions, the constancy of the intervening distance between the fixed points implies a constant space. It is this space that we abstract.[92]

No more than the space of particular bodies, is the common space thus abstracted conceived as separable from the material world. Like the internal space of any single body, it is a space internal to the material cosmos. With this conception of space, Toletus has a ready response to

157

the old question of whether the world is in a place. While he could agree with Aristotle that the world is not in an external place conceived as a physical container, the world is, nevertheless, in a place, since it is in the space that is internal to it. It is an inevitable consequence of Toletus' doctrine of space that every body, including the world itself, be in a place.[93]

Unlike his medieval predecessors, who failed to discuss or specifically elucidate their understanding of the distinction between true and imaginary space,[94] Toletus draws a clear distinction between the two. True space is the space of particular bodies. Every body possesses its own inner space that is inseparable from it and equal to its corporeal dimensions. By virtue of its inseparability from its body, this space is, as it were, directly perceptible to us. True imaginary space, by contrast, is not directly perceptible and is distinguishable into two types: a true imaginary space, which is the common space of the material world arrived at by abstraction from the true spaces of particular bodies; and a fictional imaginary space having no basis in reality, which includes all conceivable void spaces within or beyond the world.

Toletus also considered a question commonly discussed in the fourteenth century: whether place is mobile or immobile. Depending upon circumstances, external and internal places can be mobile or immobile. With regard to external place, Toletus adopts a common medieval interpretation. As the immediate surrounding surface of a body, an external place may be construed as mobile, since it can move and change. Furthermore, an external place, say a vessel, can have its distance altered with respect to different specific fixed points of the celestial circumference. External places may, however, be interpreted as immobile in species. For no matter how many different successive surfaces surround a body at rest, the distance of the body to different parts of the celestial circumference taken as fixed in position remains constant. Toletus resorts once again to a doctrine of abstraction. By abstracting a common place from the succession of changing places that might surround a body at rest, we can then abstract a common distance to any fixed point on the circumference of the heavens. With a common place and common fixed distance abstracted, the external place itself may be conceived as immobile.[95] In the Middle Ages, the immobility described here was called ''immobility by equivalence.''[96]

The same distinctions are applicable to internal space. Although

Toletus omits mention of it, the particular internal space of a body would seem necessarily mobile, since the body itself is mobile. The immobility of internal places, however, is restricted to common abstracted space, and thus to imaginary space. For bodies can succeed each other in a particular place of common space where each body bears a determinate relationship to any fixed reference point in the heavens.[97] Fictitious imaginary spaces must be conceived as immobile since bodies are deemed to move into and out of its parts. Moreover, when we conceive of a place in this imaginary space, whether the space is outside or inside the world, it will bear a fixed relationship to any point on the circumference of the heavens.[98]

Although Toletus considered place and space to have both external and internal senses, there is little doubt that three-dimensional internal space was more fundamental than two-dimensional external place as advocated by Aristotle. Nevertheless, he was genuinely pleased to have saved Aristotle and to have incorporated his views into a larger concept of place that included ''internal and external; true and imaginary; fictitious and commonly abstracted; proximate and remote; partial and total.''[99] The comprehensiveness of his approach prompted Toletus to proclaim that ''all the difficulties are resolved and almost all opinions reconciled; moreover, the opinion of Aristotle is defended so that [it can be said] he spoke only about external place. The reader is free to judge if the opinion of Aristotle is pleasing to him.''[100] Not only did Toletus find Aristotle's opinion on external place acceptable, though too restrictive, but he sided with Aristotle in rejecting the actual existence of extracosmic space and vacua within the world. Despite his radical extension of Aristotle's concept of place in ways that would have been totally unacceptable to Aristotle, Toletus sought to perpetuate and prolong his influence.

How different was the attitude of his contemporary, Francesco Patrizi (1529–97), who boldly proclaimed, what may well be true, that his was the first systematic study of space ever made.[101] As a confirmed anti-Aristotelian, he sought no reconciliation with Aristotle and indicates no interest in his scholastic predecessors. Where Toletus mentions Avicenna, Averroes, Aquinas, Albertus Magnus,[102] and other scholastics, Patrizi confines his references to Greek authors, real and fictitious, such as Plato, Posidonius, and Hermes Trismegistus, to whom the works in the Hermetic corpus were ascribed. His influences and sources of

159

inspiration are found in the non-Aristotelian Greek tradition associated with atomists, Stoics, Epicureans, Platonists, and Hermeticists. Even more than Toletus does Patrizi reveal the extent to which opinions on place and space customarily rejected in the medieval Aristotelian tradition were adopted in the sixteenth century as representative of physical reality. His break with medieval conceptions of place and space is virtually complete.

Patrizi assumes, as did the Stoics and some medieval theologians, that our finite spherical cosmos exists in an infinite void space. Assuming that an infinite magnitude could have a center, Patrizi locates the center of infinite void space in the center of the world on grounds that all radii drawn from the center of the world would be equal however far they are extended.[103]

Although infinite space is a homogeneous, immobile totality, Patrizi distinguishes two kinds of space: (1) an infinite external space that surrounds the world; and (2) the space of the world itself.[104] As one might expect, the two spaces possess many basic properties in common. If the space of the world were empty, it would be void (*inane*) in a manner identical to the infinite space beyond. It is only the creation of the world in that space which sets it apart as a *locus*. Moreover, "Neither of these two kinds of Space is a body. Each is capable of receiving a body. Each gives way to a body. Each is three-dimensional. Each can penetrate the dimensions of bodies. Neither offers any resistance to bodies and each cedes and leaves a *locus* for bodies in motion. And just as resistance (*resistentia, renitentia,* and *antitypia*) is the property of a body which makes it a natural body, so is a yielding offered to bodies and their motions the property of each kind of Space."[105]

The major difference between the two spaces derives from the fact that one is filled with body (though not completely, since Patrizi assumes the existence of interstitial and even small separate vacua)[106] and the other is totally void. Thus it is that the space or *locus* of the world is finite because the world is finite, whereas the infinite extracosmic void is both finite and infinite; finite because it terminates at the outermost boundary of the world and does not penetrate within; infinite because it recedes from the world to infinity in all directions.[107]

For Patrizi, space is prior to everything, even antedating the creation of the world. Indeed, if the world were destroyed, the whole of infinite space would remain in which God might create another world.[108] On this basis, Patrizi holds that the essence of space is emptiness; to be filled or occupied is accidental to it.[109]

But what, then, is space? It belongs to none of the categories distinguished by Aristotle. It is certainly not an Aristotelian two-dimensional surface that is exactly equal to the located body, for the place of the third dimension, depth, is not locatable in this way.[110] It is, in truth, an extension that inheres in nothing, the sort of separate space described as naturally impossible in the Middle Ages but supernaturally possible by God's absolute power. It is a substance, but not in the Aristotelian sense of matter and form. It may be described as a substance only because ''it subsists *per se* and depends on nothing else for its being; it requires nothing to sustain it, but itself provides their sustenance to substances, and sustains them in their being.''[111] Space is a mean between body and incorporeal substance. ''It is not a body, because it displays no resistance, nor is it ever an object of, or subject to, vision, touch, or any other sense. On the other hand, it is not incorporeal, being three-dimensional. It has length, breadth, and depth—not just one, two, or several of these dimensions, but all of them. Therefore it is an incorporeal body and a corporeal non-body.''[112]

In Patrizi's account, theological concepts that were the very essence of medieval considerations of infinite space are nowhere in evidence. Ignored or deliberatedly omitted are the sorts of questions that agitated his predecessors: Is the infinite space that is prior to all else a creation of the God who placed the world in that space? Is this space eternal, but related to God in some sense or other? Or is it independent of, and co-eternal with, God Himself? The medieval concern about the relations between an omnipresent God and the space He is presumed in some sense to occupy plays no role in this secularized approach that draws inspiration from Greek rather than Christian thought. In the seventeenth century, the secular and theological currents would merge for the first time, when God was deemed omnipresent in an infinite, three-dimensional, immobile, homogeneous void space. It was this God-filled space that would serve as the infinite container for the motions of bodies whose lawful relationships were described by Sir Isaac Newton in the *Mathematical Principles of Natural Philosophy.*

1. Although Pierre Duhem wrote at great length on the medieval doctrine of place (see *Le*

Système du monde, 10 vols. [Paris: A. Hermann, 1913–59], vols. 7 and 10), he did not consider this aspect of it.

2. Of those who adopted this position, we may mention Pseudo-Siger de Brabant (see Philippe Delhaye, *Siger de Brabant questions sur la physique d'Aristote* [Louvain: Editions de l'Institut supérieur de philosophie, 1941], bk. 4, *questio* 7, pp. 153–54), John Buridan (see *Acutissimi philosophi reverendi Magistri Johannis Buridani subtilissime questiones super octo Phisicorum libros Aristotelis diligenter recognite et revise a Magistro Johanne Dullaert de Gandavo antea nusquam impresse* [Paris: 1509; reprinted under the title Johannes Buridanus, *Kommentar zur Aristotelischen Physik* (Minerva G.m.b.h.: Frankfurt a.M., 1964], bk. 4, *questio* 7, fol. 73r, col. 2), and Albert of Saxony (see *Questiones et decisiones physicales insignium virorum Alberti de Saxonia in octo libros Physicorum recognitae rursus et emendatae summa accuratione et iudicio Magistri Georgii Lokert Scotia quo sunt Tractatus proportionum additi* [Paris: 1518], bk. 4, *questio* 1, fol. 43v, col. 1). Because of space limitations, references will regularly be given without the supporting texts.

3. Aristotle, *Categories,* 5a.6–14.

4. Delhaye, *Siger de Brabant,* p. 154.

5. *Commentary on the Metaphysics,* bk. 5, comment 18, in *Aristotelis opera cum Averrois commentariis* (Minerva G.m.b.h.: Frankfurt am Main, 1962; reprint of the Junctas edition of Venice, 1562–74), vol. 8, fol. 125v, col. 1.

6. *Opera,* vol. 4, Comment 43, fol. 141r, col. 2.

7. Harry A. Wolfson, ed. and trans., *Crescas' Critique of Aristotle; Problems of Aristotle's "Physics" in Jewish and Arabic Philosophy* (Cambridge, Mass.: Harvard University Press, 1929), p. 195. See also Max Jammer, *Concepts of Space: The History of Theories of Space in Physics,* 2d ed. (Cambridge, Mass.: Harvard University Press, 1969), pp. 76–81.

8. *R. P. F. Ioannis Duns Scoti . . . In VIII libros Physicorum Aristotelis Quaestiones cum annotationibus R. P. F. Francisci Pitigiani Arretini . . .* (Lyon, 1639), Tomus secundus, p. 228, col. 2. This edition of Scotus' works by Luke Wadding has been reprinted as Johannes Duns Scotus, *Opera Omnia* (Hildesheim: Georg Olms Verlagsbuchhandlung, 1968). Although ascribed to Duns Scotus, the treatise described above is actually by Marsilius of Inghen, a fourteenth-century scholastic, and was published at Lyon in 1518 under the title *Iohannis Marcilij Inguen super octo libros Physicorum secundum nominalium viam; cum tabula in fine libri posita . . . ;* reprinted as Johannes Marsilius von Inghen, *Kommentar zur Aristotelischen Physik, Lugduni, 1518* (Frankfurt am Main: Minerva G.m.b.h., 1964).

9. Probably based on Aristotle's argument against the vacuum in *Physics* IV.8.216a.27–216b.11.

10. Delhaye, *Siger de Brabant,* pp. 153–54.

11. See Benjamin Brickman, "On Physical Space, Francesco Patrizi," *Journal of the History of Ideas* 4 (1943): 231. Brickman's article contains a translation of parts of Patrizi's *Pancosmia,* which is part four of his *Nova de Universis Philosophia* (Ferrara, 1591; Venice, 1593). In Brickman's translation, the reader will find all of book 1 (*De Spacio Physico*) and part of book 2 (*De Spacio Mathematico*). Both books were probably published as early as 1587 in Patrizi's *De Rerum Natura Libri II Priores: Alter de Spacio Physico, Alter de Spacio Mathematico* (Ferrara, 1587).

12. *Physics* IV, 4.211b.19–29.

13. *Commentary on the Physics,* bk. IV, comment 37, in *Opera,* vol. 4, fol. 137r, col. 2–137v, col. 1, where Averroes is commenting on the passage cited in the preceding note.

14. An explanation similar to, but more formal than, Averroes' is offered by Hippocrates G. Apostle, *Aristotle's "Physics" Translated with Commentaries and Glossary* (Bloomington, Ind.: Indiana University Press, 1969), p. 245 n.27.

15. *S. Thomae Aquinatis In octo libros De physico auditu sive Physicorum Aristotelis Commentaria,* editio novissima, by A. M. Pirotta, O. P. (Naples, 1953), bk. 4, *lectio* VI, par. 885, pp. 197–98.

16. What follows appears in bk. 4, *questio* 2 ("Whether place is the terminus [or boundary] of the containing body") of Buridan's *Questiones super octo Phisicorum libros* (see n.2 for full title), fol. 68r, cols. 1–2.

17. A similar argument was proposed by Albert of Saxony, *In octo libros Physicorum,* bk. 4, *questio* 1 ("Whether place is a surface"), fol. 43r, col. 2 (for full title, see n. 2).

18. See Johannes Marsilius von Inghen, *Kommentar zur Aristotelischen Physik,* bk. 4, *questio* 13, fol. 55r, col. 2, or Johannes Duns Scotus, *Opera Omnia,* vol. 2, p. 270, cols. 1–2.

19. See Johannes Buridanus, *Kommentar zur Aristotelischen Physik,* bk. 4, *questio* 2, fol. 68r, col. 2.

20. Articles from the Condemnation of 1277 relevant to the history of medieval science have been translated and discussed in Edward Grant, ed., *A Source Book in Medieval Science* (Cambridge, Mass.: Harvard University Press, 1974), pp. 45–50. The Latin text appears in the *Chartularium Universitatis Parisiensis,* ed. H. Denifle and E. Chatelain, 4 vols. (Paris, 1889–97), 1: 543–55. For a general description of the background and impact of the Condemnation, see Edward Grant, *Physical Science in the Middle Ages* (New York: John Wiley & Sons, 1971), chap. 3, pp. 20–35, and chap. 6, pp. 83–90.

21. Articles 139 ("That an accident existing without a subject is not an accident except equivocally; and that it is impossible that a quantity or dimension exist by itself because it would be a substance"), 140 ("That to make an accident exist without a subject is an impossible argument implying a contradiction"), and 141 ("That God cannot make an accident exist without a subject or make several dimensions exist simultaneously").

22. *Kommentar zur Aristotelischen Physik,* bk. 4, *questio* 7 ("Whether it is possible that a vacuum exists"), fol. 73r, col. 1; see also bk. 4, *questio* 2, fol. 68r, col. 1.

23. Ibid., bk. 4, *questio* 8 ("Whether it is possible that a vacuum could exist by means of some power"), fol. 74r, col. 1.

24. See *De caelo* I.9.279a.13–24.

25. For a translation of this passage, see Edward Grant, "Medieval and Seventeenth-Century Conceptions of an Infinite Void Space beyond the Cosmos," *Isis* 60 (1969): 41.

26. The argument was often cited. See ibid., pp. 41–42 and n.15.

27. "That God could not move the heavens [that is, the world] with rectilinear motion; and the reason is that a vacuum would remain."

28. My translation is from *Nicole Oresme: Le Livre du ciel et du monde,* ed. A. D. Menut and A. J. Denomy, trans. with an introduction by A. D. Menut (Madison, Wis.: University of Wisconsin Press, 1968), bk. I, chap. 24, p. 176. See also Grant, "Medieval and Seventeenth-Century Conceptions of an Infinite Void Space beyond the Cosmos," p. 48.

29. My translation is from Menut's text, p. 176; see also Grant, "Medieval and Seventeenth-Century Conceptions of an Infinite Void Space beyond the Cosmos," p. 48.

30. My translations are from Thomas Bradwardine, *De causa Dei contra Pelagium* (London: Henry Savile, 1618; repr. Frankfurt am Main: Minerva G.m.b.h., 1964). The passage appears on p. 179. See also Grant, "Medieval and Seventeenth-Century Conceptions of an Infinite Void Space beyond the Cosmos," p. 46.

31. For a passage from Henry More's *Enchiridium metaphysicum* (1672), see A. Koyré, *From the Closed World to the Infinite Universe* (Baltimore: Johns Hopkins University Press, 1956), p. 154; for Samuel Clarke, see Clarke's Fourth Reply, par. 9, in *The Leibniz-Clarke Correspondence,* ed. H. G. Alexander (New York: Philosophical Library, 1956), p. 47.

32. *De causa Dei,* p. 177. A more extensive discussion of Bradwardine's views appears in Grant, "Medieval and Seventeenth-Century Conceptions of an Infinite Void Space beyond the Cosmos," pp. 43–47, and A. Koyré, "Le Vide et l'espace infini au xive siècle," *Archives d' histoire doctrinale et littéraire du moyen-âge* 24 (1949): 80–91.

33. *De causa Dei,* p. 177.

34. Ibid., p. 178.

163

35. Ibid., pp. 178–79.

36. Ibid., p. 177.

37. Ibid., p. 179.

38. *Le Livre du ciel et du monde,* p. 278.

39. Part VII of the eighth Dialogue quoted from the translation of Morris Ginsberg (London: Library of Philosophy, 1923), p. 211.

40. Part VIII of the eighth Dialogue, ibid., pp. 212–13.

41. *Commentarii Collegii Conimbricensis Societatis Iesu. In octo libros Physicorum Aristotelis Stagiritae. Prima pars qui nunc primum Graeco Aristotelis contextu Latino e regione respondenti aucti duas in partes ob studiosorum commoditatem sunt divisi* (Lyon: 1602), bk. 8, cap. 10, Quaestio 3, Articulus III [I], cols. 585–86. The Latin texts in support of what is said here have been quoted in my article, "Medieval and Seventeenth-Century Conceptions of an Infinite Void Space beyond the Cosmos," pp. 52–53, notes 56–66. Although the Cologne, 1602, edition is cited there, the Latin texts are virtually identical.

42. For other ways in which "imaginary" space was conceived, see my translation of Otto von Guericke's discussion in Grant, "Medieval and Seventeenth-Century Conceptions of an Infinite Void Space beyond the Cosmos," p. 54, n. 72. See also the views of John de Ripa and Franciscus Toletus below.

43. My discussion is based on the edition of this *Distinctio* in "Jean de Ripa I Sent. Dist. XXXVII: De modo inexistendi divine essentie in omnibus creaturis," édition critique par André Combes et Francis Ruello, présentation par Paul Vignaux in *Traditio* 23 (1967): 191–267, cited hereafter as "Combes and Ruello."

44. Combes and Ruello, pp. 231–32, ll. 47–68. The condemned article is 49 quoted in n. 27 above. An argument involving angels is also included.

45. Combes and Ruello, p. 233, ll. 95–96.

46. Ibid., p. 234, ll. 6–11.

47. Ibid., p. 234, ll. 16–20. Theologians such as John de Ripa, who believed that God is omnipresent in an infinite imaginary void, generally held that God was present in that space prior to the creation of the world, and not nowhere. Others, however, followed Saint Augustine and Peter Lombard and argued that God was in Himself before the world and not in any space (ibid., p. 234, n. 38). Aquinas, for example, denied the existence of any space or place before the world (see *Summa Theologiae,* part 1, question 46, article 1, p. 259a of the Latin text published by the Institute of Medieval Studies, Ottawa (Ottawa, 1941)).

48. Combes and Ruello, p. 234, ll. 20–22.

49. The four arguments below appear in Combes and Ruello, p. 232, ll. 68–78.

50. "Cuilibet substantie intellectuali simplici perfectiori essentialiter correspondet proportionaliter posse ad maiorem locum diffinitivum" (ibid., p. 221, ll. 1–3; see also p. 216, ll. 15–25).

51. "Locus enim in spatio est quod longitudine, et latitudine, et altitudine corporis occupatur" (J. P. Migne, *Patrologiae cursus completus, series latina,* 221 vols. [Paris: 1844–64], vol. 192, p. 626, col. 2).

52. For example, Combes and Ruello, p. 222, ll. 33–35; p. 232, ll. 79–80; and p. 235, ll. 26–28 and 35–37.

53. Ibid., p. 235, ll. 35–37.

54. Bradwardine, *De causa Dei contra Pelagium,* p. 131 A-C; see also Combes and Ruello, pp. 239–40, n. 49.

55. Combes and Ruello, p. 228, ll. 64–68.

56. Ibid., p. 235, ll. 26–28.

57. Ibid., p. 237, ll. 76–77.

58. Ibid., p. 221, ll. 1–3, 25–27, and p. 222, ll. 41–46.

59. Ibid., pp. 224–25, ll. 58–64.

60. Although de Ripa declares that ''many doctors assume it probable that God could produce some intensive infinite, and therefore an infinite creatable power'' (ibid., p, 224, ll.55–56), only Gregory of Rimini has thus far been identified as a believer in actual infinites (ibid., p. 224, n. 21). Buridan speaks of ''many'' who ''deny that God could make an infinite'' (see Ernest A. Moody, ed., *Iohannis Buridani Quaestiones super libris quattuor De Caelo et Mundo* [Cambridge, Mass.: Mediaeval Academy of America, 1942], Bk. 1, *quaestio* 15, p. 71, ll. 6–7).

61. Combes and Ruello, p. 238, ll. 101–4.

62. See E. A. Moody, ed., *Iohannis Buridani Quaestiones super libris quattuor De Caelo et Mundo*, Bk. 1, *quaestio* 15, p. 69, ll.30–31, and p. 71, ll. 4–19. With respect to the simultaneous occupation of the same place by two dimensions, Buridan probably had in mind Article 141 of the Condemnation of 1277 (see above, n. 21).

63. Ibid., *quaestio* 17, p. 79, ll. 1–9, and p. 82, ll.23–24.

64. Ibid., p. 79, ll.21–23.

65. *Kommentar zur Aristotelischen Physik* (for full title, see n. 8), bk. 4, *questio* 2, fol. 46 v, col. 2.

66. The supporting Latin texts from Henry of Ghent's *Quodlibeta,* Quodlibet XIII, *questio* 3, have been quoted by A. Koyré, ''Le Vide et l'espace infini au xive siècle,'' p. 63 n. 2.

67. Ibid., p. 64, n. 2.

68. Johannes Buridanus, *Kommentar zur Aristotelischen Physik,* bk. 4, *questio* 10, fol. 77v, col. 1. See my article, ''Jean Buridan: A Fourteenth Century Cartesian,'' *Archives internationales d'histoire des sciences* 16 (1963): 252.

69. Buridan did not confine this opinion to bodies beyond the world, but insisted that the material world and its parts are dimensions and therefore spaces. In the world, of course, these dimensions, which are inseparable from material bodies, are also in places. See Buridanus, *Kommentar zur Aristotelischen Physik,* bk. 4, *questio* 2, fol. 68r, cols. 1–2. Franciscus Toletus, whose strikingly similar views are described below, called the dimensions of a material body its ''internal place.''

70. What follows is drawn from *Nicole Oresme: Le Livre du ciel et du monde,* bk. 1, chap. 24, pp. 172–75.

71. Ibid., pp. 178, 179.

72. Ibid., Bk. 2, Chap. 8, pp. 368–70.

73. Par. 4 of Clarke's Third Reply to Leibniz in 1716. See H. G. Alexander, ed., *The Leibniz-Clarke Correspondence,* p. 32. I have also discussed this in Grant, ''Medieval and Seventeenth-Century Conceptions of an Infinite Void Space beyond the Cosmos,'' p. 49.

74. *Tres libri De celo et mundo* in *Questiones et decisiones physicales insignium virorum Alberti de Saxonia in octo libros Physicorum, tres libros De celo et mundo. . . .* (Paris, 1518), fol. 93v, col. 2.

75. *Leaves of Grass.* Facsimile edition of the 1860 text, with an introduction by Roy Harvey Pearce (Ithaca, N.Y.: Cornell University Press, 1961), p. 220 (number 8).

76. *D. Francisci Toleti Societatis Iesu Commentaria una cum Quaestionibus in octo libros Aristotelis De Physica Auscultatione, nunc secundo in lucem edita* (Venice, 1580).

77. Ibid., fol. 122v, col. 2.

78. Ibid., fol. 123v, col. 1.

79. Ibid., fol. 122v, col. 1.

80. ''Quaest. VIII: De concordia quadam scitu digna omnium opinionum de loco'' (ibid., fol. 122v, col. 2). The question continues to fol. 124r, which is incorrectly paginated as 121r.

81. Ibid., fol. 123r, col. 1.

82. Ibid., fol. 123v, col. 1.

83. Ibid.

84. See also ibid., fol. 124r, col. 2.

85. Ibid., fol. 123r, col. 2. Descartes says virtually the same thing in the *Principia Philosophiae*, Part 2, Principle 10 (see *Oeuvres de Descartes*, edited by Charles Adam and Paul Tannery, 8 [Paris: Vrin, 1905]: 45). Thus to Aristotle's concept of place, Toletus adds the view attributed by Aristotle to Plato (*Physics* IV.2.209b.6–16) in *Timaeus* 52, namely that matter and place are identical. In this view, place, which is dimensionality, is inseparable from body. The three dimensional concept of place on which we have concentrated in this paper differs only in the sense that it is separable from the bodies that occupy it and is therefore essentially a vacuum.

Descartes, it should be noted, appears to have adopted Toletus' distinction between internal and external place (*Principia*, Part 2, Principle 15, pp. 48–49). For a further discussion of Descartes' views on space, see the papers by Wallace Anderson and Peter Machamer in this volume.

86. Ibid., fol. 123r, col. 2.

87. Ibid., fols. 123r, col. 2-123v, col. 1.

88. Ibid., fol. 122v, col. 2.

89. Ibid.

90. Ibid., fols. 122v, col. 2-123r, col. 1.

91. Compare Descartes, *Principia Philosophiae*, Part 2, Principle 12, pp. 46–47.

92. Toletus, fol. 123r, col. 1.

93. Ibid., fol. 123v, col. 2.

94. John de Ripa distinguished between true (or positive) place and imaginary place, but offers no explicit discussion. The differences have to be inferred.

95. Toletus, fol. 124r, col. 1.

96. See Duhem, *Le Système du monde*, 7:210.

97. Toletus, fol. 124r, col. 1. Compare also Descartes, *Principia Philosophiae*, Part 2, Principle 12, pp. 46–47.

98. Toletus, fol. 124r, cols. 1–2.

99. Ibid., fol. 124r, col. 2.

100. Ibid.

101. In referring to the sciences of geometry and arithmetic as the sources of all the mathematical sciences, Patrizi sees their beginning in "Space and its study, which I am the first to set forth in these books" (Benjamin Brickman, "On Physical Space, Francesco Patrizi," p. 245). Although Gianfrancesco Pico della Mirandola and Bernardino Telesio also discussed vacuum, place, and space, and the latter is said to have influenced Patrizi, their discussions are relatively brief and omit consideration of infinite space.

102. For example, see Toletus, fol. 118r, col. 1.

103. Brickman, "Patrizi," p. 238.

104. Ibid.

105. Ibid. Professor Norman Kretzmann rightly observed that it appears redundant for Patrizi to insist that space both (1) yields to bodies and (2) penetrates the dimensions of bodies. The second alternative alone would seem to render the first incongruous. Penetration is obviously essential if space is assumed continuous and homogeneous. A yielding of space would have been appropriate only if space were deemed discontinuous, so that where body exists space is absent. Perhaps, Patrizi's "yielding" (*cessio*) is simply an aspect of "penetration" (*penetratio*). In penetrating the body, space is simultaneously yielding, i.e., allowing the body to move through it without displacing it.

106. Ibid., and p. 232.

107. Ibid., p. 238. Patrizi, a Neoplatonist, probably drew on Plato's *Timaeus* 52 for the priority

of space (Plato's "receptacle"), which furnishes a place for all things that come into being. He also cites with approval (p. 231) Aristotle's attribution of priority to place (*Physics* IV.1.208b.34–209a.1.).

108. Ibid., p. 240.
109. Ibid., p. 238.
110. Ibid., p. 229.
111. Ibid., p. 241.
112. Ibid.

Chapter Seven

CAUSALITY AND EXPLANATION IN
DESCARTES' NATURAL PHILOSOPHY

> Every Essence, created or uncreated, hath its final cause, and some positive
> end both of its Essence and Operation. This is the cause I grope after in the
> works of Natura; on this hangs the Providence of God.—Sir Thomas
> Browne, *Religio Medici,* 1635.

Descartes, like Caesar, found all divided into three: God, mind, and
matter. Concerning these three substances, he saw four different types
of inquiry that could be made: about God, about mind, about matter, and
about the union of mind and matter. In this essay I shall be concerned to
examine Descartes' natural philosophy and especially its systematic,
complex connections with the concepts and inquiries in the other realms.

Briefly and uncritically stated, Descartes conceived of his task in
natural philosophy as one whereby he would replace the explanatory
schemes of the Scholastics and natural magicians with a set of principles
or laws that would explain the inanimate world by referring only to
matter and its properties, extension and motion (and, ultimately, to
God). Extension and motion were clear and knowable concepts and, for
that reason, could serve as bases in demonstrations, proofs, and explanations about the material world. Extension and motion were simples and
served as causes in natural philosophy. This causal role was ultimately
to be justified by an appeal to the more certain principles of metaphysics.
Explanations were to be causal explanations, using only material, formal, and efficient causes. Any appeal to final causes was illegitimate for
metaphysical and theological reasons.

I shall examine Descartes' use of explanatory causes, especially
extension and motion. I hope to explain in what ways Descartes' causes
are insufficient and in what ways and why his fundamental scheme of
explanation had to be supplemented. In order to show this, I shall divide

my essay in two parts. The first elaborates in some detail Descartes' treatment of extension and motion as it occurs in the second book of the *Principles of Philosophy*. I want to demonstrate what connections actually obtain between these Cartesian concepts and ascertain the limits of their explanatory power, remarking on the causal role that these concepts play. In the second section I explicate Descartes' remarks concerning causal explanations. I shall briefly analyze his treatment of the various kinds of causes and argue that there exists an inherent insufficiency in his program as he states it. Then I shall consider some of Descartes' actual explanations and show how they reflect this inadequacy of his program. In this way I hope to shed some light on this bit of early seventeenth-century natural philosophy and to draw attention to its internal inadequacies in a way that will permit students of both earlier and later periods to see the rationale present in this portion of the history of natural philosophy.

Space permits no more than a mention of the background to Descartes' work, but it is worthwhile to do that much as the analysis I am proposing illuminates the role of Descartes as a transitional figure in the history of thought. For present purposes I wish to highlight four currents which flow into the Cartesian corpus. First is the Scholastic tradition in which Descartes was trained. His conception of the various kinds of cause was formed from this Christian-Aristotelian tradition. Likewise, the non-mathematical aspects of his explanatory scheme are of an Aristotelian type. Equally of interest are his departures from the tradition. His rejection of a comprehensible God determines many of the arguments he uses. Despite the work of Gilson and others, the rationale of Descartes' acceptances and rejections is basically unknown.[1] The second current in which Descartes is caught is his anti-magical, anti-alchemical stance. Probably theologically based, this position leads him to reject not only the world pictures of the natural magicians and alchemists of the Renaissance, but also to eschew any hint of animism that might link him with them. This is probably the basis for his rejection of final causes.[2]

The third influence is the skeptical philosophy that was revived during the fifteenth and sixteenth centuries. The newly felt influx of skeptical thought caused Descartes to spend much of his time worrying about the problem of knowledge and certainty. Specifically, many of his epistemological and metaphysical doctrines seem quite explicitly designed

169

to combat the skeptics' position and to reaffirm the truths of religion.[3] Finally, what has not been pointed out to my knowledge, is the influence of Neoplatonism on Descartes' work. This influence is seen in his early concern, in *Le Monde,* with the nature of light and in the influence of optics as exemplary of mathematical physics. As in the thirteenth-century light ontologists, Descartes seems to derive his ontological picture from a conception of the tridimensionality of body and the geometrical model of the propagation of light. The metaphysics of light and its Neoplatonic explanatory scheme is a Cartesian source much deserving of study.[4] All of these influences direct the articulation of Descartes' work. I shall now turn to the work itself.

I

Let us first look at Descartes' introduction of the notion of extension. In *The Meditations* and *The Principles*, Descartes argues that extension only must be the essential property or form of matter. The argument concerning the piece of wax in *Meditation* 2 (AT, 30–31; HR, 1.154–55)[5] and those in *The Principles,* book 2, chapters 4 and 11,[6] proceed along similar lines (though I think their purposes differ). Roughly, Descartes asks what is the one property that matter must have? The answer is the essential property of matter. But the question then takes a subtle and epistemologically suspect twist, for Descartes asks what property of matter *must always be perceived* whenever one perceives matter at all? In answer he lists many of the properties of matter that are perceived, adding an etc.-clause to ensure the list is exhaustive. One by one he goes through them to show that under certain conditions a perceiver could perceive matter but would not perceive that particular property. The only properties that must be perceived, if one is perceiving matter at all, are extension and those properties, like figure, deducible from the property of extension. Later he will add the premise that we may trust what we perceive in such cases, for God is no deceiver.

The only sense modalities he considers are sight and touch, and the argument does seem to work when restricted to these. Had he considered hearing or smell, it is doubtful that he would have felt secure about his conclusion, but he does not. This is the only unreasonable part that I can find in the argument, except for the beginning restriction to perceived properties. Even in those places where the argument smacks of sophistry, I think it is reasonable. For example, Descartes argues that color is not essential to bodies because one can perceive an extended body that has no color; i.e., is transparent. Now this looks sophistical to a modern eye; but when one recalls that, historically, transparency was opposed to color, part of the air disappears, and the rest goes when one reflects on the kind of analysis of color perception Descartes would give. The most natural suggestion is in terms of reflected light (for even transparent bodies reflect light in the visible band, elsewise they could not be seen at all (*Optics* AT, 84; O,67).[7] If then, impenetrability is deducible from extension, and light reflection depends upon the impenetrability of the object's surface, plus the particulate nature of light, it follows that the only property of the perceived object one needs to speak of is extension and what is deducible from extension. Thus, Descartes would be vindicated.[8]

In Descartes' system the ability of a body to reflect light does follow from the extension of the reflecting body plus the corpuscular nature of light and his laws of collision (though in what follows I shall ignore these latter two components). If we take a body's extension at a given moment as being its figure or shape, then to have, at a given moment, a given extension is to ensure that the body at that moment is impenetrable (*pace* Leibniz). For if one extended body were to penetrate into another, this would entail either that part of one of the bodies was annihilated—since two bodies cannot occupy the same space at the same time—or that the shape of one of the bodies was changed.[9] The latter change would be a change in its mode of extension, which *ex hypothesi* contradicts the assumption that at the given moment it had the other mode of extension. That is, the impinging body would have to push out of shape the original body. For a body to have a given extension at a given moment is to assume that at that moment its extension is constant and not annihilated, therefore that it is impenetrable. Of course, this does not preclude the possibility that at the next moment something might act upon it to change its mode of extension, i.e., its shape.

Since the reflecting body is impenetrable, when it is hit by the light particle, the light will be reflected. The spin of the reflected particle will depend upon the surface of the reflecting body. Thus, perceived color is dependent upon extension, so that one can conceive of extension without color, but not conversely.

If one combines the argument I have sketched above concerning the primacy of extension as a property of matter with an argument to the effect that sight and touch are the most important senses, Descartes is provided with a reasonable argument for the essential nature of extension in contexts where matter is perceived.[10] I think Descartes does hold, and implicitly argue for, the position that sight and touch are the primary or fundamental senses.[11] Instead of following this up, I want to take the essential nature of extension for granted and go on to examine and explore its relation to Descartes' conception of motion.

Descartes' notion of extension is somewhat unclear. It is identified with the three-dimensionality of body, i.e., a substance extended in length, breadth, and depth.[12] Such extension is internal place or the space-occupying nature of body (*Principles* 2.10). Descartes ties extension as space with its characterization as magnitude or figure. This is qualified when he discusses rarefaction: if a body were to be squeezed together so that its parts were nearer one another than previously, such a body would have the same extension as before. The change of figure, due to condensation, would not have changed the extension of the body. The reason for this, Descartes says, is that "we ought not to attribute to a body the extension of the pores or interstices which its parts do not occupy, but to the other bodies which occupy these interstices" (*Principles* 2.6). The model lying behind this reasoning is Descartes' conception of a body as made of a number of smaller particles, each of which has its own extension. The extension of the body is merely the sum of the extension of its constituent particles, irrespective of how far these particles might be from one another. The spaces between the constituent particles are filled with finer, different matter; matter that is not, properly speaking, part of the body. It is these other particles that get squeezed out when the body becomes more dense. Since the number and shape of the constituent particles do not change, the extension of the body does not change.

Elaboration of this position would require Descartes to explain the different kinds of particles, and how it is that some are properly constitutive of body whereas others are only accidentally associated in that they fill up possible voids. While Descartes will later (in *Principles* 3.52 and 4.2f.) discuss such particles, he has not as yet, so they come as a bit of a surprise here.[13] This conception of extension as the sum of the extensions of the constituent particles is somewhat at odds with the Aristotelian characterization of extension as internal place, i.e., the body's occupying of three dimensions. It would seem that strictly speaking the extension of a body would be the sum of the internal spaces of the constituent particles. But Descartes talks of the body's extension quite apart from its constitution when discussing motion. This tension is never resolved. Also, it is worth noting that the particle model of extension makes extension very close to what others will call the quantity of matter. This is misleading though, as Descartes, in his laws, only considers the quantity of matter to be a function of the "size of a body, and of its surface, which separates it from others."[14]

Making use of the premise that matter is only extension, Descartes can derive both philosophical and scientific conclusions. I have already shown above how he attempts to derive the impenetrability of bodies from their extension. From bodies as extended he also argues both that there can be no vacuum in nature (*Principles* 2.16–18) and for the cosmological conclusion that there can be but one world (*Principles* 2.21–22). It is worth spending a moment upon these arguments because they illustrate Descartes' deductive method and bring out another aspect of extension I have not yet remarked. Descartes argues that matter is extension, and that extension is a magnitude. Any distance is a magnitude, so any distance has an extension. Further, for there to be an extension there must be some *thing* that is extended. Nothing cannot be extended since it has no properties. This last point turns out, for Descartes, to be equivalent to holding that extension is both a necessary and a sufficient condition for a thing's being material. A vacuum is said to be both nothing and to have magnitude (extend over a given distance). But a vacuum cannot have a magnitude, therefore there can be no vacuum.

This argument does rely upon the questionable, crucial premise—if there is extension, there must be some thing that is extended. But it is not explicitly question-begging, for this premise seems to be an instance of Descartes' more general thesis, which can be argued for independently, that if there exists a property (either mode or attribute), there must be some substance having that property.[15] While this is not philosophically neutral, it certainly has a large degree of plausibility attaching to it. One would either have to deny this premise and hold out for pure uninstantiated properties existing in the world, or somehow argue that a vacuum is a substance. Either of these course seemed foolhardy to Descartes, and indeed the latter caused problems for many seventeenth-century thinkers.

From the conclusion that there is no vacuum, Descartes goes on to argue that since extension is indefinite, matter occupies all space. From this it follows that there can be only one world. For if there would be two, they would have to be separated from one another, and there would have to be a vacuum. Thus from extension as a necessary and sufficient condition for matter, Descartes is able to confute both the multi-world hypothesis, the atomist conception of a vacuum, and the animism with which the former doctrine was often associated.

It should be noted that in all the above arguments extension is used as a formal cause. The arguments all proceed from the premise that matter is extended and use only premises entailed by this via common or accepted definitions (or, sometimes, using premises that are independent of the concept of matter). There is no use made of efficient causes, let alone of final causes. There is no appeal to experience or experiment, though the conclusions are meant to be true of the world. These purely formal arguments (as I call such that use only formal causes) stand in contrast to the arguments concerning motion and the world that Descartes gives later in the *Principles*.

In one of its guises Descartes' conception of motion also has a purely formal role to play. In the *Principles,* Descartes contrasts the vulgar sense of motion, as the activity or action by which a body travels from one place to another, with the true (and philosophical) sense of motion: "the translation of a piece of matter (a body) from the neighborhood of bodies immediately touching it, these being regarded at rest, to the neighborhood of others" (2.25). Thinking of external forces or activities deriving from external causes, Descartes explicates his claim by arguing that his concern is only with motion as it inheres in a moving body and not with the force of body that moves it. Translation is to the force of translation as the moving body is to the body that moves it. The higher motives behind this claim are to show that motion is a mere mode (*modum*) of body and not a substantial reality in itself.

It has been noted often that this definition of motion as translation does not occur earlier in Descartes' writings. It has been suggested and, seemingly, universally accepted that Descartes' motivation for thinking up and including this definition in his *Principles* derived from his quivering reaction to Galileo's condemnation and his subsequent desire to show that the earth could truly be said to be at rest (cf. *Discourse* 6, AT, 60; 0, 49).[16]

174

While there is no doubt some truth in this charge, it should not be allowed to obscure the fact that the translational definition of motion is very close to the geometrical concept of motion that occurs in *Le Monde* (AT, 11.39–40). It too considers only geometrical projections and no efficient causes. It is also logically tied with Descartes' conception of bodies as extended. Indeed, if Descartes does want to show that motion is a mode of body, he must have something like his translational definition and not one like the vulgar. For Descartes, to talk of force or efficient causes is ultimately to refer to God, beyond body, but I shall return to this below. The logical connection between matter and extension and translational motion, as laid out in *Principles,* Book 2, especially sections 10-15, is almost that of mutual implication. The qualification is necessary because of a few additional premises which are, I think, unproblematic.

To see that argument from extension to motion, consider the following:

1. Body is an extended substance.
2. The extension of a body determines its internal place; i.e., the three-dimensional space it occupies at a given time.
3. The internal place of a body determines the boundary between that body and those that immediately surround it.[17]
4. The set of bodies that immediately surround a given body, A, determine the external place of A.
5. Change of place for a body, A, is a change of external place, and therefore a change in the set of immediately surrounding bodies constituting the boundary.
6. There are only two possible ways in which the boundary that exists between body A and those that surround it can change: (i) change A (by increase or diminution) or (ii) have the bodies which surround A at t_2 be different, all or in part, from those which previously surround A at t_1. (I believe (i) entails (ii), and I think (ii) is a sufficient condition.)
7. Assume A remains constant, does not change size (not-i).
8. Motion is change of place, change in the boundary between A and those bodies that immediately surround A by (ii).

175

9. Change of place for A by changing its immediately surrounding bodies is the definition of true motion.

This argument does not touch Leibniz' claim that Descartes cannot deduce a conception of motion as action from his conception of matter. This claim Descartes would probably agree with. The second part of Leibniz' claim that the laws of nature cannot be derived from the concept of extension I agree with.[18]

Turning to the converse derivation of extension from translational motion, there is:

1. True motion is the translation of a body from the neighborhood of bodies immediately touching it (these being regarded at rest), to the neighborhood of others. (2.25)
2. Motion is change of place.
3. External place is defined by the set of bodies surrounding a given body, A.
4. The set of surrounding bodies externally defines the extension of a given body, A, which is surrounded.
5. Therefore, to speak of the set of bodies which surround A is just to speak of the extension of A or better is what we must use to ascertain the extension of A.
6. Since the set of surrounding bodies is (what defines) the extension of A (as externally determined), the extension of A is part of the definition of motion.
7. Bodies have many properties that are true of them, e.g., extension, figure, movement, situation, divisibility, hardness, color, and so on.
8. Either hardness, divisibility, color, and so forth, are properties of body used in the definition of motion.
9. The other properties of body, e.g., figure and situation, are deducible from, or equivalent to, extension as a property of body.
10. Every property of body other than extension is either deducible from, or equivalent to, extension or is not used in defining the motion of bodies.

11. Therefore, with respect to motion, extension is the only necessary property of body.

A qualification needs to be made concerning this argument. In the argument, extension is treated wholly from an external point of view. If extension is the sum of the extensions constituent of the particles, this argument would not work. But it would be possible to construct an analogous argument concerning the motion of each particle and its extension.

So far Descartes has only treated matter in its essential or formal aspects. In order to proceed further and to be able to explain such phenomena as gravity, heat, and planetary motion in terms of matter and motion, Descartes must introduce additional causal elements into his analysis. If he were an occasionalist, as he is sometimes construed, such an introduction would not be necessary. But as Kemp Smith has pointed out, Descartes, despite his voluntaristic creationism and his atomic doctrine of time, should not be read as an occasionalist. Descartes has a conception of real causal relations, and I shall show, what Kemp Smith only stated, how these operate and what force they have.[19]

II

Descartes is quite clear about the need to use efficient causes. Also he is explicit about the necessity to reject any use of final causes. He argues for this last on theological grounds, stressing the infinity of God and the impossibility and temerity of trying to divine His purposes. He should not be astonished, he says, if his intelligence is incapable of comprehending God. He continues:

> For, in the first place, knowing that my nature is extremely feeble and limited, and that the nature of God is, on the contrary, immense, incomprehensible, and infinite, I have no further difficulty in recognizing that there is an infinitude of matters in His power, the causes of which transcend my knowledge; and this suffices to convince me that the species of cause termed final, finds no useful employment in physical things; for it does not appear to me that I can without temerity seek to investigate the ends of God. (*Med.* 4, AT 7.55; HR 1.173)[20]

The role and function of efficient and formal causes is not made clear by Descartes. In *Principles* 1.28, he states that God is the efficient cause of all things. Later (2.36) he says that God is the universal and primary cause of all motions in the universe. He replies to Gassendi that the perfection of a form can never be understood to preexist in the material [cause] but only in the efficient cause (AT, 7.374–75; HR, 2.223). In his reply to Arnauld, Descartes discusses the difference between formal and efficient causes. He agrees with Arnauld's claim (AT, 7.212–13; HR, 2.91) that one seeks efficient causes only of existence and says that efficient causes are sought and used to prove the existence of everything that is not self-derived (AT, 7.238–39; HR, 2.109–10). Formal causes are used to demonstrate what a thing is. The formal cause is the entire essence of a thing. Citing Aristotle's *Posterior Analytics* 2.11, he remarks that formal causes are those from which knowledge of any kind is derived (AT, 7.242; HR, 2.112).

A final, perplexing comment comes in his reply to the sixth objection. Descartes argues that what God does is good because he does it, and not that he does it because it is good. He then writes:

> Nor is it worthwhile asking in what class of cause fall that goodness or those other truths, mathematical as well as metaphysical, which depend upon God; for since those who enumerate the class of causes did not pay sufficient attention to causality of this type it would have been by no means strange if they had given it no name. Nevertheless they did give it a name; for it can be styled efficient causality in the same sense as a king is the efficient cause of the law, although a law is not a thing which exists physically, but is merely as they say [in the schools] a moral entity. (AT 7.436; HR 2.250–51)

From the above bits I think Descartes' position can be stated. God must be treated as both the efficient and formal cause of all things. Reference to final causes is prohibited since one cannot know the purposes of God's free creations. His intentions are inscrutable to human minds. Like a king, He freely promulgates his will, the results being (existentially) the things of the universe and (formally) their essences and the laws of nature. God as an efficient cause brings about the existence of matter and all its properties, including its modes, motion and rest.[21]

The rationale behind this position is simple. Since matter is completely passive and characterized by extension, the efficient cause of the motion of matter must come from elsewhere. In Descartes' ontology, there are only two active principles: God and mind. In the case of matter,

God must be the active principle and thus the efficient cause of motion. The introduction of God as efficient cause takes us out of the purely material realm and establishes a necessary line of activity between God and the world. God as efficient cause performs two functions in Descartes' scheme: He is first, an Aristotelian first, and only mover. In this aspect God is the activity (*actio*) or force (*vis*) in the world. Matter as passive (or, perhaps, pure potentiality) only becomes active (actual) when God acts upon it. This action of God is, in good Christian fashion, the creation and sustaining of matter with all its properties, including motion. This linkage of efficient cause with the coming into being, and preserving, of existents is the second function of God as efficient cause.

God, as efficient cause, creates things, bringing them into existence, and He continues to sustain them as they are, according to the laws of nature. To gain any knowledge concerning the things of the universe, or to come to know the laws of nature by which they act, we must look at God also as a formal cause.

As a formal cause God and his attributes are responsible for the attributes and modes of all that is created. All of our knowledge of things is derived from their formal causes. Similarly, all explanations of why things have the properties they do are to be given by referring to formal causes. Reference to an efficient cause can only show that something exists in a brute sense; to know what it is necessitates the use of formal causes. But, as Aristotle noted long ago, formal causes are tricky; specifically, very often formal causes are coincident with final causes.

Before proceeding further it will be useful to forestall a wrong interpretation of Descartes' work. When Descartes desires to give a causal explanation, he wants to explain a given state of affairs in the world by appealing to God only as the efficient and formal cause of that state. It is not enough for him merely to state that God has such and such attribute and caused to exist a state having so and so properties. Causality as constant conjunction and succession is not a viable analysis for Descartes. Descartes denies that effects must follow their causes, temporally or logically (reply to Arnauld, AT, 7.239f.; HR, 2.110f.). Further, he holds that effects are formally or eminently contained in their causes (*Med.* 3, AT, 7. 40–41; HR, 1.162 and Replies to Obj. 2, AT, 7; HR, 2.34). Thus, causality is a real relation and a causal explanation must exhibit the internal (formal) relation which holds between a cause and its effect. I shall argue that such internal relations cannot be supplied solely from efficient and formal causes as Descartes conceives them. Either he must supplement his analysis by using final causes or he must modify the traditional notion of a formal cause in such fashion that it does duty as a final cause.

Before turning to Cartesian texts, an ancient example will help to clarify the traditional roles of formal, final, and efficient causes. I shall also use the example to argue the general point I want to make concerning Descartes' failing. Classically, a formal cause is traditionally conceived as the attribute or essence of a substance (Aristotle, *Post. An.* 94a20; *Phys.* 194b26). A formal cause of a house is its shape and design (its physical structure). It is the house's shape that causes it to be a house rather than a boat or some other structure. A final cause states an end or that on account of which a thing exists. The final cause of the house is, in one sense, the dweller; in another sense the final cause is a dwelling, for it was built for the purpose of providing a dwelling (*Phys.* 194a27; *Met.* 1027b1). Because the house has the form it does, it is able to provide a dwelling for a dweller. Thus, formal and final causes collapse (*Phys.* 198a25f.). The efficient cause is the builder and his actions. The builder fashions the material according to his plan to bring it into its final form. In cases of art (as opposed to nature), the intention or plan of the artisan provides, extrinsically, the principle according to which the artifact is made. When it is (correctly) made, it achieves its proper form and fulfills its function. To completely state the principle according to which the house arises requires referring to the proper form and to the proposed function (and, thus, at least implicitly to the intention or purpose of the builder).[22]

Formally, the cause is said to be the essence of the effect. The formal cause, *per se,* refers to nothing outside of that of which it is the form. In modern terms, it does not refer to the cause at all. It is used in causal explanations when we wish to explain certain things about the thing, such as why it is a house. If we ask why the house exists, we say the builder made it. This accords with Descartes' remarks concerning the connection between the efficient cause and existence.[23] But if we ask why does this exist as a house, we must answer by referring to the final cause. This exists as a house in order to provide a dwelling. The linkage between cause and effect (in modern terms) is provided by the principle, which states the relation between this house as a dwelling and the formal cause of the housebuilder, to build houses for dwelling. In cases of art, the design or principle is provided by the artisan (his intention), and the artifact achieves its end, or proper form, when it fulfills this intention. The final cause serves as the means of stating the link between cause and effect.

For Aristotle all things happen for an end, not just those that come about by art. Everything that happens by nature happens for an end. Happening for an end is just happening according to a principle, as opposed to happening by chance. I cannot here go into what it means to happen or to be moved by a principle, but I suggest that it be read as meaning happening or moving according to a law. Reaching the state of being a mature oak tree is the final cause of an acorn. By nature, or in a law-like way, the acorn will grow into an oak if conditions are propitious. The principle or law describing oak trees' growth patterns mentioned the end of the growth process. The acorn's formal cause is its vegetative soul (for this is the principle according to which it will achieve its actuality). Here again the final cause serves as a link between two states, what the acorn is now and what it will finally come to be.

The Christian-Aristotelians, against whom Descartes was reacting, carried on this explanatory scheme (cf. below, pp. 182–83). They also added certain theological elements to it. Aquinas, for example, treated all things that happened by nature as manifestations of divine art (*Comm. Aris. Phys.,* Lectio 14, sect. 268). In one sense, he collapsed the Aristotelian distinction between nature and art and conceived of final causes in nature as the providence by which God creates and sustains the world (*Comm. Aris. Phys.,* Lectio 12, sect. 250). In the *Summa contra Gentiles,* Aquinas talks of God's reason or design for the world as realized in all natural things.[24] Everything in the world has its form and fulfills its end because it realizes the intention of God.

I suggest that it is not really final causes to which Descartes objects, but rather this medieval manner of treating all final causes as fulfilled intentions. If he had realized that final causes are not necessarily animistic and not necessarily disguised intentions, there would have been no reason for him to forbid their use. Final causes, generally construed, can be quite as physicalistic and non-animistic as any others. They are needed to state the internal relation that obtains between cause and effect. Some modern philosophers have treated final causes as semantic relations.[25]

Descartes, trained by the Jesuits and steeped in Thomism, could only

181

conceive of final causes as they were used by Thomas and the later Catholic commentators. Theologically indignant at the animistic world schemes of the magicians and alchemists, and fearful of anthropomorphism and temerity in religion, Descartes went to excess and banished all final causes from his system. Because of this, he could not provide the needed linkage between causes and their effects. In providing causes, both vertically from God to what He created, and horizontally between things in the world, Descartes was forced to reintroduce final causes, either in their classical or Christianized manner. He had to do this or else fall back upon a purely occasionalist analysis, which his own theology and metaphysics would not allow.

To show how this rejection of final causes affected Descartes' work consider the form for Cartesian explanations. In a methodological remark in the *Principles*, Descartes tells how we are to proceed with explanations of what is perceived by sense. He writes:

> But regarding Him as the efficient cause of all things, we shall merely try to discover by the light of nature that He placed in us, applied to those attributes of which he has been willing that we should have knowledge, what must be concluded regarding the effects that we perceive by sense. (1.28)

The attributes mentioned, presumably, are in the first place those that comprise the essence of God, and secondarily, those essential attributes of corporeal and incorporeal substance. The effects perceived are created by God as efficient and formal cause. For Descartes, one task of the natural philosopher is to build the bridge between the substantial attributes we know and the sensible effects. The resulting bridge is the causal explanation and constitutes, in good Aristotelian fashion, the knowledge of the builder. Ultimately, a causal explanation should lay out the internal connection that obtains between God and the observed phenomena as effect.

This picture of bridging the gap between God and his attributes and the phenomena of the world is quite traditional. As a general explanatory scheme it fits well with Neoplatonic and Christian-Aristotelian conceptions of natural philosophy. If Descartes were strictly following this form, one would expect explanations like the following: We observe a phenomenon having a certain property. We explain the phenomenon by showing that the property it has is an instance or manifestation of one of God's attributes. Since God created and sustains the world, all things

182

were given form by Him (or emanate from Him) and so cannot but reflect His mode of being.

Many of Descartes' arguments, especially those for the laws of nature, come close to this traditional form. When, in *Principles* 2.36, he demonstrates that the quantity of motion in the world is constant, one tends to read the argument according to the traditional pattern: God is constant and unchangeable. God created matter and its mode, motion. Therefore, the motion created by God is sustained in matter as he created it, i.e., the amount created remains constant. In fact, even here, Descartes' argument is more complex, in ways which I shall discuss below.

At other times Descartes abandons the traditional model and argues in a strictly epistemological fashion. The epistemological character of some of Descartes' physical arguments is rather unusual. Presumably, it derives from his reaction to the revival of skepticism and his worries about Galileo's condemnation. In these arguments, the formal attributes of God do not function as prototypes or sources for the properties and attributes of material things (as they do in the traditional arguments). They function primarily as constraints upon, or licenses to perform, certain inferences concerning human thinking about material things.

In both types of argument, Descartes uses God's attributes as the formal cause of God *and* as a final cause of something God created, either matter or man. In some cases, as fits the traditional model, the final cause of what is created is also its formal cause. Such use of final causes is not worrisome on the traditional scheme. Indeed, some Neo-platonists and Christian-Aristotelians, with whom Descartes was familiar, called explicit attention to this fact.[26] The final cause was said to be, in one sense, the form that was introduced through generation and put upon the object in the process of coming-to-be. Since what can come to be was often said to participate in its cause or have its cause as its form, the identification of formal and final causes was quite natural.

Descartes, as we have seen, does not allow himself to use formal causes in such dual fashion. To see what, in fact, he does do and to indicate by text the general limitations of the Cartesian program, I shall now examine a few arguments in detail. The arguments I shall consider cover both types, physical and epistemological.

Following the pattern of argument set down in *Principles* 1.28 (and quoted above, p. 182), Descartes should argue for the conservation of the quantity of matter as follows: We know an attribute of God, i.e., constancy. By the light of nature, we conclude that the quantity of motion in matter is constant. This conclusion in its turn would serve as a basis for further conclusions or explanations concerning observed motions. The argument actually runs: God is perfect, therefore unchangeable and constant. Therefore, His operation in the world occurs in a constant manner. Unless experiment or revelation shows us change (inconstancy) in the effects of the operation of God, it is reasonable to assume that the effects are constant and unchangeable. The creation of matter and its modes, motion and rest, was one of the effects of God's operation; "consequently, it is most reasonable to hold that, from the mere fact that God gave pieces of matter various movements at their first creation, and that he now preserves all this matter in being as he first created it, he must likewise always preserve in it the same quantity of motion" (*Principles* 2.36).

The missing premise in the last part of the argument is that experience and revelation show us no change or inconstancy in matter and motion. It is important to note that it is not always so. There does occur change and inconstancy in some cases.[27] In the occurrence of transubstantiation, for example, there is real substantial change and God does not preserve the bread as bread (Reply to Objections IV, AT, 7.248 f.; HR, 2.116 f.). How, then, in the case of motion are we assured of no change? Revelation would seem to have no part. Experience seems to argue for change since one often sees a body in motion at one time and at rest at another. Descartes is silent on this, and as a result his argument does not go through. For the argument to be sound, we would have to be assured by revelation or experience of the constancy of certain properties of the world taken as a whole. This obviously is what Descartes has in mind, for the quantity of motion is a property of the universe as a whole. The conservation of this property is quite consistent with the change attested by experience. But revelation again seems silent, and experience of such properties is not to be had by mere humans. The only way of showing that the premise is true is by an *a priori* argument that shows that God's constancy has as an instance the constant quantity of motion. This would make the argument quite Neoplatonic in character. It would also treat constancy as both a formal *and* a final cause of the universe as a whole. Descartes must do this since constancy as a property of the universe as such has no intrinsic connection with the constancy that is an attribute of God. To say that the former is an instance of the latter is to say that the creation of the universe realizes God's intentions in this particular way. Thus, to explain why the universe is such that its quantity of motion is constant, one invokes God's constant manner of creation, just as in explaining why this is a house one explains its suitability for dwelling and shows how it realizes the builder's intentions. God cannot be merely the efficient cause, for Descartes attests that efficient causes can only tell us about the existence of the universe, not about any of its properties. Only by so treating constancy can Descartes provide the link he needs between constancy as an attribute of God and constancy as a property of all matter in motion. To treat constancy as a final cause is to implicitly refer to God's purposes as exhibited in the state of the world which he created.

God chooses to maintain, or create, one state of affairs rather than another. Without recourse to His intentions or purposes (or to his attributes as final causes for created things) there is no explanation of the state that is possible (as Leibniz clearly saw).[28] This general point can be made in secular terms against any argument that uses a version of the principle of sufficient reason or its equivalent.

Descartes appeals to the principle of sufficient reason on many occasions. It is used in the proof of the third law of nature, Descartes' main law of motion (*Principles,* 2.41). In presenting the argument as to why a body rebounds from another with which it collides, Descartes says:

> In collision with a hard body, there is an obvious reason why the motion of the other body that collides with it should not continue in the same direction; but there is no obvious reason why this motion should be stopped or lessened. . . .

Likewise, in one of the laws of collision, which is supposed to follow directly from the third law of nature, again we find an appeal to sufficient reason. In *Principles* 2.51, Descartes claims to prove his sixth law of collision:

> that if the body C is at rest, and perfectly equal in size to the body B, which moves toward it, it is necessary that it will be in part pushed by B, and in part rebound from it.[29]

Descartes "proves" this law by considering the possibilities which could result from such a collision: either B will push C without rebound-

185

ing (and transfer to it half its motion—by Descartes' other laws which he takes to be established); or B will rebound completely from C, and retain all its motion; or

> finally . . . it rebounds retaining one part (of half its motion) and it pushes transferring the other part (of half its motion) (it keeping half for itself). It is evident that, since they (the bodies) are equal, and thus there is no more reason why it should rebound than push C, these two effects must be shared equally, that is, that B must transfer one (part of half its motion) to C and rebound with the other.

The form of the argument is quite clear. Since there is no reason why B should do one thing or another, it does both in part.

Since the principle of sufficient reason is used in various forms, it would be well to ask how can Descartes appeal to such a principle? He nowhere lists it as among the eternal truths that can be derived from the descriptive attributes of God, which is reasonable since, in all probability, it cannot be so derived. He cannot derive it as Leibniz does from a final cause of God, i.e., that He created the best of all possible worlds (cf. *Discourse on Metaphysics*), because Descartes has eschewed any talk of final causality. But these are the only two places from which he could get it. So he must be using it without any justification.

The principle of sufficient reason involves final causes at two levels. First, it is the principle or law used to determine the final state from a previous state. This provides the needed connection between cause (previous state) and effect (final state). But considered as formal or material cause, the description of the final state has no connection to its antecedent cause. Only when sufficient reason is invoked to show the determinate end of the process do we have an explanation, but this is to treat that end as a final cause. Secondly, the principle of sufficient reason must be justified on teleological grounds. Descartes may try to escape such a conclusion by merely saying that the principle of sufficient reason is reasonable; i.e., it is reasonable to suppose the equal division of motion during such collisions. But such reasonableness seems justifiable only by appeal either to God's manner of creation or to useful patterns of human reason that work because God created the world and man in accord with a certain design.

The general way of establishing the problems with all these physical arguments can be seen by considering another of Descartes' methodological passages. As mentioned previously, despite Descartes' explicit refusal to countenance final causes, either of God's purposes, of Aristotelian forms, or magical animism, it is clear that he still recognizes a principle of causal explanation. The principles one must use to explain the universe, the principles of philosophy, must be causal principles. Though he attempts to restrict himself to efficient and certain formal causes, Descartes is still very much under the influence of Aristotle, and his explanatory principles are to be like Aristotelian *archai* and figure fundamentally in causal explanations.

Descartes explains the causal character of explanations in the *Discourse on Method,* part 6. Descartes writes:

> First, I tried to discover the general principles, or first causes, of all that is or can be in the world, without for this purpose considering anything but God alone, its Creator, and without deriving these principles from anything but certain seeds of truth which are naturally in our souls. (52)

These "certain seeds of truth" Descartes refers to are those clear and certain ideas he arrived at through the method of doubt, and ones derived therefrom. In this context these clear and certain truths are just the eternal truths, God's existence, and the ideas of his attributes (plus, perhaps, mathematics). Descartes says that from these first principles he "discovered the skies, the stars, an earth, and even, on earth, water, air, fire, minerals, and certain other things, which are the simplest and commonest of all, and thus the easiest to understand" (52).

Now this latter passage is unclear between two interpretations: first, that these were discovered by reintroducing the material world (as he does in *Meditation* 6) and then by having clear and distinct ideas of these different things; or second, by deducing the laws of motion and then the corpuscular nature of matter (in its three forms), and then the existence of certain characteristics of those which we call stars, sky, earth, and so on. Unfortunately, what follows in the second movement of the method does not clarify this ambiguity.

To present the second part of the method Descartes writes:

> When I wanted to descend to those effects that were more particular, so many diverse ones presented themselves to me that I did not believe it possible for the human mind to distinguish between the forms or species of objects that are on earth, and an infinity of other ones which could have been if it had been the will of God to put them there. . . . the power and nature is

so ample and vast, and these principles so simple and general, that I almost never notice any particular effect such that I do not see right away that it can be derived from these principles in many different ways; and my greatest difficulty is to discover in which of these ways the effect is derived.

The problem, as put by Descartes here, has two aspects. In the first few lines, he seems to be saying that from these general principles (whatever they are) so many different conclusions can follow that there is no way, by reason alone, to tell which conclusion God, in fact, made to occur. When he talks from the effect end, he claims that by looking at the effects he can think of innumerable ways of deducing those effects from the general principles. In syllogistic parlance, the first lament concerns the possibility of too many conclusions, while the second concerns the possibility of too many middle terms (or causal chains). Let us take it then that the solution he proposes must do both jobs, viz., secure the proper conclusion and the correct middle. He proposes the means to such security as follows:

> And to do that I know no other expedient than again to search for certain experiments which are such that their result is not the same when we explain the effect by one hypothesis, as when we explain it by the other. (52)

One must, then, for Descartes find an experiment wherein the result sanctions both a given conclusion and a particular middle term of a given hypothesis. This method is avowedly causal, in that explanations proceed from the first cause, God, and that part of the search is for middle terms. There is meant to be no teleological element in such demonstrations. The bridge between God and the world is meant to be simple and, presumably, to involve only his concordance and the manifestations of his properties (cf. above, p. 182). But no experiment can provide this step and give us such a middle. Experiments, unless I misunderstand them in Descartes, can only select between conclusions. He says experiments make most of the effects certain (61). From where does one get these middle terms? They do not seem dictated by the properties of the first cause, i.e., the attributes of God, nor are they obtainable by experiment or experience simpliciter. What is missing are the connecting links describing the manner in which God works in the world and proper descriptions of the products or effects of this operation, descriptions in terms of formal and final causes. The middle terms sought for are to be essences or formal causes of the basic substances (God, mind, or matter).[30] These must then be linked up with what is experienced. The resulting links, the proper middles, are not always knowable with absolute certainty. In non-mathematical cases we only have moral certainty. It is for this reason that Descartes calls such premises, hypotheses.[31]

Descartes' program of giving metaphysical grounds for his physics always runs into the same problem. To bridge the gap between God and phenomena necessitates the use of final causes. To see how this problem arises in epistemological arguments and how epistemological considerations enter into physical arguments and into the general justificatory scheme, it is worth looking at Descartes' second law of nature. The second law of nature states: "Any given piece of matter considered by itself tends to go on moving, not in any oblique path, but only in straight lines" (*Principles* 2.39). Descartes purports to derive this from the immutability and simplicity of "the operation by which God preserves motion in matter." This would be to treat the "inclination" of matter to proceed in a straight line as a final cause of matter. Indeed, later, in *Principles* 3. 55–56, Descartes uses the terms 'inclinatio' and 'conatus' to describe the tendency of bodies moving in a circle to recede from their centers. But here Descartes does not argue this way.

Concerning an example of a stone being swung in a sling, he says, "It cannot be imagined that the stone has any definite curvilinear motion . . . none of the curvature can be conceived as inherent in its motion." But one might reasonably ask, Why not? Descartes seems to have two possible answers. First, explicitly he appeals to observation to show that if the stone leaves the sling, then it will continue in a straight line. This fits the general method set out in the *Discourse,* which I quoted above. On this reading, observation would be used to select between two possible effects of God's operation. God could have continued the stone in either a straight or curved path. Observation shows that he chose the straight path. But this does not show the *inconceivability* of curvilinear motion or its simplicity. One still needs a bridge from God's simplicity to this effect. The second possibility derives from considering straight motion versus curved motion and thinking back upon Descartes' commitment to geometry. Presumably, in analytic geometry it can be sensibly said that a straight line is simpler than a curved line, since the former only needs two points to determine it while the latter needs at least three.[32] Thus, using analytic geometry as the determiner as to what is simpler, we can say straight motion is simpler than curved. But it is problematic why analytic geometry should be simpler to God, or why he should consider two points more simple than three, given his infinite wisdom. Further, this possibility is not mentioned by Descartes in the *Principles,* and it could not be used to explain why curvilinear motion is inconceivable or unimaginable. Only upon the hypothesis that analytic geometry provides a model for human conception and correspondingly is the manner in which God has chosen to fulfill his purpose, creating man as a rational creature, does this argument succeed.[33] This again would involve us in attributing to God final causes regarding his creation of man and man's knowledge of God's effects in this world.

The assumption of purpose and design in the answering of epistemological questions is common in Descartes. One is often confronted with the implicit assumption that the world and men in it are designed so that men can come to know the world. This is said to be a manifestation of God's goodness. Let us look at some other cases.

Earlier, in the first section, I examined Descartes' argument that extension is the essence of matter. At that point I only mentioned the premise that God is no deceiver. In full, the reconstructed argument must run: Only extension is always perceived when one perceives matter. Therefore, one cannot imagine (in Descartes' imagistic sense)[34] matter without extension nor extension without matter. God is good and is no deceiver. Therefore, God would not create apart, in fact, what we cannot imagine apart (or He would be a deceiver). Therefore, by definition there is only a distinction of reason between matter and extension. But distinctions of reason hold only between substance and attribute, so matter and extension are related as substance and attribute.[35]

In the argument, God's essence (or goodness) is used to justify a claim about what was created in the world (or what was not created). This is then used to show that the world is as we perceive it. This is Descartes' general line of justification for clear and distinct ideas; because God created them in such a way, they are clear and distinct for us (elsewise things would be contrary to his goodness) (*Med.* 5, AT, 7.70–71; HR, 1.184–85).[36] Descartes is quite clear that it would be unwarranted to infer from this that God created the world in a certain way (e.g., with extension as the essence of matter) for our purposes. In many passages, he points out that "it is by no means probable that all things were made for our sake" (*Principles* 3.4).[37] But Descartes does not, and cannot, rule out that we were created by God in such a fashion that we are able to learn about the world. So while it is not true that God created the world as it is for our sake, it is true that we are the way we are (capable of knowledge) because He created us and the world according to a certain design. This design is not necessary for God, but it is necessary if Descartes is to confute the skeptic. If we are to have certain knowledge of essences of things in the world, then the world must have been created by God in such a fashion that its essences "fit" with our modes of perception and cognition. This means that the premise in the argument above concerning God's goodness implicitly includes a design component. Such design factors are aspects of the manner of God's operation in the world and serve as final causes with respect to God and ourselves as knowing beings.

I think all of Descartes' epistemological arguments are affected by the implicit recognition of God's design (with the exception of the cogito argument and the arguments for the existence of God). This design infects the pure use of formal causes and allows them to function in ways traditionally assigned to final causes. It is worth pursuing this a bit not only to make my point more effectively but because it has not always been noticed how many of Descartes' physical arguments have epistemological components.

At the end of the first section I discussed Descartes' treatment of translational motion in a purely formal manner. In that discussion, I only considered the surrounding bodies themselves and attributed rest to them in an absolute way. But in his actual definition of translational motion Descartes used the phrase "these being regarded at rest" to describe the bodies immediately touching the piece of matter in question.[38] The introduction of the mental act of regarding into the definition of motion is another kind of instance where Descartes implicitly recognizes the necessity of final causes (or a teleological, purposive component) in his mechanical explanatory model.

The reason, I conjecture, why Descartes finds that he needs the phrase "regarded at rest" is to avoid a regress. If the phrase read "at rest," then to determine the rest of the surrounding bodies one would have to look at those bodies that surrounded them. But to determine of these whether or not they were at rest, one would have to go to the next layer of surround,

191

ad indefinitum. To avoid this, Descartes introduces the mental act of regarding the surrounding bodies as being at rest. This means that the nature of motion is not solely determined by the properties of matter. Mental activity is included in the definition, for without it one would be unable to attribute to the surrounding bodies the property of rest.[39] Only in this way is it possible to attribute rest to them independently of the surrounding bodies. Regarding is an act of perception and as such is an intentional activity and requires a specification of the object of regard. This specification must be determined by reasons external to the act itself, for there is nothing in the nature of the act that specifies the object or properties of it. The object is described according to the purposes of the person doing the regarding or for some higher teleological reason. Descartes explicitly recognizes this feature of perceptual acts, as I shall show below. In sum, then, the force of this addition is to show the necessity of considering the mind and its purposes when describing the material world.

Descartes himself seems to recognize this and to accept it. In *Principles* 2.30, he discusses a case of two small bodies moving in opposite directions upon the surface of the earth and attempts to describe this case in terms of the traditional definition. He says, "It is impossible to ascribe [*nec . . . possit intelligi*] it [motion] to the Earth as a whole on account of the translation of some of its parts from the neighborhood of small contiguous bodies" (2.30). The reasons he gives for this impossibility are two, though the connection between them is unclear. First, if the two bodies were moving in opposite directions, we would have to attribute contradictory motions to the earth. It is possible that God could create things in such a way, but if He did so, we humans could not have knowledge of them. Such a contradictory state of affairs would be incomprehensible to us. Descartes always argues against such possibilities on epistemological grounds, using the premise that God is no deceiver.[40] This impossibility is inferred from the contradiction that would arise between God's goodness and our abilities as knowing beings. Ruling out such contradictions assumes a design component of the type mentioned earlier.

This design component is attested in the second reason Descartes gives for the impossibility. He says that we ascribe motion only to smaller bodies and not to the earth "to avoid too great a departure from the ordinary use of language." So, for the purpose of preserving ordinary language, and, presumably, intelligible communication and thought, we must regard the earth at rest.

Attributing such a large role to the mental act in Descartes' conception of motion may seem a bit overworked and unfair, especially since in the very next sentence of the above example he points out that "the real positive character of moving bodies . . . is also fully found in the other bodies contiguous to them [in this case the Earth] although these are regarded only as being at rest." But this seems to take back what he already has been forced to admit, i.e., that regarding and thus purpose is an essential part of the definition of motion. Without seriously taking this into account, description and knowledge are, in Descartes' own words, impossible.[41]

In very general terms, the teleological character of perceptual acts is explicitly set down by Descartes in his account of vision (in *Meditation* 6). Descartes finds he must refer to God's purposes in explaining the nature and character of perception. In *Meditation* 6, Descartes discusses "what nature teaches us" with regard to the senses. Earlier, reason caused us to set aside these teachings. Now he wants to assure us, from the fact that God is no deceiver, that we can trust—to an extent—the teachings of nature.

He remarks that perception is a passive faculty that must be caused by an active power. This causing power must be material objects for it is neither I nor God (since we are inclined to believe that material objects cause our perceptions and God is no deceiver in this regard). Again he says that, exclusive of mathematical and other clear and distinct properties of material objects, things could be otherwise than I perceive them to be. But since God is no deceiver, I can arrive at the truth about such matters.

At this point, Descartes again considers the things nature teaches us. He says he means by *nature* either God Himself or the order and position which God established in created things (AT, 7.80; HR, 1.192). It is this order and position he now explains and describes in teleological terms; i.e., Descartes finds the veracity of perception by talking of the usefulness of it to the preservation of the body-soul complex. The nature of perception is explained by Descartes by specifying the purpose for which the perceptions are placed in us. Indeed, he goes on to say that

God could have done things differently; he could have constituted man in other ways, "but none . . . would have contributed so well to the conservation of the body" (AT, 7.88; HR, 1.197).[42]

This is specifically elucidated by Descartes. He considers perceptions of properties that are non-mathematical in character, i.e., are not of extension or things entailed by extension. Why do we see these sorts of things? He answers:

> These perceptions of sense have been placed within me by nature merely for the purpose of signifying to my mind what things are beneficial or harmful to the composite whole of which it forms a part. . . . (AT, 7.83; HR, 1.194)

Here is teleology and a final cause of a rampant sort. The sensations or qualities we perceive are there because of their usefulness to the perceiving system. Further, Descartes says they were placed there with this end of utility in (God's?) mind.[43] The same point comes up again later in the *Meditations:*

> Finally . . . since each of the movements which are in the portion of the brain by which the mind is immediately affected brings about one particular sensation only, we cannot under any circumstances imagine anything better than that this movement, amongst all the sensations which it is capable of impressing on it, causes the mind to be affected by that one which is best fitted and most generally useful for the conservation of the human body when it is in health. . . . (AT, 7.87–88; HR, 1.197)

The question Descartes is implicitly answering here is, What is the purpose of God as reflected in what nature teaches us with regard to sense perception? His answer shows that he uses God's purposes, specified teleologically in terms of bodily preservation, as a middle term to provide the link between God's not being a deceiver and the fact of our perceiving heat, colors, and the like. But this is what Descartes cannot allow himself, for he has stated that we cannot know God's purposes and that we should not make use of final causes.

In the discussion, Descartes commits himself to a functional system. Since, by Descartes' principles, the regulator or creator of the system cannot be our minds and their intentions (cf. *Meditation* 6), the only possibility left for him is that God planned it that way. This does not follow logically, of course, as is shown by the possibility that one can believe in functionalism of this sort (though without the dualism) and still (a) only be committed to purely causal explanations (i.e., no emergence thesis or any such), and (b) hold a materialist position on the mind/body issue.[44] But Descartes cannot, because of his commitment to the independent substantiality of minds and his requirement that God is a first cause (in the way he uses it), opt for any of these harmless versions of teleology or final causes.

In this essay, I have shown how Descartes' work makes extensive, though unacknowledged, use of final causes and teleological factors as traditionally conceived. Historically, I think this points up the way in which Descartes was a transitional figure between the older Aristotelian or Neoplatonic conceptions of natural philosophy and those physical or mechanical views that were to follow. In the thinkers following after Descartes, science is still conceived as part of natural philosophy, but it becomes increasingly easy to isolate the more physical parts from the more theological, metaphysical axioms. A word of caution is in order here for, I believe, no science can ever be free of "metaphysics", and philosophical concerns figure prominently in later works. Descartes kept much of the older tradition, and we are only now beginning to find out what the tradition was. He also tried to throw out the parts he was unhappy with, replacing them with new conceptions, e.g., of God and epistemology. The result was not the coherent system he sought. As I have shown, he was unable to give up all that he wished, but the effort behind the breaking of the traditional patterns carried his thought in new directions and was influential for those who followed. Philosophically, there is a moral in Descartes' failure. I think it points to the necessity for using final causes in causal explanations. Today we might call such causes semantic or functional relations. I have not argued this point here, but it seems that any causal account of scientific explanation or of epistemology must take it seriously.

I wish to thank Wallace Anderson for discussing with me various aspects of Descartes' work during 1973, and Wade Robison and Peter Heimann, for critically commenting on a version of this essay. In addition, I must thank those who argued with me during the conference; especially among them I recall the taunts of Ted McGuire and the questions of Ernan McMullin.

1. Etienne Gilson, *Etudes sur le rôle de la pensée Médiévale dans la formation du système cartésien* (Paris: Vrin, 1930); and E. Gilson, *Index Scolastico-Cartesian* (New York: Burt Franklin, originally 1912); cf. N. Kemp Smith, *New Studies in the Philosophy of Descartes* (London: Macmillan, 1963), esp. chap. 7.

2. Edward Collins has a long essay on Descartes that has this point as a central theme.

3. Work on this aspect of sixteenth-century thought was done by Richard H. Popkin, *The History of Scepticism from Erasmus to Descartes* (New York: Humanities Press, 1968). Recently, very interesting work has been done by Charles Schmitt in "The Recovery and Assimilation of Ancient Scepticism in the Renaissance," *Revista Critica di Storia della Filosofia 4* (1972): 363–84. Schmitt has promised more work on the topic.

4. A start was made in my study of Kepler's explanatory scheme, "Kepler's Virtue," forthcoming.

5. When referring to the *Meditations,* I shall give in the text the reference to the volume and page of the Charles Adam and Paul Tannery standard edition, *Oeuvres de Descartes* (Paris: Vrin, 1956 ff.), cited hereafter as AT, and then the reference to the volume and page of the English translation by E. Haldane and G. Ross, *The Philosophical Works of Descartes* (Cambridge: At the University Press, 1911), hereafter cited as HR. I shall also refer to HR for many other texts.

6. When referring to *The Principles of Philosophy,* I shall cite in the text the book and section. These can be looked up in AT, volumes 8 and 9, the Latin and French editions of *Principia Philosophiae.* Often an English translation exists either in HR 1, or in E. Anscombe and P. Geach, *Descartes' Philosophical Writings* (London: Nelson & Son, 1954).

7. The edition of Descartes' optics I am using is the excellent translation found in Paul J. Olscamp, *Descartes' Discourse on Method, Optics, Geometry, and Meteorology* (Indianapolis: Library of Liberal Arts, 1965), hereafter cited as O. I shall also use this edition for the text of *The Discourse on Method.*

8. There is a better argument that he might have used against color as essential; viz., a man blind or with his eyes closed can perceive a material body by touch, but cannot perceive its color.

9. This argument is quite Cartesian, for Descartes claims that having a shape entails having extension (letter to Gibieuf, 10 January 1642, AT, 3.474); see also A. Kenny, *Descartes: Philosophical Letters* (Oxford: 1970) (hereafter cited as K), p. 124. Further, he argues that shapes are modes of the attribute extension in *Notes Directed against a Certain Programme,* AT, 8:348–49; HR, 1:435. Cf. also *Rules for the Direction of the Mind,* Rule 12, on the role of figure, AT, 10: 413 f.; HR, 1: 37 f.

10. To get from the conclusion that "if matter is to be perceived at all it must be perceived as extended" to "extension is the essential property of matter" needs further argument of an epistemological character. Roughly, one would have to argue that to be an essential property it is necessary to be a perceived property. I think Descartes does hold such a position and does at least implicitly argue for it, but I shall not attempt to ferret out the argument here. (This premise also comes up earlier in *Meditation* 2 concerning the properties of mind.) Also see below, where I take up this argument again.

11. Cf. *Le Monde* and the *Optics* (O, p. 65).

12. In *Le Monde* the discussion of tridimensionality is quite reminiscent of the light ontologists of the thirteenth century. See *Le Monde,* AT, 11:33, 39–40, 53.

13. Anderson brings out in his paper in this volume that the identity conditions for Cartesian bodies are quite problematic. Cf. also Kemp Smith, chap. 8. It is not really a surprise, for in the *Meteorology* Descartes explicitly brought up the particle hypothesis.

14. *Principles* 2.43. The other factors in the tendency to persist are the speed of the body and the kind and degrees of opposition involved when it collides with another body. Surface is an important factor when Descartes discusses motion in liquids, *Principles* 2.56.

15. See, for example, *Replies to Objections,* 4, AT, 7:222; HR, 2:98.

16. See note 7; I use Olscamp's translation of *The Discourse* for reference.

17. In his reply to the fourth set of objections by Arnauld, Descartes explicitly discusses the idea of boundary or superficies in the context of problems with the Eucharist. He defines boundaries or superficies as not being part of their substance, "nor indeed any part of the quantity of the body, nor

even part of the circumjacent bodies, but merely the limit which is conceived to lie between the single particles of a body and the bodies that surround it, a boundary which has absolutely none but a modal reality'' (AT, 7:250–51; HR, 2:118). That Descartes accords with my argument is nicely brought out in this section, for he says that when the bread changes substantially into the body of Christ, it can be said to exist in the same boundary even though ''it is not there in the proper sense of being in a place.'' I take it this means that since the substance has changed, properly speaking the place has changed.

18. ''That extension is the common nature of corporeal substance is a doctrine taught with great confidence by many, but never proved. Certainly neither motion, that is, action, nor resistance or passive force, can be derived therefrom. Nor can the laws of nature concerning motion and the shock of bodies be derived from the concept of extension, as I have shown elsewhere'' (G.W. Leibniz, ''Critical Remarks Concerning Descartes' *Principles*,'' ed. P. Schrecker [Indianapolis: Bobbs-Merrill], p. 37). Leibniz, in the *Journal des Savans*, first in 1691 and again in 1693, attacked the adequacy of Cartesian explanations based solely upon extension. As I discuss below, this is not quite fair, for Descartes himself recognizes the necessity of causal explanations. But Leibniz rightly points out that *vis inertia* cannot be identified solely with extension of bodies (or even extension plus surface) for the results contradict experiments. This argument presumably must be taken seriously by Descartes given the explanatory scheme he lays out in *The Discourse* (which I shall discuss below). I agree with Leibniz that some conception of force is necessary, as Descartes himself tacitly admits, but I cannot agree with Leibniz' remark (in the 1693 paper) that Descartes makes satisfactory use of God as general, causal hypothesis. Also Leibniz fails to remark Descartes' use of teleological elements.

19. Kemp Smith, pp. 212 f.

20. This point is repeated in *Principles* 1.28 and in Reply to Fifth Objections (Gassendi), AT, 7:374–75; HR, 2:223.

21. James Collins seems to have his finger on an aspect of this, though I am unsure exactly how it fits with what I have written. In *Descartes' Philosophy of Nature* (American Philosophical Quarterly Monograph, No. 5, 1971), he characterizes Descartes' views of nature as threefold: theistic, material, and formal or law-like. This seems analogous to the three causes other than final, for the efficient cause is God the creator, the material cause is extension, and the formal cause is the laws of motion (pp. 16 f. and 48 f.). Collins argues that it is the complex relations between these that allow Descartes to avoid a ''facile rationalistic mechanism'' (60).

22. Aristotle, *De Generatione Animalium*, 715a4.

23. Kemp Smith, pp. 109–11 and 202 f., and Gilson, *Pensée*, 224 f.

24. Cf. Clement C. J. Webb, *Studies in Natural Theology* (Oxford: 1915), pp. 249, 276; also, Kemp Smith, pp. 174 f.

25. See N. R. Hanson, *Patterns of Discovery* (Cambridge: At the University Press, 1958), chap. 3, or Gerd Buchdahl, ''Semantic Sources of the Concept of Law,'' *Boston Studies in Philosophy of Science III* (Dordrecht, Holland: Reidel, 1967), pp. 272 f.

26. For example, J. Kepler, in his *Epitome of Copernican Astronomy* (GW & 7:298–99), or Fr. Eustachio a Sancto Paulo, in his *Summa philosophica quadre partita*, 3.64–66 (quoted in Gilson, *Index*, p. 39). Both of these thinkers were read by Descartes.

27. If everything in the world were constant, this premise would not be needed, but the power of Descartes' explanation would likewise be decreased. Even so, in such a case constancy would be both a final and formal cause.

28. G. W. Leibniz writes on need for final causes in *Discourse on Metaphysics*, 21, 22, and *Specimen Dynamicam*. See also McGuire's paper in this volume.

29. The treatment of motion in the non-translational way can be seen in this passage. For on Descartes' definition of real and true motion, there would be no more reason to say that B moves toward C—which he says—than to say C moves toward B. It is just this problem that turns the two

parts of the third law of motion into contradictories. See R. J. Blackwell, "Descartes' Laws of Motion" *Isis* 57 (1966): 220 f.

30. In his letter to Mersenne, 16 June 1641 (AT, 3.382–83; K, p. 104), Descartes says that the best proofs are those in which the middle terms state essences of ideas innate in our mind.

31. Cf. *Principles* 4.204–6. For an interesting discussion of the nature of hypotheses and their uncertainty, see Gerd Buchdahl, *Metaphysics and the Philosophy of Science* (Oxford: Blackwell, 1969), pp. 119–23. Buchdahl discusses this passage from *The Discourse* from another point of view. My point remains, such middles must serve as final causes whether or not they are known with certainty.

32. Descartes does use this argument in *Le Monde,* AT, 11.44–55. It is also put forward by R. S. Westfall, *Force in Newton's Physics* (London: MacDonald, 1971), p. 59. Westfall does not consider any difficulties with the suggestion. David Hahm tells me that Proclus used such an argument for the simplicity of straight lines, but I have been unable to see the reference.

33. This seems to be the line Descartes takes in *Meditation* 5 when he discusses the role that mathematics plays in our knowledge of material things.

34. See *Meditation* 6, AT, 7:71–73; HR, 1:185–87.

35. Descartes discusses this distinction in *Principles* 1.62.

36. This point is much discussed by Descartes. It comes up in almost all the Replies to Objections, and also clearly in *Principles* 1.30, in his letter to Regius, 24 May 1640 (AT, 3.64; K, p. 73), and in his letter to Gibieuf, 19 January 1642 (AT, 3:474; K, p. 123).

37. Also see his letter to Hyperaspistes, August 1641, point 10 (AT, 3:431 f.; K, pp. 117 f.), and to Chanut, 6 June 1647 (AT, 5:53–54; K, p. 220). This point is discussed by Collins, *Descartes' Philosophy of Nature,* pp. 89 f.

38. The Latin reads "Tanquam quiescentia specantur," AT, 8:53. The French edition has "et nous considerons comme en repos," AT, 9:76.

39. I am aware that I have shifted from discussing what is true of bodies and motion to talking about what one can say about, or attribute to, bodies and their motions. I do not think this shift problematic here, for Descartes is ultimately concerned with what we as human beings can know (and, thus, properly say). While Edward Collins has argued that for Descartes there are some unknowable properties of matter (non-essential properties), these are not of use in building a science. Thus, I and Descartes talk interchangably about what is true, what is known to be true, and what is truly descriptive of bodies and their motions. The only qualification on this is that God could have chosen to create the state of affairs which we describe as a contradiction, but though this is possible Descartes says it would be unintelligible to us.

40. E.g., Letter to Mesland, 2 May 1644, AT, 4:118–19; K, p. 151. See also Kemp Smith, pp. 177–88.

41. Another instance of this impossibility that I shall not mention below is that discussed by R. J. Blackwell in his paper "Descartes' Laws of Motion." Blackwell points out that the third law of nature only makes sense if one regards one of the bodies as being really at rest. If this regarding at rest is left out, then the first and the second part of the law are straightforwardly interpretable in a way that leads to a contradiction; i.e., contradictory predictions about how the bodies in question will behave. See R. J. Blackwell, "Descartes' Law of Motion."

42. This relation between perception and material objects, for the purpose of preserving the body-soul union, is taken by Descartes as a manifestation of God's beneficence and a relation that God chose freely. The contrast between Descartes' "voluntarist" conception of the way God structured the relation between material objects and perceptions of them and the later Malebranchian rational relation is set out by Martial Gueroult, *Descartes: selon l'ordre des raisons,* 2 (Paris: Aubier Montaigne, 1968): pp. 24–25.

43. This same general argument and teleological position is repeated in summary form by Descartes in *Principles* 2.1–3, and mentioned again in a letter to Henry More, 5 February 1649, AT, 5:271; K, p. 240.

44. The way out of treating purposes as animistic is to take some sort of evolutionary, adaptive approach concerning the development of the (perceptual) system. Cf. J. J. Gibson, *The Senses Considered as Perceptual Systems* (New York: Houghton Mifflin, 1969), and D. C. Dennett, *Content and Consciousness*, chap. 2 (London: Routledge & Kegan Paul, 1971).

Wallace E. Anderson

Chapter Eight

CARTESIAN MOTION

Alexander Koyré gives a succinct account of the Cartesian program in physics. "The Cartesian universe," he writes, "is constructed with very few things. Matter and motion; or better—since Cartesian matter, homogeneous and uniform, is only extension—extension and motion; or better still—since Cartesian extension is strictly geometrical—space and motion. The Cartesian universe, as best understood, is geometry realized."[1]

Although the Cartesian universe is thoroughly geometrical, Descartes does not construct it merely as an exercise in geometry; it is intended to be an adequate representation of the actual universe. Hence it includes various kinds of celestial and terrestrial bodies organized with a due order, motions, and regular interactions, into an entire system corresponding to what Descartes calls the "visible world"—the system of fixed stars, the sun and planets, the comets, and the earth and bodies upon it. The parts and features of this system are to be explained by constructing an adequate representation of them by means of the concepts of matter and motion alone. His detailed account of the "visible world" begins, not with these fundamental concepts, but with hypotheses about particular kinds of bodies that move and affect each other in various ways. Matter and motion are called upon primarily to construct the concepts employed in these hypotheses. In this paper I shall be concerned with the way in which Descartes forms the concepts of an individual body, its particular motion, and the manner in which it affects, and is affected by, other bodies.

Of the two principles upon which Descartes bases his account, matter appears to be the simpler and more intelligible. As Koyré suggests, since it is defined solely in terms of extension, Cartesian matter is explicable in purely geometrical terms. The Cartesian concept of motion, however, is far more problematic and obscure. It was attacked at various points by Leibniz, Newton, and others in the seventeenth century. Although recent commentators have noted paradoxes in connection with Cartesian motion, the source of the difficulties remains obscure. In this paper I shall attempt to throw light on the problems associated with Cartesian motion by showing how Descartes' physical theory incorporates not one but three different concepts of motion. Nor are the ambiguities present because of inadvertence, but result from Descartes' use of the concept in three distinct roles. He appeals to motion conceived metaphysically in order to account for the existence and status of individual bodies; he accounts for motion kinematically in connection with a general theory of bodily change; and he conceives motion dynamically in developing an account of physical causation. Motion is required in all these roles, for matter considered simply as extended substance proves to be an insufficient basis for any of them. At the same time, Descartes notably fails to distinguish the various concepts of motion suited to these different roles, and so fails almost entirely to recognize problems in the relations among them. Instead, he is strongly inclined throughout his discussion to treat motion as a single, simple concept.

Descartes' metaphysics is particularly distinguished by its radical dualism. There are two kinds of substance, thinking substance and extended substance, mind and body. Each comprises a part of reality that is wholly independent of the other, though both depend immediately upon the concurrence of God. None of the attributes of body can belong to mind, nor can any of the attributes of mind belong to body. Such notions as "effort", and the distinction between the "natural" and the "violent", since they pertain to the affections of mind, are banished from Cartesian physics. Neither does body, considered as extended substance, have any of the properties that are discernible by the senses.

In 1633, in his suppressed treatise Le Monde, Descartes contrasts his conception of body with that of the Aristotelians. Divested of all specific forms and sensible qualities, they consider it to be a wholly unformed primary matter, a mere potentiality. He, on the contrary, takes matter to be a real and enformed being, having the single attribute of extension in three dimensions. As such we can suppose that matter has an actual existence. In order to secure the supposition without confusing it with our ordinary conception of matter, Descartes asks us to imagine our-

201

selves transported to a remote region of space where God has created such a matter. There we will suppose it to be

> a true substance, perfectly solid, which uniformly fills all the length, breadth, and depth of that great space in the midst of which we have stayed our thought, so that each one of its particles always occupies a portion of that space so related to its magnitude that it could not fill a greater, nor contract itself into a less, nor allow, while it remains there, any other to enter it.[2]

In this early presentation Descartes construes the extension of matter as the property by which it fills or occupies space. We are to assume no empty space; the parts of matter and the parts of space are considered to stand in a one-to-one correspondence. No part of matter can change with respect to the magnitude of the space it fills, and each part is impenetrable with respect to every other part. In the *Principia* Descartes develops the account of matter that is more commonly attributed to him, that matter *qua* extended substance constitutes space. Empty space, or extension without matter, he now holds to be logically impossible.[3] The assumption of empty space implies either that extension can exist without substance at all, or that extension is only an accidental property of material substance; and both of these consequences are taken to be logically impossible. It follows that matter is a uniform and continuous quantity, and that it is a real being, a single whole that is indefinitely extended, without any boundary or determinate size or shape.

> This world—the totality of corporeal substance—has no limits to its extension. Wherever we imagine the boundaries to be, there is always the possibility, not merely of imagining further space indefinitely extended, but also of seeing that this imagination is true to fact—that such space actually exists. And hence there must be indefinitely extended corporeal substance contained in this space.[4]

One of the most significant aspects of the Cartesian account of matter is that, though Descartes can speak of it as body, it does not as such consist in individual bodies, distinct extended things, each with a determinate size, shape, and other distinguishing properties.[5] Nor is the whole of indefinitely extended substance an individual body in its own right, since it has no determinate size or shape. Taken simply as extended substance, matter includes no properties but necessary ones,

202

which are reducible to, or are the necessary consequences of, extension. Descartes admits to Henry More that matter is tangible and impenetrable, but argues that these properties belong properly to its parts, and follow from their extension alone. Tangibility consists in the fact that every part of continuously extended substance must be contiguous with other parts. Impenetrability follows from the fact that "it is impossible to conceive of one part of extended substance penetrating another equal part without *eo ipso* thinking that half the total extension is taken away or annihilated; but what is annihilated does not penetrate anything else."[6] In any case, Descartes remarks, neither property belongs to matter conceived simply as extended substance: "Tangibility and impenetrability involve a reference to parts and presuppose the concept of division or termination; whereas we can conceive a continuous body of indeterminate or indefinite size, in which there is nothing to consider except extension."[7]

The logically prior conception of matter is thus wholly divorced from the idea of individual bodies and their distinguishing properties. It involves no accidental properties, nor even the notion of determinate parts. Even while he reminds More that it is nevertheless the concept of something real, Descartes pointedly suggests that it corresponds to our ordinary notion of empty space: "Remember nothing has no properties, and that what is commonly called empty space is not nothing, but real body deprived of all its accidents (i.e. all the things which can be present or absent without their possessor ceasing to be)."[8]

Hence, from the thesis that extension is the sole and essential property of matter, Descartes is led to a conception of matter that is properly associated with our ordinary notion of empty or unoccupied space. Matter by itself is logically insufficient to account for the existence of any individual bodies. In order to account for them, Descartes introduces motion in a metaphysical role:

> It is one and the same matter that exists throughout the universe; its one distinctive characteristic everywhere is extension. All the properties that we clearly perceive in it are reducible to divisibility and a capacity for varying motions in the various parts; from this there follows the potentiality of all the states that we see may arise from the motion of its parts. Merely mental partition changes nothing; all variegation of matter, all differences of forms of matter, depend on motion.[9]

As the final sentence of this quotation implies, the real existence of an individual body with a determinate size and shape depends upon the actual presence of motion in some part of extended substance. It is by the presence of motion in a part that it is "divided" from other parts of matter so as to constitute an individual body. The "division" of a part that has motion does not consist in its spatial separation from all other parts of matter, for this is impossible; it consists instead in the presence of a particular motion in that part as a distinguishing property of it. In *Le Monde* Descartes writes,

> Suppose that God has actually divided [matter] into many such parts, some greater, some smaller; some of one figure, others of another, whatever we may be pleased to fancy; not that, in doing so, he has separated them one from another, so that there should be any empty space between two of them; but let us suppose that the only distinction to be met with consists in the variety of motions he gives to them, in causing that, at the very instant that they are created, some of them begin to move in one direction, others in another; some more swiftly, others more slowly (or, if you please, not at all), and that they continue thereafter their motions according to the ordinary laws of nature.[10]

A comparison with contemporary seventeenth-century atomism is instructive. Both Descartes and the atomists hold that matter is extended, and they agree that extension alone is not sufficient to account for the existence of individual bodies. The atomists introduced solidity or impenetrability as a property distinct from extension and essential for individual bodies, thereby distinguishing them from the empty and penetrable extension of space. Descartes employs motion in a comparable role. Bodies are parts of matter that have motion and, consequently, other accidental properties; parts that are devoid of motion have no property but extension.

In other respects, however, Descartes' account differs substantially from the atomists'. Both speak of individual bodies resulting from a "division" of matter, but Descartes denies that any part of matter is indivisible by its own nature: "Even if God made it not to be divisible by any creatures, he could not take away his own power of dividing it."[11] And whereas the atomists supposed matter is divided into distinct bodies by spatial separation so that each body is individuated by its unique location in space, Descartes suggests that bodies are individuated by their particular motions, the motion of each differing from that of others with respect to speed and direction.

Descartes' general position, therefore, is that God has determined the fundamental topological features of the world by producing various motions in various parts of continuously extended material substance. The initial division of matter by motion, he maintains, is entirely arbitrary; we can suppose any distribution of any number of particular motions among the divisible parts that we please.[12] The presence or absense of a particular motion in any divisible part is conceived to be logically independent of the presence or absense of motion in any other part. At the same time, Descartes' own account of the division of matter appears to involve several additional assumptions which it would be well to bring out.

First, although Descartes argues that matter is divisible *in infinitum*, he does not think that God actually divided it infinitely. He assumes instead that all bodies are ultimately composed of elementary bodies or particles that, while they are further divisible, are not actually divided.[13] Second, each individual body is comprised by all the contiguous and divisible parts of matter contained within a single closed surface that share a particular motion. The matter of a body may also be divided into individual bodies, each having a particular motion of its own. "When I speak of *a* body or *a* piece of matter, I mean something that is all transferred [or moved] at once; this may, however, again consist of many parts with various motions of their own."[14] Accordingly, Descartes distinguishes between the proper motion of a part of matter, which belongs to it uniquely and distinguishes it as a single individual body, and the motion a part shares with others in virtue of which it is an integral part of an individual body. So a particular body will share in all the proper motions of all the bodies that contain it as a part; as the wheels of a watch that is in the pocket of a sailor in a ship tossing on the sea will share the motions of all these bodies as well as having its own distinct motion independently of these. "All these motions will really exist in the wheels; but since it is not so easy to conceive of so many all at once—indeed we cannot know of all of them—it will be enough to consider in each body only one motion—its proper motion."[15] Hence each body, considered as an individual body, is subject to only one motion by itself; but when considered as a piece of divisible matter containing many parts, it may be subject to many different motions simultaneously. The proper motion of a body, however, is considered to be logically independent of the motions of its component parts, and of those in which it shares by virtue of its being a part of other larger bodies.

Descartes' conception of the topology of the universe appears to involve two other assumptions, neither of which is explicitly stated. As I mentioned before, though material substance comprises a single indefinitely extended whole, Descartes does not regard this whole as an individual body. Accordingly, it would be a mistake to suppose that the whole of indefinitely extended substance could have a proper motion. Furthermore, Descartes does not seem to require that God has exhaustively divided matter into individual bodies. It is at least logically possible that a part of matter has neither a particular motion of its own nor shares in the motion of any other body as an integral part of it. On the contrary, in his developed account of the system of the physical world he seems to suppose that all the bodies it contains are included in a single closed region of extended substance. The parts of matter outside this region, he seems to assume, are not divided into bodies, and are devoid of motion.

Descartes' use of motion in this metaphysical role, to distinguish individual bodies from each other and from the parts of matter that are not bodies, involves treating motion as an accidental property. It is not a substantial reality, nor an essential attribute of matter, but only a mode or state of the parts to which it belongs. In general, a mode cannot be conceived apart from its subject, though the subject can exist and be conceived without the mode.[16] In commenting upon the status of motion, Descartes often compares it with shape, though in view of its metaphysical role motion should be considered more fundamental than shape. Yet, as the same body may have different shapes in successive times, so the same part of matter may have successively different motions. And a divisible part or quantity of matter may have no actual motion, as it may have no actual shape.

From the nature of Cartesian motion as conceived in its metaphysical role, it seems to follow that rest is the lack or privation of any particular motion, as shapelessness is the lack of any particular shape. Descartes implies this point in answering a query of Henry More, whether motion or rest is the natural property of matter. He replied, "If matter is left to itself and receives no impulse from anywhere it will remain entirely at rest. But it receives an impulse from God who preserves the same amount of motion or translation in it as he placed in it at the beginning.''[17] On the other hand, Descartes often implies the contrary view concerning rest. In *Le Monde* he attacks the scholastics for attributing to motion ''an existence much more substantial and real than they do to rest, which they say is merely a privation.''[18] Motion and rest are both to be considered qualities or modes of matter, he maintains, and laws of nature apply equally to both states. Descartes' conception of the relative status of motion and rest, as assumed in this formulation of the laws, has been the subject of much perplexity among commentators on Descartes' physics. We shall return to this problem presently.

The concept of motion as it is understood in its metaphysical role accounts for the existence of individual bodies and their properties, but this concept by itself does not account for change in the successive states of bodies. This point becomes apparent when we consider Descartes' account of motion in relation to its metaphysical cause, God. It is a central thesis of Cartesian metaphysics that what God creates exists only at the moment when he creates it; the creature cannot continue to exist by itself, but depends solely upon God's re-creating it at each successive moment that it continues.[19] The motion God creates, therefore, is only a momentary state of a part of matter, and as such does not imply continuation. It follows that the presence or absence of motion in a given part of matter at one time is logically independent of the presence or absence of motion in that part at any other moment. Change in the Cartesian world can thus be represented, not by the mere presence of various motions in its momentary states, but by differences in the motions found in the same parts of matter as God re-creates them in successive states.

This concept of change of state has a preeminent place in Descartes' account of the laws of nature, which describe how successive states of motion or rest are determined from preceding states. But although Descartes affirms that God is the sole cause of the motions in each state of the world, he denies that we can appeal to God as the cause of any changes of motion in individual bodies. Since God is immutable he is not himself subject to change, and hence he always creates by the same operation and brings about the same effect. Hence no changes in the world whatever may be explained by the manner in which God creates or conserves it. Change must rather be viewed as arising from the nature of

207

the creature, from matter and motion themselves. This position is never put forth more clearly than in a passage in *Le Monde,* in which Descartes introduces his laws of nature. It deserves quotation in full.

> By nature I do not here understand any goddess or any other sort of imaginary power, but I make use of this word to signify matter itself, in so far as I consider it with all the qualities I have attributed to it, taken as a whole, and under this condition, that God continues to preserve it in the same way that he has created it; for, from the simple fact that he continues thus to preserve it, it necessarily follows that there must be many changes in its parts, which not being, as it seems to me, properly attributed to the Divine activity—because that does not change—I attribute them to nature; and the rules in accordance with which these changes occur I call the laws of nature.
>
> In order the better to understand this, remember that among the qualities of matter we have supposed that its particles have had various motions from the instant of their creation, and, besides that, they are all in contact on every side, so that there is no empty space between any two of them; whence it follows of necessity that at the time they began to move they began to change also and to vary their movements as they encountered one another; and thus, even if God preserved them thereafter in the same manner as he created them, he does not preserve them in the same condition—that is to say, while God always acts in the same way, and consequently always produces the same effect in substance, there result, *as it were by accident,* many diversities in this effect. And it is easy to believe that God, who, as everybody ought to know, is immutable, always acts in the same way. But without involving myself further in these metaphysical considerations, I will lay down two or three principal rules in accordance with which it must be thought that God causes the nature of this new world to act, and which are sufficient, I believe, to enable you to comprehend all the rest.[20]

Descartes' evident impatience with the metaphysical difficulties involved in giving a coherent theory of change on his principles only underscores the importance of these difficulties. Change may not be attributable to God, since He is immutable. Although each successive state of the world follows directly from God's creative operation, the differences between successive states are not to be explained in the same way. Neither can change be explained by matter taken simply as extended substance. It must follow in some way from the motion that God produces in matter, and so will be seen to follow from God only indirectly, or as by accident. But again, no change follows from the mere possession of motion as a momentary state of a part of matter; this concept of motion has no further consequence than to differentiate that part as an individual body. In order to provide a theory of change, Descartes is constrained to incorporate motion in other roles. His analysis of change leads to conceptions of motion in two distinct roles, a kinematical and a dynamical, each of which differs in significant ways from the idea of motion in its metaphysical role. To this part of Descartes' theory of motion we must now turn.

The two conceptions of motion emerge as Descartes attempts to treat the nature of motion as it is exhibited by bodies distinctly from the causes of motion. Conceived kinematically, Descartes understands motion to be the change of location of a body. His definition of motion arises from an analysis of this notion as applied to the divisible parts of a continuously extended substance. In treating the causes of motion, Descartes considers motion in another way, in relation to the operation of God by which it is produced in bodies. In this context it is taken to be a property inhering in a body, and which is subject to change under certain physical conditions—when the body collides with another. Descartes usually treats these two conceptions of motion as equivalent; with regard to both he insists that motion is only a mode of a body or part of matter. Nevertheless, the two conceptions belong to quite different levels of explanation of change, each of which involves a different account of the causes and effects of change.

A central tenet of the Cartesian conception of physical causation is that all change is explainable by means of local motion. In *Le Monde* Descartes rejects the scholastics' claim that there are various kinds of motion that occur without a change of place; for example, motions as to form, as to quantity, and as to quality. He finds unintelligible the Aristotelian definition of motion provided to explain these kinds of change: "The act of a being in potency so far as it is in potency." In contrast, Descartes proposes the conception of the nature of motion assumed by the geometers, who suppose it to be "more simple than that of their surfaces and their lines, as appears in the fact that they have explained the line by the motion of a point, and the surface by the motion of a line." The clearest conception of motion, he goes on, is "the lines of the geometers, which bodies make in passing from one place to another and successively occupying all the spaces between the two."[21]

Descartes thus offers a geometrical conception of motion, as the path a body produces in successive times. The difference between matter in motion and matter at rest is that the former produces or describes a path while the latter does not. Furthermore, when the motion of a body is identified by means of the path, it follows that the body can have only one real or proper motion, even though we might imagine it to be subject to several motions at the same time. For example, a carriage wheel might be supposed to have a circular motion around its axle, and at the same time a rectilinear motion along a road; "but this distinction between motions is not a real one, as is obvious from the fact that any given point of the moving body just describes one given line."[22] Accordingly, the sense of a body's motion is to be interpreted as the shape of its path, the direction as its progress relative to a fixed starting point, and the speed as the length of segment that has been produced through a given time.

In the statement in *Le Monde* Descartes assumes that the path of motion is produced by a body's successively occupying a series of different places. As has been pointed out, in this early work he assumed a distinction between the parts of matter and the parts of space that is incompatible with his later view. From the strict identity of space and extended substance that is asserted in the *Principia,* it follows that every part of matter is identical with the space it occupies, or with its "intrinsic place." Consequently it is contradictory to speak of a body changing its place; when a body moves, its place must move in the same way.[23] Neither can motion be defined as a change of relative position, Descartes argues. The relative position of a body is determined by us with reference to other bodies that we consider to be at rest. Since it therefore depends upon which bodies we take to be at rest, there is no unique relative position or series of positions for a body.[24] But every body is surrounded by a unique set of other contiguous bodies or parts of matter, and Descartes frames his definition of motion *qua* change of location in terms of this set. Motion is "the translation of a piece of matter (a body) from the neighborhood of the bodies immediately touching it, these being regarded as at rest, to the neighborhood of others."[25]

Descartes' conception of the translation of a body is an evident attempt to accommodate the geometrical idea of motion as expressed in *Le Monde* to the doctrine that space and matter are identical. Each body can be considered to have only a single motion, and not several simultaneously, "for only one set of bodies can be contiguous to a given moving body at a given moment of time."[26] The path of translation may thus be defined as the series of sets of contiguous bodies from which a given body has been translated. Descartes also speaks of translation as the motion by which parts of matter are individuated: "When I speak of *a* body or *a* piece of matter, I mean something that is all transferred at once."[27] In thus identifying translation with motion as it is understood in its metaphysical role, Descartes also argues that translation is only a mode of a body.

> I say that motion is translation, not the force or action of transference, to show that motion inheres always in a moving body, not in the body that moves it (an accurate distinction between these two things is not ordinarily made); and to show that it is a mere aspect (*modum*), not a substantial reality (*rem subsistentem*), just as shape is an aspect of the object that has shape, and rest of the thing that rests. . . .
> . . . The body is in a different condition when it is being transferred and when it is not being transferred or is at rest. Thus motion and rest are simply two different states (*modi*) of a body.[28]

In considering both translation and its absence to be modes of a body, Descartes seems to be inconsistent with his definition of a body as "something that is all transferred at once." The nature of translation does not agree with his metaphysical conception of motion in another respect. The latter takes motion to be a momentary state of parts of matter, whereas translation consists in a body's being contiguous to different sets of other bodies at successive times. Also, in distinguishing between translation as it inheres in the moving body and its external cause, Descartes indicates in the above quotation that the cause is physical, "in the body that moves it," rather than metaphysical, in God.

In calling translation a mode of a body, Descartes also intends to distinguish the real motion of a thing from the motion we attribute to it. The definition of motion as translation requires clarification in this respect, for it makes the motion of a body depend upon us, that is, upon our taking certain other contiguous bodies to be at rest. In support of his claim that there is a difference in state between a body's being in motion and its being at rest, Descartes attempts to characterize translational motion independently of our determination of it.

> In itself, translation is reciprocal; we cannot conceive the body AB as transferred out of the neighborhood of the body CD without simultaneously conceiving the body CD as transferred out of the neighborhood of the body AB; and just the same force and activity is needed on both sides. So if we

211

wanted to assign to motion its proper, non-relative nature, we should say that when two bodies are contiguous and then undergo translation in opposite directions, there is as much motion in one as in the other.[29]

This rule for ascribing motions does not agree with our ordinary practice, Descartes admits, and it even leads to inconsistencies in certain cases. When two small bodies are translated from each other in opposite directions along the surface of the earth, each is also translated from some part of the earth's surface. But the earth could not also be transferred from each of the two smaller bodies without moving simultaneously in opposite directions. It is therefore with good reason that we consider the earth as at rest and only the two smaller bodies as being moved. Nevertheless, Descartes adds, "At the same time, the real positive characteristic of moving bodies—that which makes us call them moving bodies—is fully to be found in the other bodies contiguous to them, although these are regarded only as being at rest."[30] Descartes both does and does not recognize translation to be a relation. While he sees that it holds reciprocally for any two bodies that are separated, and that the same body may simultaneously be translated from different contiguous bodies in different ways, Descartes nevertheless continues to assert that translation is an inherent state of a body, and that a body can be in only one such state at any given time.

Although translational motion is a purely kinematic concept—it expresses the nature of change as distinct from its causes—Descartes nevertheless deduces an account of physical causation from it together with his doctrines concerning matter. Since space is identical with matter, all translation must occur within a continuous plenum of which all parts are divisible, movable, and impenetrable. It follows that a logically necessary condition for the transference of any one part of matter is the simultaneous transference of other parts along the same path. Descartes often expresses this condition in causal language: whenever a body is transferred, it must expel another from the place to which it is transferred, and must impel the other in the same direction and with the same speed that it moves.[31]

The same principles lead Descartes to the conclusion that all motion is in a closed curved path, that is, consists in the circulation of matter.

I observed above that all places are full of body, and an identical piece of matter always occupies an equal place. It follows that the only possible movement of bodies is a circulation; a body pushes another out of the place it enters, and this pushes another, and that another, till at last we come to a body that is entering the place left by the first body at the very moment when the first body is leaving it.[32]

The argument is incomplete as it stands, but it is clear that Descartes sees the conclusion following from the nature of translation in a plenum, rather than from its causes. That is, circular motion does not follow from the manner in which one part displaces another, or from the direction of the displacement, but from the fact that there must be an equal and simultaneous displacement of matter through every part of a continuous path of translation. Since all this matter is moved together, it is moved as a single body or piece of matter; and if it were not moved in the path of a closed curve, the whole quantity of matter partaking of a single translation would be of indeterminate size and shape. This is probably the reason why Descartes considers it obvious that translational motion must be circular. Motion cannot belong to extended substance as a whole, but only to determinate parts of matter.

Descartes' argument for circular motion in the *Principia* is followed by a discussion of two kinds of cases. In the first the path is a perfectly circular ring with uniform dimensions all around, and in the second the path is narrower in some parts than in others. Translation of matter in the first path, he holds, will be perfectly uniform in speed and direction in every part, but in the second case the matter must have different speeds in different parts of the path. For translation requires that equal quantities of matter be transferred through every part of the path; but this can occur in narrower segments of the ring only if the matter moves at a correspondingly greater speed there than in wider segments.[33] The example is of particular interest because Descartes treats the path of motion as a fixed physical boundary that causes changes in the speeds of the parts of matter that are transferred within it. In this way it provides a causal model for explaining changes of translational motion. By describing a boundary having different degrees of curvature as well as different dimensions in its various stretches, one could provide a model to explain changes in sense and direction of the parts of matter transferred within it, as well as changes in speed. That is, the motion of a particle passing from one to another stretch of the path determined by such a boundary would necessarily be altered in accordance with differences in the curvature and dimensions of the boundary in those stretches.

The idea of such an explanatory model is not developed by Descartes at the theoretical level, though it seems to have had a great influence in his explanation of planetary motion. As it stands, the model does not give an adequate account of the causes for changes in the motion of a body, for it assumes that the motion of matter within a closed curved boundary will remain constant through time, that is, that equal quantities of matter will pass through any given part of the path in equal times. A Cartesian warrant for this assumption can be found in his doctrines concerning God's conservation of motion through successive times. Descartes introduces these doctrines in the context of his discussion of the causes of motion; but in this discussion he abandons the above model for explaining changes of motion. Instead, he offers a quite different account, in which motion itself is given a dynamical role. Accordingly, Descartes implicitly adopts a third conception of motion.

When he undertakes to explain the causes of motion, Descartes interprets a particular motion, not as the actual translation of matter along a path, but as an instantaneous state of a body. Such a state is characterized by a speed and a direction, but these properties of the motion are not identified with the geometrical properties of the path the body actually has made. They are rather to be understood as properties of the path a body *tends* to produce when it is in that state of motion. The nature of motion, as it is incorporated in this account, is conceived in a dynamical context, in relation to the operation by which God produces it, rather than kinematically, as it is actually exhibited by the parts of matter that receive it.

The central ideas in Descartes' account of changes of motion and its causes are those of the quantity of motion and the direction or determination of motion. God produced a definite total quantity of motion in the world at the first moment of creation, and he re-creates the same total quantity at each successive moment in the history of the world. Originally this total quantity was arbitrarily distributed among various parts of matter in such a way that each part that received motion was put into a particular determinate state; it was determined in a particular direction, and had a particular speed according to the size of the part and the amount of motion given to it. Contiguous parts receiving different motions were thereby determined in different ways; that is, they were determined in different directions, or they had different speeds, or both. In addition, Descartes supposes some parts were determined by their being at rest, having no speed or direction. Successive changes in the determinate states belonging to each part are held to follow according to "laws of nature," that is, laws expressing the nature of the states of motion themselves and the relationships among them. Since the motion that is the ground of such states is produced and sustained by God, Descartes conceives that the laws expressing the relationships and successive changes among different states of motion also follow from the immutability of the divine operation.[34]

Descartes continues to maintain that the determinate states of motion and rest that belong to different bodies are only modes that cannot exist or be conceived apart from their subjects. But in comparison with the concept of translation, these prove to be modes in a significantly different way, endowed with dynamical properties and capable of affecting, and being affected by, each other. Such properties are specified by the laws of nature.

In the order given in the *Principia,* the first law, the law of inertia, expresses the tendency of any body to remain in the same state in which it is: "Every reality, in so far as it is simple and undivided, always remains in the same condition so far as it can, and never changes except through external causes."[35] Without an external cause a piece of matter will continue to have the same shape from moment to moment; so also it will continue the same with respect to the presence or absence of motion, and with respect to the quantity and direction of its motion. In opposition to the Aristotelians he maintains that motion does not cease or tend toward rest by its own nature, "for rest is the opposite of motion, and nothing can by its own nature tend towards its opposite, towards its own destruction." The tendency of motion to persist, and to persist in the same way, is the basis for the power of a body to affect, or be affected by, the motion of another.

Descartes' second law states how the direction or sense is determined in a state of motion. "Any given piece of matter considered by itself tends to go on moving, not in any oblique path, but only in straight lines."[36] By the "tendency to go on moving," he now explicitly adds, he means just the motion that God produces in the body at a given moment of time, "for he preserves the motion in the precise form in which it occurs at the moment when he preserves it, without regard to what it was a little while before.'' In *Le Monde* he had a further point of explanation:

> Among all movements, that which is in a straight line is the only one which is entirely simple, and one the whole nature of which can be embraced in a single instant; for, in order to conceive it, it is enough to think of a body actually moving in one fixed direction, which is the case in every one of the instants which can be determined during the time it is in motion; whereas to conceive circular movement, or any other that can exist, it is necessary to consider at least two of these instants, or rather two of its parts, and the relation between the two.[37]

It is evident that in the context of the second law ''straight line'' describes not the path a body makes but the direction in which it is determined at any given moment while it is in a state of motion. It will actually describe a rectilinear path only if its direction is not affected by external causes in successive times; and in that case the path will follow from the fact that it is determined in the *same* straight line in successive states. But from the concept of motion as translation and displacement of matter in a plenum, it also follows that such can never be the case.

Descartes' first two laws serve primarily to define the nature of motion in the manner in which God produces it, or neglects to produce it, in various parts of matter at particular moments of time. The third law formulates the nature of the external conditions under which the state of motion in a body is altered in successive times, either by a change in its quantity or in its direction or both. The law is specifically restricted to cases where the external conditions are the states of other bodies; changes resulting from mental causes are not to be considered. But, Descartes claims, the law covers all cases of physical causation.

> When a moving body collides with another, then if its own power of going on in a straight line is less than the resistance of the other body, it is reflected in another direction and retains the same amount of motion, with only a change in its direction; but if its power of going on is greater than the resistance, it carries the other body with it, and loses a quantity of motion equal to what it imparts to the other body.[38]

Collision is understood to occur when two bodies have contiguous surfaces and opposed states of motion. States of motion may be opposed in speed or direction or both.

We must observe that one motion is in no way opposite to another of equal velocity. Properly speaking, there are two sorts of opposition. First, motion is opposite to rest, and likewise a swift motion to a slow one, since slowness has something of the nature of rest. Secondly, the determinate direction of a motion is opposed to the body's meeting with another that lies in that direction and is at rest or is moving differently. The degree of this opposition depends on the direction in which a body is moving when it collides with another.[39]

Finally, the power of a body to continue its motion in a straight line or to resist a change of its motion, Descartes asserts, is just "the tendency of everything to persist in its present state so far as it can, according to the first law."[40] In order to determine the effects of a collision in the successive states of the colliding bodies, "we need only calculate the power of each body to move or to resist motion, and use the principle that the greater power always produces its effect."[41]

Descartes goes on in the *Principia* to give a list of seven rules of collision, to show how the calculations are to be made and the changes of motion determined from them. Although he claims, in the passage quoted above, that "slowness has something of the nature of rest," the rules show that he had a very different conception of the power to continue in motion and the power to continue at rest, and that these powers are to be calculated in different ways. The power of a body to remain in motion or resist a change of motion is calculated by the size and speed of the body. This power Descartes considered to be equivalent to the body's quantity of motion. The power of a body to remain at rest, however, is calculated by its own size and the speed of the colliding body. As a result of this conception of the power of rest, it follows that a body at rest will resist however swift a motion when it belongs to a smaller body, but will yield to however slow a motion of a larger body.[42] While rest is the absence of motion in one obvious sense, it is nevertheless to be considered a mode of a body, and not the mere lack of modal determination, as it appears to be when motion is considered in its metaphysical role. Accordingly the state of rest has a dynamical role, inasmuch as it is subsumable under the Cartesian law of inertia.

The calculable changes of motion resulting from a collision, as exhibited in Descartes' rules, consist in a reconciliation of the opposed states of the colliding bodies by alterations in their speeds or directions or both. In all cases the total quantity of motion belonging jointly to the

217

two bodies in their antecedent states is jointly preserved by them in their consequent states. Preservation of the joint quantity of motion figures as a fundamental assumption in Descartes' statement and application of the third law. Where consequences require an alteration of speeds, this is accomplished by a redistribution of the motion; motion is transferred from the body with the greater quantity to that with the lesser, so that one body has a lesser speed and the other a greater speed in the succeeding moment. Hence, in his demonstration of the third law, Descartes writes,

> Clearly, when God first created the world, he must not only have assigned various motions to its various parts, but also have caused their mutual impulses and the transference of motion from one to another; and since he now preserves motion by the same activity and according to the same laws, as when he created it, he does not preserve it as a constant inherent property of given pieces of matter, but as something passing from one piece to another as they collide. Thus the very fact that creatures are thus continually changing argues the immutability of God.[43]

The assumption that motion passes or is transferred from one body to another seems to be an essential part of the network of ideas involved in this conception of change. Yet the assumption is clearly inconsistent with the claim that motion is only a mode of the moving body. Henry More raises just this difficulty for Descartes: "In general, I cannot conceive how something that cannot be outside the subject, as all modes are, passes nevertheless into another subject."[44] In reply Descartes denies writing what More had implied. In a very puzzling passage, he writes,

> Indeed I think that motion, considered as such a mode, continually changes. For there is one mode in the first point of a body A in that it is separated from the first point of a body B; and another mode in that it is separated from the second point; and another mode in that it is separated from the third point; and so on. But when I said that the same amount of motion always remained in matter, I meant this about the force which impels its parts, which is applied at different times to different parts of matter in accordance with the [rules of collision] set out in article 45 and following of the Second Part [of the *Principia*].[45]

Here Descartes is evidently speaking of translational motion considered as a mode. Obviously, it is not the sense of a "mode of motion" to which the law of inertia applies, for translation itself is taken to entail

218

successive changes of mode. What is conserved is rather referred to as a force that impels matter to move. But in another passage in the same letter Descartes speaks of motion as being a mode of a different type from shape.

> Motion is not a substance but a mode, and a mode of such a kind that we can inwardly conceive how it can diminish or increase in the same place. Now there are notions appropriate to each type of being, which in judging about it must be used instead of comparisons with other beings; thus what is appropriate to shape is not the same as what is appropriate to motion; and neither of these is the same as what is appropriate to extended substance.[46]

A significant source of Descartes' confusions over the status of motion may be seen to lie in his failure to distinguish among, or bring together, concepts that appear to function at different ontological levels. In various ways the whole theory of motion involves God's initial creative act of dividing matter into distinct bodies, the total quantity of motion he produced in them when he did this and which he has sustained in the world ever since, the particular portions of this total quantity that were initially assigned to each body and which have since been redistributed by successive collisions, the speed and direction of each body resulting from the quantity of motion belonging to it, and finally the path produced by each body in being transferred from place to place. The last two of these levels are obviously of greatest importance in the construction of the Cartesian picture of the world: motion as an instantaneous state of a body which is characterized by an absolute determinate speed and a rectilinear direction, and motion as the translation of matter, with simultaneous displacement, along a single closed path. Anomalies in the relations between these two conceptions deserve special attention, for they seem to generate the greatest difficulties in accounting for Cartesian motion.

Descartes often assumes that the determinate states of motion belonging to bodies are expressed by the speeds and directions in which they are actually translated. The assumption does not hold in the case of direction, as he admits, for the direction of determination is always rectilinear while the direction of translation is always curvilinear. He sometimes holds that the curvilinear path results from constraint, presumably from collisions with surrounding bodies;[47] but the manner in which such collisions occur, and the reason they inevitably make the

219

path a continuous closed curve, is not apparent. There are further difficulties. Descartes formulates his laws of nature and his rules of collision primarily in connection with the nature of the divine causation, and without regard for the fact that his theory requires the application of these laws and rules to bodies that must move in a plenum. He thus speaks of collision as though it were a momentary event in the course of translation, in which two bodies approach and rebound along rectilinear paths, as if in a void. In fact, he admits, collision never occurs under such conditions: the colliding bodies are never perfectly hard, and their motions are always affected by other bodies that surround them.[48] But insofar as collision occurs whenever two parts of matter with different and opposed motions have contiguous surfaces, it necessarily follows that every body collides with some other part of matter at every moment it exists. If this consequence is admitted, it will also follow that every body is altered either in the quantity or direction of its motion at every moment. Not even the first law of nature can be applied to characterize the actual translation of a body during successive times or in successive parts of its motion.

In addition, on the assumption that a body is actually translated in a plenum of matter, it seems impossible to explain how a collision could occur. A moving body could only collide with another that lay in its path; but in the path of a translational motion all the matter must be displaced in a direction and speed which could not be opposed to the translated body. In effect, the necessary conditions for translation are contrary to the necessary conditions for collision.

In the last analysis Descartes has provided two different models of physical causation, each of which involves different assumptions about the nature of bodies and their relationships in the extended world in which they move. In practice Descartes employs both models, sometimes separately and sometimes together. Problems concerning the nature of matter are generally resolved by assuming the existence of extremely small and flexible particles that continue to move with great rapidity. Such adjustments in the explanatory apparatus, introduced as hypotheses, were themselves to become a major target for the critics of Cartesianism by the end of the seventeenth century; but these matters lie beyond the scope of this paper.

Thanks are due to Ronald Laymon, Ernan McMullin, J. E. McGuire, and especially to Peter Machamer for their comments on earlier versions of this paper.

1. Alexander Koyré, *Etudes Galiléennes* (Paris: Hermann, 1966), p. 391.

2. Charles Adam and Paul Tannery, eds., *Oeuvres de Descartes,* 12 vols. (Paris: Vrin, 1964–), 11:33. This edition is hereafter cited as AT. Translations are from Ralph Eaton, ed., *Descartes: Selections* (New York: Scribner, 1927), p. 319.

3. Descartes asserts the doctrine earlier; for example, in a letter to Mersenne dated 9 January 1639 (AT, 2:482). For a translation of the relevant passages, see Anthony Kenny, trans. and ed., *Descartes: Philosophical Letters* (Oxford, 1970), p. 62.

4. *Principia Philosophiae,* 2.21, AT, 8(1):52. Translation from Elizabeth Anscombe and Peter Geach, trans. and eds., *Descartes: Philosophical Writings,* rev. ed. (Indianapolis: Bobbs-Merrill Co., 1971), p. 207 (hereafter cited as AG).

5. Norman Kemp Smith calls attention to the difference between Descartes' use of 'body' in the sense of extension and 'bodies' in the sense of extended things which are distinguishable from each other through differences in size, shape, and motion (see *New Studies in the Philosophy of Descartes* [London: Macmillan, 1952], pp. 197–98).

6. Descartes to More, 15 April 1649, AT, 5:342; Kenny, pp. 248–49.

7. Descartes to More, 5 February 1649, AT, 5:269; Kenny, p. 238.

8. Descartes to More, August 1649, AT, 5:403; Kenny, p. 257.

9. *Prin. Phil.* 2.23; AT, 8(1):52–53; AG, p. 208.

10. AT, 11:34; Eaton, pp. 319–20.

11. *Prin. Phil.* 2.20; AT, 8(1):51; AG, p. 207.

12. This point is fundamental to Descartes' conception of the status of hypotheses in physics. In *Prin. Phil.* 3.46 he writes: "From what has already been said it is established that all bodies in the universe consist of one and the same matter; that this is divisible arbitrarily into parts, and is actually divided into many pieces with various motions; that their motion is in a way circular, and that the same quantity of motion is constantly preserved in the universe. We cannot determine by reason how big·these pieces of matter are, how quickly they move, or what circles they describe. God might have arranged these things in countless different ways; which way he in fact chose rather than the rest is a thing we must learn from observation. Therefore we are free to make any assumption we like about them, so long as all the consequences agree with experience" (AT, 8(1):100–101; AG, p. 225). Descartes also holds that the current state of the world, as known through observation, is consistent with *any* hypothesis concerning the initial state in which God created it: "The actual arrangement of things might perhaps be inferable from an original Chaos, according to the laws of nature; and once [i.e., in *Le Monde*] I undertook to give such an explanation. . . . In any case, it matters very little what supposition we make; for change must subsequently take place according to the laws of nature; and it is hardly possible to make a supposition that does not allow our inferring the same effects (perhaps with more labour) according to the same laws of nature. For according to these, matter must successively assume all the forms of which it admits; and if we consider these forms in order, we can at last come to that which is found in this universe" (*Princ. Phil.* 3.47; AT, 8(1):102–3; AG, pp. 223–24).

13. Descartes maintains there are three elements or kinds of matter in the world: "First are those which are agitated with such force that, when they meet other bodies, they are divided indefinitely into smaller parts of such shapes that they fit into all the recesses among those other bodies. Second are those which are divided into spherical particles which are very small in comparison with the bodies we see. After these we find a third form, comprising parts with such great sizes and such shapes that they have less motion than the others" (*Princ. Phil.* 3.52; AT, 8(1):105). Descartes tends to suppose, not that God initially divided matter into these three kinds, but that they emerged through subsequent divisions of the first bodies in the course of nature. The hypothesis that there are three such kinds is invoked primarily to explain light. The sun and fixed stars, which emit light, are composed of the first element; the heavens, which transmit light, are composed of the second; and

the planets, comets, and the earth, which reflect light, are composed of the third. The whole system of these bodies comprises the "visible" world.

14. *Princ. Phil.* 2.25; AT, 8(1):53–54; AG, p. 209. Anscombe and Geach translate *quae alios in se habeant motus* as "with various motions relatively to one another."

15. Ibid., 2.31; AT, 8(1):57; AG, p. 212.

16. Ibid., 1.61; AT, 8(1):29–30.

17. Descartes to More, August 1649, AT, 5.404. Kenny, p. 258.

18. AT, 11:40. Eaton, p. 324.

19. The doctrine is asserted in Meditation 3 in connection with the duration of the self (AT, 7:48–49; AG, p. 88), and defended in the Reply to the Fifth Objections (AT, 7:369). In *Prin. Phil.* 2.36 the conservation of motion in the world is held to follow from God's "ordinary concurrence" or his conservation of the world through successive moments in the same way that he originally created it (AT, 8(1): 61–62; AG, pp. 215–17). So also, in *Princ. Phil.* 2.39, Descartes argues that God "preserves motion in the precise form in which it occurs at the moment when he preserves it, without regard to what it was a little while before" (AT, 8(1):63–64; AG, p. 217).

20. AT, 11:36–38; Eaton, pp. 322–23. Italics mine.

21. AT, 11:39–40; Eaton, pp. 223–24.

22. *Princ. Phil.* 2.32; AT, 8(1):57–58; AG, p. 212.

23. *Princ. Phil.* 2.12; AT, 8(1):46; AG, p. 203.

24. *Princ. Phil.* 2.13,24; AT, 8(1):47, 53; AG, pp. 203–4, 208.

25. *Princ. Phil.* 2.25; AT, 8(1):53; AG, p. 209.

26. *Princ. Phil.* 2.28; AT, 8(1):55; AG, p. 210.

27. *Princ. Phil.* 2.25; AT, 8(1):53–54; AG, p. 209.

28. *Princ. Phil.* 2.25, 27; AT, 8(1):54, 55; AG, pp. 209, 210.

29. *Princ. Phil.* 2.29; AT, 8(1):55–56; AG, pp. 210–11.

30. *Princ. Phil.* 2.30; AT, 8(1):57; AG, p. 212.

31. *Le Monde* ch. 8; AT, 11:49; Eaton, p. 330. See also *Princ.Phil.* 2.33; AT, 8(1):58; AG, p. 213.

32. *Princ. Phil.* 2.33; AT, 8(1):58; AG, p. 213.

33. Ibid, AT, 8(1):59–60; AG, pp. 213–14. The latter motion is possible, Descartes notes, only if we admit what is at the same time inconceivable, namely, the actually infinite division of matter. *Princ. Phil.* 2.34,35; AT, 8(1):59, 60; AG, pp. 214–15.

34. *Princ. Phil.* 2.37: "From God's immutability we can also know certain rules or natural laws which are the secondary, particular causes of the various motions we see in different bodies" (AT, 8(1):62; AG, p. 216).

35. *Princ. Phil.* 2.37; AT, 8(1):6; AG, p. 216. See also *Le Monde,* chap. 7; AT, 11:38; Eaton, p. 323.

36. *Princ. Phil.* 2.39; AT, 8(1):63; AG, p. 217. This is given as the third law in *Le Monde,* chap. 7.

37. AT, 11:44–45; Eaton, p. 328.

38. *Princ. Phil.* 2.40; AT, 8(1):65; AG, p. 218.

39. *Princ. Phil.* 2.44; AT, 8(1):67; AG, pp. 219–20.

40. *Princ. Phil.* 2.43; AT, 8(1):66; AG, p. 219.

41. *Princ. Phil.* 2.45; AT, 8(1):67; AG, p. 220.

42. *Princ. Phil.* 2.46–52; AT, 8(1):68–70. Descartes' seven "rules" express the manner in which the calculation should be made and the consequent motions determined for several cases of collision. Of these, the fourth, fifth, and sixth rules pertain to collisions between a moving body and a body at rest. For a detailed discussion of these rules and the dynamical ideas they involve, see

Allen Gabbey, "Force and Inertia in Seventeenth Century Dynamics," *Studies in the History and Philosophy of Science* 2 (May 1971): 20–31.

43. *Princ. Phil.* 2.42; AT, 8(1):66; AG, p. 219.

44. More to Descartes, 23 July 1649; AT, 5:382.

45. Descartes to More, August 1649; AT, 5:404–5; Kenny, p. 258.

46. Ibid., AT, 5:402–3; Kenny, p. 257.

47. For example, in explaining the law of rectilinear tendency of motion in *Princ. Phil.* 2.39 Descartes writes, "Of course many pieces of matter are constantly being compelled to swerve by meeting with others; and, as I said, any motion involves a kind of circulation of matter all moving simultaneously" (AT, 8(1):63; AG, p. 217).

48. Ibid., 2.53; AT, 8(1):70; AG, p. 220.

Arnold Koslow

Chapter Nine

ONTOLOGICAL AND IDEOLOGICAL ISSUES OF THE CLASSICAL THEORY OF SPACE AND TIME

I would like to discuss several ontological and ideological issues that arise in a natural way upon consideration of Newton's ideas, published and unpublished, about space and time. The perspective for the present essay is adopted from that of a larger work whose general attitude is that we can learn a good deal by trying to recast empirical and philosophical theories of space and time into some canonical form in order to see more clearly what the genuine differences are between such theories. More specifically, each theory is to be cast into a canonical form which is that of an empirical relational system.[1] Roughly speaking, such a system is one expressed with the aid of standard first-order logic, in which the domain of objects is specified, the requisite predicates of various orders are listed or at least described in such a way that it is clear how all needed predicates may be generated. Certain general truths are then enunciated that use those predicates. Considerations about the domain of objects of the theory, and perhaps their individuation, belong to what I shall call, following Quine, the ontology of the theory; problems that concern the predicates, their number and type, and the general truths expressed with their aid, we shall call problems of ideology. I do not wish to maintain either that there is a very sharp division or a special dependency between these two classes of problems. I mean for the distinction to mark off in a rough way two sets of problems in a relatively neat manner. I shall not attempt to cast the Newtonian theory into such a neat form, but I shall try instead to present you with a number of considerations that I hope will persuade you that there are serious philosophical difficulties that arise en route. A discussion of these difficulties will show, I hope, that there are many philosophical riches in the theories to be discussed, which come to light in a natural way when we consider how such theories are to be cast into a preferred canonical form.

In attending to the ontological and ideological features of theories of space and time, the use of historical material can help to reshape some of the "standard" debates. Thus it is commonly believed that Newton thought that bodies are distinct from space, and that this is one major difference between Newton and Descartes. But, in a manuscript he never published,[2] Newton argued that bodies might just be special regions of space, specially endowed with certain causal characteristics by God. Moreover, Newton held this view on the basis of beliefs that he held throughout most of his life. Therefore, many new possibilities are now open, and a good deal of care has to be exercised in the description of Newton's views of absolute Space and Time. As another example of a rather standard view that may have to be discarded, or significantly modified, consider the ideological point, frequently advanced, that Newton's theory of Space and Time was different from that of Leibniz in that the latter, on his own description, held a "relational" theory, the former an absolute or non-relational theory. This contrast fails to hold up under examination. Leibniz's view of relations was such that every well-founded relation that holds between two substances is reducible to truths about the simple attributes which each of the substances has separately. Therefore, Leibniz's theory of Space turns out to be a theory about individual substances and the simple, non-relational attributes they must necessarily possess. Newton, on the other hand, far from maintaining a non-relational theory of space (and time) held the view (in the same manuscript referred to above), that although space and time are entities with parts, those parts must be understood "relationally". More exactly, he claimed that regions of space (and moments of time) are individuated solely by their relation to all other regions of space (or all other moments). Thus the differences as I see it, between Newton's and Leibniz's theories of space and time are to be understood as a difference between two distinct kinds of empirical relational systems. The difference does not concern the use or non-use of relations.

In the first part of what follows I shall discuss selected problems of determining the ontology and the ideology of Newtonian theories of space and time. In the second part[3] I shall try to muddy the waters a bit and shall suggest that at least at one stage of his thought, Newton was more pragmatic, perhaps even a bit more opportunist about ontological issues, than he is usually thought to be.

I

Absolute Space

Almost the first question we ought to ask about any empirical relational system representing Newton's theory of Absolute Space (and Time) is this: what are the entities this theory requires? A second question might then be this: what are the requisite predicates, relations, and truths relating these entities? Absolute Space from all that Newton ever said of it is at least this: an individual with parts. We shall assume that the basic entities of Newton's theory of Absolute Space is an entity or individual, call it A, together with all its parts. These parts he called *regions of space,* and I think that the bulk of ideological problems about his theory concern the relations that hold between regions of space. I shall wish to say something similar about Newton's theory of Absolute Time. Absolute Time is an entity or individual, which implies that it has parts. Newton called those parts of Absolute Time the parts of duration, or the moments of duration. It would be easy just to think of Absolute Space as some three-dimensional continuum, with or without a Cartesian-Coordinate grid superimposed, and to think of regions of Absolute Space as perhaps bounded, connected portions of this continuum. But it must be stressed that we ought not, without specific reason, to invest Absolute Space (or Time for that matter) with the trappings of modern analysis. As we shall see, Newton for a time entertained the possibility that there were minimal spatial and temporal parts. This should make us a bit more cautious; we ought to sift through the Newtonian corpus to find out what properties and relations he in fact attributed to regions of space (and moments of time).

It seems natural for us to think of parts of space as having (dimensionless) points. In most modern discussions of mechanics, it is claimed that a pointbody moves from one point to another in the course of time. But once again, it is for us to discover, rather than to assume, what views Newton had about points in Space. One thing is clear: an argument such as the one just given would not have been used by him. For one thing, Newton did not believe that there were pointbodies. Each body has mass, and he believed that mass was measured not only by its inertia but also by its quantity of matter. Since all bodies have a positive quantity of matter, they could not be punctal. Newton does of course characterize true or *absolute motion* with the aid of *Absolute Space,* but it is not the change of a body from one *point* of *Absolute Space* to another over Absolute Time; it is a change of *absolute place.* Thus in his unpublished essay "De Gravitatione et Aequipondio Fluidorum" (probably written sometime between 1664 and 1668–69) he states, in words substantially repeated in the *Principia*: "Place is a part of space which something fills evenly; Body is that which fills place; Rest is remaining in the same place; Motion is change of place." The idea is clearly that places are filled. Clearly those regions of space he calls places are not punctal.[4] As we shall see, in a work (still unpublished), the *Questiones quaedam Philosophiae,*[5] probably written between 1661–65 and perhaps over-lapping in time with the "Gravitatione" text just cited, Newton was very much puzzled over so-called points or infinitely small parts of lines (which have no proper parts). When I shall therefore speak generally of regions or parts of absolute space, I shall not assume that these parts are somehow composed or constituted from points, and I suggest that we ought not to use the notion of a dimensionless point in reconstructing the ontology of Newton's theory of Absolute Space, unless, in specific context, there is special warrant for introducing such a geometric struc-ture.

There are many features attributed to Absolute Space that I shall not discuss in great detail here but that deserve some brief discussion. These features constitute the "received view".

1. *Sensorium.* Famous or infamous is the well-known dictum that Absolute Space is the Sensorium of God. This was defended in the *Leibniz-Clarke Correspondence,* but it is now known that Newton's ideas were at issue, not merely Clarke's. The idea seems to be that God is aware of all existents, but that his awareness of them requires a special sensorium. I do not know enough about God's perceptual processes to judge whether He is unique in this way, or whether there are beings other than God who require a sensorium. But the identification of such a sensorium with Space has as a consequence a very Newtonian thesis: "*No being exists or can exist unless it is related to Space in some way.*"[6] Thus he maintains that God is everywhere and that created minds are somewhere. Elsewhere, even dreams are located—where the dreamer

is. Indeed, as we shall see below, it is this belief of Newton that sets the stage for his discussion of how moments of time, which exist, are related to space. Another aspect to the identification of Space with the sensorium of God, is that Space is not an absolute being (*sic*), independent of the existence of God. Such a notion of Space would be, in his mind, conducive to atheism. He wanted a notion of Space for which it would be impossible to conceive of extension while imagining the non-existence of God.[7] Newton also referred to God-independent notions of body as "absolute," and we shall see below how he also tried to introduce a non-absolute notion of body in this special sense of "absolute."[8]

2. *Absolute Space and Causation.* Another characteristic theme is that certain motions in absolute space give rise to certain effects, certain forces, which in their turn have their effects. True laws of motion are those that relate the action of forces to absolute motions. It is a well-known story that Newton believed that he had empirical evidence showing that certain motions were absolute. I do not propose to evaluate that argument here since I have had my say elsewhere.[9] What is worth noting is that Newton believed that certain motions, those that one would call relative motions of bodies with respect to each other, are merely *outward signs* but not the true absolute motion (or physical motion) that is taking place. This implication is clear: some twenty years before the discussion of the bucket experiment in the *Principia,* Newton seems to have held the general position that the relative motions of bodies are to be thought of as the *effects* of true causes, i.e., physical or absolute motions. Thus, in arguing against the Cartesian theory of motion as expressed in Descartes' *Principles,* Newton asserts that it has to be conceded that there is only one physical or absolute motion of a body, and that the rest of its changes of relation and position with respect to other bodies are so many "external designations". By way of example, he thinks that "only the motion which causes the Earth to endeavour to recede from the Sun is to be declared the Earth's natural and absolute motion. Its translations relative to external bodies are but external designations."[10] Newton does not say that the motion of the earth to recede away from the Sun is the true motion; he states that only the motion that *causes* the Earth to endeavor to recede from the Sun is the absolute motion. Thus, to repeat the point: absolute motions are the causes of the more patent relative motions of bodies with respect to each other.[11]

228

3. *Absolute and relative motions, reason, and synthetic a priori.* Related to the causal connection between absolute and relative motions of bodies is a third Newtonian theme that is rather attractive in its simplicity. The claim is that *if two bodies are each absolutely at rest, then they are not moving relatively to each other;* equivalently, *if two bodies are in relative motion, then at least one of them must be moving absolutely.* A related thesis states that if two bodies are at rest with respect to each other, then it cannot be that one of the bodies is in absolute motion and the other is not. The first formulation seems to provide an implication of an absolute theory for those who would like to keep to the relativistic mode. But the second, equivalent formulation (if α is moving relative to β, then either α is moving absolutely or β is) seems to bear bad tidings for those who might think that the basic relation for describing motion is the relation that attributes relative motion to a pair of bodies. For then, granting the thesis, they would also have to admit the attribution of absolute rest to at least one of the bodies. The thesis we mentioned, as well as its mates, are rather interesting statements. Of course, those who deny that attributions of either absolute rest or absolute motion have either sense or scientific value have no qualms over simply dismissing the Newtonian thesis and its mates. But those who believed them to be true seemed to feel that there was a very strong case for them. Thus Newton maintained in his study of the Cartesian relativist theory of motion that the thesis was a truth which reason was unable to deny:

> Fifthly. It seems repugnant to reason that bodies should change their relative distances and positions without physical motion; but Descartes says that the Earth and the other Planets and the fixed stars are properly speaking at rest, and nevertheless they change their relative positions.

> Sixthly. And on the other hand it seems not less repugnant to reason that of several bodies maintaining the same relative positions some one should move physically, while others are at rest. . . .[12]

While Newton found it "repugnant to Reason" to deny these theses relating absolute to relative motion of bodies, the present century has also been able to provide enthusiastic supporters and incredulous detractors. Herbert Feigl, while discussing the relational view of space and time, mentions a debate in 1920 between Oskar Kraus and Einstein over Kraus' claim that the following was a synthetic *a priori* truth: if two bodies move relatively to each other, then at least one of them moves with respect to absolute space.[13]

Feigl comments that "this illustrates beautifully the intrusion of the pictorial appeal of the Platonic 'receptacle' notion of space or a confusion of a purely definitional truth (regarding three coordinate systems) with genuinely factual and empirically testable statements regarding the motion of bodies." In a work dating from the first decade of this century it is even argued that such theses as the one mentioned above follow as a direct application of Meinong's theory of relations in general, so that they would therefore be, in a broad sense, logically based.[14]

4. *Wholes and parts: A terminological digression.* We turn next to a series of theses about absolute space (and time), which require a slight digression. These theses refer to the parts of space and time and can be expressed more clearly if we borrow some terminology and some theorems from a technical language already fashioned for talk about wholes and parts. We shall use a modified form of the calculus, originally formulated by H. Leonard and N. Goodman, and Lesniewski.[15] The terminology borrowed is from first-order logic with the addition of a two-place predicate "x o y" which has the sense intuitively of "x overlaps y" or "x and y share a common content." Using the notion of overlap, we shall say that x is a part of y, $x < y$, if and only if every part that overlaps x overlaps y; x is said to be identical with y, $(x = y)$, if and only if every part that overlaps x overlaps y, and conversely; z is the sum of the parts x and y, $Sxyz$, if and only if a part overlaps z if and only if it either overlaps x or overlaps y; that x is distinct from y, $x \# y$, if and only if x does not overlap y; that z is the negate of x, Nzx, if and only if a part overlaps z if and only if it is distinct from x; that x is an atom, $A(x)$, if and only if every part of x is identical with x; and x is a proper part of y, $x << y$, if and only if x is a part of y, but not identical to it.

The axioms will be the universal closures of the following:

$$E_1. \ x \ o \ y \equiv (\exists z) \ (z < x \ \& \ z < y)$$

$$E_2. \ (\exists z) \ S \ xyz$$

$$E_3. \ (\exists z) \ (N \ xz \equiv (\exists w) \ (w \# x))$$

$$AA. \ (x) \ (\exists y) \ (A(y) \ \& \ y < x))$$

Not all these axioms can be used to describe the Newtonian situation. Axiom E_1 can remain as it is, but E_2, the existence axiom for sums, has to be modified. Given Newton's idea that place or a region of space is a part of space which something (a body) fills evenly, it is not plausible that any two places have a sum which is also a place. Thus we adopt instead, axiom E_2':

$$(\exists\ z)\ (x\ o\ y \supset Sxyz)$$

i.e., any two overlapping regions of space have a sum. Axiom E_3, the existence axiom for negates, also seems unsuitable since there would have to be bodies almost as extensive as space itself, if the negate of a place were also a place or region of space. The fourth axiom, AA, assuring the existence of an atomic part of any part, is also not suitable. Newton's ideas about atomic parts of space and time are subtle and difficult to describe. Roughly speaking, he seems to have believed at an early period that space and time had minimal parts, and as we shall show below, he seemed unhappy with that view. Nevertheless, for rather complex reasons, he never denied outright that indivisibles existed. In the *Principia* itself he does think that Euclid has proved that certain types of magnitude cannot have atomic or indivisible parts. But he seems thereby to leave open the possibility that some parts may indeed have atoms. The assumption that *every* part has at least one atom cannot be accepted without modification. The matter is complicated, and will be discussed below when we consider Netwon's ideas about *minima naturalia*.

5. *Infinitude of space*. Newton long before the publication of the *Principia* (1687) affirmed in the "Gravitatione" that space extends infinitely in all directions. As he put it,

> For we cannot imagine any limit anywhere without at the same time imagining that there is space beyond it.

And on a related point,

> If anyone now objects that we cannot imagine that there is infinite extension, I agree. But at the same time I contend that we can understand it. We can imagine a greater extension, and then a greater one, but we can understand

231

that there exists a greater extension than any we can imagine. And here, incidentally, the faculty of understanding is clearly distinguishable from imagination.[16]

Thus, Newton seems committed to the thesis that for every proper part of space (a part which he says has ''limits''), there is a part of space, indeed a proper part, of which it is a proper part (which has greater extension). But he is prepared to go even further: he is prepared to deny that we can imagine a *space* which has all parts of space as its parts; but he also believes that such a universe or space exists, and that we can *understand* (but not imagine) that such a space exists.

More formally expressed, there is a part, U, such that

$$(1) \quad (x) \, (\exists \, y) \, (x \, \# \, U \supset y \, \# \, U \, \& \, x \, << \, y)$$

and

$$(2) \quad (\exists \, x) \, (y) \, (y < x).$$

Sometimes (2) is expressed as

$$(3) \quad (\exists x) \, (y) \, (y \, o \, x). \text{ Clearly (2) implies (3).}$$

To obtain the converse implication from (3) to (2), let x be the part whose existence is guaranteed by (3). Let $x \oplus y$ denote the sum of x and y. If $y \, o \, x$ then the sum $x \oplus y$ exists by E_2', and $y < y \oplus x$. But $y \oplus x = x$ so that $y < x$.

It should be noted that neither of these, (1) and (2), is adopted as an axiom of a general calculus of individuals, nor do either of these claims really do full justice to the idea that space is infinitely *large*. For all that (1) and (2) assert, it is still possible that each part of space, and indeed space itself, might be finite in volume. Models of the calculus can easily be supplied for which this is true. Newton himself seems to have realized that there are figures with finite area (or volume) which are infinitely open—like the finite area under a branch of a hyperbola.[17]

6. *Eternal (in time) and immutable (by nature).* We have mentioned the Newtonian claim that if anything exists at all, space does. But Newton seems to believe that space is eternal (in duration) and immutable (in nature) because

it is the emanent effect of an eternal and immutable being. If ever space had not existed, God at that time would have been nowhere; and hence he either created space later (in which he was not himself [sic]), or else, which is not less repugnant to reason, He created His own ubiquity.[18]

Thus the existence of space throughout time is a consequence of God's existence, not of the existence of bodies. But Newton held an even stronger version of this claim. In fact, in a celebrated dictum which is one of the central themes of Newtonian thought (to be found for example in Euler's works and in Kant following Euler), ''one can imagine that there is nothing in space, yet we cannot think that space does not exist, just as we cannot think that there is no duration, even though it would be possible to suppose that nothing whatever endures.''[19]

But what are we to make of such spaces emptied of bodies? Newton, at this time, seems to distinguish between space without any bodies in it and a void. As he states it, ''*something* is there, because spaces are there, although nothing more than that.''[20] The point seems to be that even without bodies in it space is not a void since there is something *in* space, namely, parts of space. Obviously there is a great struggle needed here to avoid collapse into the tautological. Perhaps we might rescue this insight as follows: There are those individuals which we have called parts *of* space. But what must also be realized is that these individuals are not only parts *of* space, they are also parts *in* space. Is this ''in-of'' distinction more than a rhetorical flourish? I think that it is. After all, parts of space are also supposed to be *in* space, if we adhere to Newton's dictum that ''any being has a manner proper to itself of being in spaces.''[21] Suppose, in locating a body in space, we thought it sufficient not merely to indicate the body but to indicate some (proper) part of space (not just space itself) in which the body was properly located. So too with parts of space. If an individual—in this case a part of space—is to be located *in* space, it is not sufficient merely to indicate which individual it is; the part of space has to be located as a proper part of some (proper) part of space—only then is it *in* space. More shortly, to be *in* space is to be a proper part of some (proper) part of space. (Note that this assertion is just the one that Newton used to argue that absolute space is infinitely large.) One might think that the analogy pressed here, between bodies and parts of space, is too stretched; that it is especially infelicitous in reconstructing Newton's thought, given his supposed contrast between body and space. I think that the importance for Newton of the contrast between body and space has been greatly exaggerated, and I shall return to this theme below.

7. *The parts of space are motionless.* Another characteristically Newtonian theme that appears at this early date (1664–68) is that parts of space are not (actually) divisible, and the related idea that the parts of space are motionless. Newton does not connect up the two themes directly, but it is not difficult to relate them. If a part of space were to move, consider what would happen at its boundary or 'limit'. The part of space that had the boundary as a part would be (actually) divisible. The motionlessness of the parts of space was an extremely important feature of space for Newton. Indeed, after criticizing Descartes' notion of place because under certain motions of bodies some places would cease to exist, he says,

> So it is necessary that the definition of places, and hence of local motion, be referred to some motionless thing such as extension alone or space insofar as it is seen to be truly distinct from bodies.[22]

Newton did consider the possibility that parts of space moved, and indeed offered some models to show how such a motion might take place.

> If they moved, it would have to be said either that the motion of each part is a translation from the vicinity of other contiguous parts, as Descartes defined the motion of bodies; and that this is absurd has been sufficiently shown; or that it is a translation out of space into space, that is, out of it self, unless perhaps it is said that two spaces everywhere coincide, a moving one and a motionless one.[23]

Popping parts of space in and out of space seems impossible without an actual division of parts of space. But think of the second model. According to it, we have two absolute spaces, one at rest and one in absolute motion. Let us call this the Double-Space Theory. Both spaces, he suggests, would coincide everywhere. What could this mean? We cannot say that each of the two regions of space, one moving, the other at rest, are parts of each other. For then, there could be only one region, since it is an easy consequence of the calculus of individuals that if α is a part of β, and conversely, then α and β are identical. Instead of exploring further the possibility of Double-Space or Double-Time theories, I want to pick up another strand in Newton's thought, as embodied in two principles that he enunciated. He used them to argue that the parts of space (as well as the moments of time) are motionless (although he did not explicitly draw this conclusion for moments).

8. *Principles of individuation for regions of space and moments of time.* The principles are principles of individuation. Here is the key passage:

> Moreover the immobility of space will be best exemplified by duration. For just as the parts of duration derive their individuality from their order, so that (for example) if yesterday could change places with today and become the latter of the two, it would lose its individuality and would no longer be yesterday, but today; so the parts of space derive their character from their positions, so that if any two could change their positions, they would change their character at the same time and each would be converted numerically into the other. The parts of duration and space are understood to be the same as they really are because of their mutual order and position; nor do they have any hint of individuality apart from that order and position which consequently cannot be altered.[24]

The principles of individuation are of the utmost importance. They bear at present in a very direct way on whether the parts of space can move. For Newton's principle of individuation of parts of space cuts across any possible mechanism or model for showing how a part of space could move. The argument is simple: suppose that by some means or other, a part of space α, moved to another part of space, β, so that where β once was, we have α, not β. How could this be? For β was related in a special way to all other parts of space; it had a special *situation* or *order*. If α is now replacing β, and so has the same situation or order as α with respect to all other parts of space, then it is not α but β. Consequently it is not α that has moved. The argument is meant to apply to absolute time, and blocks any attempt to construct a Double-Time Theory to explore the possibility of how parts of time might move. It should be noted that although Newton thinks that the temporal moments are motionless, and restates this theme in the *Principia,* he does not think that it is incompatible with another characteristic theme—that time flows evenly and uniformly. We postpone the resolution of this apparent conflict for the present.

There is some difficulty in describing these attractive principles coherently. Some content must be provided for them, so that they are not entirely empty. After all, unless a specific predicate is given that describes the ordering relation among moments, then these principles that state simply that if α and β stand in exactly the same (unspecified) relation R to all moments, then they are identical, hardly are informative. If the principle merely requires that there be *some* relation term that does the job of individuating, then the argument for the immobility of the parts of space and time still goes through; but I think that most of us would feel that until a specific relation is given, the principle has not reached the level where we would seek either to justify or to refute it.

If we call the set of those parts of time β such that $\alpha R\beta$, the (second) place of α, and the set of all those γ such that $\gamma R\alpha$, the (first) place of α, then Newton's principle of individuation for moments of time runs as follows:

> If the first place of α and the first place of β are identical, and so too are the second places of α and β, then α is identical with β.

This is but another way of stating that the places of α and β in the temporal series, their relation to all (other) parts of time, must be the same for α and β to be identical. We can provide a more formal schema by using some set-theory language:

$$(\mathrm{I}) \left. \begin{array}{l} \hat{x}(xR\alpha) = \hat{x}(xR\beta) \\[2mm] \hat{x}(\alpha Rx) = \hat{x}(\beta Rx) \end{array} \right\} \Rightarrow \alpha = \beta$$

Even if one were satisfied with the individuating principle for time, what would be a suitable relation for the principle individuating regions of space? Most of the so-called spatial relations discussed in the philosophical literature are unsuitable, e.g., "x is spatially related to y", "x and y belong to a connected portion of space", and so forth. The reason such relations are bound to fail is that they are symmetric and transitive, so that if we were to use schema (I) above, there would be only one part of space, Space. However, if we use the calculus of individuals, we can supply a "spatial relation" that does the job of individuating quite nicely. If α and β are parts of space, I suggest that we

236

use the overlap relation between parts of space as a suitable candidate for the unspecified relation R. Consider the resultant schema, upon substitution. If $\hat{x}(x\ o\ a) = \hat{x}(x\ o\ B)$, then every part of space that overlaps α overlaps β, so that $\alpha < \beta$, i.e., α is a part of β by the calculus of individuals. (Since the overlap relation is symmetric, the second clause of the schema is redundant.) We may also conclude that every part of space that overlaps β overlaps α, i.e., $\beta < \alpha$, i.e., β is a part of α. But from the calculus of individuals, it follows that $\alpha = \beta$. The result of substitution in schema (I) is therefore valid; i.e.,

$$\left\{ \begin{array}{l} x(x\ o\ \alpha) = x(x\ o\ \beta) \\[2mm] x(\alpha\ o\ x) = x(\beta\ o\ x) \end{array} \right\} \Rightarrow \alpha = \beta$$

Of course, if I have convinced you of the plausibility of Newton's thesis for space, you should then also be satisfied by a similar proposal for time. That is, if we use the overlap relation for moments of time, instead of a relation like ''x is before y'', about whose adequacy there might have been some misgivings, the calculus of individuals yields the appropriate conclusion. Of course, this proposal for parts of time only makes sense if parts of time do overlap. If all moments α overlap with nothing other than themselves, then the principle implies that $\alpha = \beta$ (since $\hat{x}(x\ o\ \alpha) = \{\alpha\}$, the premises state that $\{\alpha\} = \{\beta\}$).

9. *The diffusion of moments through space*. We mentioned the Newtonian thesis that whatever has or could have existence is related to space in its special way. That claim is intertwined with two other better-known Newtonian claims in the following seminal passage:

> . . . Any being has a manner proper to itself of being in spaces. For thus there is a very different relationship between space and body, and space and duration. For we do not ascribe various durations to the different parts of space, but say that all endure together. The moment of duration is the same at Rome and at London, on the Earth and on the stars, and throughout all the heavens. And just as we understand any moment of duration to be diffused throughout all spaces, according to its kind, without any thought of its parts, so it is no more contradictory that Mind also, according to its kind, can be diffused through space without any thought of its parts.[25]

Let us postpone discussion of the relation between body and space and consider the relationship between space and duration. According to

Newton, since everything that has being is related in its peculiar way to space, each part of time should also be related to space.

What more is there to say about that relation? The moment that is supposed to be the same at all the spatial places mentioned seems to have parts. Newton says that the moment is diffused throughout all spaces. Newton suggests that even though temporal moments, like minds, have parts and are diffused through space, we are not to pay attention to that feature. Suppose that we call those parts of a moment that are diffused through space the *spatial parts* of the moment. Each moment of time therefore has a certain internal structure. The thought seems to be that although the moment of time is the same at London as at Rome, on the earth and the stars, there are parts of time that are specifically related to each of the parts of space. We might then supplement the actual passage from the Newtonian corpus with the idea that it is the same moment of time at London as at Rome precisely because two spatial parts of that moment were respectively at the two spatial regions. Although it is the same moment at London as at Paris, it is presumably different spatial parts of that moment that are at those cities, and this raises the issue of what could it mean for a spatial part of a moment to be *at* a certain region of space. The most direct assumption is that a spatial part is nothing but a part of space itself. Time and space would thus overlap—since there would be parts of each that overlap (this uses the theorem of the calculus of individuals that parts of parts of α are parts of α.) Another suggestion would introduce a special overlap relation called 'at' such that a spatial part of time is at a region of space. None of these, and other suggestions, can be advanced on the basis of what has come to light so far in the Newtonian corpus. Nothing precise is said there about the character of the spatial parts of moments or parts of time. It does seem natural though, given his talk of diffusion, to assume that each part of time has as many spatial parts to it as there are regions or parts of space. In what follows I shall not propose any theory as to how the spatial parts of parts of time are related to parts of space. I shall simply assume that to each part of time, τ, and each region of space, ρ, there is associated a special part of τ, which will be called the spatial part of τ corresponding to the spatial region ρ, and denoted by $\tau\rho$.

Two questions arise. First, if we consider two parts of space, ρ and ρ^*, such that $\rho < \rho^*$, then for any part of time τ is the spatial part of τ corresponding to ρ a part of the spatial part of τ corresponding to ρ^*?

Second, if the part of time τ is a part of a part of time τ^*, then for any region of space ρ, is the spatial part of τ corresponding to ρ a part of the spatial part of τ^* corresponding to ρ?

Corresponding to the first question, we have two natural principles:

(I) If $\rho < \rho^*$, then $\tau\rho < \tau\rho^*$, for any regions of space ρ that ρ^* and any part or moment of time τ.

(II) If $\rho \# \rho^*$, then $\tau\rho \# \tau\rho^*$, for any regions of space ρ and ρ^* and any part or moment of time τ.

Corresponding to the second question, we have a third principle:

(III) If $\tau < \tau^*$, then $\tau\rho < \tau^*\rho$, for any parts of time τ and τ^* and region of space ρ.

The first two principles are synchronic, and the third is diachronic in character. Related to the third principle is the question of whether for two non-overlapping parts of time the spatial parts of each moment or part of time corresponding to the same region of space are "isomorphic." Clearly, such parts of time cannot be identical or even overlap, for then the two parts of time would also overlap.

Principle (III) creates a bit of mischief when taken in conjunction with (I) and (II); together they are contradictory.

Suppose that τ is any part of time. Suppose further that there are two spatial parts of τ that do not overlap. Call them $\tau\rho$ and $\tau\rho^*$. Such parts of τ exist, since there are surely two parts of space, ρ and ρ^*, that do not overlap, and Principle (II) yields the result. Thus,

$$\tau\rho < \tau, \text{ and}$$

$$\tau\rho^* < \tau.$$

Further, since $\tau\rho$ is a part of time τ, we use (III) to obtain

$$(\tau\rho)\rho^* < \tau\rho^*.$$

Furthermore, since the spatial part of a part of time is supposed to be a part of time, it follows in particular that the part of $\tau\rho$ corresponding to the part of space ρ^* is a part of $\tau\rho$. Thus,

$$(\tau\rho)\rho^* < \tau\rho$$

so that $\tau\rho$ and $\tau\rho^*$, both assumed to be non-overlapping, are seen to have the common part $(\tau\rho)\rho^*$.[26]

Faced as we are with a contradiction, Newton's warning about time being diffused throughout all space, *without any thought of its parts,* is well taken. But how can we ignore those parts and still take the notion of diffusion seriously? One obvious remedial move is to qualify (III) so as to rule out its use for any spatial part of a temporal part. Thus no application could be made for a temporal part of the type $\tau\rho$ where ρ is a part of space, and therefore the argument that led to the contradiction would be blocked. Another, more elaborate response would be to restrict (III) to just those parts of time which are *small,* where we call an item small if and only if it is a part of anything which it overlaps. That is,

$$\mathrm{sm}(\tau) \equiv (\alpha)(\tau \ o \ \alpha \supset \tau < \alpha).$$

Thus an item is small if and only if it is a part of exactly those things which it overlaps. Under this qualification, (II) becomes false, and (III) becomes patently true (but vacuous).[27]

I am not aware of any Newtonian text that would lift these suggestions from the place of the merely speculative. (I) and (II) seem natural given Newton's suggestion that parts of time are spread out through space. And we can see that these two principles imply that a good deal of the structure of space, in fact, all that can be expressed with the aid of 'overlap' or 'is part of', will be reflected in each part of time.

Although some of the suggestions we have put forward as principles may seem very strange, and quite remote from the usual account of Newton's thoughts on the matter, it is worthwhile recalling that these explications and speculations about time being diffused through space are but one strand of Newton's general belief, at this time, that all being is in space or related to space. Time and its parts are no exception for Newton, for they too have being.

But body is a kind of being also, and has its particular way of being in or related to space.

10. *Body and space.* It is a famous Newtonian theme that body is different from space. Space, at least empty space, in the "Gravitatione," is distinguished from void. Newton thought of the void

240

as nothing, whereas (as we have already indicated) he thought of empty space as still something, because parts of space are in space, and they are *something*. Body then is certainly different from void.[28]

In later writings Newton often used the absence of resistance to motion as an indication that space was empty. Regions filled with matter offer resistance to motion and conversely. And one could also in this vein remark that bodies are impenetrable to other bodies, they reflect light, have certain textures, and so forth, whereas a region of empty space has none of those features. Thus by the standard received view, body is distinct from space, and by his remarks in the ''Gravitatione,'' ''Place is a part of space which something fills evenly and Body is that which fills place.'' We should conclude that the ontology of Newton's theory of Absolute Space includes at least the empty space, its parts, and bodies. There is some ground, however, for qualifying the thesis that Newton drew a sharp distinction between body and space. I want to maintain that Newton did not wish to identify body with space. He wanted, for various philosophical and empirical reasons, to refute the Cartesian dictum that body and extension (three-dimensional volume) are identical. But in his refutation of the Cartesian dictum, he came very close to adopting it in another form. Newton offers what he terms an explanation of the nature of body. In order to show that bodies do not necessarily possess the powers or properties which Descartes ascribed to them, he asserts that since he cannot say with certainty what kinds of beings God could or could not create, he cannot speak with certainty about the nature of body. What he offers instead is to

> describe a certain kind of being similar in every way to bodies, and whose creation we cannot deny to be within the power of God, so that we can hardly say that it is not body.[29]

Newton envisions the possibility that instead of a body—say, a mountain in a given place—there is just that region of space, which God, by will, has singled out and, by his will, sees to it that other bodies cannot enter into that region of space. Light is turned back or reflected so that we witness reflection, see textures, and the like. Such a region of space would be similar in every way to a body. Moreover, if God so chooses, he can single out different regions of space and endow them with certain ''conditions''. We would then have a being similar in every way to a moving body. In this description we referred to bodies being repelled from the designated region of space, but this reference to a body can be treated in the same way, so that there can be beings which, for example, behave in every way like two colliding bodies.

If we ask whether body so understood is identical with space, the answer seems to be negative. If 'body' is understood as a count noun, then the question might read: Is it true that anything that is a body is a (part of) space? But the statement that each body is a part of space is ambiguous. On one reading, which seems to be the way Newton understood it, a body is identical with some one part of space, so that the answer to the question is surely no. It is sufficient to note that we would count something as the same body even though the specially conditioned part of space were to "shift" (i.e., be different) over time. But it is certainly true that every body is a part of space, though only in the case of those permanently at rest is it always the same part; and it is also true that every part of a body is a part of space. Moreover, if we understood 'body' as a mass noun, it would be true in this model that every part of body is a part of space. Thus, although in general no particular body is identical with one particular part of space, it is also clear that a body is just a specially "God-conditioned" part of space. In the context of refuting the Cartesian dictum and simultaneously answering the question of how body is related to space, the device or model is brilliant. Bodies are related to space in the simplest possible way: at any time, they and their parts are parts of space.

Throughout his discussion of the nature of body, Newton seems very reluctant to say definitely that his proposed explanation of body as a determined quantity of space is correct. Who could say that it was not within God's power; who is to deny that He did it? It should also be noted that in a way the Cartesian dicta such as all bodies are necessarily extended are obviously true if we use Newton's explanation of the nature of body. What is lost is Descartes' idea that extendedness is a necessary or essential feature of bodies. For Newton, the extendedness of bodies is due to the conditions endowed by God, and that they hold true is at any time a matter of His will, not a necessary consequence of something's being a body. Lest it be thought that Newton could not decide between the view usually ascribed to him—that there is a distinction between the content of the space where a body is and empty space—and the other view which he described in the "Gravitatione" that there is basically no difference: impenetrability, look, texture, and so forth, are brought about by God who is skillfully deflecting and angling off an empty region— I cite the following passage, from which it is clear that there was no doubt in his mind about where the truth lay:

> Lastly, the usefulness of the idea of body that I have described is brought out by the fact that it clearly involves the chief truths of metaphysics and thoroughly confirms and explains them. For we cannot postulate bodies of this kind without at the same time supposing that God exists and has created bodies in empty space out of nothing, and that they are beings distinct from created minds, but able to combine with minds. Say, if you can, which of the views already well-known, elucidates any one of these truths or rather is not opposed to all of them, and observes all of them.[30]

This favored view about body is not entirely compatible with other desiderata that arise from the physics of bodies. Newton attributed mass to all bodies.

Although he characterized the mass of a body by its inertia, it is not difficult to see how mass, so described, can be construed by using the resources of his favored model. However, Newton also characterized the mass of a body by its quantity of matter. Here, surely, we have the basis for a distinction between bodies and empty space. There is evidence from his correspondence with Roger Cotes and from a comparison of the difference between the first and second editions of the *Principia* that genuine conceptual and physical difficulties arose if both ways of characterizing mass were retained. It is only with the second edition of 1713, and chiefly through Cotes' insistence, that inertia rather than quantity of matter became the basic way of characterizing the mass of a body. I am suggesting that the model for body proposed by Newton in his earliest thoughts on the subject permitted an account of some of the properties attributed to bodies, such as texture, color, and impenetrability. But it is hard to see how a notion like the mass of a body, construed as its quantity of matter, could be accommodated on the basis of his model. One can only speculate here whether the attractiveness of his model and the physical problems connected with his maintaining two concepts of mass may have made his adoption of inertia as the key measure of mass easier. *In any event, these difficulties make it difficult to think of the distinction between bodies and space as a hallmark of Newton's ontological commitment.*

11. *Body and distance.* One theme very characteristic of Newton's

theory of Absolute Space is that the distance (at a given time) between two bodies is the distance between those parts of absolute space in which the bodies are at that time. One might wonder how such a statement might be checked. Do we ever have recourse to the distances between parts of space? In fact, we might think that a reasonable way to find out the distance between parts of space is to determine the distance between two bodies that occupy those spaces. Newton's thought is just the reverse.

> The positions, distances, and local motions of bodies are to be referred to the parts of space.

He refers to an earlier passage that has the same purport:

> And hence it follows that space is an effect arising from the first existence of being, because when any being is postulated, space is postulated. And the same may be asserted of duration: for certainly both are dispositions of being or attributes according to which we denominate quantitatively the presence and duration of any existing individual thing.[31]

The order is reversed. It is the distance between regions of space that determines the distances between bodies; it is lapse of absolute time that determines the duration of an individual thing. Obviously Newton is not offering practical advice on the measurement of distances and durations. Exactly how the distances between parts of space are to be estimated, checked, corrected, and confirmed is part of the story of how Absolute Space and Time are worked into the fabric of a viable empirical theory. I am not presently as concerned with that story as I am with a point that has not been sufficiently discussed: Is it meaningful to ascribe distances between parts of Absolute Space? Is there a theory that assures us that there is a distance between any two disjoint parts of space? Unless we are assured of the existence of a metric, then the Newtonian thesis that the distances between bodies and the duration of existing individuals be referred to distances and durations in Absolute Space and Time is vacuous.

Let us consider how distances might be assigned in Absolute Space. One might think of how the distance between two *points* of space is determined. But it must be recalled that in our reconstruction, we have the individual Absolute Space and all its parts, the regions of space. Are

244

there any points in Absolute Space? Surely not any arbitrary part of space is a point, since some of those parts will have proper parts. According to our current understanding, points seem to be the sort of parts that have no proper parts because they are dimensionless, and there are infinitely many in any finite line segment. If we ask whether Newton believed that Absolute Space had parts that could properly be called points, the answer, strangely enough, is not clearly in the affirmative.

Let us explore the question of Newton's commitment to any kind of entity resembling our current conception of a point, and return after that to the larger issue of whether the notion of distance is well-defined for the type of objects that constitute Absolute Space.

12. *Points and Minimalia*. Newton does mention, in the "Gravitatione," the existence of points, lines, and surfaces in space. He cites Euclid approvingly as if we were to understand these notions as they are described in Euclid's *Elements of Geometry*. But Euclid described surfaces as without width, lines as lacking breadth, and points as lacking both. In the "Gravitatione," Newton says that (all) the parts of space have limits (or boundaries) and that the limit of two parts is a surface and so forth. But he admits a possibility that is incompatible with a strict Euclidean reading.

> In all directions space can be distinguished into parts with common limits we usually call surfaces and these surfaces can be distinguished in all directions into parts whose common limits we usually call lines and again these lines can be distinguished in all directions into parts which we call points. And hence these surfaces do not have depth nor lines breadth nor points dimension *unless you say that coterminous spaces penetrate each other as far as the depth of the surface between them.* [*Italics added.*]

Newton seems to admit the possibility that surfaces and lines and points may not only have depth, breadth, and width, he also thinks it possible that surfaces and points (i.e., any boundary or part of a boundary) may penetrate each other to their full depth. One might wonder what kind of Euclidean points are these? Perhaps the following speculation may make plausible *Newton's belief that one could also treat points as (possibly) having dimension provided that such points can interpenetrate.* Euclid's geometry renders it plausible to consider a circle moving (so a circle cannot of course be a part of space) toward a line until they are tangent. Indeed, there is much use in Euclid of the motion of figures, as Newton

245

had noted (Hall and Hall, p. 122). Prior to their being tangent, both the circle and the line are complete, i.e., without 'gaps'. Once tangent, we say that they are tangent at a point. But there were two points, and now only one. One might ask, perversely, after the other. And suppose that the figures were separated? To which figure would that one point of tangency belong? An assignment of it to either figure leaves a 'gap' in the other. In flavor, how close such a puzzle seems to the following passage from Newton's unpublished *Questiones quaedam Philosophiae:*

> Extension is related to places, as time to days, years and c. Place is ye principium individuationes of streight lines and equall and like figures, ye surfaces of two bodys becoming but one when they are contiguous because but in one place (f. 91v).

How do these so-called points become one when they are contiguous? The answer seems to be that, when contiguous, these parts of space occupy exactly the same part of space and therefore must be identical. And they occupy that one part of space by "falling" into each other. In the *Questiones,* Newton developed the idea that there are least parts of space, time, and a least motion (velocity).[32] Here is a sample of his analysis of motion, using the existence of appropriate minimalia:

> *Of Motion.* That it may be known how motion is swifter or slower consider (i) that there is a least distance or least progression in motion and a least degree of time as lay two globes together so close yt they cannot come any nigher without touching yt is ye least distance, let ym be moved together. yt is ye least degree of motion and is performed in ye least part of time. There are so many parts in a line as there can stand mathematical points in a row without touching. (i.e., falling into) one another in it and so many degrees of motion along. yt line as there can be stops and stays. and there are so many least parts of time in an hower(?) as there can be $\tau \grave{o} \nu \nu \nu$'s. This proposition is proved as I proved a least part in matter. 2. these leasts have no parts for yt implies yt they are yet divisible neither prius nor posterius, nor least distance since it is passed over in an indivisible part of time and there cannot be a different time ascribed to ye entrance of a thing into yt part of space and ye leaving of it. 2 nor ye least degree of motion because too yt is performed in an indivisible pte of time and is no sooner begun yn done. . . .

The text is continued from Newton's page 10 to the top of his page 59:

ye joyning and meeting of ye two parts and posterius according to ye latter of ye two parts and so be still liable still to divisibility wch contradicts ye notion of an indivisible part. But to explaine how these leasts have no parts

And here the text ends.

Newton's page 10, *On Motion,* which we have just cited above, is lightly scratched out, though the continuation on page 59, that cliff-hanger of a last paragraph, remains untouched, and incomplete.

We cannot of course tell whether those cross-outs indicate that he rejected minimalia, or rejected his use of them in these passages, or that he rejected the arguments of Moore, Charleton, Barrow, or Gassendi. Perhaps he was unhappy with his claim that spatial and temporal minimalia could be demonstrated in the same way that the existence of least parts of matter were proved. Here is what he says about the existence of least parts of matter (*Of Attomes,* Newton's p. 3, f. 89r, again from the *Questiones*):

It remains therefore yt ye first matter must be attomes and yt Matter may be so small as to be indiscerpible. The excellent Dr. Moore in his booke of ye soules immortality hath proved—beyond all controversie y^{et} (?) I shall use one argument to show yt it cannot be divisible in infinitum and yt is this: nothing can be divided into more parts yn it can possibly be constituted of. But matter (i.e. finite) cannot be constituted of infinite parts. The Major is true for look into how many parts of a thing is divided those parts added againe make ye same whole that they were before, and so if any finite quantity were divided into infinite parts (and certainly it may if it be so far divisible) those infinite parts added would make ye same finite quantity they were before wch is againe ye Minor; and it is plaine from hence yt an infinite number of extended parts, and ye least parts of quantity must be extended, make a thing infinitely extended ([yt qua]) this you? cannot. be denied if I can prove yt things infinitely extended have fine parts. Now vacuum is infinitely extended and so may matter be fansied to be. But if ye world were removed(?) and vacuum came in ye rooms (?) of it yt very vacuum would not be be infinite. we can conceive of interspersed vacuities amongst matter but they are not infinite (though an infinite number of ym would be so.) We see ye parts of matter are finite and an infinite number of finite unities cannot be finite. To help ye conception of ye nature of these leasts, how they are indivisible how extended of wt figure &c. I shall all along draw a similitude from numbers comparing Math: points to ciphers, indivisible extension to unities: divisibility, or compound quantity to number: i.e., a multitude of attomes, to a multitude of unites. Suppose yn a number of Mathematicall points were induced with such a power as yt they could not touch nor be in one place. (for if they touch they will touch all over, & bee in one place) Then ad(?) these(?) as close in a line as they can stand together every point added must make some extension to ye length because it cannot sinke into ye formers place or touch it so here will be a line wch hath partes extra partes; another of these points cannot be added into ye midst of this line, for yt implies yt ye former points did not lie so close. But yt they might lye closer. The distance yn twixt each point is ye least yt can be and so little may an attome be and no lesse: Now yt this distance is indivisible (& therefore ye matter contained in it) is thus made plaine:

The explanation in terms of the arithmetical model with ciphers and units is given on Newton's pp. 63–65 (f.119r–120r) and almost all of those three pages bear heavy cross-outs. The argument quoted above seems to be that all parts of (finite) matter are finite, that there cannot be an infinite number of finite parts—and, since there are only a finite number of parts, there must be a least. The last part of the quotation seems to be an explanation of how one might estimate the minimal distance by trying to "pack" a line with minimalia that cannot touch. We can reconstruct the reason why they cannot touch as follows: If parts α and β touched, that would mean that there is a part γ common to both α and β. But neither α nor β have proper parts, so $\alpha = \beta$ and $\beta = \alpha$. That is, α and β would be identical. Newton maintained that these minimalia were extended, had shape and so forth, and I am inclined to interpret the cross-outs as registering disapproval of his attempt to explain via an arithmetical model just how such minimalia can be extended, have shape, and so forth.

I do not think that Newton's belief in the existence of minimalia was short-lived. We find him concerned with these tiny parts over almost half a century later.

The other aspect of points, there being infinitely many of them on a line segment, gave rise to some doubts on Newton's part. Recall that in the early *Questiones,* Newton remarked that "there are so many parts in a line as there can stand mathematical points in a row without touching (i.e., falling into) one another. . . ." Newton, I think, continued to refer to these parts as infinitely little. Thus, in his *On Analysis by Infinite Equations* (1669), he writes:

But it must be noted that unity which is set for the moment is a surface when the question concerns solids, a line when it relates to surfaces and a point when (as in this example) it has to do with lines (105) Nor am I afraid to talk of a unity in points or infinitely small lines inasmuch as geometers now consider proportions in these while using indivisible methods.[33]

And again in 1713, more than forty years later, in reflecting upon that essay, in an unpublished reply to Leibniz's Review of Wallis' collection that included Newton's *De Analysun,* he writes:

> A point is an infinitely short line. A unit multiplied by o - infinitely small moment of *quantity of time* etc. So then by a point Mr. Newton understands here an infinitely short line, & by a line an infinitely narrow space & when he calls these moments and represents them by an unit it is to be understood that this unit is multiplied by an infinitely small quantity o, or moment of time, to make it infinitely little.
>
> . . . For fluxions are finite quantities but moments here are infinitely little. Thus you see his Notation when he wrote this *Analysis* is of the same kind wth that w[ch] he uses at present. (217)

In the mid-60s Newton had reconstructed the calculus on a new basis— that of fluxions and fluents rather than the use of indivisibles. In certain writings he eschewed the use of indivisibles, and in others he used a blend of fluxions, fluents, and infinitely small quantities that he called moments. The recasting of the foundations of the calculus had, I think, very profound consequences for an understanding of Newton's later thought, especially about the concept of time, but let me document here the interesting amalgam of concepts that he offered his readers. There is, of course, the *Principia* itself:

> For demonstrations are shorter by the method of indivisibles; but because the hypothesis of indivisibles seems somewhat harsh, and therefore that method is reckoned less geometrical. I chose rather to reduce the demonstrations of the following propositions to the first and last sums and ratios of nascent and evanescent quantities, that is, to the limits of those sums and ratios, and so to premise, as short as I could, the demonstrations of these limits. For hereby the same thing is performed as by the method of indivisibles; and now those principles being demonstrated, we may use them with greater safety. There-fore if hereafter I should happen to consider quantities as made up of particles or should use little curved lines for right ones, I would not be understood to mean indivisibles, but evanescent quantities. . . .

Continuing in the same passage in the Scholium, he notes that it may also be objected, that if the ultimate ratios of evanescent quantities are given, their ultimate magnitudes will also be given: and so all quantities will consist of indivisibles, which is contrary to what Euclid has demonstrated concerning incommensurables, in the tenth Book of his *Elements*.[34]

Note that Newton does not deny the existence of indivisibles, he merely terms the hypothesis of indivisibles as "harsh," and he thinks that various principles that refer to indivisibles are somewhat unsafe, unless the same results can be secured using limit notions. Even his reference to Euclid seems indecisive. The impression given is that Euclid has shown that not all quantities consist of indivisibles. The suggestion seems to be that some quantities do consist of indivisibles, and some quantities do not. If we refer back to the Euclidean results, the indecisiveness over indivisibles is not removed. The relevent statements seem to be a definition and two propositions:

> *Definition 1.* Those magnitudes are said to be *commensurable* which are measured by the same measure, and those *incommensurable* which cannot have any common measure.
>
> *Proposition 1.* Two unequal magnitudes being set out, if from the greater there be subtracted a magnitude greater than its half, and from that which is left a magnitude greater than its half, and if this process be repeated continually, there will be left some magnitude which will be less than the lesser magnitude set out.
>
> *Proposition 2.* If, when the less of two unequal magnitudes is continually subtracted in turn from the greater, that which is left never measures the one before it, the magnitudes will be incommensurable.
>
> (Health, *Euclid's Elements,* 2d ed., vol. 3)

Of course there is no mention of indivisibles in these statements, and the relation of the three to the issue of their existence has to be supplied. Of the two propositions, only the second refers to incommensurables. How does Proposition 2 preclude the existence of indivisibles in all quantities? In short, what is the connection between the second proposition and Newton's belief about indivisibles? The Definition is of no help. If we suppose, as Proclus seems to have done (cf. Heath, *Euclid's Elements,* Book 1, 1. 268), that an indivisible is the least common measure of two magnitudes, then the relevance of Proposition 2 to the non-existence of indivisibles is apparent. For, following Proclus, we say that for any three magnitudes α, A, and B (of the same type), α is an indivisible (of A and B) if and only if it is the least common measure of A and B. However, by Proposition 2, if A and B are incommensurable, they have no common measure and, hence, no least common measure. Therefore nothing is an indivisible of the incommensurable magnitudes A and B. Following Proclus, the relation between incommensurables and indivisibles is clear via Proposition 2. However, in this case, it is also evident that nothing in the *Elements* bears on the case when A and B are commensurable. Such magnitudes might still have indivisibles as least common measure.

Thus the rejection of indivisibles is conditional upon the fact that incommensurable magnitudes have no least common measure. But readers will already have noted that Proposition 1, by itself, seems to yield a categorical rejection of indivisibles, provided that indivisible magnitudes are taken to be the smallest parts, or parts of least magnitude. Proposition 1 provides an algorithm for producing arbitrarily small sub-segments of a given segment. The proof of this proposition uses the so-called Archimedean Axiom, whose validity, as recent investigations into the mathematics of indivisibles have rightly emphasized, is crucial for the denial of the existence of indivisible quantities.

There seem to be two criteria of 'indivisible', each embedded in a proposition. One embodies the idea of a part of least magnitude, the other, the idea that an indivisible is a least common measure of two magnitudes (of the same type). The former makes more sense for geometrical quantities like length and volume, where the notion of a part has a natural interpretation. The latter characterization of indivisibles suits a quantity like velocity better. The second description of indivisible can be applied to both types of quantity, and when it is, the result, I conjecture, is a qualified rather than a categorical denial of the existence of indivisibles.

Although Newton by the 1670s had discovered an alternate foundation for analysis that required no reference to indivisibles, his writings, both mathematical and physical, are studded with references to infinitely little quantities as well as to fluxions and fluents. Here is Newton writing to Hooke almost a decade after his important mathematical investigations of the 70s:

> The innumerable & infinitely little motions (for I here consider motion according to ye method of indivisibles) continually generated by gravity in its passage from A to F incline it to verge from GN towards D, & . . .[35]

The hedging references to infinitely little quantities can be traced even into the 1690s. Consider this report on Newton's method of fluxions written in 1692, probably by Newton himself.

> For altho Flowing Quantities & their Fluxion(s) may at first sight seem difficult to Conceive (new things being somewt difficult to conceive) yet he thinks ye Notion of them will soon become more easy, than is that of Moments or parts infinitely little or differences infinitely small; because the generation of Figures and quantities by continued motion is more Natural & more easily Conceived, & ye Schemes in this Method are more Simple yn in yt of parts. Nevertheless he does not Neglect ye Theory of such parts but uses it also where it abbreviates the worke, and renders it more plain, or when it Conduces to investigate ye proportion of Fluxions.[36]

The tone of the remarks is not that of a direct rejection of indivisibles. The new "foundation" using Fluxions and Fluents is regarded more as an alternate, better method for the solution of certain mathematical problems. Even though the new method is called more natural, and simpler than that which uses infinitely small quantities, the dominant tone, if anything, seems somewhat defensive: the new methods are more difficult to conceive because they are new; almost apologetically the author states that the theory of infinitely little parts is not neglected by Newton.

Indeed, there are many reasons both theoretical and practical why such an advance would not have been best promoted by announcing the demise of the opposition. We cannot enter here into the subtle problem of how a theory can lay claim to the results of a rival and simultaneously divest itself of the conceptual apparatus of that rival. All the problems of characterizing reduction in mathematics, and clarifying the idea of a justified use of a theory, arise on the theoretical plane. But practically speaking, it should be obvious that a direct denial of a theory that had wide vogue and credibility at this time, and which was closely linked to his befriender, Sir I. Barrow, would have raised problems and initiated controversies that Newton might well have preferred to let lie. This last point, of course, speaks only to the rhetoric of the situation.

The 1692 letter from Newton to Bentley, which contains the famous disclaimer that gravity is an essential quality of bodies, also contains a passage that bears very directly upon the notion of indivisibles and their possible number and is worth quoting at length.

But you argue in ye next paragraph of your letter[3] that every particle of matter in an infinite space has an infinite quantity of matter in all sides & by consequence an infinite attraction every way & therefore must rest *in equilibrio* because all infinites are equal. Yet you suspect a parallogism in this argument, & I conceive ye parallogism lies in ye position that all infinites are equal. The generality of mankind consider infinites no other ways than definitely, & in this sense they say all infinites are equal, though they would speak more truly if they should say they are neither equal nor unequal nor have any certain difference or proportion one to another. In this sense therefore no conclusions can be drawn from them about ye equality, proportions or differences of things, & they that attempt to do it, usually fall into parallogism. So when men argue against ye infinite divisibility of magnitude, by saying that if an inch may be divided into an infinite number of parts, ye sum of those parts will be an inch, & if a foot may be divided into an infinite number of parts ye sum of those parts must be a foot, & therefore since all infinites are equal those sums must be equal, that is an inch equal to a foot. The falseness of ye conclusion shews an error in ye premises, & ye error lies in ye position that all infinites are equal. There is therefore another way of considering infinites used by Mathematicians, & that is under certain definite restrictions & limitations whereby infinites are determined to have certain differences of proportions to one another. Thus Dr. Wallis considers them in his *Arithmetica Infinitorum* where by ye various proportions of infinite sums he gathers ye various proportions of infinite magnitudes: which by way of arguing is generally allowed by Mathematicians & yet would not be good were all infinites equal. According to ye same way of considering infinites, a Mathematician would tell you that though there be an infinite number of infinitely little parts in an inch yet there is twelve times that number of such parts in a foot; that is, ye infinite number of those parts in a foot is not equal to, but twelve times bigger than ye infinite number of them in an inch. And so a Mathematician will tell you that if a body stood *in equilibrio* between any two equal and contrary attracting infinite forces, & if to either of those forces you add any new finite attracting force; that new force how little so ever will destroy ye equilibrium & put ye body into ye same motion into which it would put it were those two contrary equal forces but finite or even none at all: so that in this case two equal infinites by ye addition of a finite to either of them become equal in our ways of reconning. And after these ways we must reconn if from ye consideration of infinites we would always draw true conclusions.[37]

And finally, in a letter that must be dated slightly later than 1710,

Newton is still reluctant to say outright that there are no infinitely small parts of a line:

> May it please yo^e grace
> An Italian as I heare, has lately published a book ([in Italy]) about things being infinitely infinite & if tha . . . say that the number of points in the line AB is infinitely infinite, & by consequence that an infinite number of points will take up but an infinitely small part of a line ([AB, it would be requisite to see his Demonstration before I give my opinion in the matter]) AB, I am not able to confute him. On the contrary others may dispute that the number of points in a line infinitely long is no more than infinite & therefore ye points A & B must be at an infinite distance to be the extremes of an infinite number of points. And others may dispute whether questions of this kind are Mathematical or Metaphysical. For my part I must acknowledge them far from my understanding. I remain
>
> Your graces most h & most ob^t
S.I.[38]

Newton's attitude toward various minimalia or infinitely small quantities, whether of space, time, motion, or matter, seems equivocal to me. There is an obvious belief in their existence in the early writings. Nevertheless, despite the recasting of the calculus in the 70s, a good deal of his writing describes these small parts in terms that fall considerably short of rejection. Therefore, if we ask whether Newtonian Absolute Space had points—dimensionless entities that constitute geometric lines and form continua of sorts—then the answer is negative. The closest candidate for points seems to be minima of various kinds, but they are not dimensionless; and the constitution of geometric figures from these 'points' seems to be a problem requiring new mathematical insights. The idea expressed in his letter to Bentley, cited above, that not all infinites are equal and may not even be comparable, was quite a bold suggestion.[39] His ambivalence, even in later years, about the existence of indivisibles suggests that the reconstruction of Newton's views on space and time as an empirical relational system turns out to be a very delicate matter. We cannot assume that Newton, during this period and even later, believed that space had to be dense, Archimedean, or complete, that 'point' for him had its present sense, or that time had to consist of durationless moments. Whatever the regions of space and the moments of time may be, they cannot, without good reason be identified with the concepts so labeled, of classical analysis. In fact, the evidence seems to point away from such an identification.

Let us return then to our discussion of that most typical Newtonian thesis: the distances between bodies are to be determined by the distance between the regions of absolute space that those bodies occupy (at the same time). If the distances between bodies are to be referred to regions of space—and we do not have 'points' in absolute space; only dimensional parts or regions of space are at our disposal—then what sense is there in talk of distance between such regions?[40] We can, using the techniques of modern topology, give an analysis of absolute space that places the concept of a distance between the parts of such a space on a secure basis,[41] and thus buttress the central Newtonian theme that the distance of bodies is to be referred to the distance of space—not the other way around.

Appendix

THE TOPOLOGY OF ABSOLUTE SPACE

The question we then confront is this: If Newtonian theory of Absolute Space is a theory about a particular object and all its parts, then is this theory sufficiently rich to warrant our attributing distances between, say, any two parts or regions of space?

It is instructive to begin with the classical topological idea that distance is given by a distance or metric function on a space; that is, we assume that there is a space or set of elements, S, together with a function, ρ, that maps pairs of elements of S into the non-negative real numbers such that

1. $\rho(x,y) = 0$ if and only if $x = y$

2. $\rho(x,y) = \rho(y,x)$

3. $\rho(x,y) + \rho(x,z) \geqq \rho(y,z)$

where x, y, and z are elements of S. It is worth noting that the third condition implies the second. If (1) is replaced by

$$(1)'\ \rho(x,y) = 0 \text{ if } x = y$$

then ρ is called a *pseudo-metric function*. The ordered pair (S,ρ) of a set S and metric (pseudo-metric) function ρ, is called a *metric (pseudo-metric space)*.

It is worth noting also that any set S can become a metric space, since there is always a function Σ defined on $S \times S$ that satisfies (1)–(3): set

$$\Sigma(x,y) = 0 \text{ if } x = y, \text{ and } \Sigma(x,y) = 1 \text{ if } x \neq y \ .$$

When we turn to the set consisting of Absolute Space and all its parts, the straightforward transfer of conditions (1)–(3) presents certain difficulties. If α and β are two parts of space that overlap, then (1) seems to need modification since $\rho(\alpha,\beta)$ should surely be zero, even though $\alpha \neq \beta$. Further, if α and β overlap, and β and γ overlap, while α and γ do not, then (3) is false. Therefore, we shall think of a metric function for a set of parts as a non-negative real-valued function Δ which satisfies the following:

(1)' $\Delta(\alpha,\beta) = 0$ if and only if α and β overlap

(2)' $\Delta(\alpha,\beta) = \Delta(\beta,\alpha)$

(3)' $\Delta(\beta,\alpha) + \Delta(\beta,\gamma) \geqq \Delta(\alpha,\gamma)$, where none of α,β,γ overlap.

256

Incidentally, despite these modifications, it remains true for Δ's as well as ρ's that they can always be defined for an arbitrary set: let $\Delta(\alpha,\beta) = 0$ if α and β overlap; $\Delta(\alpha,\beta) = 1$ otherwise. (1)′ –(3)′ are clearly satisfied. The question of whether a set of parts is a metric space or not is relatively uninteresting since there always are Δ's on hand. Just as in the case of standard topology, what is interesting is not whether a certain set S is a metric space, but whether S is *metrizable*. That is, it is assumed that the set S already has been provided with the structure of a topological space with its open (closed) sets. If a metric function is provided for the set S, there is a standard method of defining a topological structure of S, called the topology induced by the metric: define the notion of a disk of radius ε (the set of all points of the space less than, or equal to, a certain distance ε from a given point).

An open set is defined as any set-theoretical union of disks. It can be shown that open sets, as just introduced, satisfy the usual conditions for a topological structure. The remaining question is whether the topological structure which S has is equivalent to the topological structure induced by the metric function (that is, precisely the same sets are open in each topology). If it is, then the set S is said to be *metrizable*. Of course, there may be many ρ's that do the trick; that is, many ρ's that are topologically equivalent; that is, any two such functions induce exactly the same topological structure for S. It should be noticed that some very different kinds of functions can still be equivalent, and their peculiarities are not relevant to the issue of metrizability.

Thus the question we asked—is the concept of distance well-defined for this set of parts of absolute space S?—has substance only if we ask whether S is metrizable, and this in turn presupposes that S has a topological structure that is "recaptured" by a metric function.

Does the set S, consisting of Absolute Space and all its regions, or parts, have a topological structure? Let us remind the reader at this point of the standard conditions for a topological structure for the set S. The ordered pair of a set S, together with a mapping K that maps subsets of S into subsets of S, is a topological space if and only if:

(1) $K(S) = S$, and $K(\phi) = \phi$

(2) $U \subseteq K(U)$, for any subset U of S

(3) $KK(U) = U$

(4) $K(U_1 \cup U_2) = K(U_1) \cup K(U_2)$

The set $K(U)$ is called the *closure of* U, and a set is closed if and only if it is equal to its own closure. We wish to add a fifth condition, proposed by Kuratowski, that is regarded as a condition on the space S rather than a condition on the function K:

(5) $K(\{a\}) = \{a\}$, for any element a of S.

The fifth condition therefore states that every unit set is closed. Such spaces are called T_1-spaces, or spaces that are T_1-separable.

We shall now try to show that the set S consisting of Absolute Space together with all its regions or parts has a natural topology, given the Newtonian dicta (chiefly to be found in the "De Gravitatione") about the parts of space and their limits. This topology, as we shall see, is T_1, and completely regular, so that according to a standard theorem,[42] such a space is pseudo-metrizable. Thus Newton's observations about absolute space almost determine a certain kind of distance function for it; not a true metric, but one that is pretty close.

The best way I know of for determining the topology of Newtonian absolute space proceeds in a roundabout fashion. Let us first introduce the notion of a *proximity space*, and via that, the Newtonian topology. Let us say that a set X, not assumed to be metric, is a *proximity space* with respect to the relation δ defined for all subsets of X, if and only if the following conditions are satisfied (read "$A\delta B$" as "A is close to B"):

1. δ is symmetric

2. $A \, \delta \, (B \cup C) \equiv (A \, \delta \, B \text{ or } A \, \delta \, C)$

3. $P \, \delta \, q \equiv (p = q)$

4. ϕ non-δ X

5. If A non-δ B, then there are sets C and D, such that $A \subset C$, $B \subset D$, $C \cap D = \phi$, and A non-δ $(-C)$, B non-δ $(-D)$ (upper-case letters indicate subsets of X, and lower-case indicate elements of X). Our strategy then is to show that there is a natural candidate for the closeness relation among regions of absolute space which satisifes the conditions for proximity spaces. Then, with the aid of this notion of closeness, we shall define a closure mapping K which satisfies the conditions for a topological structure which is metrizable: $K(A) = \{x | x\delta A\} = \overline{A}$.

It seems natural, given Newton's remarks about the parts of space and their limits in "De Gravitatione," to think of every (proper) part of space α as having a certain "environment" $E(\alpha)$ that is also a part of Absolute Space satisfying the following conditions:

1. For every (proper) part of space α, $\alpha < E(\alpha)$. That is, every part of space is a part of its environment. If the environment of α is not the whole space, then this condition would state that every (proper) part of space is also *in* space.

2. For every part α and β, if $\alpha < \beta < E(\alpha)$, then $E(\alpha) = E(\beta)$. Thus, although there may be many parts of space between a part of space and its environment, all such parts must have the same environment.

258

3. If, for any part $\gamma, \gamma < E(\alpha)$, then $E(\gamma) \circ \alpha$. Thus every part of the environment of α is such that its environment $E(\gamma)$ overlaps with α. We do not insist that every part of $E(\alpha)$ must overlap with α; only that the environment of every such part must overlap with α. In an intuitive way, this condition implies that the parts of $E(\alpha)$ never get "far" from α, since even if any part of $E(\alpha)$ fails to overlap α, nevertheless its environment will.

4. For any parts α, β, if $\alpha < \beta$, then $E(\alpha) < E(\beta)$.

And lastly,

5. $E(E(\alpha)) < E(\alpha)$.

The last condition, together with (1), implies that $E(\alpha) = EE(\alpha)$. In other words, one might extend a part of space to an even more inclusive part, its environment. This might be thought of as extending the part by including or adjoining its limits. These limits, as we have seen, might have depth or dimension, and are not to be thought of as dimensionless lines, surfaces, or points. But environments, even if they are proper parts of other parts of space, do not have more inclusive environments. Once you have "added on" the limits, those limits do not have limits in turn.

We suggest the following analysis for bringing out the proximity-space character of the set consisting of Absolute Space together with all its parts.

1. For indivisuals x and y, we set $x \delta y$ if and only if $E(x) = E(y)$.

2. For sets A and B, a notion of proximity has to be provided with some care. Obviously, if $A \delta B$ is defined so that $\{x\} \delta A$ if and only if $x \delta \alpha$ for some α in A, and we later show that $x \delta \alpha$ implies $x = \alpha$, then every set will be closed under the natural definition of closure given by $x \varepsilon A \equiv x \delta A$. The topology in this case would be the uninteresting discrete one. Thus we certainly do not want to define $A \delta B$ as holding if and only if there are elements of A and B which are close, i.e., related by δ.

3. We shall say that set A is close to set B ($A \delta B$) if and only if there is a sequence of elements of A that converge (in the environment) to an element of B, or there is a sequence of elements of B which converge (in the environment) to an element of A. A sequence of elements α_1, $\alpha_2, \alpha_3, \ldots$ converges to the element β (in the environment) provided that

   a. the sequence of α's is non-increasing environmentally, i.e., $E(\alpha_1) > E(\alpha_2) > E(\alpha_3) > \ldots$

   b. for any elements γ and γ' distinct from each other, and from β, if $\gamma < E(\beta) < \gamma'$, then there is some N such that for all $n \geq N$, $\gamma < E(\alpha_n) < \gamma'$.

259

4. We assume the uniqueness of convergence, at least up to environments. That is, if there is a sequence which converges (in the environment) to both β and β', then $E(\beta) = E(\beta')$.

It remains only to verify conditions (1)–(5) for proximity-spaces. (1) holds since δ is symmetric. (2) is the condition that $A \ \delta \ (B \cup C) \equiv A \ \delta \ B$ or $A \ \delta \ C$. From right to left is obviously true. The other direction is straightforward also. Let us defer (3), the condition that $p \ \delta \ q$ if and only if $p = q$. (4), the condition that ϕ non-δ x, is easily verified. Condition (5) states that if A non-δ B, then there are sets C and D such that $A \subset C$, $B \subset C$, $C \subset D = \phi$, and A non-δ $(-C)$, and B non-δ $(-D)$. Everything hinges upon the appropriate choice of disjoint C and D. A straightforward computation shows that C and D, as defined below, satisfy condition (5):

$$C = A \ \cup \ \{E(\alpha)|\alpha \varepsilon A\} \ \cup \ \{x|E(x) = E(\alpha) \text{ for some } \alpha \text{ in } A\}$$

$$D = B \ \cup \ \{E(\beta)|\beta \varepsilon B\} \ \cup \ \{y|E(y) = E(\beta) \text{ for some } \beta \text{ in } B\}$$

We return now to the deferred condition (3). In order to assure its satisfaction, we pass from the set S consisting of Absolute Space together with all its parts to a factor space S^*. For each α belonging to the set S, define α^*

$$\alpha^* = \{x \varepsilon S | E(x) = E(\alpha)\} \text{ and define}$$

$$S^* = \{x | x = \alpha^*, \text{ for some } \alpha \text{ in } S\}$$

By lifting the proximity relation from the set S to the set S^*, it is easy to see that $\alpha^* \ \delta \ \beta^*$ if and only if $\alpha^* = \beta^*$ by recalling that $\alpha \ \delta \ \beta$ if and only if $E(\alpha) = E(\beta)$. We therefore have a proximity-space, and hence a T_1-separable completely regular space, i.e., a Tychonov space.[43] By a standard theorem in the literature, Tychonov spaces are pseudo-metrizable.

This essay was supported in part by grants from the Ford Foundation and the Research Foundation of the City University of New York.

1. The basic concept of an empirical relational system derives ultimately from Alfred Tarski, and recently P. Suppes and D. Scott have developed a theory of measurement in the sciences with its aid. We are suggesting here that a Theory of Space, Time, or Space-Time be represented as empirical relational systems.

2. "De Gravitatione Et Aequipondio Fluidorum," University Library, Cambridge Add. 4003. Reprinted and translated into English in *Unpublished Scientific Papers of Isaac Newton*, A. R. Hall and M. B. Hall (Cambridge University Press, 1962).

3. It turned out that there was relatively little time for Part II at the original talk, and absolutely no space for it in this volume.

4. Newton added a note to his definitions: "I said that a body fills place, that is, it so completely fills it that it wholly excludes other things of the same kind or other bodies as if it were an

unpenetrable being. Place could be said however to be a part of Space in which a thing is evenly distributed; but as only bodies are here considered and not penetrable things, I have preferred to define [place] as the part of space that things fill'' (*Unpublished Scientific Papers,* p. 122). If there should be some qualm about the Halls' translation, which speaks of places being *filled,* it should be noted that Newton made the same point in an English manuscript also dating from the mid sixties: ''There is an uniform extension, Space, or expansion continued every way without bounds: in which parts of Space possessed & adequately filled by ym are their places'' (''The Lawes of Motion,'' Add 3958, fols 81–83, reproduced in *Unpublished Scientific Papers*).

5. University Library, Cambridge, Add. 3996, 87r–135r.

6. *Unpublished Scientific Papers*, p. 136.

7. One of Newton's criticisms of Descartes' concept of space refers to its God-independence or absoluteness.

8. *Unpublished Scientific Papers,* p. 144.

9. A. Koslow, ed., *The Changeless Order: The Physics of Space Time and Motion* (New York: Braziller, 1967), pp. 139–41.

10. *Unpublished Scientific Papers,* p. 127.

11. In another passage of the ''De Gravitatione'' which is almost repeated in *Leibniz-Clarke Correspondence* some fifty years later, Newton argued that ''physical and absolute motion is to be defined from other considerations than translation (meaning motion relative to other bodies), such translation being designated as merely external.'' The reason is that it seems to Newton that the relative motion of the earth and the fixed stars (heavens) can be produced in two entirely different ways depending upon whether force is exerted on the earth or upon the heavens, and in each case there are different absolute motions, taking place. Thus for him, relative motions somehow conceal the true absolute motions. What I have suggested is that these absolute motions are thought of as having a causal or explanatory role (perhaps via the forces which these motions are supposed to produce).

12. *Unpublished Scientific Papers,* p. 128.

13. H. Feigl, ''The Origin and Spirit of Logical Positivism,'' in *The Legacy of Logical Positivism,* ed. P. Achinstein and S. Barker (Baltimore: John Hopkins University Press, 1969), p. 7.

14. A. Muller, *Das Problem des absoluten Raumes und seine Beziehung zum allegemeinen Raumproblem* (Brunswick: Friedr. Vieweg & Sohn, 1911). I am indebted to David Lindenfeld of Ohio State University for identifying Muller's reference to Meinong as most probably the discussion about real and ideal relations and change in *Uber die Annahmen,* p. 844.

15. In this description of the calculus of individuals I am following the streamlined form provided by W. Hodges and D. Lewis, ''Finitude and Infinitude in the Atomic Calculus of Individuals,'' and G. Hellman, ''Finitude, Infinitude, and Isomorphism of Interpretations in Some Nominalistic Calculi,'' both in *Nous* 2 (1968).

16. *Unpublished Scientific Papers,* pp. 133–34.

17. Ibid., p. 135.

18. Ibid., p. 137.

19. Ibid., p. 138.

20. Ibid.

21. Ibid., p. 137.

22. Ibid., p. 131.

23. Ibid., p. 136.

24. Ibid. The Halls have called attention to a passage in Newton's *Principia* itself which is very similar but nowhere near as explicit. Cf. *Sir Isaac Newton's Mathematical Principles of Natural Philosophy,* ed. F. Cajori (Berkeley: University of California Press, 1947), p. 8.

25. *Unpublished Scientific Papers,* p. 137.

26. To John Josephson, of Ohio State University, I owe the inspiration for the much improved "$\tau\rho$" notation.

27. There is no suggestion in the "De Gravitatione" that parts of time or duration must be non-dimensional or point-like. In fact, in the passage about the individuation of parts of time, he readily treats Yesterday and Today as temporal parts. Even in the passage on diffusion of a moment of duration over London and Rome, Newton does not say that such a part has no parts; he says that we should understand that diffusion without thinking of, or concern with, its parts.

28. J. E. McGuire, "Body and Void and Newton's *De Mundi Systemale:* Some New Sources," *Archive for History of Exact Science* 3 (1966–67): 206–48.

29. *Unpublished Scientific Papers,* p. 138.

30. Ibid., p. 142.

31. Ibid., pp. 136–37.

32. Though most of these claims and their arguments can be traced to contemporary sources, it should be stressed that despite the sources Newton did believe them at the time, and believed in related themes even later. For an excellent description of the *Questiones,* see R. S. Westfall, *Force in Newton's Physics* (New York: American Elsevier, 1971).

33. D. T. Whiteside, ed., *The Mathematical Papers of Isaac Newton,* vol. 2, 1667–70 (Cambridge: At the University Press, 1968), p. 235.

34. F. Cajori, ed., *Newton's Principia, Mottes Translations Revised* (Berkeley: University of California Press, 1947), pp. 38–39.

35. H. W. Turnbull, F.R.S., ed., *The Correspondence of Isaac Newton* (Cambridge: Published for the Royal Society at the University Press, 1960), p. 307.

36. *Correspondence,* 3 (Cambridge: At the University Press, 1961): 222–23. For an interesting recent article which tries to show under what conditions Newton used Fluxions, Infinitesimals, and Ultimate Ratios, See P. Kitcher, "Fluxions,.Limits, and Infinite Littlenesse", *Isis* 64 (1973): 33–49.

37. I. Newton to R. Bentley, 17 January 1692–93, in I. B. Cohen, ed., *Isaac Newton's Papers and Letters on Natural Philosophy* (Cambridge: At the University Press, 1958), pp. 293–96.

38. ULC, Add. 3965.13. Expressions in double parentheses are those which were crossed out in the original text. Although the language is close to those passages on indivisibles in Galileo's *Dialogue on the Two New Sciences,* that work is now generally believed to have been unavailable in translations at this time; and anyway, even if there were an English translation in private circulation, it would hardly have counted as a publication for Newton, and certainly not as a recent publication. Thus the "Italian" was not Galileo. Inspection of the scratched-out beginning of the initial sentence seems to reveal the name of Pere Grandi. This is a reference I think to Guido Grandi (1671–1742), an Italian mathematician who published his *De infinitis infinitorum et infinite parvorum ordinibus* in Pisa (1710). The trace to Newton is fairly direct. In a letter to Newton dated 1707 (*Correspondence,* 4:506-7), Henry Newton (apparently not a relative) adds the postscript: "Father Grandi of Pisa, who is not one of ye least readers or admirers of ye same work [i.e., Newton's *Opticks*] presents likewise his humblest respects to you." Guido could very well have been "Father Grandi", having joined the Benedictine order of Camaldolites. But if Guido's first work on indivisibles was published in 1710 some three years later then Henry Newton's letter, how did Newton "heare" of the work? Most probably through Henry Newton himself, to whom Guido dedicated his *De Infinitis Infinitorum.*

39. The similar-sounding discussion in the First Day of Galileo's *Dialogue Concerning Two New Sciences* does not diminish our sense of Newton's boldness. In the *Dialogues* Galileo suggests that infinites and indivisibles are incomprehensible to humans because of their finiteness. Paradoxes drawn with the aid of these concepts simply bear witness to their incomprehensibility. Newton, on the other hand, does not think that infinites and indivisibles are incomprehensible; he believes that there are mathematically intelligible ways of handling this, and it is just false to think that all infinites are equal or even comparable.

262

40. It would be a mistake to think it impossible to introduce topological considerations if the space did not consist of points, and the subsets were not point sets. For an admirable review of the mathematical work of Moore, Milgram, Stone, Wallman, Wald, and Menger, toward eliminating the concept of point as basic in favor of pieces, lumps and so on, the reader should study the results of K. Menger, "Topology without Points," *Rice Institute Pamphlet* 27 (1940): 80–107, in which the mathematical motivation for such a program is related to a general movement to algebraize certain elementary geometries. The philosophical motivation of Whitehead and Nicod following him and the importance for physics are mentioned in passing.

41. I do not mean that we shall show that a specific metric function, say, Euclidean, must be the metric of Absolute Space. All I want to argue is that the space is *metrizable,* that some metric function preserves the topology. It seems to me that this is sufficient to justify talk about the distances between parts of Absolute Space.

42. The theorem is that any completely regular T_1-space (a Tychonov Space) is pseudo-metrizable. For a proof, cf. K. Kuratowski, *Topology,* 1 (New York: 1966): 233, Theorem 2.

43. For a proof that every Proximity Space is a completely regular T_1-space, cf. either Kuratowsky, *Topology,* vol. 1, or S. A. Naimpally and B. D. Warrack, *Proximity Spaces* (Cambridge: At the University Press, 1970).

Margaret D. Wilson

Chapter Ten

LEIBNIZ'S DYNAMICS AND CONTINGENCY IN NATURE

I

In 1699 Leibniz wrote to a correspondent:

> My Dynamics requires a work to itself . . . you are right in judging that it is to a great extent the foundation of my system; for it is there that we learn the difference between truths whose necessity is brute and geometrical, and truths which have their source in fitness and final causes.[1]

And about a decade later he remarks in the *Theodicy*:

> This great example of the laws of motion shows us in the clearest possible way how much difference there is among these three cases, first, *an absolute necessity*, metaphysical or geometric, which can be called *blind* and which depends only on efficient causes; in the second place, *a moral necessity*, which comes from the free choice of wisdom with respect to final causes; and finally in the third place, *something absolutely arbitrary*, depending on an indifference of equilibrium which is imagined, but which cannot exist, where there is no sufficient reason either in the efficient or in the final cause.[2]

The claims made in these passages for the philosophical importance of Dynamics are strong and in a sense unequivocal. Yet they are also rather mysterious. Leibniz of course wants to hold that the laws of nature *are* contingent—not true in all possible worlds; this in fact is a point he often makes side by side with the claim that the existence of any individual substance is contigent, and depends on the free choice of God to create the most perfect of the possible worlds. But in these passages Leibniz seems to say more than this. He says we *learn* from Dynamics the difference between necessary and contingent truths, that the laws of motion show "in the clearest possible way how much difference there is" between the blind or brute necessity of mathematics and contingent or "morally necessary" truths that depend on the choice of perfect reason. What is more, he speaks in the first passage specifically of "my" Dynamics, while the second passage follows a long criticism of Cartesian physics and a statement of some of Leibniz's own physical principles. The suggestion, then, is that the specific principles that Leibniz believed he had established, in opposition to the Cartesians, help make evident the difference between necessary and contingent truths.

To a twentieth-century philosopher, this notion is apt to seem very odd. The question whether the laws of nature are necessary or contingent—assuming this is a reasonable question at all—does not seem in any way dependent on what laws are found to be the true ones. (Similarly, one would have difficulty making sense of the suggestion that the question whether individual existence is necessary or contingent must be answered with reference to what individuals actually exist.) And in fact, the contingency of laws as well as the contingency of particular existents seems often to be presented by Leibniz himself as a tenet of his system resting on purely philosophical intuitions. Nevertheless (as it will be the purpose of this paper to show), Leibniz's remarks in the passages quoted do reflect an important and persistent aspect of his thought. Further, his claims for the philosophical significance of his Dynamics, though largely anachronistic today, are tied up with some issues of considerable interest from the point of view of the history of ideas.

Leibniz's doctrine of contingency has, of course, been the subject of much controversy in the critical literature of the past seventy years. However, this controversy has tended to focus on the status of propositions about particular individuals—e.g., "Adam ate the apple"—and largely to neglect any problems about contingency that might be specifically related to his views about the laws of nature. Most attention has been devoted to determining how, if at all, Leibniz's claims that propositions about particular existents are contingent, and true only because God freely selected the *best* possible world for creation, can be reconciled with his further doctrine that in *every* true proposition the concept of the predicate is "in some manner" included in the concept of the subject. For the latter claim seems to imply that *all* propositions are implicit identities. But Leibniz's standard definition of a necessary truth is a proposition the negation of which implies a contradiction—in other words, an explicit or implicit identity. Thus the theory of truth suggests that even existential propositions may ultimately have to be construed as necessary, and brings in question the sense in which alternative worlds may rightly be characterized as "possible." (That another world is in itself consistently conceivable need not imply that it might have existed.) Of course, the theory of truth presents difficulties for any claim of contingency in Leibniz's system—not just for propositions about individuals. But some of the most interesting aspects of this particular problem are tied up with the treatment of individual substances.[3]

Naturally, it has not gone unnoticed that Leibniz regarded the laws of nature as contingent. However, their status tends to be touched on only incidentally in the literature, in connection with proposed solutions to problems deriving from the theory of truth. Thus, Louis Couturat cited the contingency of laws in Leibniz's system as evidence against the proposal that Leibniz regarded only existential judgments as contingent—and meant to *exempt* such judgments from the "analytic" theory of truth.[4] And as part of his effort to reestablish a version of the latter interpretation, E. M. Curley has replied that "according to Leibniz, the laws of nature are also existential propositions, so that they do not form a distinct class of contingent truths."[5] Curley quotes a passage in which Leibniz does represent the view that the laws of nature are contingent as resting on the premise that the existence of the "series of things" depends on God's choice:

> [We said] these laws are not necessary and essential but contingent and existential. . . . For since it is contingent and depends on the free decrees of God that this particular series of things exists, its laws will be themselves indeed absolutely contingent, although hypothetically necessary and as it were essential once the series is given.[6]

Curley further clarifies his point by indicating that the laws of nature are "existential" in that they rule out certain possible states of affairs from the realm of actuality:

> . . . "All circles are plane figures," which is given [by Leibniz] as an example of an essential proposition, says that a circle which is not a plane figure is not a possible thing. But a law of nature, such as "unsupported

bodies fall to the earth,'' says only that an unsupported body which does not fall to earth is not an actual thing, i.e., does not exist.[7]

Now one may feel some hesitation about this interpretation. In particular, it seems to deny Leibniz any distinction among universal laws, local law-like generalizations, ''accidental'' generalizations, and hypotheticals true by virtue of the falsity of the antecedent. (The example Curley uses is, clearly, far from being universally true—and has the special disadvantage, in this context, of *mentioning* a particular existent.) Further, it seems that a philosopher could hold that more than one ''series of existents'' is possible, without holding that more than one set of basic laws is possible.[8] (I will suggest below that such a view can be found in the writings of Descartes.) On the other hand, Curley is clearly right in pointing out that since Leibniz *does* believe the laws of nature are contingent, and since statements of laws of nature do have negative existential import, their truth, for Leibniz, cannot be altogether independent of God's choices of particular existents. What we need now, however, is some account of why the laws of nature should sometimes be ascribed *special importance* in illuminating the distinction between necessary and contingent truth.

In what follows I shall try to provide such an account, by placing Leibniz's interpretation of his dynamical conclusions within its historical context. Fundamentally, Leibniz was concerned to oppose—for religious reasons especially—the ''geometrical'' conception of natural science exemplified (in different degrees) by his predecessors Descartes and Spinoza. That is to say, he was concerned to oppose the assimilation of physics to geometry, and of physical necessity to geometrical necessity. Leibniz shared with most of his contemporaries the view that the axioms of Euclidean geometry are among the eternal truths: within Leibniz's system this view appears as the doctrine that the axioms of geometry are true of all possible (i.e., consistently conceivable) worlds for reasons connected with the concept of space.[9] He believed he could establish that this status is *not* shared by the laws of mechanics. Leibniz believed his Dynamics yielded this conclusion in virtue of showing (1) that matter cannot be adequately conceived in purely geometrical terms—that ''the essence of matter does not consist in extension alone''; and (2) that physical laws manifest features of ''fitness and proportion,'' which not only are inconsistent with the geometrical view, but which further can only be explained with reference to the purposes of a ''wise author'' of nature. (He thus claims to have discovered the basis for a new, updated version of the Argument from Design.) That his reasoning on these issues is logically impeccable can hardly be maintained; on the other hand, the reasoning has, for the most part, a definite *ad hominem* cogency against the assumptions of his opponents.

But Leibniz's rejection of the doctrine of blind or brute geometrical necessity, on the basis of his conclusions in dynamics, has a murkier aspect as well. For he also takes his Dynamics to reveal that the underlying causes of natural phenomena must be found in immaterial or soul-like entities that are governed by final causes and may be identified with Aristotelian forms or entelechies. Such immanent purposiveness he also takes to be incompatible with the concept of determination by ''geometrical necessity'' (although, amazingly, he never makes clear that this is an entirely different point from those mentioned above). This view appears to reflect a good deal of wishful thinking about the possibility of partially defending the older philosophy of nature against the atheistical and ''materialistic'' implications of the modern view— and, one may be tempted to think, not much else. I will suggest, however, that even this conclusion, obscure as it may be from a strictly philosophical point of view, can be partially explicated in terms of the transition away from the early geometrical conception of physics. In this case a particular sort of conceptual difficulty implicit in the transition appears partially to account for Leibniz's otherwise bewildering move.

Finally, I shall point out some ways in which attention to these aspects of Leibniz's position can contribute to a balanced and historically accurate interpretation of his views on contingency. In particular I shall try to make clear that Leibniz *is* in one important sense entitled to present his system as an alternative to ''brute geometrical necessitarianism''— *even if* he does not ultimately succeed in maintaining a coherent distinction between moral necessity and absolute necessity, or necessity in virtue of the principle of non-contradiction.

II

All things, I repeat, are in God, and all things which come to pass, come to pass solely through the laws of the infinite nature of God, and follow (as I will shortly show) from the necessity of his essence.[10]

> Nothing in the universe is contingent, but all things are conditioned to exist and operate in a particular manner by the necessity of the divine nature.[11]

> Things could not have been brought into being by God in any manner or in any order different from that which has in fact obtained.[12]

These quotations from Part I of Spinoza's *Ethics* represent the strongest form of the geometricism that Leibniz wished to oppose through his Dynamics. Spinoza's extreme position can be expressed in the claim that whatever is (in the timeless sense of "is") cannot not be, and whatever is not, cannot be. Individual existents (modes), their relations to each other, laws of nature, the two accessible attributes of thought and extension, and the world as a whole are alike in this respect. Only God is self-caused, or has an essence that includes existence; however, the causal necessity by which modes come into existence is itself in no sense weaker than the necessity of a logical deduction from necessary premises:

> From God's supreme power, or infinite nature, an infinite number of things—that is, all things have necessarily flowed forth in an infinite number of ways, or always follow with the same necessity; in the same way as from the nature of a triangle it follows from eternity and for eternity, that its three interior angles are equal to two right angles.[13]

This passage and the whole deductive format of the *Ethics* epitomize the influence of the geometrical model on scientific thought of the seventeenth century.

The geometric model was of course also a primary influence on the thinking of Spinoza's predecessor Descartes, although in some respects Descartes did not take things quite so far. Descartes does not deny that other worlds are possible; on the other hand, there is evidence that he did regard the basic laws of nature as necessary—as holding in any worlds God "could have created." The laws of nature, like the axioms of the geometers, are, he claimed, innate in our minds, so that "after having reflected sufficiently upon the matter, we cannot doubt their being accurately observed in all that exists or is done in the world."[14] Further, he remarks that in his Physics he had

> pointed out what [are] the laws of Nature, and without resting my reasons on any other principle than on the infinite perfections of God, I tried to

269

demonstrate all those of which one could have any doubt, and to show that they are such that even if God had created several worlds, there could be none in which these laws failed to be observed.[15]

As this passage may suggest, Descartes in fact "deduces" his laws of motion, such as the principle that the same "quantity of motion" (mv) is always conserved, by appeal to the "immutability of God"—an appeal which may have echoes in Spinoza. (Notoriously, though, Descartes elsewhere espouses the views that the "eternal truths" or standards of logical and mathematical possibility themselves depend on God's will and that God enjoys a complete "liberty of indifference" in determining what they should be. Thus he is a source both of necessitarian thought and of the seemingly opposite tendency (also opposed by Leibniz) that views the circumstances of nature and even ordinary mathematics as "arbitrary.")

Descartes' conception of the laws of nature as *a priori* and necessary (like the axioms of geometry) is accompanied by a strictly geometrical conception of matter. This conception is advanced on purely intuitive grounds in his philosophical writings. The doctrine that what is "clearly and distinctly perceived is true" yields, in the *Meditations,* the claim that the real or objective properties of body (as distinct from mere sensory appearances) are just those properties that it possesses as "the object of pure mathematics."[16] By "pure mathematics" Descartes means, especially, geometry;[17] extension, figure, and motion or movability are apprehended by the intellect as "all that remains" when we consider a body as it is in itself, stripped of "external forms."[18] In the *Principles of Philosophy* Descartes further argues that extension is the one property of bodies that is presupposed by all the other physical properties, without itself presupposing them. On this basis he holds that extension is *the* defining or essential attribute of body.[19] These purely conceptual arguments ostensibly (at least) appeal only to ordinary intuitions, rather than sophisticated scientific understanding. This concept of body or matter is assumed by Descartes in deriving his laws of motion *a priori*. (The motion in the world, on the other hand, is not represented as itself part of the nature of body, but as a quantity imposed "externally", as it were, by the prime mover.)

Extrapolating a bit from Descartes' own statements, we might suggest the following argument as an exemplar of the moderate geometricism that views the laws of nature (if not individual existents) as obtaining in any world God could have made.

1. It is a necessary truth (true in all possible worlds) that the essence of matter consists in extension alone.

2. No world could exist that is not made by God.

3. Immutability is an essential property of God, i.e., "God is immutable" is a necessary truth.

4. The basic laws of motion, m_1 - m_n, can be derived with geometrical (i.e., logical) necessity from the assumptions that the essence of matter is just extension and that any material world is created by an immutable Being. (I take it one need not suppose the existence of motion to be a necessary truth, in order to hold that the laws of motion are true in all possible worlds.)

This series of claims is, of course, based primarily on the passage quoted above from the "Discourse." (Fortunately, we need not be concerned here with the plausibility of these propositions, and particularly of the fourth.) It yields the conclusion:

5. The basic laws of motion are necessary (are "observed", as Descartes puts it) in any world God could make.

Now someone concerned to dispute the conception of physical law implicit in this line of thought might very well wish to concentrate attention on the "laws" its propounder claims to have derived in this *a priori* manner. He might, for instance, try to show that some "laws of nature" presented as necessary either do not follow from these assumptions, or are false of the actual world (and *a fortiori* not true of all worlds that could exist). He might argue that *correct* reasoning from the geometricist's premises yields false results. None of these approaches, to be sure, would suffice to show that the laws of nature are *not* necessary. However, a sort of minimal argument against that conclusion would consist in that some particular set of laws (say that espoused by the geometricist) is both possible and false (not "observed" by nature).

In other words, one good way to refute the geometricist's conception

271

of physics is to refute the geometricist's physics (without claiming that his principles are *necessarily* false). If such an enterprise sounds somewhat farfetched today, this is, I think, largely because the geometricist's conception of physics sounds farfetched. But the quotations from Descartes and from Spinoza may perhaps serve to remind us of the grip of the geometrical model in the early seventeenth century.

Some further points about the Cartesian outlook should be mentioned before we consider Leibniz's reaction to it. First, in the minds of seventeenth-century Cartesian philosophers there seems to have been a close connection between the idea that the essence of matter consists in (the geometrical property of) extension, and the idea that the laws of nature share the *necessity* of geometrical axioms. In fact, there does *not* seem to be any direct logical route between these notions: even if we suppose that Euclidean geometry is necessarily true of the world, the doctrine that the laws of nature are necessary seems neither to entail, nor to be entailed by, the proposition that only geometrical concepts are required for the statement of them.[20] But it is understandable that these ideas should be assumed to stand or fall together as twin aspects of the notion that physics (as Descartes remarked to Mersenne) "is nothing but geometry."[21]

In the same way both necessitarianism and the Cartesian theory of matter are tied up with the exclusion of final causes from the physical world—a position shared by Descartes and Spinoza. On the one hand, it is difficult to conceive how a bare bit of extension could be endowed with a purpose or goal. On the other hand, because of the close linkage of the concept of *purpose* with that of *choice*, it might well seem natural to suppose that only efficient causality can consort with the strict necessity supposedly characteristic of geometrical axioms.[22] Thus both the necessitarianism of the geometrical conception of physics, and the attendant conception of matter, may be viewed as having *some important connection* with the denial of immanent teleology. It is understandable, therefore, that someone interested in reinstating immanent final causes should find it *necessary* to oppose the other aspects of the geometricists' view. Leibniz, as we shall see, sometimes seems to think that rejection of geometricism in physics is *sufficient* to establish that there are purposive entities throughout nature.

Finally, however, we must concede that Descartes' use of the rather inscrutable notion of *God's immutability* makes it difficult to measure with complete assurance the distance between his and Leibniz's conception of the relation between physical and mathematical truth. Thus Descartes seems to hold that the laws of nature are true in any world God could have made—on the hypothesis that God is immutable. If this hypothesis is required only for the derivation of physical laws, and not as an underpinning for geometry proper, and if, further, it is taken to introduce some reference to volition and purpose (God's immutable will), then Descartes' actual position might turn out to be much closer to Leibniz's than at first seems to be the case. It would still remain true, however, that Leibniz differs fundamentally from Descartes in his concern to emphasize rather than minimize the distinction between physical and mathematical truth.

III

In his early years Leibniz himself accepted the geometricist conception of physics. He even attempted to derive "abstract principles of motion" by reasoning *a priori* from the conception of matter as mere extension.[23] One of the propositions he "proved" in this way provides the springboard for his later attacks on the doctrine of brute geometrical necessity in nature. Briefly, Leibniz had reasoned that mere extension must be "indifferent" to both motion and rest. By this he seems to have meant that there was nothing in the purely geometrical conception of matter to provide for a force of resistance, or reaction to every action. He then proceeded to the startling conclusion that when a body in motion, however small, collides with a body at rest, however large, both bodies will then move in the direction of the original motion, and at the original speed! According to his later accounts, he was even at the time distressed by this bizarre result, and supposed that the "wise Author of nature" would not permit such a disproportion between cause and effect. Subsequently he affirms (with a rather puzzling air of informativeness) that indeed such phenomena do not occur.[24] He affirms in these later works a principle of equality of reaction, which he derives from (or perhaps equates with) the principle that there is an equality between causes and effects. This principle he seems to regard as foreign to Descartes' "geometrical" reasoning. (Elsewhere he stresses that

Descartes' disregard of the architectonic or non-geometrical principle of continuity—according to which all change is gradual—also explains fundamental errors in his physical principles.)[25]

Leibniz concludes that the Cartesian conception of matter is inadequate to account for the actual phenomena of nature, so a different conception must be substituted:

> If the essence of body consisted in extension, this essence alone should suffice to explain [*rendre raison de*] all the affections of body. But that is not the case. We observe in matter a quality which some have called *natural inertia* [read "resistance" or "reaction"], through which body resists motion in some manner. . . .[26]

> In order to prove *that the nature of body does not consist in extension,* I have made use of an argument . . . of which the basis is that *the natural inertia of bodies* could not be explained by extension alone. . . .[27]

Now we may note in passing that this argument does not depend exclusively on dynamical notions. For instance, it seems to presuppose a rather uncartesian conception of the function of an essence: Descartes does not think that even the fundamental property of motion *follows from* the essence of matter alone. On the other hand, Descartes does assume a certain conception of body in his deductions of the basic laws. And Leibniz does not make clear exactly what he means by an "account" of the properties of bodies. Therefore it is not easy to pinpoint the extent of their difference on this question; fortunately it does not seem very important that we do so.[28]

In at least one later writing, however, Leibniz seems to concede that this reasoning is perhaps not sufficient to clinch the case against Cartesianism. For his conclusion about impact might be staved off if only one assumed—as Descartes had assumed—that God conserves the same quantity of motion.[29] At this point, therefore, Leibniz brings to bear another argument. This argument, based openly on the research of Galileo, purports to prove that the true conservation principle in physics is not the conservation of motion (mv), as Descartes held, but the conservation of quantity mv^2, which Leibniz identifies as *vis viva,* active force. From this fact, too, Leibniz claims, it can be seen that there is more in nature than quantity of motion, and more to matter than is dreamed of in Descartes' Geometry.[30]

274

(The consideration that is supposed to prove this point is, briefly, as follows. A body of 4 pounds falling freely from a height of 1 foot will rise again to a height of 1 foot (assuming elastic rebound). A body of 1 pound falling from a height of 4 feet will rise again to a height of 4 feet. This shows, Leibniz says, that the "forces" operating in the two cases are equal (as the Cartesians would have agreed). However, Galileo had shown empirically that velocity in free fall from a state of rest is proportional not to the distance of fall but to the square root of the distance. Thus if force were measured in the Cartesian manner (mv), the "forces" of the two bodies on impact would be not equal but in a proportion of 2:1; i.e., the 1-pound body would rise to 2 feet, not 4 feet. So to get the correct results, we must introduce the quantity mv^2 as the measure of force. This argument, too, is held to manifest the principle of equality of cause and effect.)

Although Leibniz does use other dynamical arguments against the Cartesians, these two seem to have been his favorites. Sometimes, as I have indicated, he uses them in tandem; more often he presents them independently of each other. There are, to be sure, puzzling features in both arguments, from the point of view merely of the physical interpretation. In particular, it is difficult to understand in what way the second argument may be said to establish the *conservation* of *vis viva*.[31] What we are concerned with here, however, is Leibniz's *philosophical* interpretation of the arguments, his claim that they show the falsity of the doctrine of brute geometrical necessity in nature.

It is worth noting, first, that Leibniz's conclusions would provide the materials for a simple-minded but plausible repudiation of necessitarian claims along lines touched on above. The collision argument pretends to derive from Cartesian assumptions, "laws" that are in fact false of the world. The free-fall argument shows (according to Leibniz) that a principle Descartes himself derived is incompatible with the true "conservation" principle. Further, Leibniz in one place seems to go out of his way to indicate that Cartesian assumptions (as he interprets them), though false, are not impossible. He writes to the Cartesian de Volder:

And doubtless such a world could be imagined as possible, in which matter at rest yielded to the mover without any resistance; but this world would really be pure chaos.[32]

275

(It might be objected that Leibniz could here be saying not that a Cartesian world *is* possible but only that we could (wrongly) suppose it to be possible. One might be particularly tempted to make this objection in view of the fact that ''pure chaos'' seems incompatible with the Principle of Sufficient Reason, which Leibniz does regard as a necessary truth.[33] However, it appears from the context that Leibniz does *not* think such a world would violate the Principle of Sufficient Reason, though it *would* violate another, contingent principle, to the effect that the better we understand things, the more they satisfy our intellect. (I admit the passage is pretty peculiar.)

If the Cartesian laws are false but not *necessarily* false, it follows of course that the basic laws of nature cannot be attributed the necessity of geometrical axioms. And, given the influence of the geometric model on seventeenth-century conceptions of physical science, this way of arguing against necessitarian conceptions might well have more effect than appeals to alleged direct intuitions of alternative possibilities. However, it does not seem that Leibniz's own reasoning ever follows quite this route.

Clearly, Leibniz thinks the need to introduce the concept of reaction, and of mv^2 as a measure of force, to describe physical phenomena shows the inadequacy of the Cartesian conception of matter, and *thereby* demonstrates the inadequacy of the conception of physics as a science of brute geometrical necessity. His reasoning here appears to be, in a way, specious, since, as we have noted, there seems to be no direct logical route between the concept of matter as mere extension and the necessitarian position concerning the laws of nature. From a historical point of view, however, Leibniz's reasoning is understandable. The Cartesian characterization of physics as nothing but geometry suggests such a close association between a necessitarian conception of the laws of nature and the conception of matter as extension, that it might well be natural to view them as standing or falling together.

Up to this point the issue of teleology has not at all entered into our discussion of Leibniz's philosophical interpretation of his results in physics. We have merely considered two ways of construing the denial of necessitarianism—one manifestly present in the Leibnizian texts, and one somewhat artificially constructed out of elements provided by the texts. This negative aspect of Leibniz' position accords well enough with the conception of the status of physical laws prevalent in contemporary philosophy—so well, in fact, that some effort of historical imagination is required to understand the prominence he accords to the contention. However, as our initial quotations clearly show, Leibniz's interest in refuting the doctrine of "brute geometrical necessity" lay not in the bare demonstration of *non*-necessity but rather in showing that nature manifests "fitness and final causes." Two points seem to be involved in this claim. First, Leibniz believes his physics provides a basis for restating and vindicating the traditional Argument from Design against the view that nature is governed not by design but by geometry. Second, he believes that his concept of force requires us to assume that underlying the phenomena of physics are purposive, mind-like metaphysical entities that somehow provide a "foundation" for the phenomena.

IV

Many passages concerning the rejection of Cartesian physics reflect Leibniz's preoccupation with vindication of the concept of design in nature against the geometricist's view. The following are representative:

> *If mechanical rules depended on Geometry alone without metaphysics, phenomena would be quite different.* . . . One notices the counsels of [the divine] wisdom in the laws of motion in general. For if there were nothing in bodies but extended mass, and if there were nothing in motion but change of place, and if everything had to be and could be deduced from these definitions alone by a geometric necessity, it would follow, as I have shown elsewhere, that the smaller body would give to the greater which was at rest and which it met, the same speed that it had, without losing anything of its own speed. . . . But the decrees of divine wisdom to conserve always the same force and the same direction in sum has provided for this.[34]

> I have already asserted several times that the origin of mechanism itself does not spring from a material principle alone and mathematical reasons but from a certain higher and so to speak Metaphysical source.
> . . . One remarkable proof of this, among others, is that the *foundations of the laws of nature* must be sought not in this, that the same quantity of motion is conserved, as was commonly believed, but rather in this, that it is necessary that the same *quantity of active power* be conserved. . . .[35]

> Perhaps someone will . . . believe that a completely geometric demonstration can be given of [the laws of motion], but in another discourse I will show that the contrary is the case, and demonstrate that they cannot be derived from their source without assuming architectonic reasons.[36]

277

(Again, "architectonic reasons" means such principles as the equality of cause and effect (or of action and reaction), the principle of continuity, and certain other principles such as the law of least action, that Leibniz regards as comparable in showing the governance of a wise Author of nature.)

Leibniz himself stresses that his discernment of intelligence and a sense of perfection behind the principles that his dynamics establish (in contrast to the "chaotic" implications of Cartesian mechanics) is closely related to the traditional Argument from Design. He believes that he has advanced beyond the traditional form of the argument by showing that the general laws of nature, as well as its particular phenomena, manifest the workmanship of a beneficent intelligence. In the "Discourse on Metaphysics," for example, he first endorses the traditional form of the argument (with special reference to animals in general and eyes in particular), and makes fun of its opponents. He continues:

> Thus, since the wisdom of God has always been recognized in the detail of mechanical structure of some particular bodies, it ought also to show itself in the general economy of the world and in the constitution of the laws of nature. And this is so true that one notices the counsels of this wisdom in the laws of motion in general.[37]

In the conclusion of the passage, which has already been quoted, Leibniz cites his argument against Cartesian principles that is derived from the problem of collison. In another work he claims that the dependence of the laws of his true dynamics on the principle of fitness provides "one of the most effective and obvious proofs of the existence of God."[38]

Without wishing to endorse any version of the Argument from Design, I would like to suggest that this is a quite understandable and rather interesting move for a determinedly pious person to make in the historical situation in which Leibniz found himself. The Cartesians rejected final causes and the Argument from Design on the grounds that events in nature were determined according to a "blind" deductive system of necessary "geometrical" laws. Leibniz does not deny the lawfulness of nature, and he does not deny that physics is a deductive system. However, he holds that final causes, or considerations of fitness and proportion, may be said to enter into the system on the top level, once one recognizes that the Cartesian physics is false.

This move that Leibniz makes in defense of the Argument from Design (against, we may suppose, such ferocious critics as Spinoza)[39] has some affinity with a move later made in its defense in the face of the theory of natural selection. In the words of F. R. Tennant,

> The sting of Darwinism . . . lay in the suggestion that proximate and "mechanical" causes were sufficient to produce the adaptations from which the teleology of the eighteenth century had argued to God. Assignable proximate causes, whether mechanical or not, are sufficient to dispose of the particular kind of teleological proof supplied by Paley. But the fact of organic evolution, even when the maximum of instrumentality is accredited to what is figuratively called natural selection, is not incompatible with teleology on a grander scale . . .[40]

> . . . The discovery of organic evolution has caused the teleologist to shift his ground from special design in the products to directivity in the process, and plan in the primary collocations.[41]

Although Leibniz seems to present his reasoning as supplementing, rather than replacing, the traditional argument from the evidence of "special design in the products", he too makes use of the discovery of mechanical principles—principles at first sight inimical to teleological conceptions of nature—to argue for "teleology on a grander scale." Whereas the later teleologist finds progress and hence purpose in the process of evolution broadly considered, Leibniz insists on the evidence of wisdom in the order and proportion that are maintained throughout nature as a result of the *sort* of mechanical laws that obtain. His claim against the geometricists is that, first, these laws are *not* those of "brute geometrical necessity," and second, that they can *only* be viewed as manifesting the values and aesthetic sense of a wise Creator.[42]

Leibniz's theological interpretation of the laws or principles of his Dynamics is more asserted than argued; it could hardly be expected to carry conviction to anyone not highly sympathetic to the aims of the Argument from Design. One might, obviously, accept his negative claim—that not all fundamental principles of physics are "geometrical"—while withholding credence entirely from the teleological interpretation. For there is no need to accept as exhaustive the division of geometrical necessity on the one hand and purposiveness on the other. Leibniz's point of view is, nevertheless, readily intelligible against its historical background. Indeed, it has not been altogether absent from the science and philosophy of our century.

V

As I have indicated, however, this attempted vindication of the intrinsically teleological character of the basic principles of physics is tied up in Leibniz's own thinking with a stranger and more elusive notion. Consider the following passages, from different parts of his writings.

First, from the "Discourse on Metaphysics" (1686) following a statement of the argument for the conservation of mv^2:

> And it becomes more and more apparent, although all particular phenomena of nature can be explained mathematically or mechanically by those who understand them, that nevertheless the general principles of corporeal nature and of mechanics itself are rather metaphysical than geometrical and belong rather to some indivisible forms or natures as causes of appearances than to corporeal or extended mass.[43]

From "Critical Thoughts on the General Part of Descartes' *Principles*" (1692):

> For besides extension and its variations, there is in matter a force or power of action by which the transition is made from Metaphysics to nature, from material to immaterial things. This force has its own Laws, which are deduced from the principles, not merely of absolute, and so to speak brute necessity, but of perfect reason.[44]

And from "On the Elements of Natural Science" (ca. 1682–84):

> Certain things take place in a body which cannot be explained by the necessity of matter alone. Such are the laws of motion which depend on the metaphysical principle of the equality of cause and effect. Therefore we must deal here with the soul, and show that all things are animated.[45]

Many other similar passages could be cited from Leibniz's work. But these three should suffice to make clear that the denial of brute necessity in nature is intimately associated in his thought with the postulation of immaterial forms, entelechies, or souls as the real metaphysical basis of phenomena. Of course, Leibniz had other reasons for maintaining that extension is purely phenomenal, and that real substances are indivisible and mind-like.[46] But he seems to regard the non-geometrical nature of his "laws of force" as providing independent reason for this view.

Bertrand Russell has with justice harshly criticized this aspect of Leibniz's position.[47] As far as I know, Leibniz never fills in any of the steps that might take one from "forces" in physics to "souls" in metaphysics. He does not indicate what it might mean to say that the latter provide a "general explanation" of the former. And he does not provide elucidation of the relation between the two claims made about the "general principles of corporeal nature"; i.e., that they are "rather metaphysical than geometrical," and that they "belong . . . to some indivisible forms or natures as causes of appearances"—he merely treats the claims as if they were obviously equivalent.

I have suggested above that there would be some natural affinities for a seventeenth-century thinker among the denial that the laws of nature have the status of geometrical axioms, the rejection of the Cartesian conception of matter, and the reaffirmation of the traditional "forms" or immanent purposiveness in the (non-human) world. It is also quite clear and beyond question that Leibniz took an almost obsessive pride in the notion that his system offered a "synthesis" of traditional metaphysics and modern physics, retaining the best of both views and in particular avoiding the anti-spiritualist implications of the latter. But there is a more specific and perhaps more interesting explanation, or partial explanation, of this obscurity in his system.

Despite Leibniz's opposition to the Cartesian theory, one finds in his writings a certain tendency to assimilate the concept of the material to that of the geometrical, in just the Cartesian manner:

If nature were brute, so to speak, that is purely material or Geometric. . . .[48]

Certain things take place in body that cannot be explained by the necessity of matter alone.[49]

Of course, Leibniz uses the term "matter" in different ways in different contexts; it would be quite wrong to attribute to him without qualification the assumption that "the non-material" can be equated with "the non-geometrical.'' But we may still suppose that his transition from forces in physics to "immaterial things" reflects some implicit assumption that any entity in nature not fully describable through the concepts of geometry, and particularly anything suggesting changes not reducible to relative change of place, is by definition excluded from the realm of the material. Similarly, where the concept of efficient cause has been associated with that of geometric determination, the reintroduction of forms, entelechies, or final causes might well seem warranted or inevitable, once the geometric picture is abandoned.[50] In such ways we can make out a path from "mechanical" dynamics to soul-like purposive entities as underlying "causes."

Here someone might object that since Leibniz postulates force as part of the "essence of body" or even of the "essence of matter," he can hardly be said to conflate the non-geometrical with the non-material. And further support could be adduced for this objection: for example, Leibniz in one place explicitly characterizes the failure to provide a ground for the laws of force as an insufficiency in the "common notion of matter":

> . . . It is not possible to deduce all truths about corporeal things from logical and geometrical axioms alone, those of great and small, whole and part, figure and situation, but others of cause and effect, action and passion must be added, by which the reasons of the order of things may be preserved. Whether we call this principle Form or entelechy or Force is not important, so long as we bear in mind that it can only be intelligibly explained through the notion of force.
> But I cannot agree with the view of certain prominent men today, who perceiving that the common notion of matter does not suffice. . . .[51]

What this really shows, however, is only that Leibniz's conception of matter was ambiguous. Thus, the very next paragraph after the passage quoted begins:

> Although I admit an active principle throughout bodies which is superior to material notions and so to speak vital. . . .

Similarly, while the first quotation may suggest for a minute that Leibniz's transition from "forces" to "entelechies" or "souls" is after all a merely terminological issue, the second reminds us of the edifying talk of "higher" sources and spiritual cures for those "mired in materialist notions'' that characterizes nearly all his presentations of his conclusions in physics. ''Entelechy'' and ''soul'' invariably connote for Leibniz the unequivocally mental qualities of sense and appetition.[52] The obscure inference to these from the concepts of Dynamics is an internally important aspect of his thinking; unfortunately it is also, as Russell remarks, ''one of the weakest points in his system.''

VI

This concludes my explication of Leibniz's claim that his Dynamics teaches us ''the difference'' between necessary and contingent truths, and helps to overthrow the doctrine of brute necessity in nature. I wish, finally, to make two further observations concerning the significance of this explication.

First—to take up a very specific point—my account throws new light on a question discussed at some length in an interesting article by Leonard J. Russell: namely, how much ''community of structure'' is held to exist (in Leibniz's system) between our world and other possible worlds.[53] In the midst of some insightful and plausible analysis, Russell writes:

> . . . While the laws of physical motion may in some respects differ from those in the actual world, the fundamental principle laid down by Leibniz for the actual world, viz., the equality of cause and effect, has some claim to apply to all worlds. In GP III 45–46, and in GP II, 62, he describes this equality as a ''metaphysical'' law, and this would make it apply universally. Again he tells de Volder that if matter had no inertia the world would be a chaos (GP II, 170)—in which case it would not be a possible world—while to Remond (11 Feb. 1715, GP III, 636) he says that if there is inertia then cause equals effect. It is true that this law is here described as a rule of *covenance* (in which case it is not metaphysical), but it is clear that if the effect were less than the cause in accordance with a constant law, motion would in the end cease, and God would have to intervene from time to time to set things moving again—and Leibniz was never happy about a system which would involve God in a permanent need for miraculous action . . . I conclude then that the equality of cause and effect is likely to be maintained in all worlds.[54]

Russell has been misled, I think, by the fact that Leibniz frequently uses the expression ''metaphysical necessity'' as equivalent to ''absolute'' or "geometrical" or "brute" necessity—which he *contrasts* with moral necessity, or the necessity by which "perfect reason" chooses the best among all possibilities.[55] My exposition has shown that in calling the principle of the equality of cause and effect "metaphysical" Leibniz actually means the very opposite of what Russell interprets him as saying. In the relevant Leibnizian contexts, to say a law is metaphysical is to say that it is *not* geometrically necessary; that it depends on immaterial rather than material principles, that it has to do with form, entelechy, and purpose rather than brute necessity. Certainly this dual usage of "metaphysical" is confusing; but only until its existence has been pointed out.

Russell's mistake has further led him to take for granted a seemingly erroneous interpretation of the remark to de Volder.[56] Perhaps it has also led him to the mistaken view that Leibniz's "unhappiness" about a principle such as God's frequent miraculous intervention is the same as a tendency to believe that such a principle is false in all possible worlds. (A world that involved God's constant intervention would presumably be, to Leibniz's way of thinking, radically imperfect and therefore non-actual—but not in the least impossible.)

The role of the law of equality of cause and effect in Leibniz's philosophy certainly poses some interesting and difficult problems for interpreters. To deal with these problems effectively and accurately, however, one must take into consideration the prominent, if peculiar, line of thought I have examined here.

Finally, a more general consideration. There has been much controversy over whether Leibniz was (a) sincere and (b) justified in presenting his system as an alternative of Spinozistic necessitarianism. As we have noted above, his theory of truth seems to have the implication that all truths are such that their denial implies contradiction. Thus it is not in the last analysis clear that more than one world is *really* possible on Leibnizian principles. Further, even apart from the theory of truth, it is not ultimately clear in what sense it would have been possible that God create some world other than the maximally perfect one of those that he could conceive without contradiction.[57] These are, indeed, crucial problems, and problems that provide an interesting challenge to Leibniz's more analytically inclined interpreters. However, I think that in a certain sense they have been somewhat overstressed. Even if one should conclude that the denial of a truth of fact *must* lead to contradiction on

Leibniz's premises, there remain vast differences between his system and the necessitarianism of Spinoza. The main point can be expressed very simply: Leibniz's philosophy requires that the explanation of any existential proposition involve reference to value, purpose, perfection. As we have seen, this idea is particularly prominent in his presentation of his conclusions in Dynamics. No one could, I think, deny that in this respect Leibniz's position is indeed antithetical to Spinoza's.[58] And the antithesis is hardly trivial.[59] Granted, if Leibniz's general theory of contingency ultimately breaks down as has been claimed, he has not succeeded in providing a coherent opposition to necessitarianism *simpliciter*. Further, on one understanding of "brute geometrical necessity"—according to which a proposition is geometrically necessary if it could not, in the last analysis, have been false—difficulties for the general theory are indeed difficulties even for Leibniz's claim to have rejected the doctrine of "brute geometrical necessity" in nature. But what *can* still be said (in the face of such difficulties) is that within *Leibniz's* system—unlike the system of (at least) Spinoza—the laws of nature do have a specifically different status from that of the axioms of geometry. The latter, but not the former, are true of all worlds conceivable by God as candidates for existence. And the laws of nature that obtain in the actual world *are* the actual laws of nature *just because* a world in which they hold is the *best* of all possible (i.e., internally possible) worlds. In other words, they reflect and even (according to Leibniz) demonstrate the valuation on the part of the creator of "fitness and proportion"—however necessary *this valuation* may itself turn out to be on Leibniz's own premises. The same claim *cannot* be made for the geometry of the world—in Leibniz's system or in Spinoza's. Thus, there is a clear if limited sense in which Leibniz's attack on the Spinozistic doctrine of brute geometrical necessity in nature may be said to survive even the most serious objections to his general theory of contingency.

Members of the philosophy faculties of Ohio State and Rutgers universities, and several other individuals, have greatly influenced the present form of this work through their comments on earlier versions. I am particularly indebted to Robert Turnbull, Wallace Anderson, and Norman Kretzmann for specific comments and suggestions, and to Fabrizio Mondadori and James F. Ross for discussion of some of the problems dealt with in the paper. I learned of some of the articles cited from a bibliography prepared by E. M. Curley.

1. Leibniz, *Die philosophischen Schriften*, ed. C. J. Gerhardt (Berlin: Weidmannsche Buchhandlung, 1875–90; repr., Hildesheim: Georg Olms, 1965), 3:645 (hereafter cited as GP).

2. GP, 6:321. In general, translations throughout the paper are my own unless otherwise indicated. However, I quote directly from the Elwes translation of Spinoza's *Ethics* (which is accurate for the passages cited), and from the admirable Lucas and Grint translation of Leibniz' "Discourse of Metaphysics." My versions of other passages have sometimes been influenced by the published translations cited in the notes.

3. For example, its connection with the claim that every substance has a "complete concept" from which all its properties—past, present, and future—are somehow derivable.

4. "Sur la métaphysique de Leibniz," *Revue de métaphysique et de morale* 10 (1902): 12.

5. "The Root of Contingency," in H. Frankfurt, ed., *Leibniz: A Collection of Critical Essays* (Garden City, N.Y.: Doubleday, 1972), p. 91.

6. Ibid. The original source for this quotation is L. Couturat's edition of *Opuscules et fragments inédits de Leibniz* (Paris: Alcan, 1903), pp. 19–20. I have altered Curley's wording slightly.

7. Ibid., p. 92.

8. It might be questioned whether this idea would be congenial to Leibniz. A statement in one of his letters to Arnauld could be read as suggesting that he thought no two possible worlds have a law in common: "For as there is an infinity of possible worlds, there is also an infinity of laws, some proper to one, others to another [*les unes propres à l'un, les autres à l'autre*], and each possible individual of any world includes in his notion the laws of his world" (GP, 2:40; this passage was brought to my attention by James Alt). However, I doubt that Leibniz really means to imply that every law is peculiar to some particular possible world. Earlier in the paragraph, for instance, he indicates that the decree to create a particular substance (Adam) was distinct from the "few free primary decrees capable of being called laws of the universe."

9. Cf. J. Moreau, "*L'Espace et les verités éternelles chez Leibniz,*" *Archives de Philosophie* 29 (1966): 483 ff.

10. Spinoza, *Ethics,* Part I, Proposition 15, Note, translated by R. H. M. Elwes (New York: Dover Publications, 1951), 2:59.

11. Ibid., Proposition 29, p. 68.

12. Ibid., Proposition 33, p. 70.

13. Ibid., Proposition 17, Note, p. 61.

14. "Discourse on the Method of Rightly Conducting One's Reason and Seeking Truth in the Sciences," part 5, *Oeuvres de Descartes,* ed. C. Adam and P. Tannery (Paris: Leopold Cerf, 1897–1910), 6:41 (hereafter cited as AT). Cf. *The Philosophical Works of Descartes,* trans. E. S. Haldane and G. R. T. Ross (Cambridge: At the University Press, 1967; 1st ed. 1911), 1:106 (hereafter cited as HR). There is a non-trivial error in the HR translation, however.

15. AT, 6:43; HR, 1:108.

16. Meditation VI; AT, 7:80; HR, 1:191.

17. Descartes's seventeenth-century French translator renders "*in purae Mathesos objecto*" as "dans l'object de la géométrie speculative." Cp. AT, 9:63.

18. Meditation II; AT, 7:30–31; HR, 1:154–55.

19. *Principles of Philosophy* 1:53 (AT, 8:Pt. 1, 25; HR, 1:240).

20. See A. Quinton, "Matter and Space," *Mind* 73 (1964): 347–49, for an elaboration of this point. Quinton, surprisingly, seems to go along with the view that geometry *does* provide a body of necessary truths about "spatial qualities in the external world." But perhaps I have misunderstood him.

21. To Mersenne, 1639 (AT, 2:268).

22. Spinoza, in fact, makes clear in the Appendix to Part I of the *Ethics* that he regards the claim that "everything in nature proceeds with a sort of necessity" as incompatible with the supposition of final causes (Dover ed., 2:77).

23. GP, 4:228–32. This paper of 1671 is translated by L. E. Loemker in *Leibniz: Philosophical*

Papers and Letters, No. 8, 2d ed. (The Hague: Martinus Nijhof, 1972), pp. 139–42 (hereafter cited as Loemker). Leibniz makes many allusions to it in his later writings.

24. See *Specimen Dynamicum* (1695) in *Leibnizens mathematische Schriften*, ed. C. J. Gerhardt (Berlin-Halle: Ascher, Schmidt, 1849–63), 6:241–42 (hereafter cited as GM) and letter to de Volder (1699), GP, 2:170; Loemker, no. 55, pp. 516–17.

25. See, e.g., *"Tentamen Anagogicum"* (ca. 1696), GP, 7:279; Loemker, no. 50, p. 484; "Critical Thoughts on the General Part of Descartes's *Principles"* (1692), II, *ad art.* 45, GP, 4:375, Loemker, no. 42, p. 398.

26. Letter to *Journal des savans* (18 June 1697), GP, 4:464; *Leibniz Selections*, ed. P. P. Wiener (New York: Scribner's, 1951), p. 100 (hereafter cited as Wiener).

27. (1693) GP, 4: 466, Wiener, p. 102. As Gerd Buchdahl has pointed out, Leibniz fails clearly to distinguish *inertia* (a fundamental concept of Descartes' physics, taken up by Newton and reexpressed in his First Law) from *reaction* (the concept of Newton's Third Law), (*Metaphysics and the Philosophy of Science* [Cambridge, Mass.: MIT Press, 1969] pp. 421–22). However, it does not seem to me that Leibniz's argument against the Cartesian conception of matter depends in any important way on this confusion. In one letter to de Volder, further, Leibniz very explicitly distinguishes a version of the law of inertia ("each thing remains in its state unless there is a reason for change")—which he says is "a principle of metaphysical necessity"—from the law of reaction (24 March/9 April 1699, GP, 2:170, Loemker, no. 55, p. 516).

28. It is rather interesting, however, to note the contrast between Descartes' *a priori* and Leibniz's *a posteriori* approach to this problem. For an illuminating account of the history of the concept of the essence of matter, see Ivor LeClerc's essay, "Leibniz and the Analysis of Matter and Motion," in his *The Philosophy of Leibniz and the Modern World* (Nashville: Vanderbilt University Press, 1973). Oddly, Leibniz sometimes suggests that the independence of motion from the concept of body proves the necessity of postulating God as the cause of nature (see Loemker, p. 639).

29. GP, 4:465, Wiener, p. 101.

30. Ibid. See also "Discourse on Metaphysics," pp. xvii–viii, GP, 4:443–44, trans. P. G. Lucas and L. Grint (Manchester: Manchester University Press, 1953), pp. 28–32 (hereafter cited as Lucas and Grint).

31. This point is made by Carolyn Iltis, "Leibniz and the *Vis Viva* Controversy," *Isis* 63 (1970): 26 ff. (Iltis provides a very helpful critical analysis of several of Leibniz's dynamical arguments against Cartesianism.) One might suppose that in speaking of "conservation" Leibniz must mean that the force of a body *immediately after* impact is the same as the force *acquired* in free fall, and that this is expressed in the quantity mv²—not that mv² is constant *throughout* the fall and rebound event. Certainly some passages lend themselves to this interpretation: cf. "A Brief Demonstration of a Notable Error of Descartes and Others . . . " (March 1686), GM, 6:117, Loemker, no. 34, p. 296; "Critical Thoughts," GP, 4:370, Loemker, no. 42, pp. 394 f. But R. C. Taliaferro seems to suggest a different view in *The Concept of Matter in Descartes and Leibniz*, Notre Dame Mathematical Lectures, no. 9 (Notre Dame, Ind.: University of Notre Dame Press, 1964), p. 29.

The controversy on this issue between Leibniz and the Cartesians is apt to seem utterly mystifying to laymen—and apparently to many physicists as well. Some of the important things to note are these.

1. The *principle* that Leibniz is particularly concerned to reject is the false principle that mv is conserved as a scalar quantity, not the true principle that it (understood as the "momentum" of post-Newtonian physics) is conserved as a vector quantity. In later writings he seems to endorse the latter principle (though his concept of mass may not be identical with that of a contemporary physicist). However, he still maintains that mv², not mv, is the "true measure" of force in nature.

2. Whether force should be expressed as proportional to the Cartesian quantity mv, or the Leibnizian quantity mv² depends on whether one is concerned with force acting through time or

through distance: a body with twice the velocity of another will (in Mach's words) overcome a given force through double the time, but through four times the distance. Thus Leibniz' talk of the "true measure" of force is not defensible.

3. Kinetic energy (represented today as the quantity $1/2 \ mv^2$) is indeed conserved in perfectly elastic collisions. Leibniz of course was aware of the problem posed by inelastic collisions, and dealt with it by postulating that all collisions involve perfect elasticity on the micro-level.

32. GP, 2:170, Loemker, no. 55, pp. 516–17.

33. As he says in this letter.

34. "Discourse on Metaphysics," p. xxi, GP, 4:446, Lucas and Grint, pp. 36–37.

35. "On Nature Itself" (1698), Sec. 3–4, GP, 4:505–6, Loemker, no. 53, p. 499.

36. "Tentamen Anagogicum" (ca. 1696), GP, 7:279, Loemker, no. 50, p. 484. Taliaferro comments: " . . . The Tentamen Anagogicum is written for the sole purpose of showing the necessity for architectonic principles in mechanics and the insufficiency of geometry alone" (Concept of Matter, p. 31).

37. "Discourse of Metaphysics," p. xxi. See also "The Principles of Nature and of Grace" (1714) Sec. 11, GP, 6:603, Loemker, No. 66, pp. 639–40.

38. Ibid. (He adds the qualification, "pour ceux qui peuvent approfondir ces choses.")

39. Ethics, Part I, Appendix. Spinoza's criticism, though vehement, is somewhat diffuse.

40. In The Existence of God, ed. John Hick (New York: Macmillan, 1964), p. 126; excerpt reprinted from Philosophical Theology (Cambridge: At the University Press, 1930), vol. II, chap. 4, pp. 79–92.

41. Ibid., p. 127.

42. Nicholas Rescher is therefore quite wrong in stating that, with respect to the Argument from Design, "the only characteristic touch Leibniz adds to the classic pattern of reasoning has to do with the pre-established harmony [among substances]" (The Philosophy of Leibniz, [Englewood Cliffs, N.J.: Prentice-Hall, 1967], p. 151).

43. Sec, xviii, GP, 4:444, Lucas and Grint, p. 32.

44. II, ad art. 64, GP, 4:391, Loemker, no. 42, p. 409.

45. Loemker, no. 32, p. 278. This work was translated by Loemker from an unpublished Latin manuscript.

46. For example, the paradoxes of division; also the argument that since extension involves plurality and repetition, while substance is by definition a "true unity," we must postulate indivisible, unextended entities whose repetition somehow accounts for the well-founded phenomena of extension. Leibniz also has another sort of dynamical argument for force, derived from the relativity of motion: see Bertrand Russell, A Critical Exposition of the Philosophy of Leibniz (London: Allen & Unwin, 1900; 2d ed. 1937), chap. 7, esp. Sec. 41. This argument seems so confused that I have ignored it in the text.

47. Ibid., p. 87.

48. "Tentamen Anagogicum," GP, 7:279, Loemker, p. 484.

49. See n. 42.

50. See n. 38.

51. "Specimen Dynamicum," GM, 6:241–42, Loemker, no. 46, p. 441.

52. See especially "New System of the Nature and Communication of Substances" (1695) Sec. 3, GP, 4:478, Loemker, no. 47, p. 454.

53. "Possible Worlds in Leibniz," Studia Leibnitiana 1 (1969): 161–75.

54. Ibid., 166.

55. See e.g., Fifth Letter to Clarke, Sec. 4, GP, 7:389, Loemker, no. 71, p. 696; Theodicy, "Discourse of the Conformity of Faith and Reason," Sec. 2, GP, 6:50.

56. See above, p. 275. My interpretation, incidentally, accords with that of Bertrand Russell,

who also interprets the remark to de Volder as showing that Leibniz regarded resistance as a contingent feature of bodies (*Critical Exposition*, p. 79). However, for independent metaphysical reasons Leibniz probably *should* have held that the Cartesian physics is impossible (see above, n. 38).

57. For a lucid defense of this claim see A. O. Lovejoy, *The Great Chain of Being* (Cambridge, Mass.: Harvard University Press, 1957), pp. 172 ff. I have stated some reservations about Lovejoy's position in "On Leibniz's Explication of 'Necessary Truth'," in Frankfurt, ed., *Leibniz*. See also N. Rescher, "Contingence in the Philosophy of Leibniz," *Philosophical Review* 6 (1952): 26–39; and *The Philosophy of Leibniz*, chaps. 2, 3, and 5. This problem can be derived from the theory of truth, but also arises independently of it if "God chooses the best world" is taken to be a necessary truth.

58. In Note II to Proposition 33 of *Ethics* I, Spinoza does speak of the "perfection" realized by God in the creation of things; elsewhere, however, he explains that perfection is to be understood as the same thing as reality. Understood otherwise, perfection and imperfection are "merely modes of thinking": the judgment that things in nature are "well-ordered" reflects merely the fact that they happen to conform to our imagination. See the Appendix to Part I, and the Preface to Part IV.

59. Some participants in the discussion at Ohio State did object to this claim. To the extent that their objection may have resulted from a feeling that necessitation by final causes is still necessitation, and that is all that really matters, I am not inclined to concede it. (I do not see why this is *all* that really matters, although I agree it is one of the things that really matter.) However, their protest also made me aware of some purely logical difficulties in the conclusion of the paper as originally presented, which I have attempted to correct.

J. E. McGuire

Chapter Eleven

"LABYRINTHUS CONTINUI": LEIBNIZ ON SUBSTANCE,
ACTIVITY, AND MATTER

Of the problems of the continuum Leibniz wrote that "no one will arrive at a truly solid metaphysics who has not passed through that labyrinth".[1] This was no idle remark. As this paper will seek to show, the continuum in its mathematical, perceptual, and metaphysical aspects goes to the bottom of Leibniz's theories of matter, motion, substance, and dynamics. It shaped his dynamics through the law of continuity, which was most fully formulated in 1687.[2] Leibniz uses this law as a criterion to expose what he saw as erroneous consequences in the Cartesian laws of impact. (G. 3:51–55, G. 4:354–92). In his analysis of the perceptual continuum, Leibniz argues that material phenomena, which he holds to be aggregational and divisible in nature, presuppose indivisible simples (G. 6:548, 607). The solution of the metaphysical continuum also forms an important part of Leibniz's theory of the action of simple substances. This, along with the law of continuity, supports his attack on Cartesian and Newtonian theories of space, time, and matter. In general the continuum is a *"principle of general order"*[3] according to which the processes of nature can be conceived as a harmonious and intelligible system. Of singular importance, then, is Leibniz's unraveling of the labyrinth of the continuum. It will be of central concern in the analysis to follow.

The themes of this paper can best be introduced by placing them in their seventeenth-century intellectual context. Leibniz thought that the Mechanical Philosophers had failed singularly in their attempt to provide a theory of physical activity (G. 4:504–16). The Cartesians and the occasionalists, he argues from about the middle of the 1670s (G. 4:504–16; G. 1:330–31), conceived matter as passive and devoid of any intrinsic principle of change. Rejecting the existence of active principles extrinsic to matter, the occasionalists made all occurrences occasions of divine will. The Cartesian program also denied the existence of extrinsic agents in nature. Moreover, the Cartesian doctrine that God actively preserves finite things from moment to moment[4] afforded Leibniz ground for claiming that they had retreated to a form of trans-creationism in order to explain physical activity (G. 2:168 and 192). For Leibniz both positions placed a heavy burden on the miraculous. But the Newtonians fared no better. Though they recognized activity in nature and developed an ontology of force, Leibniz regarded Newtonian gravitation as a reintroduction of occult qualities into natural philosophy.[5] For him, as for many contemporaries, the ontological status of Newtonian forces was vexatiously obscure.[6] Since the Newtonians conceived them as neither derived from, nor reducible to, material phenomena, forces seemed to be arbitrarily superimposed upon the latter. Which of the traditional categories would force fall under? As an active agent it could not be an attribute or mode of passive matter. On the other hand, it could not be a substance, as it appeared to depend on matter for its existence.[7] As an account of activity, the ontological status of force was in a conceptual limbo. It was not surprising to Leibniz that the Newtonians held God to intervene periodically in nature to sustain activity.[8]

Within this context, Leibniz's program in natural philosophy takes on meaning. In the 1680s he began to articulate a theory of motion based on a measure of force different from that of the Cartesians.[9] Leibniz's conception of the ontology of force was that of an intrinsically active agent, which was categorically different in nature from the contract action view proposed by his contemporaries. Only in the 1690s, largely in reaction to Newton's *Principia* of 1687, did Leibniz develop his ideas on force and motion into a "science of dynamics".[10] But he did so within the framework of his metaphysics of active substances, which after 1695 he called monads. Also he endeavored to show how this metaphysics could legitimate physical activity in a manner that avoided premature appeal to Providence. This meant that the Leibnizian program would attempt to explicate concepts like power, force, action, change, and motion in terms of principles intrinsic to a plurality of active substances. The practice common to the Mechanical Philosophers of grounding concepts in the nature of God, though not to be rejected, had to be reinterpreted within a general scheme of reality. By failing to develop an adequate theory of physical activity, the Cartesians, the Occasionalists, and the Newtonian interventionists were "throwing open sanctuaries for ignorance."[11] And their claim that nature could not provide a sufficient basis for activity abrogated the right of the mind, so far as it is able, to explain the self-regulating and self-determining traits of reality.[12]

Within the context of this intellectual background, the general thrust of this study will be to analyze Leibniz's theories of substance and matter. The order of business is as follows. Section 1 deals with Leibniz's general conception of substance, the notion of *haecceitas,* the 'predicate-in-notion' doctrine, the metaphysical thesis that substances contain at once all their attributes, and the related doctrine of the dissimilarity of the diverse. The doctrine that *actiones esse suppositorum* and the general theory of change and activity are discussed in section 2. In section 3 Leibniz's theory of extended pheonomena is related to his metaphysics of substance, and especially to his doctrine that there can be no compounds without simples. Section 4 deals with the notion that the continuum is not actual but ideal, and considers difficulties that it raises for the law of continuity and the theory of monadic action. In this context monadic time also presents a difficulty. In section 5 the doctrine of activity is linked with Leibniz's dynamics through his use of the categories of substance, attribute, and mode. In connection with this, his negative arguments against the ontologies of the Mechanical Philosophy are briefly discussed. It will be seen that *Materia Prima*—monadic resistance—presents a difficulty for Leibniz's dynamics and for his theory of substance. In fact, it is shown that Leibniz's emphasis on activity led to a totally inadequate account of passivity.

I

It has become a commonplace to hold that Leibniz's metaphysical theory of substance is not an important part of his dynamics. Russell holds that the relationship between Leibniz's dynamical principles and his theory of active substance is obscure.[13] Couturat and Cassirer maintain the more radical position that Leibniz's dynamics is more mechanistic and deterministic than that of the Cartesians;[14] and therefore that his theory of substance is not part of the same conceptual picture as his dynamics and theory of force. Gueroult has attempted to turn this interpretation on its head.[15] He maintains that the mature doctrine of substance is largely engendered by Leibniz's dynamical theories.[16]

These interpretations are extreme. With regard to physical explanation Leibniz was, of course, an advocate of the Mechanical Philosophy. Such explanation must pertain to classes of events and should be formulated in terms of the universal quantities of mathematics and mechanics. But if it is to be intelligible, explanation must rest on sound constitutive principles. Leibniz sought to show that abstractions like space, time, matter, and motion were adequate only for understanding the realm of appearance as given in sensory experience. To purchase ultimate intelligibility, however, concepts of scientific explanation are to be grounded in the infrastructure of reality, the realm of metaphysical substances.[17] So it was that from the beginning of the 1670s Leibniz became an uncompromising critic of the ontologies of the Mechanical Philosophy. And it is the theory of substance that largely generates both the criticisms of the Mechanical Philosophy and Leibniz's science of dynamics.

The key notion pertaining to the general theory of substance is the term 'individual'. The basic problem of individuation for Leibniz is the question, What constitutes the difference between two individual things at a given time, as distinct from the problem of what constitutes the identity of an individual thing over time?[18] In Leibniz's thought these problems are kept clearly apart, though as we shall see they are connected.

The Leibnizian doctrine of individual substance involves two related theses. The distinction between complete and incomplete concepts, and the claim that substances have a 'thisness' or *haecceitas*. In the *Discourse on Metaphysics* (1686) and *The Nature of Truth* (1686),[19] Leibniz argues that there are complete and determinate concepts of individual substances, but not of abstract concepts such as extension and kingship. The general argument is that abstract concepts, even of great complexity, do not necessarily individuate a particular thing. Abstract concepts, being essentially general, apply equally to a number of individuals. Should they succeed in the task of individuating, this would be a contingent fact. Nor can abstract concepts capture the way in which individuals exist. Leibniz wants to affirm something positive about the reality of each individual substance: *individuum ens positivum*. Thus a complete concept of *this* or *that* particular thing is necessary for individuation.

But what constitutes a *complete* concept—something that can only be uniquely instantiated by an actual individual substance? Leibniz tells us that the complete notion of "an individual substance is a subject which is not in another subject, but others are in it, and so all the predicates of that subject are the predicates of the individual substance."[20] A concept is thus complete if, and only if, it cannot be contained in a concept more complex than itself. For example, if we know all that can truly be said of Alexander, we know that, among other predicates, the abstract predicate kingship is contained in the concept of Alexander. But the concept of Alexander that involves his being the son of Philip and the conqueror of Darius is not predicable of the concept kingship: *subjectum non predicatur.*[21] Just as the general concept "sphere" does not support an inference to the diameter of any actual sphere, so from the concept of kingship no particulars about Alexander can be inferred, e.g., that he is the son of Philip and the conqueror of Darius. As an abstract, incomplete concept does not contain all its particular and possible instantiations, it cannot be predetermined to all actual temporal circumstances. Leibniz does not claim that we have any complete concepts. For he is concerned not with our names for individual substances but with the way in which God names possible and actual individuals. A divine proper name, for Leibniz, includes a complete description of the individual named. Moreover, unlike Descartes and the empiricists, he makes notions like existence and possibility fundamental to the general structure of thought, rather than problems in the theory of knowledge. He is thus not primarily concerned with the conditions in experience that would make any assertion about an individual substance true. In any event, "individuality involves infinity, and only he who is capable of understanding (infinity) can have knowledge of the principle of individuation of such or such a thing" (G. 5:268. *New Essays*). Of this, only God is capable. When Leibniz's criticisms of the ontology of the Mechanical Philosophies are considered below, the distinction between complete and incomplete concepts as it relates to his theory of substance will take on important light.

An individual substance, then, falls under a complete concept. This brings us to the notion of *haecceitas*. The Leibnizian analysis of complete and incomplete concepts indicates that he is not only interested in unambiguously pinning down conceptually what actually exists, but concerned also with the 'thisness' or *haecceitas* of substances. And here Leibniz seems concerned to give a meaning to judgements of individuality such as 'this is near that', even though 'this' and 'that' are each applicable to any existing thing. Four of Leibniz's chief doctrines are advanced to elucidate *haecceitas,* a term that he borrows from Duns Scotus, whose theory of individuation was discussed by Leibniz in his earliest philosophical work, *Disputatio metaphysica de principio individui* (1633):[22] (1) the subject-predicate form is basic to true affirmative propositions; (2) all predicates are contained in their subject terms; (3) the doctrine of sufficient reason; and (4) the doctrine of the dissimilarity of the diverse. All four doctrines were adumbrated in the early 1670s.[23] Only in the 1680s were they forged together as a metaphysical unit.

It is best to begin with the doctrine of sufficient reason in its widest sense, as it relates to the doctrines of creation and compossibility. Leibniz's general position is this: Possible substances only qualify for actualization as independent particulars when their possible and individuating attributes are intrinsic to them, and compatible with other possible particulars. From the 1680s onward, and especially in the *Discourse* (1686) and *On the Radical Origination of Things* (1697), Leibniz made a clear distinction between the role of the divine understanding and will (G. 4:438; *Discourse,* Section 13, G. 7:302–8). Eternal truths, construed much in the manner of Augustine, exist in the divine mind and in no way depend on divine will. But with contingent, individual particulars, the case is different. They depend for their existence on the fully actualized, free decrees of the divine will. Within the possibility of an individual substance is expressed its possible connections with other possible individual substances. And this also involves the possible purposes of God. To think of a possible substance in conjunction with other possibles is to think of possible divine decrees, as to think of an actual substance in conjunction with actual circumstances is to think of God's actual decrees. But the definitions of abstract truths like the concepts of kingship and triangularity make no reference to divine decrees. Whether of contingent or necessary truths, then, clear, distinct, and adequate knowledge can only be grounded in God's existence, apart from which there can exist nothing (G. 7:302).

But what determines that a system of possible individual substance will exist? At this point the distinction between the divine will and understanding is crucial. All attributes that relate to the free exercise of God's will are contingent, as, for example, his choice in creating the world that actually exists (G. 7:307). It was never a necessary truth in the mind of God that this present, actual world exist. Of all possible worlds that might exist, there must be a sufficient reason for the one in actual existence. God had a sufficient reason in creating this world in that it uniquely expresses the principle of the best. Only those possible individual substances that can form a compatible or compossible world are candidates for existence. For "not all possibles are compossible" (G. 3:573). The distinction is this: A possible individual is that which does not imply a contradiction, while a world of compossible individuals is such that there is the possibility of joint coexistence. Thus a collection of possible individuals does not form a cosmos as the individuals do not necessarily form mutual connections among themselves. Now the compossible world that actually goes into existence can *only* be that which satisfies the principle of the best. This means that a world can come into existence because each of its individual substances harmoniously manifests its perfections, or quantity of existence, to a maximum degree. Such a world contains the greatest variety of jointly possible or compossible individuals. God thus actualizes the possible world in which perfection is at a maximum among all compossible substances. The principle of perfection, then, provides God with a means by which to decide among the infinite and mutually exclusive systems of compossibles. There is therefore a sufficient reason why the actual world exists. For possible individuals have a claim to exist only because God has chosen to act in selecting a compossible world, according to the archetype of compossibility and perfection. That God freely acts as a moral agent in selecting the present world by means of the principle of perfection follows from God's necessary nature. But such acts, Leibniz maintains, do not follow from God's nature with Spinozistic necessity, but through an infinity of steps that converge on God's essential nature. For "the ground of contingency is in the infinite."[24] The appeal to the infinite, however, is little more than an evasion.[25]

Individual substances go into existence, then, because they have the highest degree of individual perfection, yet are mutually compatible. Now to know the *haecceitas* of a substance is to know that it is "at the same time the foundation of and reason for all the predicates which can truly be stated of (it)" (G. 4:433; *Discourse,* section 8). That the *haecceitas* of a substance is a "foundation" for all its predicates is explained by the thesis that a substance contains its attributes. The first thing to note is the fact that the Latin term *subjectum,* 'that which lies beneath,' has for Leibniz, as it had for medieval thought, the connotation both of a propositional subject and of a substance (G. 2:486, to Des Bosses, April 1714). The term can thus refer to both an individual substance and a logical subject, and has ontological and logical import. Leibniz, of course, keeps these two senses separate. In his analysis of subject-predicate propositions, he seems to have interpreted the Greek term *huparchei,* as in "*x* belongs to every *y*" in an ontological sense. Parkinson has suggested that he may well have thought of this notion through its standard Latin equivalent *in-esse,* 'to be in'.[26] Hence, the Aristotelian doctrine that to say truly *x* is *y* is to say *x* belongs to *y* came to be construed by Leibniz as sufficient only when *x is contained in y.* The Aristotelian doctrine of predication is thus given a strong ontological interpretation. It would be wrong, however, to suppose that the formal 'predicate-in-notion' model alone engendered the ontological theory that substances "contain" their attributes. The *predicatum inest subjecto* thesis is at bottom a piece of metaphysics. As we have seen, possible substances only qualify for actualization as complete individuals when their possible individuating attributes are intrinsically inherent and compatible with other possible individuals, and, moreover, when they satisfy the principle of the best. The containment thesis with regard to actual substances is therefore generally conditioned by the principle of perfection and the doctrine of compossibility.

Actual substances thus contain all their attributes, and these attributes completely express the individual *haecceitas* of a given substance. Leibniz develops this position in the thesis that diversity implies dissimilarity, that two things cannot have the same nature (G. 2:249, to De Volder, 20 June 1703). From this, two things are clear: that the doctrine of the diversity of content is essential for Leibniz's general conception of substance, and that the principle of indiscernibles, "that there are not in nature two indiscernible real absolute beings," (G. 7:393) is logically subsequent. Since it is impossible to have diversity without dissimilarity, the nature of a substance expresses completely what the substance is. A complete expression of what a substance is cannot be true of another substance. If there are really two numerically distinct substances, then their content must be diverse in at least the minimal sense that the

297

one has an attribute that the other has not. Hence the *haecceitas* of an individual substance is such that it "contains", independently of all else in existence, a diversity of content. It is the "foundation" therefore of all the attributes that inhere in it.

It might be claimed, however, that Leibniz's notion of *haecceitas* commits him to the view that substances can possess a diversity that is independent of dissimilarity. That is to say, it might be claimed that a substance's individual *haecceitas* shows it to have an individuality *independent* of its attributes, since judgments using ostensive terms (such as 'this is near that') have a clear meaning. It could then be said that two substances are numerically diverse without being diverse in content. If this is so, is there not sufficient ground for rejecting the doctrines of containment and the dissimilarity of the diverse? But the 'thisness' of a substance is itself a quality. As such it supposes the existence of 'something' of which it is an attribute. And the fact that a substance can be individuated independently of its attributes by means of judgments like 'this here' does not prove that its individuality is independent of its attributes. Therefore Leibniz is not committed to the view that substances can be diverse without being dissimilar.

In what sense does *haecceitas* give a "reason" for the attributes contained in an individual substance? What Leibniz seems to mean is that each substance has a constitutive basis that provides a mediating link between subject and predicate. Hence what appears to perception as a merely *de facto* association of related qualities is anchored in the internal reality of a substance's individual nature. Leibniz in a polemic against the Cartesians and Occasionalists elaborated his position in *On Nature Itself* (1678). Without 'intermediaries' or mediating links, through which "what is remote in time and space can operate here and now . . . anything can be said to follow from anything else." Accordingly there must be "some connection operating here and now between cause and effect." This mediating link is "a certain efficacy residing in things, a form or force such as we usually designate by the name of nature" (G. 4:507). This provides a ground of truth that consists in the link that binds the predicate with the subject. The principle of sufficient reason follows from this theory of truth (G. 7:199–200, *Veritates absolute primae*). This link constitutive of a substance's *haecceitas* supports all change in its successive states. There is thus a reason in terms of a substance's nature by which the manifestation of its attributes over time is determined. An individual nature is such that, were it known by the finite mind, all separate episodes in the substance's history would be seen as grounded in its existence, and would not be conceived merely as being hypothetically connected with antecedent conditions related to the substance's behavior.

II

Leibniz's general doctrine of activity can now be introduced. As early as 1668 he linked the idea of substance with the notion that a substance is an entity that has an intrinsic principle of action. An entity that subsists in itself is a *suppositum,* and is "denominated by action; hence the rule that actions belong to *supposita.*"[27] This rule he attributed to medieval thought.[28] What acts is thus a *suppositum* or individual substance. The *actiones esse suppositorum* doctrine—that actions belong to substances—is discussed later in the *Discourse* (1686) and in *On Nature Itself* (1698) (G. 4:432, *Discourse,* sect. 8; G. 4:509). There, as in his earlier writings, Leibniz uses the term *suppositum,* rather than the more neutral *subjectum,* to link action with the nature of substance.

With the development in the 1680s of the doctrine that predicates are contained in their subjects, and in the 1690s with the systematic articulation of his "science of dynamics", Leibniz was able to advance an explanation of change and physical activity. One of the important aims of the mature theory of substance is to develop a theory of dynamics that would explain these notions (G. M. 6:234, *Specimen Dynamicum*). Along with his contemporaries, Leibniz held that change is simply the transition from one state of something to that of another (G. M. 6:238). After all, this conception was one of the principles underlying the laws of impact as understood in the seventeenth century. Activity, however, is a teleological notion, the completed result of some determined action. The Mechanical Philosophers, of course, had denied that activity could in any sense truly exist in nature. In their explanatory program final causes were restricted to volitional acts. Leibniz took the unusual course of attempting to unite in the same theory of substance, change, activity, and final causes. In this way he was able to place the source of change and activity in nature, rather than arbitrarily grounding it in divine efficacy (G. 4:504–16, *De Ipsa Natura*).

It is important to realize that the theory of predication provides an abstract account of change, which is also the general framework for Leibniz's theory of physical activity. It is best to consider change first. As has been shown already, Leibniz holds that each substance contains, independently of any other, a complete set of predicates and attributes. And as each substance generates its own history, change must be founded in its individual nature. But if a substance has all its attributes from its beginning such that every predicate of that individual now refers to some determined moment in its total history, can Leibniz account for change over time? Leibniz, however, can distinguish between the state of a substance at a particular time, and the fact that it has an attribute such that it is in that state at that time. The latter is a permanent attribute of the substance, the former a temporary state of the substance. A particular state exists at a particular time and not at the next. Therefore it cannot be a permanent attribute of the substance. Change is thus accounted for because each substance is the subject of different states that occur at successive times. And since each substance contains all its attributes and generates its own history, there can be no real change that is not an intrinsic modification.[29] As the permanent nature of each individual substance is the subject of a succession of temporary states, it assures a sufficient reason for each mode of change that occurs in nature. For as Leibniz says, "the concept of an individual substance involves all its changes and all its denominations" (G. 2:56, to Arnauld, 14 July 1686). The Leibnizian theory of substance thus provides an alternative explanation of change to those embodied in the ontologies of the Mechanical Philosophy.

But a theory of change does not explain activity. Nor does it explain Leibniz's dogma "that not only whatever acts is an individual substance, but also that every individual substance acts without intermission" (G. 4:509, *De Ipsa Natura*). Now it is incontestible fact that Leibniz makes a distinction between the source of activity, and activity itself. As we shall see later, this distinction is important for his analysis of motion and force in terms of the theory of substance. The distinction is clarified somewhat in *On Nature Itself.* " . . . There can be no action without a power of acting, and conversely, a power which can never be exercised is meaningless. Yet activity and power are different things, the former a matter of succession, and the latter permanent" (G. 4:509). These passages throw light on the doctrine that what acts is an individual substance. In the first place it is clear that a substance acts because it has an inherent power or capacity to do so. Moreover, a substance exercises its power, so that fully actualized change is manifested continuously in successive states. A permanent attribute of an individual substance, then, is the power to act. As the inherent power of substances is at all times exercised, there is present a preexistent tendency to change. This tendency is continuously actualized, and every state of each substance ineluctably passes to the next state. Hence activity is a general consequence of the power of substances, and arises from the intrinsic tendency of successive states to change. Since activity results from an inherent tendency to change which always strives to actualize itself, this is tantamount to ascribing final causes to nature. And Leibniz invokes both the source and end of action in his analysis of substance. There is thus a sufficient reason for the action of an individual substance, and for the end to which it strives, completed action. Moreover, substances mutually satisfy the principle of perfection, the end for which they were created.

Leibniz's notion of active power is not that of potentiality in the modern sense a "what would happen" *if* something were affected by external circumstances. Rather, power is a permanent attribute of substances by which they continuously act. As Leibniz says in *The Concept of Substance* (1694): "Active force . . . contains a certain act or entelechy and is thus midway between the Faculty of acting and the act itself and involves a conatus. It is thus carried into action by itself and needs no help but only the removal of an impediment" (G. 4:469). Leibniz's account of activity and power will have to be considered later in the context of his dynamics. For the action of substance is connected essentially with a doctrine basic to Leibniz's dynamics, that a cause expends itself just in proportion to the resultant effect that it produces (G. 4:426). The theory of substance, however, will be seen to raise difficulties for the law of continuity and Leibniz's conception of time.

III

Within the framework of this general theory of substance and activity, Leibniz articulated a specific conception of active substance in the 1680s. This conception goes to the bottom of his "science of dynamics". And it is intimately connected with the doctrine of the continuum.

In his letters to Arnauld (1686) (G. 2:111–29), and at the beginning of both the *Monadology* (1714) (G. 6:607) and the *Principles of Nature and Grace* (1714) (G. 6:598), Leibniz states that there can be no compound things without the existence of simples. Put briefly, Leibniz's position is that material, extended things are always divisible either actually or logically. Anything divisible is a compound, and Leibniz argues that from a logical point of view what is compounded must ultimately be made of uncompounded elements (G. 2:111–29). So the concept 'compound' presupposes the concept 'simple', or put the other way, the concept 'divisible' is a derivative concept, presupposing the concept 'indivisible'. If what is material is compounded and divisible, the indivisible must be simple and nonspatial.

It is the purpose of Leibniz's theory of substance to show that these logical connections between compounds and simples represent the nature of reality and to show that the ontologies of the Mechanical Philosophy do not. Substances are simple, indivisible, partless, and intrinsically active (G. 2:268–69, to De Volder, 30 June 1704). What does this claim involve? The Leibnizian position is summed up in an apt aphorism from the correspondence with Arnauld (1686–87): " . . . That which is not truly *one* being is not truly a *being*" (G. 2:97, to Arnauld, 30 April 1687). This is the ancient maxim to the effect that *ens et unum convertuntur*. So as to fall into the category of an ontological primitive, an entity must satisfy the criterion: *nec vere ens, nec vere unum*. Atoms and compounded entities cannot satisfy this criterion. They are divisible and hence not one entity. Only an entity that is indivisible can satisfy it, as such an entity is spatially partless and thus one entity. Unlike compounds it cannot be reduced to anything other than itself. It is primitive. Such entities Leibniz called *monads,* or simple unities.

The view that being and unity are one presents a difficulty for Leibniz's general theory of substance. It seems to oppose the doctrine of the dissimilarity of the diverse. If substances are truly simple, how can they contain diverse contents? The notion of simplicity is intimately connected with the claim that monads are non-spatial. They are not divisible in the dimension of space: they are partless. Yet this does not preclude their having an internal organization, for predicates and attributes are not "parts" in a strict spatial sense. Yet they make up the distinct internal organization that gives substances the reality and individuality that they possess. Simplicity, then, does not deny to substance a unique diversity of content. It denies only that they are divisible in the dimension of space. If Leibniz is concerned to deny the simultaneous differentiation of monads into spatial parts, what of successive differentiation? Are they also undivided in the dimension of time? This question raises deep problems. For if monads are simple in the sense that, besides having no simultaneous parts, they are undivided in the dimension of time, they will have no histories, will exist only for one simple and indivisible moment of time, and will not change. A full treatment of these problems must await the analysis of the continuum.

Does Leibniz offer any specific arguments for the claim that monads are simple, partless, and indivisible? He does not. He merely states the traditional view that that which is indivisible is immaterial (G. 4:561–65, *Reply to the Thoughts on the System of Preestablished Harmony*, 1702). Besides God, only souls or soul-like entities were regarded as immaterial. And being without parts, they were considered to be simple and perduring (G. 6:548, and G. 7:552, to Sophia, 12 June). In the early 1670s Leibniz had designated only human minds as true substances or indivisible unities. By the 1680s he had extended this view to all actual things, whether human minds or otherwise (G. 4:459, *Discourse*, Section 34). Reality had become irreducibly spiritual.

Simple substances, being partless, cannot be made or unmade. That is to say, they do not begin or cease to exist naturally: they begin by creation, end by annihilation. Compound things can be reduced to their constituents. Monads cannot be thus reduced or destroyed (G. 4:433–34, *Discourse*, Section 9). They are like neither Cartesian extended corpuscles nor Newtonian atoms. As the notion that substances are indivisible is taken over from traditional thought, so too is the doctrine that they are intrinsically active. We have seen how change and activity are a consequence of the containment of predicates thesis. This thesis does not explain the fact that substances have, as a principle attribute, the power to act and to generate activity. And activity no more follows from their simplicity or indivisibility than does motion from extension. That the capacity to act is a principle attribute of spiritual substance, Leibniz accepts without argument from traditional theology.[30] Like Berkeley, he locates the source of action in volition and perception. This he extends to all actual things. As he says: "This principle of action is most intelligible, because there is something in it analogous to what is in us, namely, perception and appetition. For the nature of things is uniform, and our nature cannot differ infinitely from other simple substances of which the whole universe consists'' (G. 2:270, to De Volder, 30 June 1704).

The doctrine that compounds presuppose simples and that the latter are intrinsically active is a direct consequence of Leibniz's conception that only the spiritual is ultimate and real. The problem which must be faced is this. How is Leibniz's theory of substance related to his conception of the perceived nature of phenomena, and to the related arguments that attempt to show that the ontologies of the Mechanical Philosophy are based on incomplete and abstract concepts? Does the Leibnizian claim that extended phenomena presuppose the doctrine of simple active substances arise from a positive metaphysical commitment? Or is this doctrine merely derived from Leibniz's critical and negative examination of the Mechanical Philosophy. Russell holds the latter view.[31] And Couturat maintains that Leibniz's dynamics depends on the negative arguments against the Mechanical Philosophy, but that its validity is independent of the metaphysics of substance.[32] Although it is true that a close examination of the science of his contemporaries sharpened Leibniz's own dynamics, that science was developed entirely within the framework of his theory of substance. It is clear that the doctrine of *actiones esse suppositorum* (1668) antedates the destruction analyses of the ontologies of Cartesianism and Newtonianism. And the earliest expression of the view that the essence of matter is not extension occurs in *The Theory of Abstract Motion* (1671), where Leibniz argues that matter is compounded of a series of momentaneous conatuses (G. 4:228–30). Leibniz thus discarded the notion that extension is the essence of matter a decade before his systematic and critical analysis of the Cartesian philosophy. In what follows an attempt will be made to show that his theories of substance and matter naturally form part of the same conceptual picture. And to show also that the theory of substance provides the primary motivation for Leibniz's dynamics and for his rejection of atoms and extended phenomena as ontological primitives.

We must begin with extension. This has a peculiar, but important status in Leibniz's thought. A proper understanding of the nature of extended phenomena shows an intimate connexion between Leibniz's conception of simple substances, and his contention that extension is not a primitive notion. He put his position to De Volder as follows:

The notion of extension is thus relative, or extension is the extension of something, as we say that multitude or duration is the multitude or duration of something. But the nature which is presupposed as diffused, repeated, continued, is what constitutes the physical body (*corpus physicum*), and can only be found in the principle of action and passion, since nothing else is suggested to us by phenomena (G. 2:269, 30 June 1704).

Notice first that Leibniz claims to derive his analysis of extension from the perception of phenomena: "I believe that perception is involved in extension" (G. 2:183). This means for Leibniz that matter and motion *qua* extended beings "are not so much substances or things as they are the phenomena of percipient beings" (G. 2:270). Thus he is here concerned with the perceptual continuum and the perception of change, not with the mathematical and temporal continuum. The extended continuum is not a complete notion as it depends on something that continuously acts, as in numbers where aggregate parts are discerned "something is necessary to be numbered, repeated and continued" (G. 2:164, to De Volder, 3 April 1699). Not only is extension an attribute of the extended, as duration is an attribute of the enduring, but its existence is based on a principle of action. For it is only something that has the power to act that can produce repeated activity. At once the theory of simple substance and the doctrine of activity is linked with the perceived phenomena of extended things.

But why should the perceptual continuum be based on the nature of something that acts continuously? Is not extension an absolute and invariant attribute of the extended? According to Leibniz, the Cartesians were led to this view because "they were content to stop where their sense perceptions stopped" and had failed "to distinguish sharply between the sensible and intelligible realms" (G. 2:269, to De Volder, 30 June 1704). If the mind seeks for intelligibility, it finds that extension "may be analyzed into plurality, which it has in common with number; into continuity, which it has in common with time; and into coexistence in common even with things which are not extended. I should not have thought that plurality is to be denied in the extended thing, especially if we admit that it has actual parts; we should in that case also have to deny plurality to a herd or army; in other words, everywhere" (G. 2:183, to De Volder, 23 June 1699). Extension is thus resolvable into plurality, continuity, and "the coexistence of parts at one and the same time" (G.

2:169, to De Volder, 3 April 1699). Only extension is resolvable into all three. Time and space have continuity alone. Time like duration is successive, and so "perishes continuously" (G. 7:408, fourth letter to Clarke, 1715). But extended things have a plurality of coexisting "actual parts" that they have in "common even with things which are not extended." This is the crucial claim. It involves two things. That extended things are *phenomena bene fundata,* and that they are simply a plurality of coexisting and active substances. Thus the "actual parts" of extended things are non-extended substances. This means that extended things are well-founded phenomena in perception, and that they are in reality a plurality of substances that *appear* to be continuous and extended.[33] What is real in the perception of extended things is only the reality of substances taken one at a time. And it is substances, by virtue of their nature, which act so as to give a "nonsuccessive (unlike duration) but simultaneous diffusion or repetition" (G. 2:269, to De Volder, 30 June 1704), of their existence. Just as in a rainbow an aggregate of colorless particles is perceived as colored, so a collection of unextended substances is perceived as continuous and extended. The exteriority of phenomena is translated into the interiority of diverse substances, as the spatially distributed colors of a rainbow come together at the unextended center of while light. Thus well-founded phenomena are not truly continuous in nature, they simply appear to be so. And Leibniz appropriately observes of phenomenal change that "the same holds of changes, which are not truly continuous" (G. 2:279 to De Volder, 11 October 1705).

Leibniz thus comes to a radical conclusion regarding the perceptual continuum. It has a totally different nature from other continua. If there can be no true continuity of the extended, in perception the physical world comes into existence in discrete chunks.[34] For in the perceptual world repetition is discrete and determinate, rather than being continuous as with ideal entities whose parts are indeterminate. And if phenomenal change is not continuous, there can be in the phrase of Whitehead "no continuity of becoming."[35] This means that experience of extension is made up of a plurality of discrete 'parts' that coexist simultaneously, but not successively. These 'parts' are in reality the diverse states of substances that exist at one time but not at the next. As a plurality, however, they form a coherent set of perceived characteristics at any given moment. Thus extension and change involve a succession of extraneous 'parts' that demand a momentary state, a state that consists of a force striving toward change, a true quality inherent in simple substances. So the appearance of continuity in the extended continuum is in reality the repetitive action of a plurality of discrete substances. Entities of perception are nothing but *aggregatum per accidens,* which derive their unity from genuine unity, a proper *unum per se,* namely, simple substances, for "multitudes exist only through true unity" (G. 4:478, *Système nouveau de la nature,* 1695). It is the unity of being of simple substances with their diverse contents that, through continuous action, well-found the appearance of continuity in the perception of aggregational entities and their changes. And it is thus that extended phenomena are perceived as unities by the mind (G. 2:96, to Arnauld, 30 April 1687).

IV

Yet the question has not been asked: Why does Leibniz wish to deny that ultimate reality is continuous? Partly, this conclusion arises from his treatment of the extended continuum. There as we saw, he denied that phenomenal extension and change are truly continuous; they are merely an appearance of indivisible substances. However, at the bottom of the doctrine of phenomena, lies Leibniz's solution to what he called the "labyrinth of the continuum."

But it also involves other categories of entities. Leibniz divided the central categories of natural philosophy into three distinct groups. Space, time, and motion are *entia rationis:* phenomenal extension and change are well-founded appearances: and substances and their attributes are actual existents (G. 7:552–53, to Sophia, 12 June 1700). Leibniz's picture of the physical world is this: Analysis of the phenomena of perception shows them to be extended aggregates; extension is essentially plurality; plurality is actually infinite, as there is no possible or actual limit to the divisibility of the extended. Since all extension is composite, if there is to be ultimate unity in reality, the plurality of the extended must be a result of entities which are indivisible and non-extended. Hence what gives rise to the appearance of matter is not material, if what is material must be extended. If the extended arises from a repetition of plurality, it must result from the activity of something intrinsically active, since it cannot be repetition of itself. Nor can it be the repetition of mathematical points: these are mere modalities, or the limit of a process of subdivision. As such, though simple, they are not *real*. Only something endowed with a power to act, like a soul, can be real and simple, and can account for our perception of the extended. Simple substances alone satisfy this criterion. Since the nature of what actually exists is to act, only simple individual substances truly exist. Hence extended phenomena are in reality only the simultaneous action of a plurality of actual and coexisting substances. On the other hand, the notions of space and time, with motion conceived as a mathematical function of these, form a conceptual grid that the mind imposes upon phenomenal change. Everything at the phenomenal level has its own extension and duration, which it is perceived to carry with it, but not its own space and time, which are merely *entia rationis* or possible orders. Space is a possible order of coexistence, time a possible order of change and succession. As with extension and duration, however, space and time, when applied to the phenomenal ordering of perceptual entities, have their *fundamenta* in the realm of substances, and are hence well-founded phenomena. For then space arises from the perceived ordering of simultaneous phenomena, and time is the perceived ordering of successive events. This view of the physical world, with its distinctions between ideal, phenomenal, and real entities, is a direct consequence of Leibniz's analysis of the continuum.

Leibniz's views on the continuum are complex. Consider the following two passages. The first is from a letter to Des Bosses in 1709:

> Space, like time, is a certain order—which embraces not only actuals, but possibles also. Hence, it is something indefinite, like every continuum whose parts are not actual, but can be taken arbitrarily—Space is something continuous but ideal, mass [*massa*] is discrete, namely an actual multitude, or being by aggregation, but composed of an infinity of units. In actuals, simples are prior to aggregates, in ideals the whole is prior to the part. The neglect of this consideration has brought forth the labyrinth of the continuum. (G. 2:379)

In 1706 he wrote to Des Bosses:

> A continuous quantity is something ideal which pertains to possibles and actuals, insofar as they are possible. A continuum that is, involves indeterminate parts, but, on the other hand, there is nothing indefinite in actuals, in which every division that can be made, is made. Actuals are composed as a

number is composed of unities, ideals as a number is composed of fractions; the parts are actual in the real whole but not in the ideal whole. But we confuse ideal with real substances when we seek for actual parts in the order of possibles, and indeterminate parts in the aggregate of actuals, and so entangle ourselves in the labyrinth of the continuum and in inexplicable contradictions. (G. 2:282)

A proper understanding of these passages goes to the heart of the Leibnizian philosophy of nature. Notice that they embody a trichotomy of the ideal, the phenomenal, and the actual. These are analyzed first in terms of a distinction between wholes that are prior to parts and wholes to which simples are prior. Second, wholes are distinguished into those that involve indeterminate parts and those that are determined, in traditional terms the distinction between the potential and actual infinite. But since those "who have philosophised—have become confused, for want of distinguishing between resolution into notions and division into parts," (G. 3:583) wholes must be analyzed further into distinctions between what is resolvable, composable, and divisible. These sets of distinctions have a direct bearing on Leibniz's views concerning the mathematical and metaphysical continuum, and also illuminate further his reduction of the non-substantial to *phenomena bene fundata*. For by means of these distinctions Leibniz attempts to demarcate the ideal, the phenomenal, and the actual.

Leibniz denies that the geometrical continuum is composed of actual parts for "in ideals the whole is prior to the parts." Geometrical lines thus contain an infinite number of points potentially, but not actually, since "an infinite number of points collected together [cannot] compose an extended line" (G. 2:97). A given line, of course, can be divided into a finite number of parts. Geometrical lines are thus not composed of actual parts, but are resolvable into notions or modalities and divisible into finite parts. The notion of the spatial continuum is different from the geometrical, since it involves the idea of a possible order of coexistence. As such, it does not involve actual distances that are extended and presuppose division into parts. Nor is it composed of parts. It is, however, resolvable into ideal modalities such as a point "which is nothing but a mode, namely an extremity" (G. 2:347). Since time is also an ideal principle of ordering, similar reasoning applies to it. It is neither composed of, nor divisible into, instants, as these are "not properly a part of time'' (G. 3:591). Being truly continuous it contains its points potentially and is arbitrarily resolvable into them.

The passages above speak of actuals that are fully determined and definite. The only entities that satisfy this criterion are individual substances. These, as we have seen, fall under complete concepts and are determined to all possible circumstances, unlike abstract indeterminate concepts such as kingship. Now Leibniz says that ''in actuals, simples are prior to aggregates''. Since actuals are simple, individual substances, they are presumably indivisible in all dimensions. So that simple actuals can be neither composable nor resolvable nor divisible. Such, for Leibniz, are true unities. And since unity and being are one, only simples can be actual and *real*. In line with this reasoning the existing world is for Leibniz a composite actual. For it is composed of simple individuals, and is resolvable and divisible into a plurality of coexisting individual substances. In terms of his distinction between the determinate and indeterminate, and the criteria of composability, resolvability, and divisibility, Leibniz thus demarcates ideal wholes from actual wholes.

But he can also demarcate the composite and the simple. This brings us again to the phenomenal continuum. Unlike simple actuals, the phenomenal continuum is composable, resolvable and divisible. It is thus not a true whole, but merely an *unum per accidens*. Aggregational unities that manifest degrees of ordering of their parts, such as an army or a machine, are no more than the mode of existence of that which composes them (G. 2:100, to Arnauld, 30 April 1687). Every entity by aggregation thus ''presupposes entities endowed with a true unity'' (G. 2:97, to Arnauld, 30 April 1687). This can only be entities that are neither composable, nor resolvable, nor divisible, namely, individual substances. Thus phenomenal and aggregational entities have an ordered coherence in perception to the extent to which they are ''modes of genuine substances'' (G. 2:97). The point is put succinctly in the *New Essays* (Bk. 2, chap. 13, p. 149n.6): ''This unity of the idea of aggregates is very true, but at bottom it must be confessed, this unity of collection is only a congruity or relation, whose foundation is in what is found in each simple substance by itself. And so these beings by aggregation have no other complete unity but what is mental; and consequently their entity also is in some way mental or phenomenal, like that of the rainbow.''

Leibniz is now able to provide a solution to the labyrinth of the continuum. Ideal wholes are truly continuous and contain their parts potentially. Actual wholes are indivisible and simple. Hence the question, How can that which is continuous consist of indivisible elements?, cannot properly arise. Moreover, the actual cannot be continuous, as it is simple and indivisible "and not formed by the addition of parts" (G. 5: 144). Being composable, resolvable, and divisible, the perceptual continuum cannot be truly continuous. Yet is is not actual, since these notions do not apply to what is truly simple—the existence of individual substances.

Leibniz's reasoning is ingenious, yet it bristles with difficulties. In the first place, it seems that he must restrict the law of continuity, so important in his dynamics, to the continuity of space, time, and motion. These alone he allows to be truly continuous. For the perceptual continuum is not truly continuous, but comes into existence in a series of discrete chunks. Moreover, Leibniz explicitly denies that actual substances are continuous in the sense of containing potential and indeterminate parts. On both these scores it seems that his philosophy of nature is a denial of the continuous. But the law of continuity is a "principle of general order" that, as the second passage above indicates, is applicable to the possible as well as to the actual. In the second place, how is continuity of action in the states of individual substances to be understood, if continuity is applicable only to ideal entities? For Leibniz says that one of his great principles is "that *nature never makes leaps:* which I called the *law of continuity*—I have remarked also that, in virtue of insensible variations, two individual things cannot be perfectly similar, and must also differ more than numerically" (G. 5:49). Not only is the law of continuity considered valid for gradual variation in monadic attributes, but it is clear that Leibniz saw a close relationship between the diverse and dissimilar contents of substances and the law of continuity. Two things seem clear. That the internal generation of monadic attributes is continuous. This means that monadic activity is itself a manifestation of continuity, for the unity of change in individual substances is gradual and non-arbitrary. Second, there is a continuous ordering of actual substances such that there exists a continuous series of substances each of which differs infinitesimally from the next. (*New Essays,* p. 712n.6, and G. 2:168, to De Volder, 3 April 1699). Continuity thus maintains that the series is everywhere dense: the dissimilarity

311

of the diverse asserts that no two things in the universal series are the same. If only the ideal can be truly continuous, in what sense are the internal relations of monadic attributes and inter-monadic relations continuous? The last major difficulty is this. If time is an ideal notion, it does not apply to the actual. But actual substances have expressed states, are expressing states, and will express states. Moreover, as they are states of one and the same individual substance, that substance is programmed to unfold a unique history. But such action implies not only some notion of continuity but some conception of monadic time.

In answer to the first difficulty, it must be said that the law of continuity covers a number of distinct but related principles of order. In Leibniz's dynamics it is properly restricted to the laws of impact. It states that when the difference between two cases diminishes without limit, the difference in their results also diminishes without limit (G. 4:375, *Critical Thoughts on the Principles of Descartes,* 1692). For example, Leibniz shows that the first two of the Cartesian rules of motion taken together violate this principle. They allow a gap or leap to exist between the initial and resultant velocity of two hard bodies even when their difference in size progressively becomes infinitesimally small (G. 4:370–75). Leibniz admits that mathematical continuity is never exemplified in any exact way by phenomena since the abstract statements of mathematics do not describe the physical world. Yet he felt justified in applying the law of continuity to phenomena because it enables the mind to order and make experience intelligible (G. 4:568, *Reply to Thoughts on Preestablished Harmony,* 1702). Moreover, he believed that there was no inconsistency between coherent notions and the actual determinations of phenomena. Though in strict terms, ideal notions are continuous, yet they are applicable to phenomena insofar as they are consistent, and lead to correct empirical results. Even though Leibniz held phenomenal change and extension to be discrete in character, the law of continuity applied. For the repetition of discrete but simultaneous states creates the perception of continuity. The law of continuity can then be employed in the analysis of change in phenomena. Leibniz held that the same reasoning applied to spatio-temporal continuity. Though it embodies ideal notions that are mathematically continuous, he held it to be consistent with the ordering of actual phenomena, and to be well-founded in the perception of change (G. 4:568).

312

If the mathematical notion of continuity can be applied to the perceptual continuum, can it also be applied to monadic action and to the continuity of individual substances? Since Leibniz expressly denies that anything actual is truly continuous, it cannot be applied. On the other hand, he unambiguously held that the internal action of substances is continuous, and that coexisting substances form in terms of their attributes a continuous series of variations. In this context the notion of continuity is intuitively based on the principle of plenitude. This states that, as a principle of metaphysical ordering, nature is maximumly full of diverse existents. This principle is directly related to the principle of perfection according to which God created the most perfect and compossible world. In such a world there is the greatest 'quantity of reality' and the greatest variety of kinds. And then, the internal organization of the attributes in individual substances must be continuous, so that all aspects of their diversity are expressed. This means that as principles of change they exfoliate their attributes continuously, not continually. The same reasoning was applied by Leibniz to the intuitive notion that individual substances form a continuous series (G. 5:287–88).

This solution to the continuity of monadic action is vague. Moreover, it is not clear how Leibniz could have reconciled this aspect of his thought with his crisp analysis of the continuum. The import of that analysis does not seem to illuminate the intuitive notion of continuity embodied in the principles of perfection and plenitude. It is not only that continuity is an ideal notion and thus not applicable to actual substances: there is another difficulty. If monadic action and inter-monadic relationships are manifestations of continuity insofar as they exemplify the principles of plenitude and perfection, and as the perceptual continuum results from this action, what grounds are there for supposing it to be atomic in structure? Certainly, construing the action of monads as continuous cannot be squared with the doctrine that our perception of phenomenal becoming is discontinuous. But there is a more pressing difficulty. Since the perceptual continuum results from the action of monadic substances, there is presumably a causal relation between the two levels. However, if one is discontinuous and the other continuous, the relation between them seems inconsistent with a principle basic to Leibniz's thought: that the entire effect is proportional to the total cause. From these difficulties, there seems no exit for Leibniz.

But the third difficulty, that of monadic time, is more intractable. If monads have as one of their principal attributes an intrinsic power to act, activity will arise continuously from the exfoliation of predicates. This, however, involves some notion of temporal direction. But Leibniz has denied that monads are divisible in the dimension of time as they are truly simple substances (G. 2:258, to De Volder, 10 November 1703). Moreover, he also holds that time is an *ens rationis* that can function as a principle of order only at the level of phenomena. Time, thus construed, does not apply to what is actual and real.

To begin, recall two Leibnizian distinctions that were discussed earlier: the distinction between the successiveness of activity and the permanent nature of power inherent in individual substances; and the distinction between the state of a substance at a given moment and the invariant attribute that the substance is in such and such a state at that given moment. Now fully actual activity is a phenomenal category comprising states of substances that exist at one moment and not at the next. In this sense activity is a phenomenon of succession. For Leibniz succession is the result of repetition. And repetition arises from an inherent tendency in substances to pass from one state to the next. He tells us that "all repetition is either discrete as where parts are discriminated . . . or continuous when the parts are indeterminate and can be assumed in infinite ways" (G. 4:394). Phenomenal activity is successive in the former sense. As we have seen, it arises from the continuous action of a plurality of discrete and individual substances. Thus the plurality of states that arise from substances, as they coexist at a given time, is the foundation of the direct perception of extension. But the repetition of these states is discrete. The perceptual continuum and the perception of becoming are thus atomic in character. But we have seen how, in Leibniz's view, temporal continuity applies to the ordering of actual change in perceptual phenomena. By means of his distinction between the successive character of activity, and power as a permanent attribute of substances, Leibniz can apply temporal notions to the continuity of changes at the phenomenal level in states of substances. It is clear that Leibniz is here drawing on the traditional conceptual picture regarding change and non-change. There can be no meaningful notion of something's changing in the absence of a related notion of something that is not simultaneously undergoing change. It is not at all clear, however, that Leibniz can attribute to individual substances, which contain at once all their attributes, the contrasting principles of change and non-change. This will be discussed in the next section with respect to the notions of passivity and resistance.

Analysis in terms of successive states and permanent attributes, however, does not throw light on the problem of monadic time. It only clarifies activity that results from the phenomenal perception of successive states of substances. Since states exist at one moment and not at the next, activity is irredeemably temporal. And like all that exists in time and duration, it is successive and perishes continuously. But if simple perduring substances are the "reason for and foundation of" all their contained attributes, and hence a ground for changes in their past and future states, in some sense, time must be inherent in the nature of substances, and not merely characteristic of their states. As will be expected from the analysis of states and attributes, an account of propositions about states of substances in terms of a timeless mode of 'is', and indexes marking the time of utterance of sentences using these propositions will not help. For then there is an irreconcilable confrontation between the notion of timeless facts dated by such utterances and Leibniz's ontological theory that substances contain at once in their nature the ground for all separate modifications in their ongoing activity. In a passage in which he discusses his theory of substance and the phenomenality of extension, Leibniz claims that "time is neither more or less a being of reason than space, but coexistence, and existing before and after, are something real" (G. 2:183, to De Volder, 23 June 1699). Now the only entities that are real and explicitly held to coexist are simple substances. There are good grounds, therefore, for supposing that Leibniz is referring to these entities. But if time does not apply to actuals, how can Leibniz speak of simple substances in terms of simultaneity, before, and after? If they are simple, how can they be divided in the dimension of time? For strict simplicity involves a denial of successive differentiation.

Leibniz's attempt at clarifying the nature of monadic time in the face of these difficulties involves a further claim. Besides the view that actual substances can coexist, and exist before and after, is the claim that "nothing of time does ever exist, but instants; and an instant is not even itself a part of time" (G. 7:402, to Clarke, fourth letter). Referring to succession, in a letter to Bourguet, Leibniz says that instants are "the

315

foundation of time." And that "the preceding instant has, over the succeeding instant, the advantage of priority not of time only but also of nature" (G. 3:581). Apart from difficulties of understanding the notion of a non-temporal instant, Leibniz seems to be making a distinction between a real, though *discontinuous, succession* of instants and temporal continuity that is applicable only at the phenomenal level.

Leibniz's conception of simple substances that can actually coexist and exist before and after can be interpreted in the following way. Since instants are real and hence indivisible, the complex predicate organization of simple substances, as they coexist or exist before and after, can be characterized as being in a series of momentary states. Moreover, Leibniz seems to hold that instants are endowed by 'nature' with qualities of simultaneity, and of preceding and succeeding, a line of reasoning reminiscent of the Scotist doctrine of *instantia naturae*. Thus simple substances can be said to form a *discontinuous* series of momentary states that are characterized by intrinsic temporal qualities. As the extension of phenomenal objects is a result of the repeated action of coexisting substances, so phenomenal duration is founded in the actual existence of momentary states of these substances. Leibniz can thus say that phenomenal duration is "the multitude of momentary states" (G. 7:408).

Although this conception of discrete, momentary instants that are endowed with temporal qualities shows Leibniz to have an atomistic picture of duration and becoming, it is incompatible with the unity of being that he attributes to real, substantial existence. Unless substances can be truly said to have successive differentiation, namely, to be divisible in the dimension of time, they can only exist for one simple and indivisible moment of time. But in no sense could they then be said to endure. Leibniz's doctrine of discrete instants, in which simple substances are said to exist, leads to this difficulty. For if change in the temporal nature of substantial existence can only be characterized by means of a discontinuous series of momentary states, simple substances cannot truly endure. They can only exist for a durationless moment, or from one durationless moment to the next. But this abrogates the thesis that substances maintain their individuation through time. On the other hand, Leibnizian substances must endure, since they have careers and are programmed to exfoliate attributes. Yet an existent can only be said to endure if it is coextensive with the dimension of time. Substances as they have histories are coextensive with time, and temporal predicates thus apply to them. But again the inescapable conclusion is that they are not truly simple in the sense of being indivisible in all dimensions.

Though he would have hedged, Leibniz must accept this conclusion. It is instructive to see why. From the ontological point of view, Leibniz attempts to unite two inconsistent pictures of the mode of existence of monads. One picture concentrates on the unique set of attributes that each simple monad possesses. Given that each monad contains at once all its attributes, Leibniz is quite naturally seduced into construing the mode of existence of these on the model of a set of timeless facts. And as mere progress of monads in time is not a kind of change, he is further confirmed to accord them timeless existence. The other picture construes monads as having a predestined history that, like the purposive actions of a soul, will unfold *in time*. And the mischief arises from his attempt to apply these different pictures to one and the same kind of entities. At best, then, monads are sempiternal entities, entities that exist at all moments of time. As they were created, it could not be otherwise.

In any event, the whole conception of momentary, discrete instants is far from coherent. In his attempt to construe the nature of real instants of coexistence and before and after, Leibniz presupposes a notion of temporal priority that he can only import from the phenomenal realm. The whole conception flounders on an elicit appeal to an intuitive sense of time direction. But only thus has it got coherence. Nor does Leibniz's attempt to sketch a causal theory of time fare better.[36] He tells us, "If of two elements which are not simultaneous one comprehends the cause of the other, then the former is considered as preceding, the latter as *succeeding*" (G. M. 7:18). This means that an asymmetrical causal connection is to be directly invoked by which to define temporal priority. In terms of the action of simple substances, Leibniz could only appeal to a direct awareness of the increasing complexity of monadic attributes, the less complex being defined as *earlier than* the more complex. But this would violate the principle of perfection, which says that in a compossible world all substances mutually contain at once their attribute organization in a completely actualized form. Nor does the distinction help between the permanent attribute of substances and the state of a substance at a given moment. States being successive are by nature inextricably temporal. They cannot, therefore, be used to define the direction of time at the monadic level. Leibniz's exit from the labyrinth of the continuum creates more difficulties than it solves.

V

Leibniz's theory of activity can now be related to his conception of attributes and modes. Through an understanding of how he related these notions to his theory of simple substances, connections between Leibniz's conception of primitive and derivative forces can be illuminated, as well as the doctrine of the proportionality of cause and effect so important in the "science of dynamics."

Leibniz's conception of the relationship between modes, attributes, and substance was one that he shared with most seventeenth-century thinkers. It can be found in his writings from the time of his penetrating analysis of Spinoza in 1676, through to his letters to De Volder, Des Bosses, and Clarke. Attributes and modes are properties of a substance, and cannot be conceived as existing apart from it (G. 2:458, to Des Bosses, 20 September 1712). An attribute is a general property that can be essential to a substance, as, for example, in Descartes' system, thought and extension. A mode, on the other hand, can only be conceived through an attribute, is liable to variation and thus non-essential, such as somebody thinking that they exist, or the specific shape of a particular object. The first is a modification of the attribute of thought, the second a modification of extension. Leibniz also held that no real distinction existed between substance and attribute. For a real distinction can only obtain between entities capable of existing apart from each other, such as substances. When one of two entities, such as an attribute, is incapable of separate existence, any distinction between them is either one of reason or of modality. Nevertheless, such distinctions are true and not illusory.

Given this framework, Leibniz's criticism of the status of Newtonian forces, and his rejection of the Cartesian attempt to conceive passive extension as a true subject of motion, both have meaning. He avers:

> It must above all things be considered that the modifications which can attach to a subject naturally or without miracles, must come to it from the limitations or variations of a real genus, or of an original nature which is

318

constant and absolute. This is how we distinguish in philosophy the modes of an absolute being from that being itself; for instance we know that size, shape, and motion are manifestly limitations and variations of corporeal nature. For it is clear how a limitation of extension gives figures. . . (*New Essays,* Preface, pp. 60–61 n. 6)

The central thrust of this is clear. Changes in a substance must be natural and conceptually proper to it. As we cannot conceive how matter can naturally think, without having recourse to the miraculous, so we cannot conceive how extension alone can account for change without recourse to divine efficacy (*New Essays,* Preface, pp. 56–62). So, too, with Newtonian forces. We can conceive neither how they naturally exist independently of matter, nor how they are ontologically related to matter in proportion to its mass. In terms of extension the only modes of change conceivable are size, shape, and motion construed as a form geometrical translation. These modes are proper to extension, and to none other. While extension can support a notion of geometrical change, it cannot be a basis for a full-fledged conception of activity. Leibniz declared to De Volder: '' . . . Derivative or accidental forces are mere modifications and an active thing cannot be the modification of something passive (extension), since a modification is merely varying limitation, and modes merely limit things but do not increase them, and hence cannot contain any absolute perfection which is not in the thing itself which they modify'' (G. 2:257, 10 November 1703). Activity, therefore, cannot naturally arise from passive matter and/or extension. It must, then, have another basis. Since activity is an undoubted datum of experience, it must be a mode of something that has as a principal attribute the capacity to act, and that acts by its very nature. This can only be an individual substance, a *suppositum.* Only such entities are true unities that have the intrinsic power to act. Extension is merely an incomplete abstraction, which itself depends on the active repetition of the power of individual substances.

But how is Leibniz's dynamics related to his theory of substance and activity? At the phenomenal level Leibniz characterizes changing effects and our perception of them as derivative active forces (*conatus* and *vis viva*) and derivative passive forces (*impenetrability* and *inertia*) (G. M. 6:234–54, *Specimen Dynamicum,* 1695). The former arise from the primitive active power of substances, and the latter from their primitive passive force, *materia prima* or resistance. Consider first the relationship between primitive active force and derivative active force. We have seen how Leibniz's theory of substance conditioned his conception of the source and nature of physical activity. Since all change and activity arise from the intrinsic nature of simple substances, they can be conceived as modes of these active substances. Now in terms of Leibniz's theory of substance there is a parallel between his conception of how modes are related to attributes and how effects are comprehended through their causes. Both effects and modes can only exist in, and be conceived through, their causes and attributes. Since modes are modifications of their attributes, and effects derive from their causes, Leibniz can identify causes and attributes in the light of his theory of active substance. It is in terms of this conceptual framework that he conceived his dynamics. Derivative active forces are modes or effects of the primitive attribute of substances, active power. And as effects depend on their cause, active derivative forces, as modications of this attribute, cannot transcend the nature of substances. Leibniz could maintain that this argument concerning the relationship of modes and attributes to simple substances goes through, even though our perception of compound things gives no reason to believe that there are simple substances. For it is possible to maintain, as he does, that the perceived differentiation of phenomenal things does not destroy the possibility that they result from unperceived simple substances. As Leibniz repeatedly insists, we must not confound the objects of imagination with those of the intellect (G. 6:499–508, *On What Is Independent of Sense and of Matter,* 1702).

Against this conceptual background, an axiom that Leibniz placed at the center of his dynamics assumes significance: that the total cause must be proportional to the entire effect. This is a principle that Leibniz's contemporaries accepted, though his interpretation of it is characteristically his own. If the foundation of the modifications that can occur in a substance is intrinsic to its nature, the cause of these modifications is also intrinsic. Hence there is a sufficient reason within substances themselves that accounts for their effects. No occurrence in nature can be merely arbitrary. All events, then, are in principle explicable. Now since effects are to be understood through their causes, active derivative forces, conceived as modes of substances, are modifications of their principal attribute, active primitive power. And as modifications, they cannot transcend their genus or attribute. The principle of the commensurability of cause with effect is thus an intrinsic consequence of the nature of substance. And there cannot occur in nature circumstances in which an entire effect could be greater than the total cause. Thus a sufficient reason exists that prohibits perpetual motion.

It is clear from this analysis that the basic relationship between cause and effect is that of derivation. Effects are derived from their causes. Since the source of change and activity springs from within each simple substance, and since the cause must be commensurate with the derived effect, an effect will be the same whether it is manifested over a long or short period of time. As Leibniz put it:

> It is still appropriate to remark that the force can be measured without having time enter into consideration. For a given force can produce a certain limited effect that it will never surpass regardless of the time allowed it. And whether a spring recoils all at once or little by little, it will not raise more weight to the same height, nor the same weight higher. And a weight which raises by virtue of its speed will not go any higher, whether it raises perpendicularly, obliquely (inclined plane) or in a curved line.[37]

In Leibniz's interpretation of the causal axiom, the derived effect of an intrinsic force is dependent on a unit of distance, not time. Thus a force is to be estimated and measured by means of the quantity of its completed effect, not by the velocity that it can impress upon an object. The latter was the Cartesian measure, which makes time the essential parameter. If this analysis is correct, the principle of metaphysical conservation intrinsic to the theory of substance provided the conceptual foundation that could rationalize and legitimize Leibniz's measure of phenomenal force. Thus *vis viva* (mv^2), as a principle of dynamical conservation, can be seen as a phenomenal analogue of the theory of substance. It is not motion (mv) that is conserved, as the Cartesian's maintain, but *vis viva* or active force. It seems reasonable to suggest that the theory of substance played a more fundamental role in Leibniz's reasoning to mv^2 as the quantity conserved than the negative arguments levied against the conceptions of force held by his contemporaries.

If Leibniz's conception of modes and attributes links his doctrine of phenomenal or derivative forces with the theory of substance, the situation is not so happy with the attempt to legitimize phenomenal resistance. The difficulty is simply put, and is a sequel to an earlier difficulty in his treatment of the terms "action" and "passion". In 1678

Leibniz attempted to define these terms, as indeed he must, in a way that avoided the notion of the interaction of substances.[38] As Martha Kneal has pointed out, he failed.[39] These notions are correlatives, and if a meaningful activity-passivity contrast is to be made, interaction between two distinct things must be assumed. If, as Leibniz does, the notion of passion is introduced by speaking of a passion as a state of substance in which something arising from its own nature is prevented (action), the absurdity follows of something by its own nature preventing it from proceeding according to its own nature. It seems that Leibniz returned to this conception in his developed monadic theory. He says to De Volder that as in phenomenal matter its mass limits the velocity that another body impresses upon it, so "in things that are limited we need a principle of limitation, just as we need a principle of action in acting things" (G. 2:257). As a correlate of phenomenal mass and resistance Leibniz endowed simple substances with *materia prima*—monadic resistance. It is to inertia and resistance what primitive action force is to *vis viva*. This is a dark conception Leibniz conceives as metaphysically necessary, the principle that limits or modifies the essential activity of substances. In nature it acts as a balancing principle to unbridled activity. This conception is open to the same difficulty as the earlier theory of action and passion. How can a simple substance that is by nature a principle of action at the same time limit or modify its activity by that same nature? Leibniz's conception of combining within the nature of one entity two such correlative principles leads to another difficulty. The attempt to legitimate inertia and impenetrability in terms of a primitive passive force threatens the very conception of simple substance. What is needed are two separate but correlative notions—an active and a passive principle. But together they cannot be grounded in the unity of Leibnizian simple substances. Leibniz needs a theory like Aristotle's that conceives correlatives like form and matter as contrary differentiations of a primary substance. Given his theory of substance, this means is not open to Leibniz. Nor does the conception of activity and passivity that he outlines in the *Monadology* afford a way out (G. 6:609). A theory that correlates a conception of activity and passivity among phenomena with a simultaneous, mutual increase and decrease in the clarity of perception of pairs of monads will not illuminate the conception of two opposed principles—active force and *materia prima*—as intrinsic to the nature of *one* and the same simple substance. And this conception makes more puzzling Leibniz's strange claim that *materia prima* is the principle that explains how it is that nonextended monads result in "extended" phenomena. If the essence of a simple substance is in principle activity, it is difficult to understand not only how *materia prima* can be an aspect of its nature, but also how the repetition of such a quality explains the 'extension' of a collection of monads (G. 4:394). For it is not clear how monadic reality can be extended in any sense, since the extended is a phenomenal concept.

In conclusion I should like to make some observations on Leibniz's criticisms of the ontologies of the Mechanical Philosophy. His arguments against these are not in the nature of the 'aporetic arguments' of a Zeno, Berkeley, Bradley, or McTaggart. He does not argue, for example, that space, time, and motion are merely illusory, or that the individuating principles of material phenomena are incoherent. On the contrary, he accepts the Mechanical Philosophy as the only system that can explain nature rationally. Although he interprets mechanical and geometrical explanations otherwise than his contemporaries, insisting that they be restricted to well-founded phenomena, Leibniz holds with his opponents that science must be concerned with classes of things and abstract descriptions. If his arguments are not designed to show that these explanatory notions are illusory or incoherent, they are meant to show that ultimately they cannot be constitutive of reality. For they demand if they are to be intelligible, an ontology that can support a theory of inherent action and change.

We have seen how Leibniz attempted to link extension and duration with his theory of simple substance by means of his treatment of continuum problems. The same approach is evident in his consideration of phenomenal motion. Like time, it does not truly exist for Leibniz, because it never exists as a true whole, with all its parts coexisting. It is "but a change of situation" and thus "a mere relationship" (G. M. 6:251, *Specimen Dynamicum,* 1695). What is real *in* phenomenal motion is a momentary perceptual quality that is constituted by substantial power repeatedly striving for change. Power is thus real, complete, and the originator of activity. Activity has motion as its recurring effect. And motion is thus a relation of successive times and of successive positions. As we have seen, times and positions are solely phenomenal orders. From the kinematics of rest and change of position, Leibniz is able to draw the following conclusion. As change of position *is* motion, at any given moment a position is one and one only. Thus at every moment, and therefore continually, there can be no change of position and hence no motion, as there is no intrinsic mark whereby a moment of rest is distinguishable from a momentary change of position. The indiscernibility of spatio-temporal coordinates puts them into the category of incomplete concepts. Such concepts being abstractions apply indifferently to actual circumstances. As such they cannot individuate the diversity of actual things, nor can they discern the true motion of these. For a change in a thing's spatio-temporal coordinates is no more an actual change *in* it than is a rise in the price of meat. Moreover, as a change of position is successive, it is the recurrent effect of something real and repetitive. Thus motion, like extended things, is solely the appearance of something that gives rise to activity, and therefore of the appearance of simple, indivisible substances that are the "foundation and reason" of motion.

Leibniz's theory of substance and of the continuum harbors many difficulties. It is, however, a systematic attempt to clarify some of the deepest problems in natural philosophy—matter, change, and activity. These all inextricably involve the "Labyrinth of the Continuum."

I should like to thank the National Science Foundation, Grant G5-38080, P351013, which made possible the research for this paper. Also I should like to thank, for their encouragement and criticism, Professor Peter Machamer and Dr. George Molland.

1. C. I. Gerhardt, ed., *Leibnizens mathematische Schriften,* 7 vols. (Berlin and Halle: Weidmannsche Buchhandlung, 1849–63), 7:326 (hereafter cited in text as G.M.).

2. C. J. Gerhardt, ed., *Die philosophische Schriften von Gottfried Wilhelm Leibniz,* 7 vols. (Berlin: Weidmannsche Buchhandlung, 1875–90), 3:52. (Hereafter cited in text as G.).

3. G. 3:52. Of this principle of order Leibniz says, " . . . The actual phenomena of nature are arranged, and must be in such a way that nothing ever happens which violate the law of continuity, which I introduced into philosophy. . . ."

4. E. Anscombe and P. T. Geach, *Descartes: Philosophical Writings,* (London: Nelson, 1954), Third Meditation, p. 88.

5. For a discussion of Newtonian Forces and Leibniz's objections, see J. E. McGuire, "Force, Active Principles, and Newton's Invisible Realm," *Ambix* 15 (1968): 154–208.

6. A. G. Langley, trans., *New Essays Concerning Human Understanding* (Chicago and London: Open Court, 1916), Preface, pp. 54–62; G. 4:50–61.

7. For unambiguous statements of the dependence of force on matter, see Isaac Newton, *Philosophiae Naturalis Principia Mathematica* (London: 1687), book 3, prop. 6, theor. 6, corollaries 1, 2, 3, 4, pp. 408–11, and *Opticks* (New York: Dover Publications, 1952), Query 21, pp. 351–52.

8. Langley, *New Essays,* Preface, p. 60, and *Correspondance Leibniz-Clarke,* ed. A. Robinet (Paris: Presses universitaires de France, 1957).

9. *A Brief Demonstration of a Notable Error of Descartes and Others Concerning a Natural Law* (1686) (G.M., 6:117–19), *Discourse on Metaphysics* (1686), sec. 17(G. 4:427–63), *Letter of Mr. Leibniz on a General Principle Useful in Explaining the Laws of Nature through a Consideration of the Divine Wisdom* (1687) (G. 3:51–55).

10. *Critical Thoughts on the General Part of the Principles of Descartes* (1692) (G. 4:354–92), *Specimen Dynamicum* (1695) (G.M. 6:234–54 (Leibniz uses the phrase "science of dynamics" in this work, p. 234), *A New System of the Nature and the Communication of Substances* (1695) (G. 6:477–87, *Ten tamen Anagogicum* (1696) G. 7:270–79).

11. Langley, *New Essays*, Preface, p. 61.

12. G. 4:504–16. Leibniz puts it thus in *Specimen Dynamicum:* " . . . And yet I think that there is no natural truth in things for which we must find the reason in the divine action or will but that God has always put into things themselves some properties by which all their predicates can be explained."

13. Bertrand Russell, *A Critical Exposition of the Philosophy of Leibniz* (London: Allen & Unwin, 1971), chaps. 4, 7.

14. Louis Couturat, "Sur la métaphysique de Leibniz, " *Revue de métaphysique et de morale* 10 (1902); Ernest Cassirer, *Leibniz' System in seinen wissenschaftlichen Grundlagen* (Marburg: N. G. Elwert, 1902); repr. (Hildesheim: Olms, 1962). Both writers hold uncompromisingly that Leibniz's metaphysical principles are really reducible to his logic, and that therefore the main tenets of the mechanics are logical rather than metaphysical. They also conclude that Leibniz's notion of Force is similar conceptually to the "mechanical" ideas of the Cartesians. This is true; but only at the phenomenal level. See section V below.

15. Martial Gueroult, *Leibniz: Dynamique et métaphysique* (Paris: Aubier-Montaigne, 1967).

16. Ibid.

17. *Discourse on Metaphysics*, sec. 10 (G. 4:427–63).

18. Ibid., sec. 8 (G. 4:430).

19. Ibid., and G. H. R. Parkinson, ed., *Leibniz: Philosophical Writings* (London: J. M. Dent & Sons, 1973), pp. 93–95.

20. Ibid., p. 95.

21. *Discource on Metaphysics,* sec. 8.

22. G. 4:18–24. In this early work Leibniz rejected the Scotist notion of *haecceitas*. By 1668, however, in a piece on transubstantiation he had accepted it. There he says that a substantial form, which he identifies with the notion of *suppositum,* "is the principle of individuation" (LeRoy E. Loemker, trans. and ed., *Gottfried Wilhelm Leibniz: Philosophical Papers and Letters* [Chicago: University of Chicago Press, 1956], p. 117). In the correspondence with Arnauld, Leibniz makes the same point: " . . . I am very much persuaded of what St. Thomas had already taught regarding intelligences and which I consider to be generally true, namely, that it is not possible for there to be two individuals entirely alike, or differing in number only" (H. T. Mason, trans., *The Leibniz-Arnauld Correspondence* [Manchester: Manchester University Press, 1967], p. 61).

23. Ibid., Introduction by G. H. R. Parkinson, pp. xxxix–xliii.

24. Louis Couturat, *Opuscules et fragments inédits de Leibniz* (Paris: F. Alcan, 1903), p. 212.

25. See Nicholas Rescher, *The Philosophy of Leibniz* (Englewood Cliffs, N.J.: Prentice-Hall, 1967), chaps. 2 and 3, for a discussion of these problems.

26. *The Leibniz-Arnauld Correspondence,* p. xii.

27. G. W. Leibniz, *Sämtliche Schriften und Briefe,* ed. Preussischen Akademic der Wissenshaften (Darmstadt and Leipzig: O. Reich, 1923–70), 6,1:509, *On Transubstantiation.*

28. Ibid., p. 510.

29. For a discussion of this and related topics, see C. D. Broad, "Leibniz's *Predicate-in-notion Principle* and Some of Its Alleged Consequences," *Theoria* 15 (1949): 54–70. See also the

perceptive study of Wilfred Sellars, "Meditations Leibniziennes," *American Philosophical Quarterly* 2 (1965): 105–18.

30. McGuire, "Force, Active Principles," and Samuel I. Mintz, *The Hunting of Leviathan* (Cambridge: At the University Press, 1962), chaps. 4 and 5, for a general account of this theme. Further research is needed to determine Leibniz's indebtedness to sixteenth- and seventeenth-century theories of *Spiritus*.

31. Russell, *A Critical Exposition of the Philosophy of Leibniz,* chapts. 7, 8.

32. Louis Couturat, *La Logique de Leibniz* (Paris: Alcan, 1901), chap. 6.

33. Limitations of space will not permit a discussion of Leibniz's phenomenalistic tendencies in relation to the phenomenal continuum. Ultimately he disavows phenomenalism, as the principles of plenitude and perfection demand that a plurality of individual substance exist. On this topic see Montgomery Furth, "Monadology," *Philosophical Review* 76 (1967): 169–200.

34. For an excellent discussion of this theme, see Wesley C. Salmon, *Zeno's Paradoxes* (New York: Bobbs-Merrill, 1970).

35. A. N. Whitehead, *Process and Reality* (New York: Macmillan, 1929), p. 53.

36. For an excellent discussion of this problem, see Adolf Grünbaum, "Relativity and the Atomicity of Becoming," *Review of Metaphysics* 4 (1950–51): 142–86.

37. Pierre Costabel, *Leibniz et la dynamique: Les textes de 1692,* (Paris: Hermann, 1960), p. 105. For useful discussions of the technical aspects of Leibniz's dynamics, see Carolyn Iltis, "Leibniz and the *Vis Viva* Controversy," *Isis* 63 (1970): 21–35, and Gerd Buchdahl, *Metaphysics and the Philosophy of Science* (Oxford: Blackwell, 1969) chap. 7.

38. G. Grua, ed., W. Leibniz, *Textes inédits* (Paris: Presses universitaires de France, 1948), 2:512–37.

39. Martha Kneale, "Leibniz and Spinoza on Activity," in Harry G. Frankfurt, ed., *Leibniz: A Collection of Critical Essays* (New York: Doubleday, Anchor Books, 1972), pp. 215–37.

Thomas L. Hankins

Chapter Twelve

ALGEBRA AS PURE TIME: WILLIAM ROWAN HAMILTON
AND THE FOUNDATIONS OF ALGEBRA

INTRODUCTION

On 10 August 1835 the British Association assembled in Dublin for its annual meeting. As secretary to the association, William Rowan Hamilton, royal astronomer of Ireland, read the annual report at the evening meeting. George Ticknor, visiting from America, declared it to be a "beautiful and eloquent address . . . exactly hitting the tone of the occasion," and added that Hamilton was the "great man" at the meeting—pleasant, warmhearted, one of the first mathematicians in Europe, but also a fine Greek scholar, with an extremely metaphysical mind and an ability to write good poetry.[1] Ticknor had further opportunity to hear Hamilton expound on the foundations of algebra, the solution of equations of the fifth degree, and the wave nature of light. He witnessed him deep in conversation with a German visitor on the subject of German metaphysical idealism, and saw him knighted by the lord lieutenant of Ireland at the closing ceremonies of the meeting. Ticknor might have been even more impressed if he had known that Hamilton had written most of his address on the morning of the day that he delivered it, and had arrived late even for this important occasion, partly because he found it impossible ever to get anywhere on time, but mostly because he had wanted to make the final corrections on a paper that he was anxious to have published during the meeting. This was his "Theory of Conjugate Functions, or Algebraic Couples; with A Preliminary and Elementary Essay on Algebra as the Science of Pure Time." In writing to his friend Aubrey De Vere he called it "the first installment of my long-aspired-to work on the union of Mathematics and Metaphysics."[2]

With metaphysical notions about algebra foremost in his mind, it is not surprising that they appeared at least fleetingly in the annual report that evening.[3] They also represented a certain commitment on Hamilton's part, because he expected them to be unpopular with other mathematicians, and he was now for the first time presenting them in a scientific journal and at a major scientific meeting.

My purpose in this paper is to describe the background to Hamilton's writing of this "Essay" and to indicate how his metaphysics, particularly his use of the concept of time, was important for his work in algebra.

MATHEMATICAL CONTENT OF THE ESSAY

The paper that Hamilton completed for the meeting was in three parts. The last part, the "Theory of Conjugate Functions, or Algebraic Couples," was the first to be finished, and had been read to the Royal Irish Academy on 4 November 1833. The middle and longest section was completed in the spring of 1835 and was read to the Academy on 1 June with the title "Preliminary and Elementary Essay on Algebra as the Science of Pure Time." The first part, entitled "General Introductory Remarks," was finished some time during the summer. Although portions of the work had been read as early as November 1833, the entire "Essay" was first published in the volume of the *Transactions* for 1837.[4]

Hamilton was pursuing three related directions of mathematical research in his paper. The first was an attempt to place the algebra of real numbers on a more secure logical foundation. The second was to give an algebraic definition of complex numbers that avoided the concept of imaginary numbers, and the third was to confirm the general expression for the logarithm of a complex number taken to a complex base, an expression that had been proposed by his friend John Graves. The search for a better logical foundation for algebra was increasingly attracting the attention of British algebraists at the time, and Hamilton had been thinking about it for ten years—that is, ever since his university study of mathematics. He believed that negative and imaginary numbers had no *meaning* in ordinary algebra, and that until they could be adequately defined, algebra rested on a very shaky base. In the middle section of the

"Essay" Hamilton attempted to state the properties of the real number system and to define negatives by the use of steps in the ordered relations of time. Thus he attempted to base algebra on the *ordinal* character of the real numbers. Evaluated by the standards of rigor demanded in modern algebra, Hamilton's attempt was highly intuitive and badly flawed. Yet his attempt to define the natural numbers by appealing to our intuitive sense of progression in time has been revived in this century by the Dutch "intuitionists" and still provides subject for debate among philosophers of mathematics.

The entire "Essay" is an effort to get away from the intuitive concept of number and to base algebra on another intuitive concept, that of Pure Time, which Hamilton believes is more immediate and also more consistent. Therefore he resists the temptation to introduce the ordinal and cardinal integers until he has developed the operations on time steps. As he says, he wants to "treat these spoken and written names of the integer ordinals and cardinals, together with the elementary laws of their combinations, as already known and familiar."[5]

In developing his notion of algebra as the science of Pure Time, Hamilton presents us with one of the earliest attempts to list systematically the properties of the real number system.[6] He comes surprisingly close to defining an algebraic field. There are certain points of confusion where he passes from steps to cardinal numbers expressed as multipliers of time steps, but the paper as a whole is a remarkably successful beginning. The commutative and distributive properties are described, definition of zero, additive and multiplicative inverse, and the law of closure are all there. What he misses is the associative rule, probably because it is more subtle than the others, and because it did not occur to him that there might be an algebra that did not follow it. Later on when he studied systems of hypercomplex numbers, he found that the associative law did not always hold, and he stated it for the first time in 1843 as an important property of real numbers, complex numbers, and quaternions.[7]

The second direction of research in Hamilton's "Essay" is his effort to define complex numbers as ordered pairs of real numbers, thereby avoiding the concept of imaginary numbers.[8] For some historians of mathematics Hamilton's representation of complex numbers by number couples is his greatest achievement in algebra, even more significant than his later discovery of quaternions.[9] It meant breaking away from the real number line as the basis for algebra and constructing new elements defined in a different way.

The third direction of research in Hamilton's "Essay" is his attempt to confirm the theory of logarithms developed by his friend John Graves. It was his interest in this problem of logarithms that led him to the other problems treated in the paper. Ever since the seventeenth century there had been controversy among mathematicians over the possible existence of logarithms of negative and complex numbers.[10] The difficulty lay in the fact that the logarithm of a complex number has many values. John Graves had shown that in the most general case, that of the logarithm of a complex number taken to a complex base, two arbitrary integer numbers must be specified to determine the solution, since the complex number of the logarithm and the complex base each introduce an infinite number of possible solutions. In his defense Hamilton derived Graves's results by expressing complex numbers as number couples.

Hamilton's paper showed great mathematical imagination. By using time steps and number couples he demonstrated that algebra could be more than the ordinary algebra of real numbers, and that real meaning could be given to negative and imaginary numbers. At the end of his paper Hamilton stated that he hoped "to publish hereafter many other applications of this view [of algebra as the science of Pure Time]; especially to Equations, Integrals, and to a Theory of Triplets and Sets of Moments, Steps and Numbers, which includes this Theory of Couples," a hope that he finally fulfilled in 1843 with his discovery of quaternions after a long and fruitless search for triplets.[11] With the quaternions Hamilton created an algebra that did not obey the commutative law. It was a bold step that opened the way for a rash of new algebras.[12] Although the quaternions appeared as a startling discovery, Hamilton claimed that they were merely "a continuation of those speculations concerning algebraic couples, and respecting algebra itself, regarded as the science of Pure Time, which were first communicated to the Royal Academy in November 1833. . . . The author has thus endeavored to fulfill, at least in part, the intention which he expressed in the concluding sentence of his former Essay. . . ."[13] As Hamilton saw it, the couples, the ill-fated triplets, the quaternions, and all other new hypercomplex numbers were part of the same speculation that he grounded in his metaphysical notion of Pure Time, a notion that had been growing in his mind for many years. Once we understand the metaphysical and mathematical background to the "Essay," it is easier to see why Hamilton brought together such diverse subjects into a single paper.

SETTING THE HISTORICAL PROBLEM

Since the publication of L. Pearce Williams's biography of Michael Faraday, historians have debated his assertion that crucial figures in nineteenth-century British science, especially Davy and Faraday, were strongly influenced in their work by the idealistic philosophies of Roger Boscovich and the German *Naturphilosophes,* or at least by an early version of *Naturphilosophie* that can be found in Immanuel Kant's *Metaphysische Anfangsgründe der Naturwissenschaft.*[14] The route by which Kantian philosophy reached Davy and Faraday was through the poet Samuel Taylor Coleridge, who studied Kant during his trip to Germany in 1798 and discussed his new enthusiasm with Davy when the two first met at Thomas Beddoes's Pneumatical Institution.[15]

Hamilton's only meeting with Davy was not particularly successful, and his first philosophical discussion with Faraday came only after his own philosophy was well established; but it is not surprising that Hamilton's name has been mentioned in the debate.[16] In the preface to his famous paper on a *General System of Dynamics* (1834) Hamilton professed adherence to Boscovich's philosophy, and in the preface to his *Lectures on Quaternions* (1853) he claimed that his reading of Kant's *Critique of Pure Reason* had encouraged him to present his views on algebra as Pure Time.[17] Thus Hamilton's commitment to Boscovich and Kant appeared in his most widely-read publications. Moreover, Hamilton's own philosophical inclination was to idealism as can be seen in his correspondence and in many of his astronomical lectures.[18] He also fancied himself a poet, and enjoyed a close friendship and extended correspondence with William Wordsworth.[19] Coleridge was for him the philosopher *par excellence*. He studied Coleridge's poetical and philosophical works in great detail and met him twice, in March 1832 and June 1833. These meetings were crucial. Hamilton read his paper on algebraic couples in November 1833, five months after the second meeting. Coleridge died 25 July 1834. Discussion of Coleridge and

Kant filled many pages of Hamilton's correspondence during these crucial years from 1832 through 1835. The period of his greatest interest in their philosophies coincided with his work on the foundations of algebra.

The manuscripts give ample evidence of his progress in algebra and in philosophy. They also illustrate the difficulties facing the historian who wishes to find the inspiration for discoveries in pure mathematics outside of mathematics itself. Mathematics is an abstract, logical structure that does not require physical phenomena for its justification. In some cases where physical problems require new mathematical techniques for their solution, such as the development of the calculus to solve problems in mechanics, the demands of the physical problem lead directly to new mathematical methods. But in the foundations of algebra this is certainly not the case, and historians of mathematics have not dealt kindly with Hamilton's ideas. His concept of Pure Time has been called a "metaphysical speculation without foundation in history or in mathematical experience,"[20] and he has been criticized for loading his mathematics with "philosophical and even metaphysical speculations, which makes the reading of his publications sometimes excessively tiresome."[21]

But the route of discovery is not always the straightest one, and the historian of science should be interested in the process of discovery as well as in the logical structure of science. It is impossible to *prove* that Hamilton could never have obtained his couples and quaternions without the concept of Pure Time, but I can hope to suggest where that concept was useful to him in the process of discovery and to trace the development of his mathematical and metaphysical ideas.

MATHEMATICAL BACKGROUND

To find the origin of the "Essay," it is necessary to go back to Hamilton's college years. He entered Trinity College, Dublin, in 1824 as a recognized prodigy. His college career was a continued succession of awards and medals received for academic excellence in classics, science, and even poetry (although one is led to conclude that this last talent was considerably less than the previous two). His first major scientific achievement was in geometrical optics, where he discovered a single "characteristic function" that would fully describe any optical system. His *Theory of Systems of Rays* made a sufficient impression at the college for the fellows to feel justified in appointing him astronomer royal and Andrews Professor of Astronomy, while he was still an undergraduate. Hamilton continued to develop his theoretical optics in three supplementary papers, the last of which was a major treatise in itself and concluded with his prediction of conical refraction. This third supplement was completed in 1833. Also in 1833 Hamilton began to apply the characteristic function to dynamics and published the two famous papers on a "General Method in Dynamics" in 1834 and 1835.

I mention Hamilton's work in optics and dynamics in order to indicate that algebra was not his major concern before 1835, and that he worked on it in short intervals between the study of other subjects. His great success in optics and mechanics might have distracted him completely from algebra if it had not been for constant needling by his classmate John Graves. Hamilton was close to the three Graves brothers: Robert Perceval Graves, an Anglican clergyman and author of the enormous three-volume biography of Hamilton; Charles Graves, fellow and later professor of mathematics at Trinity College, Dublin, and eventually bishop of Limerick; and John Graves, Hamilton's closest friend of the three. Hamilton and John Graves went through Trinity College together, where they both showed an enthusiasm for mathematics. After their graduation in 1827 John Graves moved to England, where he was called to the English Bar in 1831. The two friends continued their discussions of mathematical subjects in correspondence, which, fortunately for the historian, also records the development of their ideas.

The first extant letter of their correspondence is one from Hamilton to Graves dated 5 April 1825 on the subject of logarithms.[22] Some time in 1826 Graves concluded that the general logarithm must necessarily contain two arbitrary integers—the discovery that Hamilton would later defend in his "Essay on Conjugate Functions, or Algebraic Couples." He and Hamilton continued to discuss the subject through 1827, and in 1828 Graves submitted a paper to the Royal Society entitled "An Attempt to Rectify the Inaccuracy of Two Logarithmic Formulae." Graves feared that his discovery might have been anticipated; Hamilton assured him that John Herschel would certainly know if it was new, because he read so widely in current research in algebra.[23] Herschel did not know of any precursor to Graves's paper, but both he and George Peacock were unconvinced by Graves's arguments.[24] Herschel had

333

hoped to avoid commenting on the paper, because, as he candidly admitted to Graves, he could not determine if it was correct.[25] But the Council of the Society, responsible for the *Transactions,* appointed him judge, and he concluded that the only honorable thing to do was to ask Graves to withdraw his paper.[26]

Hamilton intervened at this delicate and sensitive point in the negotiations, without obtaining Graves's prior consent, and tried to convince Herschel of the paper's worth. It was published the following year in the *Philosophical Transactions.*[27] Peacock continued to oppose the results in his report to the British Association of 1833, but De Morgan was able to show that Graves's results were correct if one accepted Graves's definition of the logarithm.[28]

The controversy over the logarithms of complex numbers was stimulated in part by the possibility that a new kind of imaginary number might be discovered, that is, that the complex numbers might not have the property of closure under some of the ordinary operations of algebra, a possibility that John Graves had urged on Hamilton in 1829.[29] De Morgan admitted later that he had half-expected the same thing.[30] Hamilton doubted that such a new imaginary number was likely to arise from the study of logarithms (a surmise that proved to be correct), but, as we have seen, he did not hesitate to consider new hypercomplex numbers *defined* as such; and it was the search for such hypercomplex numbers that eventually led to his discovery of quaternions in 1843.

Another reason for the great interest in complex numbers during the 1820s and 1830s was the realization that such numbers could be represented geometrically by a pair of axes in what is now called the complex plane. This discovery was first published by Caspar Wessel in 1797 and was rediscovered independently by several others, but Hamilton knew nothing about it until 1829 when he read it in a book by John Warren entitled *A Treatise on the Geometrical Representation of the Square Roots of Negative Quantities.*[31] Typically it was John Graves who called Hamilton's attention to Warren's book and stimulated him to search for an algebra of hypercomplex numbers that could be represented geometrically in three-dimensional space.[32]

Throughout the debate over logarithms of complex numbers Hamilton apparently believed that the difficulty lay not in the particular problem at hand but in the inadequate logical foundations of algebra. When he first received in 1828 a copy of the paper that Graves intended

334

to submit to the Royal Society, he extended his congratulations and thanks, but added:

> I have often persuaded myself that the whole analysis of infinite series, and indeed the whole logic of analysis (I mean algebraic analysis) would be worthy of [ra]dical revision. But it would be for a person who should attempt this to go to the root of the matter, and either to discard negative and imaginary quantities, or at least (if this should be impossible or unadvisable, as indeed I think it would be) to explain by strict definition, and illustrate by abundant example, the true sense and spirit of the reasonings in which they are used. An algebraist who should thus clear away the metaphysical stumbling-blocks that beset the entrance of analysis, without sacrificing those concise and powerful methods which constitute its essence and its value, would perform a useful work and deserve well of Science.[33]

Hamilton urged on Graves the task of reformulating the foundations of algebra and later made the same request of Baden Powell.[34] His appeal to others did not mean that he gave up the subject himself, and his theory of algebraic couples and conjugate functions was obviously the result of continued thinking on Graves's paper and on the problem of defining imaginary quantities in a clear and consistent way.[35]

We have seen that the mathematical questions that Hamilton raised in his paper on algebra as the science of Pure Time have their origin in 1825 when he first began discussing logarithms of complex numbers with his classmate John Graves. This is important because it indicates that there were specific mathematical issues in addition to the metaphysical questions that led him to criticize the foundations of algebra. In order to determine whether Hamilton's metaphysics helped direct his mathematical research, it is necessary to inquire how the development of his metaphysics fits chronologically with his mathematical ideas.

PHILOSOPHICAL BACKGROUND

In the preface to his *Lectures on Quaternions* Hamilton stated that he had been "encouraged to entertain and publish [his] view [that algebra is the science of Pure Time], by remembering some passages in Kant's *Critique of the Pure Reason,* which appeared to justify the expectation that it should be *possible* to construct *a priori*, a Science of Time, as well as a Science of Space."[36] Notice that Hamilton says his reading of Kant *encouraged* him to publish his view; he does not say that reading Kant led him to that view. Hamilton first began to read the *Critique of Pure Reason* in October 1831. His first mention of algebra as Pure Time appears in an "Account of a Theory of Systems of Rays" presented to the Royal Irish Academy on 23 April 1827 more than four years earlier. In this early abstract Hamilton wrote: "The sciences of Space and Time (to adopt a view of algebra which I have elsewhere ventured to propose) become intimately intertwined and indissolubly connected with each other."[37] If Hamilton had already proposed his view of algebra as time "elsewhere" I have been unable to find it, but certainly the idea was familiar to him in the spring of 1827.

There is another short mention of time as related to algebra in a manuscript dated 15 November 1827, with the title "Consideration on some points in the Metaphysics of Pure Mathematics."[38] In this manuscript Hamilton used the concept of time "to give greater precision and simplicity to our notion of ratio." There is nothing further in the manuscripts until 14 December 1829, when he wrote another short sketch entitled "Principles of Mathematical Reasoning" in which he introduced time again.[39] The first real attempt to use time steps to define the operations of algebra appears in a manuscript dated 9 June 1830.[40] Finally in a notebook entry dated February 1831 and titled "Metaphysical Remarks on Algebra" Hamilton states a metaphysical position that appears with increasing frequency in his manuscripts. It is worth quoting at least in part:

In all Mathematical Science we consider and compare relations. In algebra the relations which we first consider and compare, are relations between successive states of some changing thing or thought. And numbers are the names or nouns of algebra; marks or signs, by which one of these successive states may be remembered and distinguished from another. . . . *Relations between successive thoughts thus viewed as successive states of one more general and changing thought, are the primary relations of algebra*. . . . For with Time and Space we connect all continuous change, and by symbols of Time and Space we reason on and realise progression. Our marks of temporal and local site, our *then* and *there*, are at once signs and instruments of that transformation by which thoughts become things, and spirit puts on body, and the act and passion of mind seem clothed with an outward existence, and we behold ourselves from afar. And such a transformation there is when, in Algebra, we contemplate the change of our own thoughts as if it were the progression of some foreign thing, and introduce Numbers as the marks or signs to denote place in that progression.[41]

When Hamilton wrote this passage he still had not been able to locate a copy of Kant's *Critique*. It is not surprising, however, that he was very anxious to find one, for he must have realized by that time that his ideas had great affinity to those of Kant.

Kant's philosophy was not well-known in England until 1830, when it began to find able translators and expositors.[42] Hamilton was exposed to a confused version of it in the works of Coleridge, although his early interest was more in Coleridge's poetry than in his philosophy. His first contact may well have been in Madame de Staël's *Allemagne*, which he read early in 1826.[43] Another definite source was Dugald Stewart. Hamilton was reading Stewart's *Elements of the Philosophy of the Human Mind* in the spring of 1824, and the *Philosophical Essays* in 1826.[44] Stewart's references to Kant are unsympathetic and often inaccurate, but Stewart was an important source for Hamilton in spite of his opposition to the idealism that Hamilton greatly admired.[45] Hamilton was probably inspired to read Stewart by his contact with the Edgeworth family. He became very friendly with the Edgeworths in 1824, and spent a great deal of time at Edgeworthstown with Maria, the novelist, and her much younger stepbrother Francis Beaufort Edgeworth, who fancied himself to be a poet and metaphysician. Another of Maria's brothers, Henry, had studied at Edinburgh under Stewart, and Hamilton read Stewart's *Philosophical Essays* from a much annotated copy borrowed from the Edgeworth library.[46]

Before Hamilton began any serious study of Kant he had plunged into the idealism of Berkeley and Boscovich, probably as a result of reading the second of Stewart's *Philosophical Essays,* "On the Idealism of Berkeley," which treats and compares the views of Berkeley and Boscovich on the nature of matter.[47] Hamilton obtained a copy of Boscovich's *Theoria philosophia naturalis* in 1834. Before that time his knowledge of Boscovich came largely from reading Stewart and other Scottish philosophers.

It was this study that led him to a theory of matter described as points of "unific energies," not an unusual position at the time, held by those who wished to deny contact action between atoms. Hamilton's "unific energies" had a stronger metaphysical flavor than most of the models of forces in space created at the time, and they illustrate an inclination toward idealism that was also characteristic of his poetry and philosophical writing.

It is not clear why Hamilton turned from Berkeley and Boscovich to Kant during the 1830s. In part it was a result of his great interest in Coleridge's philosophical works, which he read and discussed in great detail with his friend Aubrey De Vere. The manuscripts are of little help. There is, however, one long extract from a book review that indicates the direction of his thought. The review was of Novalis's *Schriften,* and it appeared in the *Foreign Review and Continental Miscellany* for 1830.[48] It is less a review of Novalis than a review of all idealist philosophy during the previous one hundred years, marking Kant as a crucial figure. It puts together the contributors to idealism in just the right way to catch Hamilton's fancy. There is great praise for Coleridge, strong support for Berkeley and Boscovich, emphasis on immaterialism as a defense against atheism, and the conclusion that the domain of Reason contains the "pure, ultimate light of our nature, wherein . . . lies the foundation of all Poetry, Virtue, Religion; things which are properly beyond the province of the Understanding to *contradict* Reason."[49] As a description of Kant's philosophy the review left much to be desired; but it piqued Hamilton's curiosity, and he copied at length those parts of the review that dealt with metaphysics. He decided about this time that in order to complete his education in the philosophy of metaphysical idealism, he would have to go to the source itself and read Kant's *Critique of Pure Reason,* especially since Kant's doctrine of space and time spoke directly to his concerns about the foundations of mathematics.

Kant's *Critique* was not easy to find in Ireland, and Hamilton enlisted the aid of William Wordsworth to secure a copy. Wordsworth was not a philosopher himself, but he was willing to help Hamilton in his philosophical quest. His son William was studying in Germany, and he tried in 1830 to get Hamilton a copy of the *Critique.*[50] On 29 October 1831 Hamilton wrote to him: "I have got Kant's *Kritik der Reinen Vernunft.*"[51] Whether it came from Wordsworth is not clear, but Hamilton set to work trying to read it. In spite of the great linguistic skill that supposedly allowed him to master thirteen languages by age thirteen, Hamilton had great trouble reading the *Critique* as have many since his time. His manuscripts contain no less than six attempted translations. The dates run from October 1831 to some time in 1845. Four are translations of the first few pages of the introduction, one is a translation of the preface, and the sixth is a translation of the table of contents.[52]

Progress was slow, and he had completed only a third of the book by August 1834.[53] His efforts had been frustrated in 1832 when he lost his copy of the *Critique* on a visit to Oxford in June. Hamilton always carried books with him wherever he went, and while seated on his luggage atop a Birmingham omnibus, he saw his books dribbling out of his portable library, which, for this journey, was a pillowcase. He grabbed Laplace's *Calculus of Probabilities,* but Kant's *Critique* slipped away. He was not able to replace it until 1834.[54]

In the meantime he met Coleridge. Hamilton had hoped to obtain a letter of introduction to him from Wordsworth, but Wordsworth was strangely reticent to write one, either because he knew that Coleridge was ill, and did not want to impose on him by sending unknown visitors to his door, or because he was somewhat annoyed by Hamilton's great enthusiasm for Coleridge's philosophy.[55] Coleridge, in the later years of his life, rambled on to visitors in an almost incoherent fashion, usually about his *Opus Maximus* which he had outlined in 1828. Visitors, including Hamilton, were submitted to long harangues on the logic, which was the most nearly completed part of the work and was taken largely from Kant.[56]

In the letters that he wrote to his friends, Hamilton said little about his visits to Coleridge. It is difficult to discover what transpired between them. Certainly there was a discussion of Kant, because Coleridge offered to loan Hamilton his copies of Kant's *Miscellaneous Essays* and wrote later that he was searching for the volumes that he had not been able to find in his library.[57] After the 1833 visit Coleridge also sent Hamilton his copy of Kant's *Critique of Judgment,* and gave his copy of the *Critique of Practical Reason* to Hamilton's pupil, Lord Adare.[58] Hamilton's previous reading of Kant had been superficial and second-hand, but after his visits with Coleridge he began to read the *Critiques* more seriously.

When Hamilton first met Coleridge in 1832 he was completing his work on the characteristic function as applied to optics and had possibly considered applying it to mechanics as well. Much of the conversation must have turned on Hamilton's theory of matter, but it is also likely that he talked with Coleridge about the nature of science, particularly Coleridge's distinction between Reason and Understanding as it had appeared in his *Aids to Reflection.* Hamilton wrote a memorandum on the *Aids* in June 1831 in which he concerned himself with precisely this

339

question.[59] Coleridge claimed that the distinction was badly understood, and that as a result, great strides had been made in furthering the Understanding, while Reason had fallen into "utter neglect."[60] Following his own interpretation of Kantian thought, Coleridge argued that the term *science* could only rightfully be applied to the product of the Reason.[61] Reason was "the power of universal and necessary convictions, the source and substance of truths above sense, and having their evidence in themselves."[62] A science was "any chain of truths which are either absolutely certain, or necessarily true for the human mind, from the laws and constitution of the mind itself."[63]

This definition of science is one that Hamilton took very much to heart. It appears clearly stated in the "Introductory Remarks" to his "Essay," and it separated him from the more practical-minded British scientists and mathematicians of his time. A visit from George Biddell Airy left him horrified, especially Airy's comment that the Liverpool and Manchester Railway was the highest achievement of man. To Adare he declared that Airy's mind was an instance of "the usurpation of the understanding over the reason, too general in modern English Science," and then echoing Coleridge's lament, "When shall we see an incarnation of metaphysical in physical science. When shall the imagination descend, to fill with its glory the shrine prepared for it in the Universe, and the Understanding minister there in lowly subjection to Reason."[64] The following year and before his visit to Coleridge he confessed his intellectual loneliness to his friend Aubrey De Vere: "I differ from my great contemporaries, my 'brother-band', not in transient or accidental, but in essential and permanent things: in the whole spirit and view with which I study science."[65] It is clear that Hamilton's isolation at the Observatory was not only a physical one. He felt philosophically isolated as well.

Whether Hamilton talked with Coleridge specifically about his idea of "Algebra as the Science of Pure Time" is not certain. There is some evidence, however, that he did and that Coleridge might have been a sympathetic listener. One thing that Hamilton did mention in his letters describing the meeting was Coleridge's enthusiasm for one of his own poems entitled "On Time, Real and Imaginary."[66] The poem is an allegory on the difference between time as real and absolute, and time as felt or experienced.[67] It does not contain any reference to mathematics, but it may well have come up as a result of a discussion over Hamilton's meaning of "Pure" Time. In 1819 and 1820 Coleridge had held a position close to that of Hamilton. In the marginal notations added to several of his presentation copies of the *Friend* he anticipated Hamilton by claiming that arithmetic, at least, was derived from our intuition of time. In this passage Coleridge claimed that the Science which transcends the evidence of physical phenomena should properly be divided into two parts, metaphysics and mathematics. Mathematics "has for its department the acts and constructions of the *necessary* Imagination and is subdivided into Geometry as the correspondent to Space or the Outer Sense, and Arithmetic, correspondent to Time, or the Inner Sense: while Algebra may [be] considered as the conversion of the one into the other by principles of equation and compensation."[68] Coleridge was completely ignorant of mathematics and apparently mistook algebra for analytical geometry, but the reference to Geometry as the Outer Sense, and Time as the Inner Sense is significant, since it indicates that Coleridge used the notions of Time and Space as understood by Kant to give an intuitive basis for geometry and algebra. With Coleridge and Kant as authoritative support for his position, Hamilton was able to summon sufficient courage to present his metaphysical ideas in spite of the certain opposition to them by most mathematicians.

WRITING THE ESSAY

Hamilton broached his theory of Algebra as Pure Time with caution. First he tried it out in November 1832 in the first lecture of his astronomy course at Trinity College. The lecture was published with Hamilton's permission, and possibly at his instigation, in the *Dublin University Review* for January 1833.[69] Then in a bizarre hoax he wrote a review of his own public lecture, which he sent to Viscount Adare and to Aubrey De Vere to test their reactions. The review was amusingly critical, but attacked severely his notions about algebra and his taste for "the ravings of the German school, and the unintelligible mysticism of Coleridge."[70] In July of the following year he told Adare that the passage that he had submitted to the greatest ridicule "might almost be taken as a summary of Kant's view of Space and Time."[71] After these self-deprecating remarks Hamilton said nothing more publicly about his theory until just before he completed the "Essay" in 1835. In order to clarify his own ideas and to instruct Adare, he wrote a series of letters on the subject

341

beginning 4 March 1835. The letters contain an exposition of Kant's philosophy and an explanation of his own ideas about algebra. The seventh and last letter in the series was never sent. Hamilton expanded it with the inclusion of some material from the first letter to form the "General Introductory Remarks" to his "Essay."[72]

ALGEBRA AS THE SCIENCE OF PURE TIME

In his "General Introductory Remarks" Hamilton asks how it might be possible to create "a Science of Algebra properly so called; strict, pure, and independent; deduced by valid reasonings from its own intuitive principles; and thus not less an object of [a] priori contemplation than Geometry, nor less distinct, in its own essence, from the Rules which it may teach or use, and from the Signs by which it may express its meaning."[73] He concludes that the intuition of Time is the rudiment from which such a science may be constructed.

Hamilton distinguishes between three different schools of algebraists, the Practical, the Philological, and the Theoretical. The Practical school sees Algebra as an instrument providing rules for the solution of specific problems. The Philological school sees it as a set of formulas for the manipulation of symbols, symbols that have no meaning except as marks on paper. Hamilton wishes to go further, and this is the aim of the theoretical school—to go beyond the signs to the things signified and to arrive at the primary intuition from which the concept of number is created.

Kant's influence is conspicuously present in all the "General Introductory Remarks." The one passage in the *Critique of Pure Reason* that especially caught Hamilton's eye was that part of the Transcendental Aesthetic where Kant states: "Time and Space are, therefore, two sources of knowledge, from which bodies of *a priori* synthetic knowledge can be derived. (Pure mathematics is a brilliant example of such knowledge, especially as regards space and its relations.) Time and space, taken together, are the pure forms of all sensible intuition, and so are what make *a priori* synthetic propositions possible."[74] Space and time are the only content of the Transcendental Aesthetic, and are therefore the necessary forms for all sensible intuition. It is in these pure forms of sensible intuition that Kant seeks the foundation of mathematics. Kant states explicitly that "all thought must, directly or indirectly, by way of certain characters, relate ultimately to intuitions, and therefore, with us, to sensibility, because in no other way can an object be given to us."[75] We cannot think of appearances except as they are ordered in space and time, nor can we have knowledge of space and time without the experience, or possible experience, or ordering perceived objects.

Space and time, are therefore, pure *forms* of sensibility or pure intuitions.[76] They are *a priori* and do not exist in the objects themselves, but only in our manner of perceiving them. Space is the form of outer sense that allows us to represent to ourselves objects outside of ourselves. Time is the form of inner sense by which we order all perceptions or sensible intuitions as existing simultaneously or successively.[77] Kant states clearly that time is unidimensional, that in its original representation it is unlimited, and presents us with the notion of order or succession. Time is more general than space. "Space, as the pure form of all *outer* intuition, is so far limited. . . . Time [on the other hand] is an *a priori* condition of all appearances whatsoever."[78]

Hamilton follows the Kantian notion of time very closely in his writings on the foundation of algebra. Since the inner sense of time is more general than the outer sense of space, Hamilton concludes that algebra is a more general and fundamental branch of mathematics than geometry.[79] Moreover time is not merely one of many ways to illustrate the rules of algebra. The intuition of pure time (corresponding to Kant's use of the term "pure intuition")[80] "will ultimately be found to be co-extensive and identical with Algebra, so far as Algebra itself is a Science."[81] Hamilton takes care to explain why he uses the expression *pure time*. Pure Time is to be carefully distinguished from apparent time. It is the pure form of sensible intuition described by Kant, not the objects of the intuition or the order of events. When it is thus purified from all sensations leaving only the *a priori* form of sensibility, time becomes the ground for a mathematical science. It can be used as the rudiment for algebra when it is "sufficiently unfolded, and distinguished on the one hand from all actual Outward Chronology (or collections of recorded events and phenomenal marks and measures), and on the other hand from all Dynamical Science (or reasonings and results from the notion of cause and effect)."[82]

In spite of this debt to Kant, Hamilton believed that he had gone beyond the *Critique* in asserting that algebra is *the* science of Pure Time.

He believed that Kant had recognized that there *might* be, or *ought* to be, a Science of Pure Time analogous to geometry, but never suspected that such a science lay ready at hand in algebra.[83] His reading of Kant is essentially correct. Only in the *Prolegomena* did Kant openly assert that arithmetic was the science of time.[84] Elsewhere, and especially in the *Critique of Pure Reason,* he was much more reticient. While time as the form of inner sense was of greater general importance than space, which provided material for the science of geometry, Kant did not take the additional step of making time the ground for a pure science.

In the Transcendental Analytic, where Kant presents the schematism of the Pure Concepts of Understanding, the intuition of time appears in a crucial role. The transcendental schemata link the categories of understanding with sensory intuitions, that is, the schemata mediate between the pure concepts of understanding and sensations. According to Kant the schema of a pure concept is "a product which concerns the determination of inner sense in general according to conditions of its form (time)."[85] Thus the schema of each of the twelve categories "contains and makes capable of representation only a determination of time."[86] Of particular importance is the category of magnitude: ". . . The pure schema of magnitude, as a concept of the understanding, is *number,* a representation which comprises the successive addition of homogeneous units. Number is therefore simply the unity of the synthesis of the manifold of a homogeneous intuition in general, a unity due to my generating time itself in the apprehension of the intuition."[87]

Thus for Kant, number is constructed by the generation of time in intuition much as Hamilton later argued. Not all of Hamilton's contemporaries agreed that magnitude must necessarily be generated in time. In his paper "On the Foundation of Algebra" De Morgan asserted that it is possible to conceive of a line segment as generated instantaneously, "no portion of it coming into thought before or after another."[88] Hamilton objected. Quoting Kant as an authority, he insisted that we cannot even think of a line without drawing it in thought.[89] This conviction, that every extensive magnitude is apprehended in space and time by a succession or continuous sequence of acts of perception, persuaded Hamilton that without the pure intuition of time we should have no concept of magnitude or number, and it confirmed his belief that algebra had to be based on the ordinal property of number. At the urging of "mathematical friends" he had considered basing algebra on continuous progression alone, without reference to time in order to avoid the criticism of De Morgan and the other formalists. But as he realized that progression itself had its intuitive basis in the notion of time, he felt that he could not exclude this primary intuition from his presentation.

Two notions that are crucial for Kant's mathematics and important for Hamilton as well are those of "intuition" and "construction". Both of these notions are less ambiguous when applied to geometry than to algebra, and this is probably one reason why Kant preferred geometrical illustrations to arithmetical ones.[90] Kant argued that the great advantage of mathematical reason over philosophical reason is the fact that in mathematics we construct the concepts we need from pure intuitions, whereas in philosophy we must confine ourselves to reasoning on universal concepts.[91] The intuition required for geometry is that pure intuition of space that appears in the Transcendental Aesthetic. From this pure intuition we can construct "mental images" and actually display our constructions mentally or on paper to reach conclusions regarding these figures.[92] But what is the analogous case for arithmetic and algebra? The reference of arithmetic is not necessarily to objects in space and time. For example, mathematical objects, which we can count, do not appear to have an existence in space and time.[93] What are the pure intuitions that we need for arithmetic and algebra, and how do we construct numbers or other mathematical objects of algebra from these intuitions? There is no doubt in Kant's mind that time is the one essential intuition for the construction of number, but it is not clear what intuitions we need—i.e., instants in time, portions of time, succession in time—or how the construction is to be accomplished.[94] Until Kant could remove these ambiguities he could only suggest, but not create, a science of time.[95]

Hamilton's "Essay" attempts to make precise the construction of number where Kant left it ambiguous, and it is this emphasis on the *construction* of numbers that brings him closest to the arguments of the modern intuitionists.[96]

Hamilton begins his construction by saying: "If we have formed the *thought* of any one moment of time, we may afterwards either *repeat* that thought, or else think of a different moment."[97] By comparing two such moments we can create the notion of a "step" either forward or backward in time. A time step may be greater than, equal to, or less than another given step. It is part of the original intuition that time steps can

345

be equal by analogy even if they are not actually coincident. Number is obtained from a sequence of equal time steps, either positive or contrapositive, taken from an arbitrarily chosen zero moment. A "contrapositive" step is a step backward in time. It allows Hamilton to define negative numbers by the concept of opposite direction in time, rather than by the meaningless notion of negative magnitude. The concept of positive number is originally ordinal—that is, one denotes the first, second, third, and so on, time step in the progression. It is also proper to call these numbers (or "multipliers" of time steps) cardinal numbers, since one may ask "how many" steps there are between two given moments in time. The answer is one, two, three, and so on; but these cardinal numbers are obtained by counting in progression the steps from the zero moment, and therefore they are dependent on the ordinal relationship of the steps.[98]

Any multiple step, or number of steps in progression, may be treated as a new base or new unit step for a different system. A new series of multipliers specifies the order of these steps in time according to the new base, thereby defining multiplication. If, for example, three steps of base (a) are counted off and specified as the unit base (b) of a new system, and five of these new unit steps (b) are counted off to reach a specific moment in Time (A), then the number of time steps according to base (a) from the zero moment to moment (A) will give 15, or the "product" of 3 and 5.

In his earlier manuscripts and letters Hamilton had frequently argued that number should be defined as the *ratio* of steps in a progression of time.[99] For the integers this amounts to the ratio of a time step to a unit step by which it is counted, but to define the real numbers and to provide for division, Hamilton has also to consider ratios of incommensurable time steps. This he does in a lengthy discussion of ratio. He is particularly concerned to show that the intuitive idea of the continuity of the progression from moment to moment in time also provides the idea of a *continuous progression* in ratio, thereby assuring the continuity of the real number system.

When Hamilton comes to consider complex numbers, he constructs his number couples in a manner very similar to his construction of the integers. He begins by comparing couples of moments in time (A_1, A_2) rather than single moments as was the case for ordinary algebra. This "comparison" of two moment couples $(B_1, B_2) - (A_1, A_2)$ is a complex relation that Hamilton chooses to define as

$$(B_1, B_2) - (A_1, A_2) = (B_1 - A_1, B_2 - A_2)$$

thus generating a couple of time steps much as he did in the case of ordinary algebra.[100] Numbers (or in this case number couples) are again multipliers of time steps, and Hamilton derives a "reasonable" and not wholly arbitrary definition of multiplication of number couples that is equivalent to the multiplication of complex numbers.[101]

Hamilton's construction of numbers and number couples is highly imaginative, but his argument is weakened at the outset by his use of the ambiguous concept of "moments" of time. It is from "moments," considered as objects of thought, that Hamilton constructs time steps, which in turn provide the basis for number. By the word "moment" he means either a position in time or a very small, possibly infinitely small, segment of time. Hamilton's constant reference to moments *of* time, rather than moments *in* time would indicate that he probably means the latter. But if his moments are to be taken as pieces of time, then they would have to be durationless instants in order to define adequately the boundaries of his time steps, and it is not at all obvious that we can intuit such instantaneous moments.[102]

Hamilton may have been led into this confusion by his insistence that moments of time must have an objective reality. In a manuscript from 1832 he argued that there was a difference between the intuition of time and the demands of algebra. Our knowledge of time seems to contain much more than pure order or progression because it refers to events happening, while algebra is grounded in a purely mental idea of order *that can be changed at will*. Hamilton noted that the entire theory of functions was based on this notion of changing relation or continuous progression altered at will. In a short memorandum he wrote:

> . . . The idea of *time* is that of an *order objective*. *Subjectively* viewing it, that is endeavoring to attend to thoughts rather than to things, we form the *nearest approach* to the idea of time when we think of one order as the mental basis of another, and consider the latter arrangement, which in this view resembles the course of events, as reducible to a mental dependence on the former arrangement which corresponds to the course of time. But though

we may thus approach . . . the idea of time, we do not quite attain nor adequately express it thus, nor by any other method which excludes the ascription of objectivity, or makes the order of time a voluntary arrangement of the mind. Do I then reject the doctrine of Kant that time is subjective, and in the mind? No, if I rightly believe the doctrine to be that time is a form of human thought, a result of the mechanism of our Understanding. But surely, even according to this doctrine, thus interpreted, our thought of time is involuntary, and the arrangement of order of time is so far objective. I am aware that most persons would call all this a play upon words, but I am of opinion with Mme de Staël and with others, that when men dispute about words there is always some difference of ideas, too, and the question whether time be not in part objective, serves at least to make more clear the meaning attached by the inquirer to the word objective, if not [to] their idea of time.[103]

The relations of algebra depend on the notion of continuous order in succession, which can be found in time, but is the order of time something that we can vary at will? In 1832 Hamilton thought that it was not. Even if we remove from our intuition of time all actual events and all notions of cause and effect, there would still remain an *objective* content not subject to will. In time we recognize a past, present, and future—the notion of an inexorable progress beyond human control. By contrast, in arithmetic and more especially in algebra, the mind creates its progressions at will, and to this extent mathematical time is a more subjective creation than time as it is known through the act of perception.

Hamilton also worried about the *existence* of mathematical objects, a problem closely related to the objective character of time. A notebook entry dated 24 April 1835 contains the following passage:

> When we consider any moment in the indefinite succession of time and regard it as an *object* of thought, we must think of it as having some certain place of its own in that succession by which it is distinguished *as an object* from all other moments of time. . . . The possibility of thus treating moments in algebra, and points in Geometry as *objects,* seems to be an essential postulate of these sciences and a condition of their possiblity as such.[104]

The existence of moments is guaranteed if they are objects of thought, that is, if they are intuited directly by a mental act and held in the memory. They must have some kind of objective existence and permanence (albeit an existence in the mind) if we hope to construct numbers from them.

Hamilton's arguments on the objectivity of time show an interesting similarity to some of Kant's arguments in the Analogies of Experience. The purpose of Kant's analogies is to prove that experience is possible only through the representation of a necessary connection of perceptions, thereby providing a *unity* of all perceptions in time. Of the three analogies the second is the most important for algebra, because it deals with the *succession* of appearances in time. Kant argues that "experience itself . . . is . . . possible only insofar as we subject the succession of appearances, and therefore all alteration, to the law of causality."[105] This law or formal rule is an *a priori* condition for ordering appearances in time. Kant gives the famous examples of a man viewing the facade of a large house and a man viewing a boat as it moves downstream.[106] In both cases the manifold of appearances must be generated in the mind successively. The appearance of the house is obtained by synthesizing a succession of apprehensions. The eye wanders over the facade of the house at will as the mind synthesizes the apprehensions received into the appearance of a house. Kant asks if the order of apprehensions is in the manifold of the house. Clearly not. This order of apprehensions must be a subjective succession, for the observer might well view the parts of the house in a different order and still obtain the same appearance of a house.

The case of the boat moving downstream in quite different. In this case the sequence of apprehensions is "bound down," as Kant says, to a determined order, because the observer is perceiving a series of *events*. He recognizes that this order of events is a *necessary* order, determined by the formal *a priori* rule of cause and effect that unifies experience. Without this rule the "succession of perceptions would be . . . merely subjective, and would never enable us to determine objectively which perceptions are those that really precede and which are those that follow. We should then have only a play of representations, relating to no object; that is to say, it would not be possible through our perception to distinguish one appearance from another as regards relations of time."[107] In the case of the boat moving downstream, it is important to make a careful distinction between the subjective succession of apprehensions (the kind of subjective succession that took place in viewing the house) and the objective succession of appearances. Kant says that in the case of the boat, we must derive the subjective succession of apprehensions from the objective succession of appearances.[108]

Kant's distinction between subjective and objective succession in time is similar to Hamilton's distinction between subjective and objective order. The subjective order is arbitrary and subject to the will while the objective order must conform to a determined rule that the observer does not himself create. Algebra requires the subjective order and cannot be created from a notion of time bound down to the law of causality. In his memorandum of 1832 Hamilton expressed the concern that if time could not be separated from the objective order of appearances, it would not be an adequate basis from which to construct mathematical objects.

Of course the objective order of appearances in time that Kant discusses in his second analogy is an order of *events,* whereas the doctrine that he presents in the Transcendental Aesthetic identifies time as the form of inner sense. Time as the form of inner sense is an *a priori* condition of any appearances. He says: "Appearances may, one and all vanish, but time (as the universal possibility) cannot be removed."[109] Whether his statements about time in these two parts of the *Critique* are consistent is a matter that need not be argued here.[110] It is enough that for Hamilton, Kant's argument in the Transcendental Aesthetic brought conviction. In the "General Introductory Remarks" he does not raise the objections that appeared in his memorandum of 1832, and he concludes that Pure Time can be distinguished from all physical events and from all reference to cause and effect. Pure Time becomes "coextensive and identical" with algebra. He retains the notion that moments of time must be "objects of thought." To that extent time remains objective, but he relinquishes the notion that it is in any way determined by an objective order of events.

Hamilton did not necessarily believe that Kant had resolved all the metaphysical problems regarding time, but he did believe that his doctrine was adequate for the creation of algebra. As a mathematician this was his real concern. "There is something mysterious and transcendent in the idea of Time," he wrote, "but there is also something definite and clear: and while the Metaphysicians meditate on the one, Mathematicians may reason from the other."[111] From the "definite and clear,'' he proceeded to create what he called the Mathematical Science of Time.

In closing his ''Introductory Remark'' Hamilton wrote that he had struggled long against the idea of Algebra as Time and had long forborne to make it public, ''on account of its departing so far from views now commonly received.''[112] He was quite right, because it went directly against the rising enthusiasm for symbolical algebra as it was taught by the Cambridge school of Augustus De Morgan, George Peacock, and Duncan Gregory. To that group would have to be added Hamilton's old friend John Graves, who had attracted him to the fundamental problems of algebra in the first place.

It is amazing that in his long correspondence with Graves, Hamilton had not even hinted at his new ideas about algebra. While reading the proof-sheets of his ''Essay,'' he mentioned casually in a letter to Graves the title and remarked: ''Perhaps I have not ever talked to you about this crotchet of mine, for I know that with all our personal and intellectual ties we belong to opposite poles in algebra; since you, like Peacock, seem to consider Algebra as a 'System of Signs and of their combinations'.''[113] Graves became enthusiastic about Hamilton's ideas after he understood them better, but Hamilton was correct in his supposition that he would find few supporters. Augustus De Morgan, a great friend and correspondent, took a swipe at algebra as time in his papers on the foundation of algebra, and Arthur Cayley insisted that the concept of time was totally foreign to algebra.[114]

Peacock, De Morgan, and Graves were members of the school that Hamilton called ''philological''. We would now characterize them as ''formalists''. For them the symbols of algebra were ultimate; there was no ''signification behind the signs'' as Hamilton claimed. Algebra was a set of permitted operations or manipulations of given signs. At a later date when Hamilton had come to regard the philological school more favorably, he confessed that when he had first read Peacock's *Treatise of Algebra* (1830) he had felt that Peacock had wanted to ''reduce algebra to a mere system of symbols, and *nothing more;* an affair of pothooks and hangers, of black strokes upon white paper, to be made according to a fixed but arbitrary set of rules; and [he] refused . . . to give the high name of *Science* to the results of such a system.''[115] Although he became more tolerant of symbolical algebra later on,

Hamilton never lost the conviction that the concepts of algebra had to be based in intuitions, and that number had its origin in the intuition of time.

The question that then presents itself is, "Why did Hamilton's intuitionist approach to mathematics allow him to create new algebras of couples and quaternions, while the formalists could not see beyond ordinary arithmetical algebra?" G. J. Whitrow sees a real paradox in this history:

> Thus, the ultimate conclusion of his train of thought was the revelation that algebra is not unique, a view very difficult to reconcile with his Kantian conception of its nature and a powerful argument in favor of the formalist philosophy of mathematics to which he was so resolutely opposed.[116]

If our concept of time is unique and primitive, Whitrow argues, then the algebra created from it should also be unique. Thus it is contradictory for a supporter of the intuitionist school to create new algebras—but that is just what Hamilton did. In fact, he was the first to do it.

This inconsistency in Hamilton is more apparent than real. As we have seen, the intuition that Hamilton requires for algebra is not that of number *per se* but the intuition of *countinuous progression,* which he finds in time. From this primitive concept of progression in time, Hamilton *constructs* numbers—real numbers, number couples, and finally quaternions. Hamilton calls it a *synthesis* of algebra, the "building of it up anew (in its most essential parts) from the Idea of Pure Time."[117] Whether or not the notion of continuous progression comes from time is, he believes, a *metaphysical* question.[118] The mathematical problem is one of *constructing* algebra from this basic concept of progression, and more than one algebra can be so constructed, as Hamilton demonstrates.

In spite of his anti-intuitionist position De Morgan understood what Hamilton had done in creating new algebras of couples and quaternions. Both men had been searching for an algebra of triplets in October 1843 at the time Hamilton discovered his quaternions, but it had never occurred to De Morgan to create such an algebra. In a letter to Hamilton he wrote:

> [John] Graves gave me some extracts from your letter now published. . . . He never dropped a hint about *imagining* imaginaries. On such little things do our thoughts depend. I do believe that, had he said no more than

"Hamilton *makes* his imaginary quantities," I should have got what I wanted.[119]

John Graves had the same problem:

> . . . There is still something in the system [of quaternions] which gravels me. I have not yet any clear view as to the extent to which we are at liberty arbitrarily to create imaginaries, and to endow them with supernatural properties. You are certainly justified by the event . . . [but] what right have you to such luck, getting at your system by such an inventive mode as yours?[120]

In contradistinction to Whitrow's criticism, it was the intuitive approach that gave Hamilton an advantage over the formalists in creating new algebras. By grounding algebra on the idea of succession in time rather than on the idea of number, he was able to construct new algebras while De Morgan and Graves remained tied to the rules of arithmetic and arithmetical algebra.

CONCLUSION

As is frequently the case when one attempts to penetrate into the details of a historical problem, the questions that should have simple answers have to be qualified to fit the historical situation. Did Hamilton's concept of Algebra as the Science of Pure Time come from his reading of Kant and Coleridge? We have seen that in a very direct way his "Essay" was a product of his intense interest in their philosophies during the four years prior to its composition. And yet the idea was at least a glimmer in his mind eight years earlier in 1827. At that time he knew of Kant from reading Dugald Stewart and Mme de Staël. More than that we cannot say, and it is doubtful whether Hamilton himself could have said at any time precisely where his idea came from.[121] Nevertheless it is clear that Kant and Coleridge were responsible for turning this glimmer into a burning interest, and their authority gave him the confidence to publish his ideas. Probably we should take at face value his claim, made to William Wordsworth, that when he read Kant in 1834 he found himself to be "at least as much *recognizing as learning.*"[122]

As to the significance of Hamilton's metaphysics for his algebra, I believe that the metaphysics was very important. Of course the influence of metaphysics can work two ways, both for and against discovery in mathematics. There is great irony in the following passage from the "Essay": "No candid and intelligent person can doubt the truth of the chief properties of Parallel Lines, as set forth by Euclid in his Elements, two thousand years ago. . . . The doctrine involves no obscurity nor confusion of thought and leaves in the mind no reasonable ground for doubt."[123] Hamilton made this statement to contrast the clarity of geometry with the confusion and obscurity of algebra. But just because he was a Kantian, he could not appreciate non-Euclidean geometry, which was being developed at just this time in the first half of the nineteenth century. However, the concept of Algebra as Pure Time, even with its unsatisfactory metaphysics, gave Hamilton a freedom that was not available to De Morgan, Peacock, and the other algebraists of the Cambridge school. By using the pure intuition of time he was able to consider operations on objects other than the real and complex numbers. In this he went beyond Kant, who believed that the natural numbers had a uniqueness comparable to the Euclidean space of geometry. Thus he was able to consider operations on time steps that were only later given algebraic meaning. This way he was able to define negative number and arrive at an adequate algebraic definition of complex numbers in terms of number couples. Most dramatic, of course, was his rejection of the commutative law in the algebra of quaternions.

Peacock and De Morgan attempted to free algebra from just that intuitive character that Hamilton inserted into it. But they were unable to recognize that there might be some profit in considering systems that did not follow all the rules of ordinary algebra. Once Hamilton had opened the way—and I believe his metaphysics helped him to open the way—new algebras came fast and furiously. The path to discovery is often a devious one. In this case it was also a strangely metaphysical one.

Research for this article has been supported by grants from the National Science Foundation and the Graduate Research Fund of the University of Washington.

1. Robert P. Graves, *Life of Sir William Rowan Hamilton*, 3 vols. (Dublin: Hodges, Figgis & Co., 1882), 2:154 (hereafter cited as "Graves").

2. Hamilton to Aubrey DeVere, 4 Oct. 1835; Graves, 2:164.

3. Graves, 2:151.

4. *Transactions, Royal Irish Academy* 17 (1837): 293–422.

5. W. R. Hamilton, *The Mathematical Papers,* vol. 3, "Algebra," ed. H. Halberstam and R. E. Ingram (Cambridge: At the University Press, 1967), pp. 15–16.

6. *Math. Papers,* 3:xv.

7. *Math. Papers,* 3:233. I learned of Hamilton's discovery of the associative law from an unpublished undergraduate thesis by Barbara Underwood Cohen, "From Algebra to Algebras: A Study of Changing Conceptions of Mathematics," Harvard University, April 1966.

8. This is a modern idea. See Ross A. Beaumont and Richard S. Pierce, *The Algebraic Foundations of Mathematics* (Reading, Mass.: Addison-Wesley, 1963), pp. 286–98.

9. For instance C. C. MacDuffee, "Algebra's Debt to Hamilton," *Scripta Mathematica* 10 (1944): 25.

10. An excellent history of this controversy is Florian Cajori, "History of the Exponential and Logarithmic Concepts," *American Mathematical Monthly* 20 (1913): 5, 35, 75, 107, 148, 173, 205.

11. *Math. Papers,* 3:96.

12. See Eric Temple Bell, *Development of Mathematics,* 2d ed. (New York: McGraw-Hill, 1945), p. 189.

13. "Researches on Quaternions," read 13 Nov. 1843; *Math. Papers,* 3:159. Hamilton was even more explicit in a letter to De Morgan, 7 May 1847, Graves, 2:574.

14. L. Pearce Williams, *Michael Faraday: A Biography* (New York: Simon & Schuster, 1971), p. 62.

15. Ibid., 67.

16. Graves, 1:211, 215, 2:95–96.

17. *Math. Papers,* 2:104; 3:117–18 note.

18. See Robert Kargon, "William Rowan Hamilton and Boscovichean Atomism," *Journal of the History of Ideas* 26 (1965): 137–40.

19. George Dodd, "Wordsworth and Hamilton," *Nature* 208 (1970): 1261–63.

20. Eric Temple Bell, *Men of Mathematics* (New York: Simon & Schuster, 1937), p. 358.

21. C. Lanczos, "William Rowan Hamilton—An Appreciation," *University Review, National University of Ireland* 4 (1967): 156. In a similar vein C. C. MacDuffee apologizes for Hamilton's apparent need to justify his abstract notions by "appealing to the physical universe" ("Algebra's Debt to Hamilton," *Scripta Mathematica* 10 [1944]: 27).

22. Trinity College, Dublin, MS. 1493–2171, no. 10.

23. Hamilton to John T. Graves, 20 Oct. 1828; Graves, 1:303–4.

24. Graves 1:307–8. J. F. W. Herschel to J. T. Graves, undated, TCD MS. 4015, no. 114.

25. J. F. W. Herschel to J. T. Graves, undated, TCD MS. 4015, no. 113.

26. J. F. W. Herschel to J. T. Graves, 24 Jan. 1828, TCD MS. 4015, No. 112.

27. Hamilton to J. F. W. Herschel, 25 Feb. 1829, TCD MS. 1493–2171, no. 35. The paper appeared as "An attempt to rectify the inaccuracy of some logarithmic formulae. By John Thomas Graves, of the Inner Temple, Esq. V. P. Read 18 December 1828," *Philosophical Transactions of the Royal Society* 209 (1829): 171–86.

28. George Peacock, "Report on the Recent Progress and Present State of Certain Branches of Analysis," *3rd Report* of the British Association for the Advancement of Science (London, 1834), p. 266 n. See Cajori, "History of the Exponential and Logarithmic Concepts," p. 178.

29. Hamilton to John T. Graves, 11 June 1829, TCD MS. 1493–2171, no. 54.

30. Augustus De Morgan, "On the Foundation of Algebra," *Transactions, Cambridge Philosophical Society* 8 (1844–49): 141.

31. (Cambridge: At the University Press, 1828). On Hamilton's knowledge of the geometrical representation of complex numbers, see his "Preface" to *Lectures on Quaternions, Math. Papers,* 3:135 note.

32. Hamilton to De Morgan, 25 June 1864; Graves, 3:189.

33. 20 Oct. 1828; Graves, 1:304.

34. Hamilton to Baden Powell, 5 April 1833; Graves, 2:40.

35. The couples first appear as "couples of curve surfaces" from which Hamilton derives the "equations of conjugation," in a letter to John T. Graves 5 Aug. 1829, TCD MS. 1493–2171, no. 58. See note 97.

36. *Math. Papers,* 3:117 note.

37. Graves, 1:228–29.

38. TCD MS. 1492 Misc. Papers, Box VI, These papers have not been individually catalogued.

39. Ibid.

40. Ibid.

41. TCD MS. notebook 1492–24.5, fol. 49. Hamilton incorporated these remarks in his introductory lecture on astronomy of 8 November 1832 (Graves, 1:643).

42. René Wellek, *Immanuel Kant in England, 1793–1838* (Princeton: Princeton University Press, 1931).

43. Graves, 1:209, and TCD MS. notebook 1492–16, fol. 49.

44. Graves, 1:208, and Hamilton to his cousin Arthur Hamilton, 14 April 1824, TCD MS. 1493–3804, no. 79.

45. Wellek, *Immanuel Kant in England,* pp. 40–46.

46. TCD MS. 1492–16, fol. 28; also Richard Lovell Edgeworth, *Memoirs,* 2 vols. (London: 1820), (facsimile ed., Shannon: Irish University Press, 1969), 1:293.

47. Dugald Stewart, *Collected Works,* ed. Sir William Hamilton, 7 vols. (Edinburgh: Thomas Constable & Co., 1854–60), 5:87.

48. TCD MS. 4015, no. 37, copied from *The Foreign Review and Continental Miscellany* 4 (1830): 97ff. The author was Thomas Carlyle, although the work appeared anonymously.

49. Ibid., p. 117.

50. William Wordsworth, Sr., to Hamilton, 9 Sept. 1830, Graves, 1:393, and 26 Nov. 1830, Graves, 1:401.

51. Hamilton to Wordsworth, 29 Oct. 1831, Graves, 1:478.

52. Of these six attempted translations four are in the uncatalogued Miscellaneous Papers, one in TCD MS. notebook 1492–24.5, fol. 101, and a second is in notebook 1492–70, not foliated. Hamilton had the first edition in 1831; he read the second edition in 1834.

53. Hamilton to Aubrey DeVere, 29 Aug. 1834, Graves, 2:103.

54. Hamilton to Aubrey DeVere, 3 July, 1832, ibid., 1:585; and to his sister Eliza, 15 March 1832, ibid., 1:535.

55. Wordsworth to Hamilton, 24 Jan. 1831, ibid., 1:425.

56. Alice D. Snyder, *Coleridge on Logic and Learning with Selections from the Unpublished Manuscripts* (New Haven, Conn.: Yale University Press, 1929), p. 68.

57. S. T. Coleridge to Hamilton, 4 April 1832, Graves, 1:545.

58. Hamilton to Viscount Adare, 19 July 1834, ibid., 2:96; 20 Aug. 1834, ibid., 2:100. Hamilton to Aubrey DeVere, 29 Aug. 1834, ibid., 2:103–5. Hamilton to Viscount Adare, 19 April 1842, ibid., 2:364. Hamilton to Rev. William Lee, 2 July 1859, ibid., 3:115–17.

59. Ibid., 1:437.

60. S. T. Coleridge, *Aids to Reflection, and the Confessions of an Inquiring Spirit* (London: George Bell & Sons, 1901), p. xvii.

61. Ibid., pp. 114, 117 n, 195.

62. Ibid., p. 143.

63. Ibid., p. 195.

64. Hamilton to Adare, 23 Aug. 1831, Graves, 1:444.

65. Hamilton to Aubrey De Vere, 9 Feb. 1831, ibid., 1:519.

66. Hamilton to Aubrey De Vere, 27 March 1832, ibid., 1:541.

67. S. T. Coleridge, *Complete Poetical Works,* ed. Ernest Hartley Coleridge, 1 (Oxford, 1912): 419–20.

68. S. T. Coleridge, *Collected Works,* ed. Barbara E. Rooke (Princeton, N.J. and London: Princeton University Press and Routledge & Kegan Paul, 1969), vol. 4, pt. 1, p. 440.

69. Graves, 1:639–54.

70. Ibid., 2:32.

71. Hamilton to Adare, 19 July 1834, Graves, 2:96–97.

72. Graves, 2:138, and Hamilton to Aubrey De Vere, 13 May 1835, ibid., 2:141. The entire series of letters was later published by Robert P. Graves: "Sir W. Rowan Hamilton on the Elementary Conceptions of Mathematics," *Hermathena* 6 (1879): 469–89.

73. *Math. Papers,* 3:5.

74. Immanuel Kant, *Critique of Pure Reason,* trans. Norman Kemp Smith (London: Macmillan & Co., 1961), (A38–39 = B 55–56). All passages from Kant's *Critique of Pure Reason* are from the above translation. I cite the page numbers in the first and second editions as they are given in Smith's translation.

75. Ibid., (A 19 = B 33).

76. Ibid., (B 35).

77. Ibid., (A 30 = B 46, A 200 = B 245).

78. Ibid., (A 34 = B 50).

79. *Math. Papers,* 3:7.

80. *Critique* (B 35).

81. *Math. Papers,* 3:7.

82. Ibid.

83. Hamilton to Aubrey De Vere, 13 May 1835, Graves, 2:142.

84. "Arithmetic produces its concepts of number through successive addition of units in time" (quoted from Norman Kemp Smith, *A Commentary* to Kant's 'Critique of Pure Reason,' 2d ed. [London: Macmillan & Co., 1923], p. 129).

85. *Critique* (A 142 = B 181).

86. Ibid., (A 145 = B 184).

87. Ibid., (A 142–3 = B 182).

88. De Morgan, "On the Foundations of Algebra," *Cambridge Philosophical Society, Transactions* 7 (1839–42): 174.

89. Hamilton to De Morgan, 8 May 1841, Graves, 2:342. He quotes Kant's *Critique* (A 162–3 = B 293–4).

90. Kant's use of these notions in arithmetic is a current topic of debate. See Jaakko Hintikka, "Kant on the Mathematical Method," *Monist* 51 (1967): 352–75, and Charles Parsons, "Kant's Philosophy of Arithmetic," in *Philosophy, Science, and Method,* ed. Sidney Morgenbesser et al. (New York: St. Martin's Press, 1969), pp. 568–94.

91. *Critique* (A 713 = B 741, A 724 = B 752).

92. Hintikka believes that Kant was thinking specifically about the Euclidean method of proof, where the construction or "display" is called the *ecthesis* or *setting-out* ("Kant on the Mathematical Method," p. 361).

93. Parsons, "Kant's Philosophy of Arithmetic," pp. 581–82.

357

94. Kant is not even consistent in his claim that only the inner sense of time is involved in arithmetic construction. In the schematism he derives number from ordered succession in time; but in the Transcendental Doctrine of Method he generates number from an intuition of "the universal element in the synthesis of one and the same thing in time and space" (A 724 = B 753). Hintikka argues that for Kant the important aspect of intuition is not its immediacy, but the fact that all intuitions are individual and singular. Both aspects are closely related, because Kant realizes that in order for intuitions to be immediate, they must be of singular objects. (Hintikka, "Kant on the Mathematical Method," p. 354ff.).

95. N. K. Smith, *Commentary,* p. 131.

96. For L. E. J. Brouwer and other intuitionists the existence of mathematical entities is synonymous with the possibility of their construction, whereas their opponents, the formalists, emphasize the static and timeless character of mathematical objects. Brouwer announced his intuitionist position by rejecting Kant's *a priority* of space, but adhering the more resolutely to the *a priority* of time (L. E. J. Brouwer, "Intuitionism and Formalism," in *Philosophy of Mathematics,* ed. Paul Benacerraf and Hilary Putnam [Prentice-Hall, 1964], p. 69). He constructs the natural numbers from "the intuition of the bare two-oneness" which he calls the basal intuition of mathematics. This bare two-oneness is intuited from the division of time, which he in turn claims is "the fundamental phenomenon of the human intellect" (ibid., p. 69).

97. *Math. Papers,* 3:9.

98. Ibid., 3:29. Hamilton speaks specifically of counting as an ordinal relation in a letter to the Viscount Adare, 9 Aug. 1843 (Graves, 2:417).

99. Hamilton to Viscount Adare, 16 May 1833, Graves, 2:46. TCD MS. 1492 Miscellaneous Papers, Box VI, Feb. 16, 1830 "Triads," December 1830 "Triads," and undated "Fundamental Idea."

100. *Math. Papers,* 3:76–77.

101. The definition of multiplication is not wholly arbitrary, because Hamilton has already defined addition of number couples in the obvious way:

$$(a_1, a_2) + (b_1, b_2) = (a_1 + b_1, a_2 + b_2)$$

and he wants the following distributive law to hold:

$$(a_1, a_2) [(b_1, b_2) + (c_1, c_2)] = (a_1, a_2) (b_1, b_2) + (a_1, a_2) (c_1, c_2).$$

A second restriction is that division of number couples should be determinate as long as the divisor is not (0,0). These two restrictions do not completely determine the process of multiplication, but they limit the possible choices (*Math. Papers,* 3:80–83).

Hamilton apparently came to the number couples in the following way. Consider a function of complex numbers of the form

$$(u + v \sqrt{-1}) = \phi \ (x + y \sqrt{-1})$$

If this relationship holds, then the following "conjugate functions" must also hold:

$$\frac{du}{dx} = \frac{dv}{dy}; \frac{du}{dy} = \frac{dv}{dx}$$

These are the Cauchy-Riemann equations, first used by d'Alembert in the eighteenth century, but rediscovered in this context by Hamilton. These "conjugate functions" determine a relation between the variables (u, v) and (x, y) that does not involve the imaginary $\sqrt{-1}$. Hamilton then pursues the possible representation of complex numbers by number couples (see *Math. Papers,* 3:97).

102. In the Analogies of Experience Kant denies that point-like instants in time are identifiable by events (*Critique*, A 209 = B 254, A 215 = B 262). Kant states explicitly that there is no smallest part of time A 209 = B254). He also uses the word "moment" in a way different from Hamilton's usage. For Hamilton a "moment" is a durationless instant; for Kant it is a continuous action of the causality (A 208 = B 254), the most common example being the "moment of gravity" (A 168 = B 210). Kant's use of the word has a definite dynamical character. It is related to the notion of an instant through the concept of "intensive magnitude." The real in the field of appearances can have *extensive* magnitude only through succession in time. Sensation, however, takes place in an *instant;* yet it also possesses degree, or *intensive* magnitude. The intensive magnitude is entitled a "moment". It is possible that Hamilton had Kant's use of "moment" in mind when he wrote his "Essay." He carefully avoids talking about instants of time and refers only to moments.

103. TCD MS. 1492 Misc. Papers, Box VI, 21 April 1832.

104. TCD MS. notebook 1492–35.5, unpaginated, underlining by Hamilton.

105. *Critique* (A 189 = B 234).

106. Ibid., (A 190 = B 235 − A 193 = B 238).

107. Ibid., (A 194 = B 239).

108. Ibid., (A 193 = B 238).

109. Ibid., (A 31 = B 46).

110. N. K. Smith says in his *Commentary* that they are not consistent: "There are for Kant, two orders of time, subjective and objective. Recognition of the latter (emphasized and developed in the *Analytic*) is, however, irreconcilable with his contention that time is merely the form of inner sense" (p. 137).

111. *Math. Papers,* 3:7.

112. Ibid., p. 7 note.

113. Hamilton to John T. Graves, 11 July 1835, Graves, 2:143.

114. Augustus De Morgan, "On the Foundation of Algebra, "*Transactions of the Cambridge Philosophical Society* 7 (1839–42): 175–76; G. T. Whitrow, *The Natural Philosophy of Time* (New York: Harper & Row, 1963), p. 118.

115. Hamilton to Peacock, 13 Oct. 1846, Graves, 2:528.

116. Whitrow, *The Natural Philosophy of Time,* p. 118.

117. Hamilton to John T. Graves, 11 July 1835, Graves, 2:144.

118. Hamilton to Augustus De Morgan, 12 May 1841, ibid., p. 342.

119. De Morgan to Hamilton, 16 Dec. 1844, ibid., p. 475.

120. John T. Graves to Hamilton, 31 Oct. 1843; ibid., p. 443. Why indeed? Graves set out at once to create an algebra of "octaves." Barbara Cohen has described this creative aspect of Hamilton's approach to algebra in her thesis "From Algebra to Algebras" (see note 7 above). She points out that Peacock's "principle of permanence of equivalent forms" is a statement that whatever symbols may be chosen for the creation of an algebra, the rules of operation in that algebra must be consistent with the rules of arithmetic and arithmetical algebra. The rule denies the possibility of a really "new" algebra.

121. In later years Hamilton could not remember whether he had read Kant's *Critique of Pure Reason* before he had the number couples. He wrote to De Morgan: "I did know the outlines of double algebra, not only before I thought of the quaternions, but also (I *think*) before I had formed, partly with the help of Kant, any definite views about pure time" Hamilton to De Morgan, 3 Jan. 1852, Graves, 3:307).

122. Hamilton to Wordsworth, 20 July 1834, Graves, 2:97–98.

123. *Math. Papers,* 3:4.

Erwin Hiebert

Chapter Thirteen

AN APPRAISAL OF THE WORK OF ERNST MACH:
SCIENTIST-HISTORIAN-PHILOSOPHER

Ernst Mach (1838–1916) stands out as a seminal figure in the development of those areas of nineteenth- and early twentieth-century conceptual thought that encompass the study of the origins, nature, methods, and function of the exact sciences. Apart from any intrinsic merit that may be attributed to Mach as scientist and nonconformist thinker, it is evident that he was born at an opportune time and within an environment and intellectual locus that made it possible for him to become a key figure in the evolution of a critical and analytical approach to the philosophy of science. By as early as the 1870s Mach had become notoriously hypercritical, even censorious, about a number of trends that were visible within the natural sciences in general. The more mature period of his life coincides with the deeply felt need on the part of the European scientific community to construct bridges from the classical to the new physics. An appraisal of Mach's role in that construction is the main theme of this paper.

For almost fifty years Mach was actively involved in the philosophical and historical examination of scientific puzzles and problems that he felt were in need of elucidation and purification. Given these facts, it may seem odd to discover that although, at times, Mach was able to unmask, pinpoint, and clearly formulate significant issues of the day, he invariably was reluctant to accept the most successful directives that were being explored and implemented by his contemporaries. For example, criticism and rejection characterized his position toward the atomic-molecular theory of matter, certain aspects of thermodynamics connected with the second law, and relativity theory.

If one looks at Mach's career circumspectly in connection with his scientific work and if one examines his ideas alongside the most noteworthy scientists who rubbed shoulders with him intellectually toward the end of the century, then it becomes obvious that Mach's critical posture within the professional discipline of physics is not an utterly idiosyncratic one. That is, among Mach's scientific contemporaries we recognize a considerably wider consensus about the inadequacies of classical theory than about what might constitute acceptable resolutions of the inadequacies. Among academics and intellectuals, such as Mach, Boltzmann, Ostwald, Planck, and Einstein, for example, discontent with the conceptual foundations of the physical sciences almost became a way of life. Now and then agreement was achieved on how to build bridges to avoid scientific incompatibility—but at what a price. At least, that was the way Mach saw it: too high a price.

It is only in retrospect, now that the dust has settled, that the older physics appears to have been assimilated so effectively into quantum, atomic, and relativity theory. This is because most of what was thought and said and written down at the time has since been swept under the rug. It makes history much easier that way. As a consequence, the so-called revolutions in science now stand out much clearer in bold relief—mostly, in my opinion (to adapt a favorite expression of Mach's), thanks to the economy of historical thought. But is that what historians want?

An attitude of intellectual discontent among scientists who are engaged in the critical revision of the conceptual foundations of a discipline is hardly an abnormal state of affairs. Productive, creative scientists, it seems, invariably acquire a conservative temperament with respect to new ideas—especially where the novelty does violence to personal contributions. Max Planck said in a lecture delivered in 1933 at the society of German engineers: "An important scientific innovation rarely makes its way by gradually winning over and converting its opponents. It rarely happens that Saul becomes Paul. What does happen is that its opponents gradually die out and that the growing generation is familiarized with the idea from the beginning—another instance of the fact that the future lies with the youth."[1]

Satisfactory explanations of the world of natural phenomena are sufficiently complicated that scientists, at all times and notably over the last several centuries, have learned to live, more or less, with an untidy world picture. They have acquired the aptitude for tolerating a modicum of internal theoretical difficulty and inconsistency. A rock-bottom level of explanatory expertise has seldom in recent times been claimed confidently for a domain of scientific investigation—at least not by a majority of practitioners over an extended period of time. Nevertheless, which particular scientific difficulties contribute to an untidy picture, and which in turn are tolerated, often has been a matter of genuine debate among scientists. Apparently the choices that are made comprise an element of personal judgement. If the matter of judicious personal decision, *viz.*, to tolerate and learn to live with some difficulties and untidiness, were not operative among scientists, then their research programs would degenerate into a wild and silly chase after insignificant side-issues that result in the fabrication of scientific air-castles.

An examination of Mach's life work, I suggest, reveals that he was unusually critical about the world of physics into which he was born, but also that he was overly cautious about the direction in which physics was being thrust during his lifetime. As a potential theoretical scientist, he was, in fact, too negative about his scientific inheritance and too taciturn about the future direction of science to become a Boltzmann, Ostwald, Planck, or Einstein. Mach did manage, however, to raise considerable intellectual commotion that opened the way for strong responses on the part of both scientists and philosophers of science. As for the history of science, he used it wherever he could to illuminate his philosophy of the sciences—although he disliked intensely being associated in any way with the discipline of philosophy and the profession of philosophers. Mach wanted to be looked upon as a scientist who knew how to use history in order to buttress his arguments within the sciences.

What was it that perturbed Mach so profoundly about the physics of his day? What intolerable difficulties threatened to ruin his world picture? Or did he have one? How did his contemporaries treat the same problems? How did they react to Mach's suggestions? The analysis and interpretation of the scientific and intellectual career of Mach, and the study of the acceptance (and rejection) of his ideas furnishes a manageable *terminus a quo* in the domain where science, history, and philosophy criss-cross. What I have in mind is to provide an overall contextual evaluation of Mach as scientist, historian, and philosopher; and in particular, to indicate how these three faces of Mach are related to one another in his work and in the intellectual climate of scientific opinion of his day.

MACH'S EDUCATION

Ernst Mach lived and worked, all but the last three years of his life, in the Austro-Hungarian empire during a time of increasing national self-determination, the evolution of linguistic identity, and the physical and intellectual emancipation of various ethnic groups. During the formative years of his life, the Mach household provided an environment in which unrestrained, critical, and stubborn scientific inquisitiveness and skepticism were nurtured.

At the University of Vienna, Mach studied mathematics, physics, and philosophy, and in 1860 completed a doctoral dissertation on electrical discharge and induction. Working as privatdocent in the laboratory of his teacher, Professor Andreas von Ettingshausen (Doppler's successor in the chair of physics), Mach carried out experimental investigations devised to furnish theoretical support for the controversial law of Doppler; *viz.*, the law that relates the changes of musical pitch and optical frequency to the relative motion of signal and receiver. At the same time he delivered several papers at the Academy of Sciences in Vienna in which the idea of intermolecular vibration was used to interpret gaseous spectra. "Molecular functions" were introduced, as well, to account for resonance in mechanically vibrating systems, the behavior of fluids in relation to density and viscosity, capillary phenomena, and calculation of the principal radii of curvature of liquid surfaces.

While in Vienna, Mach drew most of his income from popular scientific lectures on optics, musical acoustics, and psychophysics. In addition, he gave university lectures on the principles of mechanics and organized a special physics course for medical students. The latter formed the basis for his *Compendium der Physik für Mediciner* of 1863. As in his work on gaseous spectra, this textbook reveals Mach's early adoption of the atomic-molecular theory, the kinetic theory of gases, and a mechanistic interpretation of natural phenomena in general—at least as a working model and hypothesis.[2] At this time in his career Mach obviously was following the atomistic philosophy of physics that was current among most contemporary physicists. However, in the preface and conclusion of the *Compendium,* the inadequacies of the mechanical atomic theory were critically reviewed by Mach.

Even before leaving Vienna, Mach's scientific interests had begun to turn from physics to the physiology and psychology of sensation, and to the new discipline of psychophysics. In the reflections on his life that appear in his *Leitgedanken* (1910) Mach relates that he was motivated to take up the study of the psychology of sensation because he lacked the experimental means for physical investigations. He wrote: "Here, where I could observe my sensations, and against their environmental circumstances, I attained, as I believe a natural *Weltanschauung* freed from speculative, metaphysical ingredients."[3]

As Mach's investigations shifted to the exploration of psychophysical problems associated with vision, audition, and variations in blood pressure, he became convinced that his initial mechanistic and atomistic strategy had served him poorly. A noteworthy reorientation in Mach's scientific research program already is visible in the "Vortrage über Psychophysik" that appeared in print toward the end of 1863. For Mach this new direction in research represented an escape from metaphysical questions and a retreat from mechanism and physical reductionism. He wrote: "For the value of an hypothesis consists mainly therein, that by a kind of *regula falsi* it always leads closer and closer to the truth."[4] By 1863 already, the atomic hypothesis was for Mach a kind of *regula falsi*, but no more than that.

PROFESSOR IN GRAZ, PRAGUE, AND VIENNA

In his twenty-sixth year, in 1864, Mach accepted a professorship in mathematics at the University of Graz, but without the promise of an institute or financial resources for scientific equipment. He therefore drew on his own private funds to secure the research facilities he needed. In 1866, although he was given the title of professor of physics, he continued to pursue physiological and psychophysical problems on aural accommodation, the sense of time, and spatial vision.

Notably important during the years at Graz was Mach's discovery of what later came to be known as "Mach's bands"—a phenomenon that relates the physiological effect of spatially distributed light stimuli to visual perception. A Mach band is seen by an observer when a spatial distribution of light results in a sharp change in illumination at some point. A negative change corresponds to a band brighter than its surroundings in the region of sharp change. A positive change corresponds to a band darker than its surroundings in the region of sharp change. This phenomenon is a physiological effect that has no physical basis. It was the subject of five papers that Mach published between 1865 and 1868. A final paper appeared in 1906. The Mach band effect was rediscovered in the 1950s and has been the subject of considerable investigation ever since.[5]

In 1867 Mach accepted a call to the Charles University in Prague. The years of Mach's professorship in the chair of experimental physics in Prague (1867–95) constitute the most productive period of his life. During twenty-eight years at this teaching post Mach published over a hundred scientific papers. Deeply involved during these years in the theoretical reformation of his views on mechanics and thermodynamics, his journal publications, nevertheless, chiefly exhibit experimental researches.

The most celebrated among Mach's psychophysical investigations carried out in Prague are the studies on the changes of kinesthetic sensation and equilibrium associated with physical movement, acceleration, and change of orientation in the human body. In a steady stream of research papers Mach continued, as well, to examine questions that he had already promoted in Vienna and Graz: experiments on spatially distributed retinal stimuli (referred to above as Mach band studies), monocular stereoscopy, the anatomy and function of the organs of auditory perception, and aural accommodation.

Mach's contributions within the area of more conventional physics certainly were credible. He carried out a wide variety of experimental studies on refraction, interference, polarization, and spectra. He investigated the wave motion associated with mechanical, electrical, and optical phenomena, and notably clarified the longitudinal wave propulsion characteristics of stretched and unstretched glass and quartz rods. He studied the mechanical effects resulting from spark discharge within solids and on surfaces.

Over a period of twenty years, from 1873 to 1893, Mach and various of his collaborators, including his son Ludwig, devised and perfected optical and photographic techniques to study sound waves and the wave propulsion and gas dynamics of projectiles, falling meteorites, explosions, and gas jets. Stimulated by remarks made by the Belgian artillerist Henri Melsens in Paris, Mach in 1881 undertook to study the flight of projectiles by means of photographic techniques that he already had devised for other experiments in his Prague laboratory.[6]

Mach's celebrated 1887 paper on supersonics was published jointly with Professor P. Salcher of the Marine Academy of Fiume in the *Sitzungsberichte* of the Academy of Sciences in Vienna.[7] The experiments described in this classic paper were carried out in Fiume with the support of the Royal Austrian Navy. In this paper the angle, α, that the shock wave surrounding the envelope of an advancing gas cone makes with the direction of its motion, was shown to be related to the velocity of sound, ν, and the velocity of the projectile, ω, as $\sin \alpha = \nu / \omega$. This relationship holds for cases in which $\omega > \nu$. Following the work of Ludwig Prandtl at the Kaiser Wilhelm Institut für Strömungsforschung in Göttingen, in 1907, the angle α was called the "Mach Angle."[8] Recognizing the fact that the value of ω / ν (the ratio of the speed of an object to the speed of sound in the undisturbed medium in which the object is traveling) was becoming increasingly significant in aerodynamics for high-speed projectile studies, J. Ackeret in 1929 in his inaugural lecture as privatdocent at the E.T.H. in Zurich suggested the term "Mach Number" for this ratio.[9] The "Mach Number" was introduced into the English literature by the late 1930s and since the end of World War II has taken on considerable importance in theoretical and fluid dynamics. We may conjecture that the work that has come to be most closely associated with the name of Mach, in all probability, represents contributions that Mach would have conceived to be almost inconsequential when compared with his criticisms of classical mechanics and his experimental innovations in psychophysiology.

While in Prague, Mach published numerous popular scientific lectures and essays of historical and educational import, in addition to major treatises and monographs on conservation of energy (1872), spectral and stroboscopic investigations of musical tones (1873), the theory of the sensation of motion (1875), a critical history of mechanics (1883), and a volume on the analysis of sensations (1886). From 1882 to 1884 Mach was rector of the University of Prague during difficult days while the university split into a German and Czech faculty. In 1887 Mach and Johann Odstrcil published the first of a series of physics textbooks. These textbooks, of which eventually some twenty editions appeared, were written with the help of a number of collaborators, and were adopted with more or less success in Germany and Austria for about four decades. Likewise, demonstration apparatus designed in Mach's laboratory was in use in Prague, Vienna, and Leipzig.

Mach moved to Vienna in 1895 to take a teaching position in

philosophy at the university under the titular head of Professor of the History and Theory of the Inductive Sciences. His *Popular Scientific Lectures,* a later edition of which was dedicated to William James, was first published in English in 1895. His *Principien der Wärmelehre,* dedicated to another American, J. B. Stallo, appeared in 1896.

After only two years at the University of Vienna, Mach suffered a stroke that paralyzed the right side of his body. He managed, after a time of recuperation and with the help of self-imposed strenuous discipline, to continue both to lecture and to write. Mach officially retired from his professorship in 1901, the year in which he was appointed to the Upper Chamber of the Austrian Parliament. His *Erkenntnis und Irrtum* appeared in 1905; *Space and Geometry* in 1906. In 1913 Mach moved with his wife to the country home of his son Ludwig in Vaterstetten, a village in the town of Haar outside Munich. *Kultur und Mechanik* was published in 1915, the year before he died; *Die Principien der physikalischen Optik* appeared posthumously in 1921.

ROLE OF EXPERIMENT AND THEORY

Ernst Mach was a physicist by training and wanted to be acknowledged as such. Nevertheless, he was actively involved as well, for most of his life, in the investigation of problems in physiology, psychology, the history of science, and the philosophy of science. Above all, he was singularly alert to the significance of exploring problems in areas where physics, physiology, and psychology overlap. He was, in his day, acknowledged as an outstanding and ingenious experimentalist. Although his status as a theoretician was never acclaimed in that way, there is no doubt that he was a keen and inventive *critic* of scientific theories. Concerning theories, we might say that Mach the positivist was predominantly a negativist, i.e., consistently and formidably critical.

From early on, Mach's interests and efforts were permeated with a yearning inquisitiveness to observe natural phenomena, free from abstraction, in their essential unadulterated complexity. At the University of Vienna he discovered that he wished to become a physicist, but a physicist whose activity would not be limited by the academic straitjacket of conventional physics. Nor was Mach attracted to large syntheses and over-arching reductive world views. He did subscribe to a monism of scientific method, but as for subject matters and scientific disciplines and the use of theoretical principles to describe those subject matters and disciplines, Mach lived with, and fostered, a thoroughgoing pluralism of conceptual ideas. There are two sayings that would seem to fit in especially well with Mach's attitude toward intellectual lumping and splitting—the first from a source unknown to me, and the second from Schiller. First: "Stupidity must ponder the whole; intelligence can afford to confine itself to the relevant." Second: "*Mit der Dummheit kämpfen Götter selbst vergebens.*"

In 1916, in his obituary for Mach, Einstein wrote: "The unmediated pleasure of seeing and understanding, Spinoza's *amor dei intellectualis,* was so strongly predominant in him that to a ripe old age he peered into the world with the inquisitive eyes of a carefree child taking delight in the understanding of relationships."[10] Einstein asserted that even when Mach's scientific investigations were not founded on new principles, his work at all times displayed "extraordinary experimental talent." Where Mach's philosophy seemingly got in the way of his science, colleagues nevertheless praised his scientific intuition and skill. In 1927 Wilhelm Ostwald wrote: "So clear and calculated a thinker as Ernst Mach was regarded as a visionary [*Phantast*], and it was not conceivable that a man who understood how to produce such good experimental work would want to practice nonsense [*Allotria*] which was so suspicious philosophically."[11] Arnold Sommerfeld referred to Mach as a brilliant experimentalist but a peculiar theoretician who, in seeking to embrace the "physiological" and "psychical" in his physics, had to relegate the "physical" to a less pretentious level than physicists were accustomed to expect from a colleague.[12]

Whenever Mach encountered theoretical statements that in any way struck him as implausible—and that happened rather frequently—he became a detective, the Sherlock Holmes of science. In *The Adventure of the Priory School,* Watson says to Holmes: "Holmes, this is impossible." Holmes to Watson: "Admirable! A most illuminating remark. It *is* impossible as I state it, and therefore I must in some respect have stated it wrong."[13]

"Being stated wrong," as Newton for example had it wrong, was for Mach the *incipit* for critical analysis. To be in the wrong here was, in a sense, the best whetstone of ingenuity. What is more, Mach was actually dubious, leery, uncomfortable about "theoretical" statements in general—the more general,.the more uncomfortable. Mach felt at home with facts, ideas, the adaptation of ideas to facts (i.e., observation), and the adaptation of ideas to other ideas (i.e., theory)—but not much more.[14] *Tatsachen, Gedanken, die Anpassung der Gedanken an die Tatsachen (Beobachtung), die Anpassung der Gedanken aneinander (Theorie)*—that was Mach's world. He was in the habit of treating facts in the style of Sherlock Holmes, who said in *A Scandal in Bohemia* (1891): "I have no data yet. It is a capital mistake to theorise before one has data. Insensibly one begins to twist facts to suit theories, instead of theories to suit facts."[15]

Among Mach's principal contributions and observations that relate to theory are to be mentioned his studies on visual, aural, and kinesthetic sensation, his views on mechanics, thermodynamics, optics, and molecular spectroscopy, and the analysis of wave propulsion in various media. Mach's theoretical discussions on such topics often were seen by others to be unnecessarily interwoven with comments extraneous to physics proper. Indeed, both the subject matter of his inquiries into physics, i.e., his conception of the discipline, and his approach to physics were generally too far off the beaten path to merit the acclaim of his contemporaries.

For example, Mach's obstinate repudiation of the atomic theory went far beyond that of his contemporaries, who, for the most part, even when noncommittal about atoms and molecules as existential entities, seldom questioned the remarkable value of the theory as a powerful hypothesis. We know that Mach, openly at least, became less denunciatory about the theory as he became older, but it is not certain that he ever abandoned his dogmatic anti-atomistic mood even after the new discoveries at the turn of the century had furnished rather convincing evidence that he was dead wrong. At most Mach treated the atomic theory and "the artificial hypothetical atoms and molecules of physics and chemistry" as "traditional intellectual implements" of the discipline. He wrote: "The value of these implements for their special, limited purposes is not one whit destroyed. As before, they remain economical ways of symbolizing experience. But we have as little right to expect from them, as from the symbols of algebra, more than we have put into them, and certainly not more enlightenment than from experience itself."[16]

Although Planck at one time was attracted to Mach's historico-philosophic writings, he became the severest censor of his critical positivism. The origins and flowering of this Mach-Planck antagonism

369

can be followed in the literature in connection with their respective views on thermodynamics.[17] For Planck the principle of conservation of energy was a true law *in* nature, a reality independent of man's views or man's existence. For Mach the principle assumed the role of a maxim or convention useful for correlating a large class of natural phenomena. It was rooted in anthropomorphic reasoning and was seen to be related biologically to economy of effort that is conducive to survival. For Mach "energy" was no more than a plausible, and powerful, concept, like force, space, and temperature.[18]

Ludwig Boltzmann felt that neither Mach nor Planck had comprehended the statistical interpretation of the concept of entropy and the second law of thermodynamics. After 1900 Planck came around, at first slowly and reluctantly, then with a vengeance, to adopting Boltzmann's views on the subject. Mach persisted in questioning the merits of atomic, mechanical, and statistical explications of the second law. He chose in his own distinctive way to stress the more simple and purely postulational footing of the laws of thermodynamics. In 1895 in his *Wärmelehre* Mach wrote:

> The mechanical conception of the Second Law through the distinction between *ordered* and *unordered* motion, through the establishment of a parallel between the increase of entropy and the increase of unordered motion at the expense of ordered, seems quite *artificial*. If one realizes that a real analogy of the *entropy increase* in a purely mechanical system consisting of absolutely elastic atoms does not exist, one can hardly help thinking, that a violation of the Second Law—and without the help of any demon— would have to be possible, if such a mechanical system were the *real* foundation of thermal processes. Here I agree with F. Wald completely, when he says: "In my opinion the roots of this (entropy) law lie much deeper, and if success were achieved in bringing about agreement between the molecular hypothesis and the entropy law, this would be fortunate for the hypothesis, but not for the entropy law."[19]

CRITICAL REFLECTIONS ON MECHANICS

The historical and critical analysis of the science of mechanics, in relation to other sciences, constitutes the hard core of ideas that belong to Mach's philosophy of science. Two of his treatises, notably the youthful monograph of 1872 on the conservation of energy, and the more mature and comprehensive history of mechanics of 1883, gave rise

370

to highly controversial discussions on the scientific, historical, and philosophical foundations of classical physics. Among the mathematicians, physicists, and philosophers who were drawn into these debates on mechanics toward the end of the nineteenth century, we may cite here such persons as Heinrich Hertz, Karl Pearson, Ludwig Boltzmann, August Föppl, A. E. H. Love, J. B. Stallo, W. K. Clifford, Emile Picard, Henri Poincaré, Pierre Duhem, Hugo H. Seelinger, Giovanni Vailati, and Philip E. B. Jourdain. As late as 1916, however, Einstein could still assert that the brilliant ideas in Mach's *Die Mechanik* had not yet become the common property of physicists.

Newton's mechanical views were presented and analyzed in considerable detail in *Die Mechanik*. Mach recognized in Newton two outstanding characteristics of scientific greatness—an imaginative grasp of the cardinal elements of experience of the world, and truly extraordinary intellectual powers of generalization. Although Mach indulged in lavish praise for the clarity of presentation of ideas in the *Principia,* he also emphasized the need for critical, rational reformulation of a number of well-entrenched mechanical concepts. For example, Mach proposed that the concept of inertial mass be identified not as an intrinsic property of objects but as a quantity defined by the dynamical coupling of objects with the rest of the universe. Newton's conception of mass, as quantity of matter, he suggested, could be replaced by an "arbitrarily established definition," namely, that "all those bodies are bodies of equal mass, which, mutually acting on each other, produce in each other equal and opposite accelerations."[20] According to Mach, the adoption of the new definition of inertial mass would relegate Newton's principle of reaction to a position of redundancy—a conclusion Mach would have promoted for the sake of economy of thought.

Mach's aim was to eliminate all scientific propositions from which observables cannot be deduced. Thus he criticized Newton's views on absolute space, time, and motion on the grounds that such quantities could not be related in any way to experimental observations. He proposed, instead, that the motions of bodies be referred to all observable matter in the universe at large. Mach wrote: "When we reflect that we cannot abolish isolated bodies . . . , that is, cannot determine by experiment whether the.part they play is fundamental or collateral, that hitherto they have been the sole and only competent means of the orientation of motions and of the descriptions of mechanical facts, then it will be found expedient provisionally to regard all motions as determined by these bodies.''[21] Accordingly, on Mach's terms, a body in an empty universe can have no inertia, since the inertia of any body or system of bodies is a function that relates that body or system of bodies to the rest of the universe, including the most distant parts of the interacting material system.

Einstein drew attention to this principle in 1918 in a short paper on general relativity,[22] where he introduced the expression ''Mach Principle'' (*Machsches Prinzip*) to accentuate a generalized form of Mach's claim that the inertia of an isolated body is without meaning; that inertia is a function of the reciprocal action of bodies; that the inertial frame is defined by the mass distribution in the universe; and that the inertial force on a body corresponds to the gravitational interaction of distant matter on the body. This choice, even if framed conditionally, implies a material system that mathematically approximates absolute space.

The interpretation that Einstein placed upon Mach's critique of Newtonian mechanics, as seen within Riemannian field theory, thus became a motivating factor for the development of gravitational theory; this, in spite of the fact that Einstein eventually discovered that the Mach principle did not hold for his new theory. Einstein saw that the metric of space-time could be specified by the apportionment of matter and energy; that the curvature of space, and from this the motion of bodies, could be determined from the matter in space. The motion of bodies thus was seen to depend on the surrounding masses (including the stars), and not, as Newton had assumed, on the tendency of bodies to maintain their direction of motion in absolute space.

Einstein discovered that he was unable to furnish a satisfactory mathematical expression for the Mach principle. In that his field equations proved to have no solution, no metric in matter-free space, he had found some support for the Mach principle. While, therefore, he had succeeded in absorbing the compatible part of the Mach principle into the general theory of relativity—based on the equivalence principle and the conception of covariance under general transformations of space-time coordinates—he was not willing to adopt the specification, the limiting factor, of finite space-time boundary conditions implied in the Mach principle. Actually, he found that he could develop field equations that gave correct solutions only by adding the so-called cosmological term.

As is well known, the literature on the scientific and philosophical explications and reformulations of the Mach principle is large and growing. To this day, proofs of its consistency with gravitational theory and other cosmological models continue to be put forward, are revised, and are waved aside. Efforts to demonstrate where the principle, or some part or particular formulations of the principle, is or is not applicable persist. Indeed, there is poor consensus on the mathematical formulation of the principle. The expression "Mach principle", anachronistic in its use, serves as a collective term that has little in common with Mach's 1883 formulation. What is historically significant is the logical connection between Mach's critique of mechanics, including the formulation of the "Mach principle," and Einstein's intellectual route to the theory of relativity. In the *Autobiographical Notes* of 1946 Einstein mentioned that he saw "Mach's greatness in his incorruptible skepticism and independence"; that these traits in Mach had drawn him to Mach's thought in his early years. But Einstein also alluded in the *Autobiography* to Mach's epistemological position as one "which today appears to me to be essentially untenable."[23]

All this is not to show that Mach ever supported the theory of relativity. On the contrary, he rejected it categorically. In *Die Principien der physikalischen Optik,* written in 1913 but not published until 1921, Mach remarked in the preface that he was "gradually becoming regarded as the forerunner of relativity" and that philosophers and physicists were carrying on a crusade against him. He wrote:

> I have repeatedly observed that I was merely an unprejudiced rambler, endowed with original ideas, in various fields of knowledge. I must, however, as assuredly disclaim to be a forerunner of the relativists as I withhold from the atomistic belief of the present day. The reason why, and the extent to which, I discredit the present-day relativity theory, which I find to be growing more and more dogmatical, together with the particular reasons which have led me to such a view—the considerations based on the psychology of the senses, the theoretical ideas, and above all the conceptions resulting from my experiments—must remain to be treated in the sequel.[24]

The sequel—apparently an attack on Einstein's theory of relativity—was never committed to print.

Although the criticisms of Newtonian dynamics are the most frequently discussed sections of Mach's treatise on mechanics, there are other noteworthy parts of this work that admirably illustrate his use of the history of science as a point of departure for critical reflections on mechanics. Mach believed that the critical, historical, and psychological examination of the roots of science would expose metaphysical ambiguities, objectionable anthropomorphisms, and narrow, artificial mechanical interpretations of science. He also believed that the history of science would provide insights concerning the function of so-called instinctive knowledge, the role of memory, the comparative evaluation of the origins with the mature status of scientific ideas, the contrast between psychology of discovery and logic of discovery, and the authority that induction and deduction derive from experience.

Mach sought to show that what constitutes an acceptable proof at any one time is neither universally agreed upon nor unambiguously clear. For example, in the analysis of Simon Stevin's treatment of the principle of the inclined plane (based on the condition of equilibrium of an endless uniform chain on a triangular prism), Mach indicated that the mental steps of proof that were adopted rested upon an argument that proceeds from the general to the specific case. By contrast, in his treatment of the principle of the lever, Archimedes' proof moves from the specific to the general case. Again, when Daniel Bernoulli claimed to have proved the proposition of the composition of forces as an experience-independent geometrical truth, Mach recognized this ''proof'' as no more than a questionable reduction of what is easy to observe from what is more obscure and more difficult to observe—a strange model of proof that nevertheless makes it possible to cover up the precise steps in the proof. Mach thought it would be even more valuable to investigate the historical circumstances as to why Bernoulli's demonstration had been considered acceptable as a proof in his day. Mach supposed, in this case as in many others, that the demonstration had been founded on an appeal to so-called instinctive knowledge that is supposed to be distinguished from experimental knowledge.

Mach was always on the lookout for historical examples that would support the decisive role of experience in scientific investigations. Thus he maintained that though instinctive knowledge could serve a useful heuristic function, it invariably contained important concealed elements of experience. He was intent on eliminating concepts that had no counterpart in experience. In the literature this idea was later characterized as the *Mach Criterion,* according to which only such propositions enter into theory from which statements about observable phenomena can be deduced. Mach maintained that scientific proofs that are not bound to experience serve either as a screen to conceal spurious rigor or derive their sanction from the high authority of purely instinctive cognition.

One other example will suffice to illustrate the strong empirical component in Mach's view of theoretical science. Mach demonstrated to his own satisfaction that a whole group of statical and dynamical principles—for example, exclusion of the *perpetuum mobile,* principle of virtual work, d'Alembert's principle, Gauss's principle of least constraint, Maupertuis's principle of least action, Hamilton's principle—each can be derived from any other, provided sufficient perseverance and mathematical ingenuity are exercised. That is, Mach showed that the above-mentioned principles are connected in such a way that if any one of them is accepted as true, then all the others can be deduced from that one by mathematical reasoning alone. The point he wanted to drive home was that there was no possible way to establish the validity of any one of them except through an appeal to experience: for example, to the experimental discovery of the impossibility of designing a machine to supply unlimited quantities of mechanical work without the exertion of effort. To summarize Mach's point of view: the behavior of nature cannot be predicted solely on the grounds of so-called self-evident propositions unless the propositions are themselves drawn from experience.

The history of the treatment of problems in statics (using both static and dynamic arguments) and the history of the treatment of problems in dynamics (using both dynamic and static arguments) are analyzed at some length by Mach. Mach saw that, historically considered, statics preceded dynamics, whereas from a logical or conceptual point of view, statics could be reduced to dynamics. In this case, what clear meaning, if any, was to be attached to saying that one scientific principle is more fundamental, more basic, than another? The wider generality of one principle over that of another, clearly, was no sufficient warranty to accept it as more basic. Historical priority in the enunciation of one principle over that of an alternative formulation should not necessarily determine logical status. When severed from the context of the scientific problem, the terms "basic" and "fundamental" simply had no fixed significance for Mach. The practical solution to the understanding of mechanical principles, Mach felt, was to learn by experience in scientific problem-solving.[25] Mach wrote: "The most important result of our reflections is . . . *that precisely the apparently simplest mechanical theorems are of a very complicated nature: that they are founded on incomplete experiences, even on experiences that never can be fully completed; that in view of the tolerable stability of our environment they are, in fact, practically safeguarded to serve as the foundation of mathematical deduction; but that they by no means themselves can be regarded as mathematically established truths, but only as theorems that not only admit of constant control by experience but actually require it.*"[26]

Let us summarize Mach's critical position in relation to the science of mechanics. His position, to say the least, is an ambiguous one. Mechanics plays a central role in physics. The critical examination and purification of its concepts is therefore a crucial matter for which the professional physicist is responsible. On the other hand, from a logical, economical, and practical point of view, there is no credible reason for patronizing either the historical legacy of mechanics or the pertinence of mechanics for the growth of other scientific disciplines. The grandiose position of mechanics in history does not guarantee the relevance of mechanical principles for the other natural sciences. Mach felt that mechanics had little to offer to heat theory, electricity, and optics; but its influence on psychology, he conjectured, was decidedly pernicious. The restless, defensive, and resolute thrust that symbolizes Mach's attempts to exploit experimental physiology and psychology—in part, by means of physical techniques—is suggestive of the authenticity with which he was searching for novel problems and perspectives beyond the conventional physics of his day; nonetheless, without ever being able to relinquish his hold on professional physics or being able to lay down experimental physics as such. In all of this he seems to have valued philosophy—or rather, critical reflective thought—for its utility in helping him to formulate, but not to initiate, his scientific endeavors.

HISTORY OF SCIENCE

In my opinion all of Mach's historico-critical writings support the thesis that the history of science was for him a tool to interpret and illuminate epistemological problems in the philosophy of science that he was perplexed about as a physicist.[27] The extent to which Mach puzzled about those particular scientific issues into which he was impelled by his own methodological assumptions and preoccupations is of course a fascinating question; it is not one that is easily settled from a study of his published works.[28]

It is evident that Mach did not engage in historical investigations with the aim of reconstructing the history of science *per se,* or *Wie es eigentlich gewesen.* His historical studies were problem-oriented. He made no effort to write history in systematic fashion. The analysis of scientific revolutions, and the organizational, institutional, and sociocultural framework of science were not a prominent feature of his histories. For Mach, history was not conceived chronologically, biographically, or even topically. The subject matter that served as support for his historical analyses was virtually limited to original scientific writings. He had a good command of Latin, Greek, French, Italian, and English, and he examined the primary sources in the original languages of the authors. He was acquainted, besides, with a secondary literature of broad scope, and carried out—as an examination of his personal correspondence reveals—an exchange of ideas with an international constellation of scholars. Although he put forward no systematic philosophy of science, he was, nevertheless to become one of the writingest scientists of his time.

Mach's opinions about many things in science were implemented in his historical writings by the belief that the most important step toward having definite and worthwhile opinions was to have expressed them clearly and to have done so with an eye to their historical, conceptual context. The epigraph to Boltzmann's lectures on the principles of mechanics reads: *"Bring' vor, was wahr ist; Schreib' so, dass es klar ist; Und verficht's, bis es mit dir gar ist!"*[29] That might have been an appropriate motto for Mach's approach to the history of science.

Introspection and puzzlement about the nature of physics, and serious thoughts about the responsibility of the physicist toward newly developing areas of scientific investigation—these were at the heart of the concern of the youthful Mach. The least that he hoped to gain from his historical studies of the foundations of physics was to acquire some perspicuity about how to find a meaningful direction for his own life's work; and about how to approach the teaching of physics so as to instill in the minds of students a critical attitude toward their chosen professional discipline. It is doubtful that Mach saw himself playing the role of a scientific investigator whose work was meant to serve as a practical testing ground for any preconceived philosophical point of view. Rather, I wish to argue that the kinds of epistemological questions that Mach thought he was pursuing were the result of a stubborn curiosity about the inherited scientific tradition of physicists.

As a young man, in 1863, Mach was convinced that his scientific inquisitiveness and his effectiveness as a teacher might be sharpened by the study of the history of science. He argued that students should not be expected to adopt, as self-evident, propositions that had cost several thousand years of thought. In 1872, in his first major historical work— *Die Geschichte und die Wurzel des Satzes von der Erhaltung der Arbeit*—he conjectured that there is "only one way to [scientific] enlightenment: historical studies!" The investigation of nature, he believed, should be backed up with a special classical education that "consists in the knowledge of the historical development . . . of science." It was not the logical analysis of science but the history of science that would encourage the scientist to tackle problems without engendering an aversion to them. There were two paths that the scientist might follow in order to become reconciled with reality: "Either one grows accustomed to the puzzles and they trouble one no more, or one learns to understand them with the help of history and to consider them calmly from that point of view."[30]

In *Die Mechanik* Mach tells us again why it was that he undertook to study the history of science:

> . . . Not only a knowledge of the ideas that have been accepted and cultivated by subsequent teachers is necessary for the historical understanding of a science, but also—the rejected and transient thoughts of the inquirers, nay even apparently erroneous notions, may be very important and very instructive. The historical investigation of the development of a science is most needful, lest the principles treasured up in it become a system of half-understood prescripts, or worse, a system of *prejudices*. Historical investigation not only promotes the understanding of that which now is, but also brings new possibilities before us, by showing that which exists to be in great measure *conventional* and *accidental*. From the higher point of view at which different paths of thought converge we may look about us with freer vision and discover routes before unknown.[31]

PHILOSOPHY OF SCIENCE

In Mach's small volume on the history of the principle of conservation of energy of 1872 we are confronted with problems that were to become a relatively permanent part of his intellectual baggage. Mach unpacked and repacked these problems a number of times in his major historico-critical treatises on mechanics, optics, and heat theory, as well as in his *Analyse der Empfindungen* and *Erkenntnis und Irrtum*. The issues that were of perennial importance in Mach's thought, and that we encounter repeatedly in his published writings and in his private correspondence, would be the following: the utility and terms of reference of scientific theories; the importance of the physiology and psychology of sensation for the epistemological analysis of the natural sciences in general; the role of economy of thought; the internal ambiguities connected with Newtonian mechanics; the sterility of the atomic theory; and the difficulties associated with classical causality, physical reductionism, mechanism, materialism, and all forms of metaphysical speculation.

It is informative to identify the leading kinds of questions that Mach tried to elucidate historically: How had scientists acquired their current concepts and theories? Why had the concepts and theories been received in their given form rather than in a form that was logically more plausible or mathematically and conceptually more elegant? What historically fortuitous circumstances had entered into the choices that had fixed the modes of representing science? What patterns of analogical or reductivist reasoning or concept borrowing had contributed to advances in the sciences? Had there been, at different times in history, different views and practices as to what constitutes evidence, verification, conclusive proof for a scientific theory?

Mach's attempts to elucidate such questions with historical examples constitutes the main thrust of his histories of mechanics, heat theory, and optics. The general lines of reasoning that are most consistently pursued are such as the following: the discipline of physics includes but a small part of the total body of the natural sciences, the intellectual tools of physics—concepts and methods of investigation—are highly specialized; the various branches of science are so inexpressibly rich in content, and so fruitfully explored from the special points of view of the different sciences, that no set of intellectual tools borrowed from physics can do justice to them. Accordingly, the history of the different sciences (i.e., different ways of investigating the phenomenal world) are inexhaustibly diversified. On examination of the evolution of any particular branch of science, it becomes evident that the metamorphosis and not the fixity of concepts is the rule.

Scientific concepts and theories are perennially obsolete. It therefore becomes meaningless to speak of their absolute truth status. The great value of scientific history is that it brings into focus the circumstances, training, peculiarities, and biases of the scientist *vis-à-vis* his contribution. Compare, for example, the points of view and behavior, within a given scientific context, of physicist, physiologist, and psychologist.

Mach's historical analysis suggests that the form a scientific concept takes hinges in large part on the caprices of history and the contextually conditioned standards of cognitive behavior that have been adopted by scientists at a given time and place and by a given discipline. Biologically annexed to the cognitive faculties of the scientist is the desideratum for abstraction and the need for economy of thought. Recognizing no conceivable rock-bottom level of knowledge about the nature of things, and alive to the many potential options for shaping and functionally relating phenomena, Mach sees the scientist as one necessarily driven to drastic abstraction and wide generalization—in order to achieve economy of effort, *viz,* thought.

Mach rightly perceived that scientists often deliberately aim in the course of their work to realize logical congruity, simplicity, polished formulation, completeness, and all the other subtle traits that commonly are supposed to characterize science at its best. Mach took special pains to emphasize, however, that the history of science normally does not so visibly exhibit the elegant, logical, and clean-shaven face of the orthodox scientific model. Scientific extrinsicality, in fact, invariably shows up as a prominent component of scientific concepts, laws, hypotheses, theories—when examined within the setting of their historical genesis. Within a given situation over a period of time the historically acquired comes to be philosophically affirmed, reified, given existential status. According to Mach, it is the duty of the scientist to recognize and counteract the cunning ways by which scientific constructs acquire the status of philosophical necessity in place of historical contingency.

Mach's emphasis—arbitrariness, fortuitousness, historical contingency, the special points of view of different scientific investigators, and the many open-ended alternatives for mapping the world of nature—did not sit well with many of Mach's readers. Small wonder that he was attacked by realists from all sides—Planck, Boltzmann, Einstein, Lenin—and that he was called many names: positivist, empiricist, phenomenalist, sensationalist, operationalist, atheist, subjective idealist.

ANALYSIS OF SENSATIONS

We have seen that Mach was motivated by training and predilection to mount a campaign that would expose the epistemological roots of science. He believed *ab initio* that such a campaign could not be promoted exclusively by analyzing the work of the scientist in isolation from his behavior. Where this bias shows up most prominently is in the use to which he put the physiology and psychology of sensation in order to offset the mechanistic physicalism of his day.

Mach's accentuation of scientific laws as compendious, descriptive statements of observation gave rise to a barrage of criticisms directed toward his "narrow-minded" conception of science, especially on its theoretical side. One of the harshest, most notorious, most influential, and most commented-upon critiques of Mach's empirical psychologism was Lenin's *Materialism and Empiriocriticism,* published in Moscow in 1909.[32] The work was a scalding condemnation of Ernst Mach, Richard Avenarius, Bishop Berkeley, and all the rebellious young Russian Machists who were toying with a philosophy for integrating some of the doctrines of Machist positivism into Marxist dialectical materialism. Thus had they planned to keep step with the current philosophical attempts to resolve the *fin de siècle* "crisis in physics"— the *banqueroute de la science,* so-called. Criticisms of Mach's positivism, by Lenin and others, were directed in large part at the subjective idealism implied in Mach's way of analyzing sensations, and in his support of a theory of knowledge based on the study of biological behavior. This we must explain.

The views exhibited in Mach's *Beiträge zur Analyse der Empfindungen* of 1886 and *Die Analyse der Empfindungen und das Verhältnis des Physischen zum Psychischen* of 1900 developed from the deep persuasion, as he tells us in the prefaces to these works, "that the

381

foundations of science as a whole, and of physics in general, await their next greatest elucidations from the side of biology, and especially from the analysis of sensations.''[33] In various revisions of the 1900 treatise, the 1901 edition of which was dedicated to Karl Pearson, Mach interpreted his ideas on how experience as a composite of sensations may be analyzed. The elements of sensation—physical, physiological, and psychological—constitute man's only store of knowledge about the world. Colors, sounds, temperatures, pressures, spaces, times—these are the sensations. They are functionally related, and come to be associated with mind, feelings, volition. Certain associations of sensations are more visible, become stamped on the memory, and thus enter into expressions conveyed by language. Actually, events, observed phenomena, are received as a unity of sensations that embrace both object and subject of the sensations, sensed and senser.

The problem of object-subject dualism vanishes in Mach's analysis of sensations. In experience, the elements of sensation are not received as independent quanta, so to speak. They are interrelated, but for the sake of analysis Mach differentiates the elements that belong to three overlapping categories. The external (*äussere*) elements, A, B, C . . . , represent physical bodies and define the discipline of physics. The internal (*innere*) elements, K, L, M . . . , represent human bodies and define the discipline of physiology. The interior (*innerste*) elements α, β, γ . . . , represent the ego, memory, volitions, etc. and define the discipline of psychology. The elements A, B, C . . . , are modified and conditioned by K, L, M . . . , and α, β, γ . . . , so that the sense organs in different situations receive different sensations and perceptions. Thus the differences that distinguish physical from physiological or psychological investigations cannot be attributed merely to differences in subject matter. Rather, these disciplines represent different methods of investigating a given subject matter. For example, to focus by means of the tools of the discipline of physics on a given situation represents but one of several methods of cognitive organization of the elements that are given in nature.

What distinguishes physics from physiology and from psychology is not the facts but the points of view employed in investigating the facts. The ''facts'' are not unvarnished ''givens'' but a complexity of elements from among which the senser chooses those that he wishes to describe and examine. It was in Mach's opinion only in the logical part of the mind, not in nature, that one thing can be seen so definitely to exclude another. The arena of Mach's action was located in nature and not in logic.

Psychophysiologically structured, the process of cognition that Mach subscribes to consists of the adaptation of thoughts to facts (i.e., *Boebachtung*) and the adaptation of thoughts to other thoughts (i.e., *Theorie*). For the percipient the given world is totally bound up with the sensations. The Kantian *Ding-an-sich,* that is assumed to persist after all of the qualities of a body are removed, is an illusive figment of the imagination. So-called objects of our experience, such as things, bodies, matter, are all mental symbols that represent groups of elements, complexes of sensations. Objects are nothing apart from the sum of perceived attributes in a given experiential situation. To present a rich description of phenomena, it suffices to supply a more or less complete statement of the most relevant functional dependencies that can be recognized and studied. Whereas a scientist may engage in a variety of different modes of analysis of his subject matter, he cannot from the manipulation of the empirical data predict with certainty what nature will be like in a novel situation. This is not to suggest that the scientist refrain from making predictions about what nature is like in the empirically unknown, but rather that experience is ever the final arbiter concerning the nature of things.

The principle of psycho-physical parallelism, invoked by Mach, follows on the supposition that all sensations can be studied from the point of view of physics, physiology, and psychology. There are no sensations exclusively physical or physiological or psychological. Thus an investigation from a specific (discipline) point of view is specified by the choices that are made from among the virtually limitless connections between the elements that are conceivable. Given psycho-physical parallelism as a principle, and formulated to correlate with the analysis of sensations, how was Mach to stimulate the fruitful cooperation of physical/biological research programs? His answer, at least provisionally, was that teleological motives were of great heuristic value in formulating significant physics-biology–linked questions.

In this case Mach did not have in mind what now goes under the heading of biophysics. To understand Mach's position here we must once more examine his doctrine of sensations, but now from the standpoint of how they are to be related to animal behavior, and how explained in terms of biological evolution. According to Mach, the source of all science lies in the demands of life. The animal body is a comparatively steady organism in terms of experienced sensations, primarily of touch and sight. Physical knowledge, at most, sketches, but sometimes anticipates, knowledge about combinations of elements of sensation. This is because the phenomenal world is incredibly heterogeneous and interrelated.

Biological organisms cannot adapt to this complex environment, i.e., cannot survive, without in some way transforming the heterogeneous sensations into a convergent, manageable, homogeneous environment that promotes self-preservation. Biological organisms, over long time spans, have managed to develop means of scanning heterogeneous stimuli and integrating them to the advantage of purposeful preservation-conducive behavior. Mach postulates that this takes place by a sensory mechanism—for example, tactile, aural, thermal—that excites only a single nerve path, which in turn communicates to the brain and thus initiates voluntary behavior. Through habituation the voluntary behavior becomes involuntary. The net reward is that unorganized stimuli are replaced by anatomically and symmetrically organized responses, thus contributing to the organism's evolutionary acquisition of control over the stimuli that nature provides.

This brings us back again to Mach's principle of economy of thought, for it too, we see, is buttressed by evolutionary arguments that smile on acts of survival related to the minimum exertion of effort. The principle of economy of thought (*Denkökonomie*) was first introduced in Mach's 1868 lecture on "The Form of Liquids," and then reiterated in his *Geschichte und Wurzel* of 1872. In the former, Mach compares the principle of least surface for liquids to the miserly, but intelligent, mercantilist principle of a tailor working with the greatest saving of material. And he adds: "But why, tell me, should science be ashamed of such a principle? Is science . . . as a maximum or minimum problem . . . itself anything more than a business? Is not its task to acquire with the least possible work, in the least possible time, with the least possible thought, the greatest possible part of eternal truth?"[34] Again, in his lecture of 1882 "On the Economical Nature of Physical Inquiry," and in a chapter of *Die Mechanik,* Mach discusses his principle of *Denkökonomie* in connection with mathematical reasoning, causality, teleology, evolution, and psychic phenomena.[35] Charles S. Peirce (who liked the expression, too, but gave it a different meaning) referred to the diaphanous ubiquity of "economy of thought" in Mach's writings as follows: "Dr. Ernst Mach who has one of the best faults a philosopher can have, that of riding his horse to death, does just this with his principle of Economy in science."[36]

CONCLUDING REMARKS

Mach was a physicist. He endorsed psycho-physiology as the science of the future. To judge from contemporary reactions, as well as from more recent ones, Mach's posture in these matters was not a persuasive one. Can we reconstruct Mach's rationale for adopting so gauche a position? Was Mach promoting the role of psycho-physiology as bene-factor to the physics discipline? Was he suggesting that psycho-physiology was pregnant with ideas, concepts, and methodological cures that would help to initiate advances in physics? Or was it that problems characteristically psycho-physiological were waiting to be solved by the traditional methods of physics? The first alternative, *viz,* psycho-physiology in the service of physics, or the psycho-physiological reconstruction of physics—whatever that may entail—is an extraordinarily revolutionary, if not subversive, option, concerning which Mach had too little to say specifically to be informative. On the other hand, could Mach possibly have chosen the latter alternative, a position in which physics is called upon to reinforce, shore up, or rescue psycho-physiology? This emphasis either encourages physical reduc-tionism, toward which Mach was ill-disposed, or else promotes psycho-physics in the style of Gustav Fechner. Alas, Fechner treated the physical and psychical as two completely different (natural) sides of the same reality. This was a view that Mach could not condone. He treated the physical and the psychical as one reality seen from different points of view. At any rate, Mach apparently adopted different approaches to suit the occasion. That is, he played up the role of psycho-physiology in suggesting new problems and directions for physics, i.e., physics *qua* physics but enriched by stimuli and hints coming from psycho-physiology. On the other hand, he played up the role of physics to decorate and improve psycho-physiology by making use of the intellec-tual and scientific equipment borrowed from traditional physics. To be more specific, Mach was inclined philosophically toward the former (psycho-physiology to the rescue of physics), but he leaned experimentally toward the latter (physics to the rescue of psycho-physiology). Mach's contributions to the study of aural, optical, and kinesthetic sensation, as mentioned in the above section on theory and experiment, would support this conclusion.

Though a mountain of literature touches upon Mach's scientific philosophy, he made no claim to being a philosopher. In 1886 he wrote in his *Analysis of Sensations:* "I make no pretension to the title of philosopher. I only seek to adopt in physics a point of view that need not be changed immediately on glancing over into the domain of another science; for, ultimately, all must form one whole."[37] Again, in the preface to *Erkenntnis und Irrtum* Mach mentioned in 1905 that without claiming to be a philosopher, the scientist has a strong urge to satisfy an almost insatiable curiosity about the origins, structure, process, and conceptual roots of science. Opposed to being called a philosopher, Mach nonetheless was not frightened away by methodological and epistemological questions. The shift in Mach's emphasis from physics to the philosophy of science, by way of the history of science as the intermediary stimulus, is clearly in evidence in his *Erkenntnis und Irrtum.* This volume contains the most illuminating exposé of Mach's philosophical position. Unfortunately it is seldom read nowadays, and almost never cited by historians of science.

We know that Mach's critics and opponents on occasion drew attention to "*die Machsche Philosophie.*" In reply Mach wrote:

Above all there is no Machist philosophy. At most [there is] a scientific methodology and a psychology of knowledge [*Erkenntnis-psychologie*]; and like all scientific theories both are provisional and imperfect efforts. I am not responsible for the philosophy which can be constructed from these with the help of extraneous ingredients. . . . The land of the transcendental is closed to me. And if I make the open confession that its inhabitants are not able at all to excite my curiosity, then you may estimate the wide abyss that exists between me and many philosophers. For this reason I already have declared explicitly that I am by no means a philosopher, but only a scientist. If nevertheless occasionally, and in a somewhat noisy way, I have been listed among the former then I am not responsible for this. Of course, I also do not want to be a scientist who blindly entrusts himself to the guidance of a single philosopher in the way that Molière's physician expected and demanded of his patients.[38]

Mach conceived of science primarily as "the compendious representation of the actual." In 1872, early in life, he wrote: "The object of
science is the connection of phenomena; but the theories are like dry
leaves which fall away when they have long ceased to be the lungs of the
tree of science."[39] This sterile conception of theory was never abandoned by Mach, and it is at this point that his thought reveals so gross an
obliquity and misjudgement. He mightily underestimated the potency
and compass inherent in the synthetic conceptual tools (theories) that
scientists draw on to study the nature of things. If this was true in Mach's
day, it has only become more conspicuous in recent times as scientists
become more occupied with "natural phenomena" that are the product
of their own creation.

Against the background of Mach's ambitious program designed to
examine the historical and philosophical dimensions of physics, the
reader of Mach's works is never able to escape his strong-minded,
inflexible determination to unmask the theological, animistic and
metaphysical trappings of science as he saw them. The implications of
this negativism toward causality, mechanism, and atomism gave birth to
strong feelings, both blame and praise, on the part of scientists, historians, and philosophers.

As Mach dug deeper into historical studies, he was smothered by
philosophical questions that were not easily answered. He disavowed
the title of philosopher, but he became one. He rejected metaphysics,
but he could not maneuver without one. He distrusted theories, but he
surreptitiously fabricated his own. What his contemporaries discovered
to be impossible was to ignore him. Einstein wrote: "Ich glaube sogar,
das diejenigen, welche sich für Gegner Machs halten, kaum wissen,
wieviel von Machscher Betrachtungsweise sie sozusagen mit der Mut-
termilch eingesogen haben."[40]

It is a pleasure to acknowledge the support of the National Science Foundation for my study of the
life and work of Ernst Mach.

1. Max Planck, "Ursprung und Auswirkung wissenschaftlicher Ideen," in *Physikalische
Abhandlungen,* (Braunschweig: Friedr. Vieweg & Sohn, 1958), 3:245.

2. Erwin Hiebert, "The Genesis of Mach's Early Views on Atomism," in *Ernst Mach:
Physicist and Philosopher,* ed. R. S. Cohen and R. J. Seeger, Boston Studies in the Philosophy of
Science, 6 (Dordrecht, Holland: Reidel, 1970), pp. 79–106.

3. Ernst Mach, "Die Leitgedanken meiner naturwissenschaftlichen Erkenntnislehre und ihre Aufnahme durch die Zeitgenossen," *Scientia* 9 (1910): 234.

4. Ernst Mach, "Aus Dr. Mach's Vorträgen über Psychophysik," *Oesterreichische Zeitschrift für praktische Heilkunde,* 9 (1863): 366.

5. Floyd Ratliff, *Mach Bands: Quantitative Studies on Neural Networks in the Retina* (San Francisco: Holden-Day, 1965).

6. Ernst Mach, "On Some Phenomena Attending the Flight of Projectiles,". *Popular Scientific Lectures* (La Salle, Ill.: Open Court, 1943).

7. E. Mach and P. Salcher, "Photographische Fixirung der durch Projectile in der Luft eingeleiteten Vorgänge," *Sitzungsberichte der Akademie der Wissenschaften in Wien,* 95 (1887): 764–80.

8. Ludwig Prandtl, "Neue Untersuchungen über die strömend Bewegung der Gase und Dämpfe," *Physikalische Zeitschrift* 8 (1907): 23–30.

9. J. Ackeret, "Der Luftwiderstand bei sehr grossen Geschwindigkeiten," *Schweizerische Bauzeitung* 94 (1929): 179–83.

10. Albert Einstein, "Ernst Mach," *Physikalische Zeitschrift* 17 (1916): 101–4.

11. Wilhelm Ostwald, *Lebenslinien, Eine Selbstbiographie,* 2 (Berlin: Klasing, 1927): 171.

12. Arnold Sommerfeld, "Nekrolog auf Ernst Mach," *Jahrbuch der bayerischen Akademie der Wissenschaften* (1917), pp. 58–67.

13. Arthur Conan Doyle, *The Annotated Sherlock Holmes,* ed. by W. S. Baring-Gould, (New York: Clarkson N. Potter, 1967), 2:620.

14. Ernst Mach, "Anpassung der Gedanken an die Tatsachen und einander," *Erkenntnis und Irrtum,* (Leipzig: J. A. Barth, 1905), pp. 162–79.

15. Doyle, *The Annotated Sherlock Holmes,* 1:348–49.

16. Ernst Mach, *Beiträge zur Analyse der Empfindungen und das Verhältnis des Physischen zum Psychischen* (Jena: Gustav Fischer, 1886), pp. 142–43; English ed. (New York: Dover Publications, 1959), p. 311.

17. Erwin Hiebert, *The Conception of Thermodynamics in the Scientific Thought of Mach and Planck,* (Freiburg im Breisgau: Ernst-Mach-Institut, 1968), pp. 1–106.

18. It is erroneous to include Mach among energeticists such as Ostwald and Helm, as is often done. See Erwin Hiebert, "The Energetics Controversy and the New Thermodynamics," in Duane H. D. Roller, ed., *Perspectives in the History of Science and Technology,* (Norman, Okla.: University of Oklahoma, 1971), pp. 67–86.

19. Ernst Mach, *Die Principien der Wärmelehre. Historisch-kritisch entwickelt* (Leipzig: J. A. Barth, 1896), p. 364.

20. Ernst Mach, *Die Mechanik in ihrer Entwickelung historisch-kritisch dargestellt,* 2d ed. (Leipzig: F. A. Brockhaus, 1889), p. 203; English ed. (LaSalle, Ill.: Open Court, 1960), p. 266.

21. Ibid., pp. 215–16 and 283 respectively.

22. Albert Einstein, "Prinzipielles zur allgemeinen Relativitätstheorie," *Annalen der Physik* 55 (1918): 241–44.

23. P. A. Schilpp, ed., *Albert Einstein: Philosopher-Scientist,* (New York: Tudor Publishing, 1949), p. 21.

24. Ernst Mach, *Die Principien der physikalischen Optik. Historisch und erkenntnis-psychologisch entwickelt* (Leipzig: J. A. Barth, 1921), pp. viii–ix; English ed. (New York: Dover Publications, n. d.) pp. vii–viii. See also Joseph Petzoldt's essay "Das Verhältnis der Machschen Gedankenwelt zur Relativitätstheorie" in the eighth edition of Mach's *Die Mechanik,* 1921; and Ludwig Mach's preface to the ninth edition of *Die Mechanik,* 1933.

25. For Mach's views on the role of physical experiments and thought experiments in scientific problem-solving see Erwin Hiebert, "Mach's Conception of Thought Experiments in the Natural

Sciences in Y. Elkana, ed., *The Interaction between Science and Philosophy* (Atlantic Highlands, N. J.: Humanities Press, 1975), pp. 339–48.

26. Mach, *Die Mechanik* (1889), pp. 221–22; English ed. of 1960, pp. 289–90.

27. Erwin Hiebert, "Mach's Philosophical Use of the History of Science," in Roger Stuewer, ed., *Historical and Philosophical Perspectives of Science* (Minneapolis: University of Minnesota Press, 1970), pp. 184–203.

28. The study of Mach's extensive correspondence in connection with this issue is reserved for another paper.

29. Ludwig Boltzmann, *Vorlesungen über die Principe der Mechanik* (Leipzig: J. A. Barth, 1897), part 1, p. 5.

30. Ernst Mach, *Geschichte und Wurzel* (Prague: Calve, 1872), pp 1–4; English ed., 1911, pp. 15–18.

31. Ernst Mach, *Die Mechanik,* 1889 ed., pp. 237–38; English ed., pp. 316–17.

32. Vladimir I. Lenin, *Materializm i empiriokrititsizm. Kriticheskiya zametki ob odnoy reaktisionnoy filosofii* (Moscow: Zveno, 1909). [*Materialism and Empirio-criticism. Critical Comments on a Reactionary Philosophy.* Vol. 14 of Lenin's *Collected Works.*]

33. Ernst Mach, *Beiträge,* 1886 ed., p. 5; English ed., pp. xxxv–xxxvi.

34. Ernst Mach, "The Form of Liquids," in *Popular Scientific Lectures,* p. 16.

35. Ernst Mach, "On the Economical Nature of Physical Inquiry," ibid., pp. 186–213; and "Die Oekonomie der Wissenschaft," in *Die Mechanik,* 1889 ed., pp. 452–66; English ed., pp. 577–95.

36. *Collected Papers of Charles Sanders Peirce,* ed. C. Hartshorne and P. Weiss (Cambridge, Mass.: Harvard University Press, 1931), 1:122.

37. Ernst Mach, *Beiträge,* 1886 ed., p. 21; English ed., p. 30.

38. Ernst Mach, *Erkenntnis und Irrtum. Skizzen zur Psychologie der Forschung* (Leipzig: J. A. Barth, 1905), pp. vii–viii.

39. Ernst Mach, *Geschichte und Wurzel,* 1872 ed., p. 46; English ed., p. 74.

40. Einstein, "Ernst Mach," p. 102: "I even believe that those who consider themselves to be adversaries of Mach scarcely know how much they, so to speak, have been born and bred to Mach's outlook [literal translation: 'have absorbed with their mother's milk']."

Chapter Fourteen

THE METHODOLOGICAL FOUNDATIONS OF MACH'S
ANTI-ATOMISM AND THEIR HISTORICAL ROOTS

I. INTRODUCTION

In those annals of history that record the noble espousal of lost causes, the name of Ernst Mach is often linked with the opposition to atomic and molecular theories, along with such other figures as Ostwald, Stallo, and Duhem. And rightly so, for Mach's writings over a fifty-year period from 1866[1] until 1916 reveal an extreme suspicion of, bordering often on a hostility to, most of the atomistic theories of that period. Moreover, the running dispute between Mach and Boltzmann and later between Mach and Planck centered squarely on the efficacy of atomic and molecular approaches to the exploration of natural phenomena. But though the fact of Mach's opposition to atomic/molecular theories is well known and widely cited, Mach's specific argument strategies against such theories have been less fully explored and understood. Still less well documented is the relation of Mach's stand on atomism to his other work in philosophy, especially in the area of the logic and epistemology of science. Finally, the relation of Mach's critique of atomism to the views of his scientific and philosophical contemporaries is almost completely unexplored terrain.

It will be the aims of this paper to make a little clearer (1) Mach's specific criticisms of atomism; (2) to suggest, albeit tentatively, the extent to which those criticisms do, and the extent to which they do not, link up with other portions of Mach's general views on the nature of scientific knowledge; and (3) to explore, in a preliminary fashion, the intellectual traditions and affiliations to which Mach's critique of atomism reveals him to be allied. Perhaps the best place to begin is by surveying briefly the two major extant views about the origins and rationale for Mach's anti-atomism, in order to indicate why I find them unsatisfactory before I move on to state the positive case for the view of Mach I shall be defending. I shall label these positions the *sensationalist* explanation of Mach's anti-atomism and the *scientific* explanation of Mach's anti-atomism.

The Sensationalist Account

On this, the most common approach to the matter, it is argued that Mach's reservations about atomic and molecular theories arise from his sensationalist epistemology.[2] Mach, it is claimed, believed that the knowable world consisted solely of sensations (or, as he preferred to call them, 'elements') and their spatio-temporal contiguities and interconnections. Any reference to entities beyond sensation, to *Dinge an sich,* was illegitimate. All talk about external objects is just a shorthand way of speaking about our perceptions, actual or possible. Because atoms and molecules are, in principle, beyond the reach of our senses, because in short they are radically imperceptible, no theory that refers to atoms or molecules is meaningful.

There is much that is appealing in this approach to Mach. It provides immediately a neat linkage between his abstract philosophical concerns and his concrete physics, by thrusting Mach's *Contributions to the Analysis of Sensations* of 1886 to the center stage in any exegesis of Mach's thought. (It also, incidentially, makes it possible for recent philosophers to dismiss with a wave of the hand Mach's reservations about atomic doctrines by pointing out that those reservations rest on a discredited epistemological cliché: that the world consists entirely of sensations.)

But, for all its initial plausibility, this approach simply will not do as a *general* explanation of Mach's stand on atoms, nor will this interpretation bear the exegetical weight that it is forced to carry by those who see sensationalism as the philosophic cornerstone of Mach's approach to theory construction. Let me survey very briefly some of its weaknesses:

1. Chronologically, it makes little sense because Mach raises serious doubts about the viability and efficacy of atomic theories long *before* he becomes a convinced sensationalist. Almost a quarter of a century before his *Analysis of Sensations,* and fully a decade

before he develops the phenomenalistic epistemology which that work adumbrates, Mach was voicing the most serious reservations about atomic/molecular theories. Those who see Mach's sensationalism as the underpinning for his views on atomism must claim, in the absence of any substantial evidence, that he had adopted a sensationalist approach long before he published anything whatever on epistemological or generally philosophical questions.[3]

2. Of much greater concern is the fact that relatively few of the specific criticisms that Mach directs against atomic theory are couched in terms of a sensationalist theory of perception. As I shall show below, the bulk of Mach's reservations about atoms and atomic theories have nothing whatever to do with the irreducibility of atoms to sensations. His arguments come from different quarters, elsewhere in his philosophical system, and neither stand nor fall with the fate of his sensationalism.

3. Most telling of all, however, is the fact that the sensationalist account of Mach's stand on atomism explains far *too much,* for it fails to indicate why it was atomic and molecular theories in particular that offended Mach's natural philosophical sensibilities. As Mach himself persistently points out, virtually *every* scientific theory postulates objects that transcend our sensory experience. To speak of oxygen, or heat, or a center of gravity or a gravitational force or even ordinary matter is to super-add or super-induce a conception of the mind onto our sensory experience. Even to refer to tables and chairs in the customary way as permanently enduring material objects is to involve oneself in the postulation of theoretical constructs that go well beyond the sensuously given.[4] Yet Mach does not recommend that we abandon or even seriously qualify our theories of chemical elements, or the sciences of statics or celestial mechanics. He has *no* general ax to grind against theorizing as such, nor against most of the scientific theories of his day, despite the fact that virtually all of them go well beyond what a sensationalist account of knowledge would legitimate. Clearly, if we are to understand what it was that specifically rendered atomic/molecular theories otiose in Mach's

eyes, we must look elsewhere than to his sensationalism for the source of the anxiety.[5]

The Scientific Account

At the other extreme of Mach historiography are those scholars who are inclined to dismiss any attempt to find the source of Mach's stand on atomism in his philosophical terms and who point, instead, to certain important facets of Mach's scientific career that allegedly serve to explain his diffidence about theories of the micro-realm. Thus, in an important study of Mach's thought, Erwin Hiebert has suggested that Mach abandoned the atomism of his youth for two straightforwardly scientific reasons: (1) because Mach discovered certain empirical phenomena (especially concerning the spectra of elements) that seemed to defy atomistic interpretation and (2) because, after 1865, most of Mach's own scientific work, on acoustics, on resonance and on psycho-physical problems, was concerned with macro-level problems to which atomic/molecular theories seemed irrelevant.[6] Indeed, Hiebert goes so far as to claim that if Mach had been working in other domains of physics, "Mach very likely would have been more tolerant about atomistic conceptions."[7]

There is doubtless much in this approach to Mach, not least because it rightly stresses the developmental dimensions in understanding Mach's work. But it seems to me that it still, like the sensationalist account, fails to throw much light on Mach's basic reasons for opposing atomism. Limitations of space preclude a full rehearsal of the apparent limitations to this approach, but let me here simply outline what seem to be its major deficiencies:

1. If Hiebert is right that Mach's hostility toward atomic theories coincides with a shift in research concerns, then it seems difficult to understand why Mach, even when he is utilizing the atomic theory in his early work, voices serious doubts about its scientific credentials. As Hiebert himself has pointed out, Mach's early *Compendium der Physik für Mediciner* (1863) is as striking for the reservations it voices about atomic doctrines as it is conventional for its extensive use of such doctrines. Under such cir-

cumstances it seems implausible to explain Mach's aversion to atomism by invoking a shift from traditional theoretical physics to macro-level or psycho-physics, since Mach's reservations about atomic/molecular theories are voiced even in contexts where he is doing 'micro-physics'.

2. If Mach's objections to atomism had been primarily experimental objections—as Hiebert repeatedly stresses they were—it is curious that on most occasions where Mach discusses atomic/molecular theories he fails to rehearse most of the known experimental weaknesses of such theories. In sharp contrast to most of the other opponents of atomism in the late nineteenth century—figures like Stallo, Helm, Brodie, Ostwald, and Mills—Mach rarely if ever goes through a detailed inventory of the empirical anomalies confronting the microphysical theories of his day. That is, I believe, because Mach does not care to show merely that atomic theories are false. Rather, *he wants to show that they are inappropriate and dangerous as a species of theorizing and he needs more than a few refutations to establish that general thesis.*

As a final caveat concerning Hiebert's view that Mach's stand in the atomic debates might well have been different if he had been working in other areas of physics or chemistry, we should remember that views on the propriety of atomic theories in the nineteenth century seem to have been as divided among those scientists working in fields where those theories were prominent as they were in fields far removed from atomic and molecular concerns. Still more to the point, there seems to have been as much epistemic skepticism about the atomic theory from those very scientists who utilized it as from those who had nothing to do with it.[8] The fact that figures with scientific interests as diverse as those of Kelvin, Duhem, Kekulé, and Poincaré did, at various stages in their careers, voice grave doubts about atomic and molecular theorizing suggests that the sources of doubt had rather less to do with the problems on which a scientist was working than Hiebert allows.

II. THE METHODOLOGICAL ROOTS OF MACH'S SKEPTICISM

But we now find ourselves in an interpretative vacuum. If it is to neither the heady philosophical heights of sensationalism nor the scientific bustle of the laboratory that we are to look for the sources of Mach's skepticism about atomism, what remains? If it is neither abstract epistemology nor concrete physics that generates Mach's anxieties, where do they come from? It will be one of the central claims of this paper that the answer to that question is that Mach's reservations arise not from his sensationalism, but rather from his methodology—that they come less from his theory of knowledge and more from his theory of science.

Put another way, my claim will be this: that it is Mach's views about the *aims* and *methods* of scientific inquiry that provide the framework within which Mach develops most of his criticisms of atomic/molecular hypotheses. A more general corollary of this thesis is that if we wish to understand what was really at stake in the atomic debates that raged from 1860 to 1910, we must pay much more attention to *Wissenschaftstheorie*, particularly as it developed among Mach's scientific contemporaries, and rather less attention to *Erkenntnistheorie* as it developed among late nineteenth-century philosophers, including Mach himself.

Nor should this be surprising, particularly not in the case of Mach himself, who wrote approximately ten times as much on scientific methodology as he did on sensationalist epistemology. Indeed, almost all his major works are written either as explicit tracts on the methodology of science or else they contain lengthy sections on the appropriate methods and procedures of scientific inquiry. From *Die Geschichte und die Wurzel des Satzes von der Erhaltung der Arbeit* (1872), through *Die Mechanik* (1883), on to *Die Principien der Wärmelehre* (1896), *Populär-wissenschaftliche Vorlesungen* (1896), and finally to the magnificent but much neglected *Erkenntnis und Irrtum* (1905), Mach's continual preoccupation is with the nature of scientific inference, the relation of theory to experiment, the role of abstraction in theory construction, and other issues in the philosophy and methodology of science.

However, it might be thought that an approach such as mine runs counter to the chronology of Mach's thought. After all, someone might say, we know that most of Mach's methodological writings date from the period 1872 to 1905, whereas he voices opposition to atomistic modes of theorizing well before that. There are at least two points that mitigate such criticisms. (1) It is not primarily my purpose here to identify the initial reasons that led Mach to be skeptical of atomism. We know too little about that period of Mach's intellectual development (roughly from 1860 to 1866), to make confident assertions about what *initially* led him to his skeptical position. It is my aim here to explore the reasons and arguments he gives for his skepticism once it is well defined. (Obviously those may or may not be his original reasons for suspicion of the atomic theory.) (2) Moreover, there is abundant evidence that demonstrates that Mach's interest in methodology dates from a very early point in his career. As Wolfram Swoboda has shown in his study of the early development of Mach's thought, he was acutely concerned with questions of scientific methodology while still a young *Privatdozent* in Vienna. For instance, as early as 1860 (when he was 22) Mach became involved in a heated controversy with Josef Petzval about Doppler's study of the relation between frequency and pitch. In the course of that controversy he came to the conclusion that Petzval's criticism of Doppler and of Mach hinged on a confused notion of the nature of analogical inference. Mach stated a resolve at that time to look more carefully into the logic of scientific inference. There is much evidence to suggest that he did just that, including a reference to Mill's *System of Logic* in a work of his in 1863 and a course of lectures he gave in Graz in 1864 of "The Methods of Scientific Research."[9]

My thesis, then, will be that it is in Mach's methodology, not his general epistemology, that we will find the clue to Mach's opposition to atomism. So far as his stand on atomism is concerned, I want to take Mach at his word when he warns us "above all, there is *no* Machian philosophy, but at most a scientific methodology and a psychology of knowledge."[10]

III. MACH'S GENERAL THEORY OF METHOD

Before surveying Mach's specific arguments against atomism and their rationale, it would be helpful to recount some of the salient features of Mach's theory of science, particularly in the context of its ancestry through the nineteenth century. Even allowing for Mach's considerable eclecticism, his views on scientific knowledge and methodology can nonetheless be usefully classified as *positivistic,* provided that we take that term in its nineteenth- rather than its twentieth-century signification.[11] For Mach, as for the positivists generally, the aims of science were descriptive and predictive. An ideal theory was one that, for the

396

least labor, allowed one to represent as many known facts as possible and to anticipate or to predict correctly as many yet unknown states of the world as possible. Contrary to popular mythology, there was nothing in nineteenth-century positivism that was hostile to speculative theory construction.[12] All the major positivists from Comte to Mach, Poincaré, and Duhem enthusiastically accepted Kant's point about the active knower and were thoroughly contemptuous of that eighteenth-century brand of empiricism and inductivism which imagined that theories would somehow emerge mechanically from the data. Indeed, virtually all the nineteenth-century positivists, including both Comte and Mach, stressed that theories and hypotheses were a precondition for the coherent collection of experimental evidence.[13] In a very perceptive passage in his *Science of Mechanics,* Mach formulates the point this way: after remarking that Galileo had certain "instinctive experiences" in his mind before he actually experimented on inclined planes, Mach stresses that

> for scientific purposes our mental representations of the facts of sensual experience must be submitted to *conceptual* formulation . . . this formulation is effected by isolating and emphasizing what is deemed of importance, by neglecting what is subsidiary, by *abstracting,* by idealizing. . . . Without some preconceived opinion the experiment is impossible. . . . For how and on what could we experiment if we did not previously have some suspicion of what we were about?[14]

But, as positivists kept stressing throughout the nineteenth century, even if theories are preconditions for the assimilation of data, theories are nonetheless *about* the data, about the facts, and it is those facts that provide the ultimate touchstone for choosing between theories. However, to put the point that way is to phrase it too mildly, for one of the most persistent themes in all of nineteenth- and early twentieth-century positivism was the *conservative* nature of theorizing. Given that the aim of a theory is to correlate the facts, given that the process of correlation must take one beyond the facts, it should do so without postulating any other entities or processes than are *necessary* for that task of correlation. To develop a theory or hypothesis that, in order to codify and interrelate known phenomena, makes use of mechanisms or entities that are in principle beyond the reach of experimental analysis is to confuse science with pseudo-science. This requirement, that we should postulate no more than is necessary to explain the data, was called by Comte "the fundamental condition of hypothesis" and became, at the hands of Mach, the cornerstone of his doctrine of the economy of scientific thought.[15]

Closely linked to this ontological conservation was a thesis of positivistic elimination—a kind of Comtist razor. The thesis of elimination, as formulated by Mach and his predecessors, insisted that any theoretical entities that were not themselves subject to experimental analysis and control, that were not *verae causae,* were either to be eliminated from science altogether or else were to be treated as fictions, prophylactic devices that were themselves ontologically sterile. Thus, as Comte argued in the 1830s, the Huygensian construction of a wave front was purely a calculational device, a *façon de parler,* for reconstructing how light would move through a refracting interface. Secondary wavelets, so essential to the model, were purely ficticious. This conservatism, coupled with the doctrine of elimination of fictions, made it natural for the positivists, including Mach, to distinguish between what we might call *purely observational theories* and *mixed theories*— the latter being those which referred, at least in part, to entities or properties that were in principle beyond the reach of experimentation. Positivists, let it be stressed, were not necessarily opposed to either type of theory; for there were certain circumstances in which even the most orthodox positivist was willing to endorse the use of hypotheses involving imperceptible, or purely theoretical, entities. (Comte, for instance, was sympathetic to the atomic theory of matter; Mach and Duhem to the undular theory of light.) But they objected strongly to efforts to put both types of theory on the same epistemic and methodological footing. To treat Fourier's or Carnot's theory of heat as being just like the kinetic theory of gases in all relevant respects was, they felt, to ignore the fact that the theories of Fourier and Carnot refer to nothing beyond what can be measured, whereas the kinetic heat theories of a Bernoulli, a Boerhaave, or a Herapath make continuous reference to entities beyond the reach of experimental analysis. Thus all the major positivists stress an important *distinction* among the class of acceptable theories—a *distinctio* that has profound implications for what kinds of things one regards a theory as establishing the existence of.

Within Mach's own writings this distinction is most often put in terms of a distinction between universal theories that are *direct* descriptions (i.e., that contain only terms that give an abstract description of what is observable) and theories that are (or contain) *indirect* descriptions. Mach is willing to concede that either type of theory is scientific, although he himself has a decided preference for theories that only involve direct descriptions. His program moreover envisages the gradual replacement of indirectly descriptive theories by directly descriptive ones: it is, he asserts,

> not only advisable, but even necessary, with all due recognition of the helpfulness of theoretic ideas in research, yet gradually, as the new facts grow familiar, to substitute for indirect descriptions direct description, which contains nothing that is unessential and restricts itself absolutely to the abstract apprehension of facts.[16]

There is still another important positivist strain that is prominent in Mach's writings and that bears directly on his controversy with the atomists. From the time of Comte onward, the question of the *classification of the sciences* had loomed large in positivistic and empiricist writings. Comte himself, Ampère, Mill, Cournot, and others had addressed themselves to the question of the logical and conceptual linkages between the various sciences. The standard positivist line, and one to which Mach usually subscribes (especially after 1865), is that the sciences do *not* generally stand in relations of logical deducibility to one another. Chemistry is *not* reducible to physics; biology does not follow from physics and chemistry; the social sciences are not theorematic consequences of the natural sciences.[17] This doctrine that each of the sciences has its own domain, its own concepts, and even its own methods is a very important one for Mach, and important at several levels. As a psycho-physicist, he is keen to avoid the Fechnerian line that all psychical phenomena are intrinsically physical. Equally, as a physicist, Mach is adamant that everything physical is not exhausted by the science of mechanics. As we shall see below, Mach's views on the respective domains of the various sciences do much to condition his approach to the so-called atomic debates.

IV. MACH'S CRITICISMS OF ATOMIC/MOLECULAR THEORIES

Much has been written in the last decade concerning Mach's views toward the so-called atomic-molecular hypothesis.[18] Some of this literature is flawed by a failure to distinguish a number of subtle but important differences in Mach's approach to this issue. Mach himself distinguishes between *physical* atomism and *chemical* atomism and has rather different observations to make about each one. More crucially, scholars have generally failed to distinguish between "the rejection of an atomic hypothesis" and "the stating of reservations about an atomic hypothesis," too readily assuming that the latter entails the former. Again, there has been a tendency to assimilate Mach's views on the scientific weakness of the so-called *molecular* approach to his views on the metaphysical problems involved in an ontology of discrete entities.

But the ambiguity about the specific *objects* of Mach's attack have done less scholarly mischief than a failure to distinguish the various types of argumentative strategies that Mach deploys against atomic/molecular theories. Basically, Mach's arguments fall in four distinct groups:

1. *The aim-theoretical argument:* objections or strengths here would be determined by showing the degree to which a theory was, or was not, conducive to, or compatible with, the accepted aims for scientific inquiry.

2. *The interpretative argument:* a theory may be criticized, not for its intrinsic structural or evidencing weakness, but rather for what its interpreters or partisans take it to have established or proved.

3. *The programmatic argument:* a theory may be criticized or endorsed because it is part of a larger program of scientific research that is ill or well-conceived.

4. *The inferential argument:* a theory may be criticized or commended by examining the extent to which it conforms to inferential procedures for which there is a sound methodological rationale.

These distinctions are still rough-hewn, and in problematic cases it may well be difficult to determine into which category a specific argument fits. But I believe that they are sufficiently clear-cut to permit us to begin to understand the various levels at which Mach does, and does not, see weaknesses in the atomic and molecular theories of his day.

400

Aim-Theoretic Objections

As is well known, Mach held deep-seated views on the aims of the scientific enterprise. In a phrase, Mach took the view that the aim of science was to describe and predict the course of nature as economically as possible. That meant, among other things, constructing theories that explained the known laws of nature and led to the discovery of new laws of nature by making the fewest possible existential commitments over and above those that the evidence warrants (or could conceivably warrant). This statement, however, needs a good deal more unpacking, for too often Mach's commentators and critics have assumed that his so-called descriptivist aim of science leaves the whole question of theoretical heuristics and theoretical growth untouched. In arguing that the aim of science is to find an economical description of nature, Mach is *not* suggesting that we must simply describe those natural regularities that we already know to be the case. It was of central importance for Mach that a theory must anticipate the future as much as recapitulate the past. Any theory worth its Machist salt will lead the scientist to the discovery of new modes of interconnection between the appearances (i.e., to discover new laws). In order to achieve this aim, a theory may well go significantly beyond what we already know of the world and may make assertions about connections between phenomena that we have not yet explored. Far from viewing this process of "going beyond the data" as contrary to the aims of science, Mach stresses time and again the need for such theories.[19]

There are, in Mach's view, only two constraints on such theorizing. One is that such theories should lead us correctly to anticipate connections between observable data, i.e., they should predict some laws that later experiment may confirm. This I shall call the *weak constraint*. The second and *strong constraint* on such theories is that, if they are to be a relatively permanent part of our scientific ideology, we must be able to get some direct evidence for the existence of all the entities that such theories postulate.

Mach is by no means completely consistent concerning which of these constraints he will utilize in appraising micro-theories. In his most extreme moments he espouses the strong claim that a necessary condition for any satisfactory theory of physical processes is that *all* the connections postulated between entities in the theory must correspond to verifiable connections between physical objects or properties. As he formulates the thesis:

> In a complete theory, to all details of the phenomenon details of the hypothesis must correspond, and all rules for these hypothetical things must also be directly transferable to the phenomenon.[20]

Atomic/molecular theories, if assayed against this strong requirement of isomorphism, turn out to be counterfeit for, as Mach notes, "molecules (and atoms) are merely a valueless image,"[21] in the sense that there are no experimental analogues for them or for their postulated modes of interaction.

But this extreme position, which would have literally legislated all atomic and molecular theories outside the domain of the sciences, is by no means Mach's most persistent or most characteristic stance. Much more often, Mach will allow that micro-theories are in principle acceptable, provided that they lead to the anticipation of new phenomenal or experimental laws; that is, *so long as they exhibit strong heuristic capacity.*[22] Mach is quite well aware of the fact that some of the most predictively powerful theories in the history of science have involved the postulation of entities that were regarded as unobservable in principle. Far from unequivocally condemning the use of such theories, Mach has much to say on their behalf and frequently points out that the construction of theories of that kind is often one of the best ways of discovering new facts.[23]

Thus when Franklin sought to explain what was happening in the Leyden jar, he postulated the existence of an imperceptible, elastic electric fluid whose particles mutually repelled one another according to a l/r repulsion law. His ideas about this (unobservable) electrical fluid led Franklin to anticipate the fact that there are two modes of electrification (positive and negative), and that any change in one is accompanied by an equal and opposite change in the other. However, and Mach is insistent on this point, once we have established the functional dependence at the level of appearances, the theoretical model has served its purposes and can and should be discarded. Above all, Mach stresses, we must not confuse the tool with the job by pretending that the model does anything more than establish functional relationships between the data. As he put the point in 1890:

> The electric fluid is a thing of thought, a mental adjunct. (Such) implements of physical science (are) contrived for very special purposes. They are discarded, cast aside, when the interconnection of *ABC* . . . has become familiar; for this last is the very gist of the affair. The implement is not of the same dignity, or reality as *ABC* . . . and must not be placed in the same category.[24]

Thus, for Mach, theoretical entities may play an important but essentially temporary role in natural science. Once they have suggested those empirical connections that are the warp and woof of scientific understanding, they can be discarded as so much unnecessary scaffolding. Insofar as the atomic theory helped scientists discover connections between the appearances (as Mach concedes it did with respect to the laws of definite and multiple proportions), it was of great heuristic use. But Mach considers that atomism long ago outlived its usefulness and is now positively redundant.

Like other positivists before him, Mach maintains that theoretical models are essentially temporary. In the course of time, either the models themselves become verifiable matters of fact or else they remain untestable and, having served to reveal whatever empirical connections they are capable of drawing attention to, are dropped in favor of the lawlike, empirical connections themselves. Either way, they do not (or should not) remain as models for any significant length of time.

To the best of my knowledge, Mach *never* argues against the use of atomic and molecular theories, *provided* that such theories continue to lead us (as Dalton's did) to the discovery of new modes of connection between data (in Dalton's case, to the laws of definite and multiple proportions). That Mach does not object to the use of such theories does not mean that he is always happy with certain interpretations that certain atomists put on their theories (as we shall see below). But we cannot begin to understand Mach's reservations about atomic and molecular theories until we realize that he appreciates the heuristic potential that such theories have sometimes had and that he does not view the utilization of such theories as necessarily incompatible with the aims of science (so long as they anticipate new discoveries).[25]

Interpretative Objections

If Mach is relatively liberal and undogmatic about the *use* of atomic theories, he is substantially more adamant about the interpretation of such theories, particularly about interpretations that invest such theories with ontological significance above and beyond the data that they correlate. Like Comte before him, and like Braithwaite and Campbell in our own century, Mach is concerned about the ontology of theoretical terms and concepts. In brief, Mach's position is that theories only make existential claims about those entities and properties which can be experimentally determined or measured. He develops this point at length in an important essay of 1890:

> A perfect physics could strive to accomplish nothing more than to make us familiar beforehand with whatever it were possible for us to come across (experimentally); that is, we should have knowledge of the interrelation of *ABC*. . . . A motion is either perceptible by the senses, as the displacing of a chair in a room or the vibration of a string, or it is only supplied, added (hypothetical), like the oscillation of the aether, the motion of atoms, and molecules, and so forth. In the first instance the motion is composed of *ABC* . . ., it is itself merely a certain relation between *ABC* . . . In the second instance the hypothetical motion, under especially favorable circumstances, can become perceptible by the senses. In which case the first instance recurs. But as long as this is not the case, or in circumstances in which this *can never happen* (the case of atoms and molecules), we have to do with a *noumenon*, that is, a mere mental auxiliary, an artificial expedient, the purpose of which is solely to indicate, to represent, after the fashion of a model, the connexion between *ABC*. . . .[26]

I take Mach to be arguing in this important passage that so long as it seems in principle impossible to get direct experimental evidence for the entities and modes of interaction to which a theory ostensibly refers, then we should limit the existential scope of that theory solely to its (in principle) observable entailments and view all other terms and concepts therein as purely fictional, as convenient algorithms for correlating data, which *mean* and *assert* nothing more than the observational connections that they entail.

It is crucial to distinguish this line of argument from Mach's views on the *use* of atomic/molecular theories. As we have seen, so long as such theories lead to the discovery of new empirically testable connections,

Mach is more than willing to regard them as scientifically useful, regardless of whether we can "observe" all the entities that the theories postulate. If, however, we cannot see any way to get direct empirical evidence for the existence of some of the entities to which the theories refer, then we should regard those entities as convenient fictions, not as laying bare a true ontology of nature. Atoms and molecules, in Mach's day, clearly fell into this category. He stresses this point at some length in a lecture in 1882 where he argues for the purely algorithmical character of atoms and molecules.[27]

It is important to stress again that Mach's anxieties are interpretative ones about ontology, not pragmatic worries about utilization. So long as atomic/molecular theories continue to inspire discoveries of new macroscopic connections between observables, Mach would be the first to concede their value. It is only when scientists assume that atoms and molecules exist, even in the absence of any "direct" evidence for their existence, that Mach feels the interpretative machinery of science has gone astray.

Programmatic Objections

In Mach's view, specific attempts to formulate atomistic or molecular theories were part of a more general reductionist program for the natural sciences. Indeed, Mach thought that much of the appeal of such doctrines to many theorists was precisely that they were inexorably linked with a program for reducing all of science to mechanics and all of the so-called secondary qualities to primary ones.

At this level of argument, Mach is not attacking atomic/molecular theories because they postulate unobservable entities but rather because such theories are inextricably bound up with what Mach regards as an outmoded and defunct research program for reducing all of nature to mechanics. His arguments here are fairly subtle, both historically and philosophically, so we must attend to them carefully.

As Mach points out, atomism has, at least since the middle of the seventeenth century, been tied up with two related *reductionist* doctrines. On the one hand, atoms have often been seen as the vehicle for effecting the reduction of all of nature to the science of mechanics. Wherever mechanical philosophers observed any phenomenon that could not be directly reduced to the equations of motion (say, heat or

405

chemical change), they immediately assumed that such phenomena—despite their apparently non-mechanical behavior at the macroscopic level—were the result of the motion of unseen atoms behaving as mechanical systems in miniature. Atoms thus functioned as *ersatz* equations of motion in the absence of the real thing. On the other hand, to look at the other tradition to which atomism was wedded, it has traditionally been associated with the primary-secondary qualities distinction and with the effort to reduce secondary qualities to primary ones, atoms again serving as the vehicle whereby the reduction was to be effected. As a historian of science, Mach saw clearly that atomism was the not very thin end of the wedge whose other apices asserted the primacy of mechanics and the universality of the primary qualities. As he saw the science of his own time, atomism was being used to mechanize chemistry and gas theory and to deny the reality of that world of sensible qualities which Mach took to be fundamental.

Not surprisingly, therefore, many of Mach's arguments against atomic/molecular theories are directed against their mechanistic and reductive dimensions. As he argues at length in the *Science of Mechanics,* "the view that makes mechanics the basis of the remaining branches of physics, and explains all physical phenomena by mechanical ideas, is in our judgment a prejudice."[28] Mach urges his critics not to be confused by the historical accident that mechanics emerged first among the empirical sciences. Equally, he urges them to attach no more objective significance to "our experiences concerning relations of time and space than to our experiences of colors, sounds, temperatures and so forth."[29]

To seek to reduce everything to the behavior of mechanical atoms is prematurely to commit oneself to the general thesis that all change is mechanical and that all qualities are exclusively mechanical ones. Apart from the fact that such a thesis is unproven (and presumably unprovable), it suffers from other defects. As Mach points out in *Science of Mechanics,* if it is part of the aim of science to explain the unfamiliar in terms of the more familiar, then it scarcely shows good sense always to seek to reduce the so-called secondary qualities of macroscopic bodies to conjectured primary qualities of hypothetical imperceptible bodies. Heat, light, work, pressure, and the like are palpable qualities of sensory bodies (and systems of bodies); and rather than explain them away by reducing them to atoms shorn of all such properties, Mach maintains that we should accept them as ultimately given and relate the behavior of such secondary qualities to others equally familiar in experience. Mach felt particularly strongly about the limits of mechanistic/atomistic science when it came to the biological and psychological sciences. The attempt by scientists such as Fechner and others to explain mental phenomena in atomic/molecular terms seemed particularly outrageous, even to involve what Mach calls a "flagrant absurdity."[30] A fuller understanding of the relations between physical and biological systems would, Mach believed, make one much less sanguine about the viability of reducing everything to a congeries of mechanical atoms. In his words:

> an overestimation of physics, in contrast to physiology . . . a mistaken conception of the true relations of the two sciences, is displayed in the inquiry whether it is possible to *explain* feelings by the motion of atoms.[31]

What, in Mach's view, had made the mechanistic research program so attractive was the "misconception" that mechanical processes are more comprehensible, more intelligible, more clear than non-mechanical ones. Mach has two arguments against such Cartesian wish-fulfillment. For one thing, he maintains that most partisans of mechanistic reduction confuse intelligibility with mere familiarity. Because the science of mechanics has been around a good deal longer than (say) electrical theory or thermodynamics, people are more accustomed to utilizing it and to thinking of the world in terms of pulleys, levers, and inclined planes. But Mach argues that we no more understand what is happening when one body hits another than we understand what is going on when a body cools or when we have a pain. All these processes are conceptually *unintelligible*. To seek to reduce everything to mechanics is merely to substitute what Mach calls a more "common unintelligibility" for a less common one. Mach underscores this point by stressing, in a more general vein, that the fundamental principles of any science, its "unexplained explainers," are always conceptually opaque, for they are the things posited as being beyond further analysis or comprehension. So, if we cannot render any of our basic conceptions intelligible, we must at least strive to make them factual by basing them on warranted abstractions from what we do know about the world. Atomic/molecular theories fail on this score, for they postulate entities that are neither intelligible nor experimentally verifiable. In his classic study on the *Conservation of Energy,* he puts it this way:

The ultimate unintelligibilities on which science is founded must be facts, or, if they are hypotheses, must be capable of becoming facts. If the hypotheses are so chosen that their subject can never appeal to the senses, . . . and also can never be tested (as is the case with the mechanical molecular theory), the investigator has done more than science, whose aim is facts, requires of him—and this work of supererogation is an evil.[32]

Clearly, if Mach is any sample to go by (and there is much to suggest that he is very typical in this regard), one cannot begin to come to terms with the so-called atomic debates in the nineteenth century unless one realizes that those debates were as much about the viability of the mechanical philosophy as they were about the existence of atoms.

Inferential Objections

Mach has two objections he often voices against the inferential moves that partisans of atomic and molecular theories frequently make. On the one hand, Mach points out, they often assume that the fact that their theories work is presumptive evidence for the truth of such theories. Mach observes that such a fallacious affirmation of the consequence is generally egregious, but particularly so in the domains of physics and chemistry, where we already know of the existence of non-atomic and non-molecular theories that are as well-confirmed by the data as atomic/molecular ones.

As early as 1863, Mach was using this argument against the claims of the atomists for the unique empirical adequacy of their particular *Weltanschauung*. In his *Compendium der Physik für Mediciner* he characterized the atomic theory—which he utilized extensively in that work—as simply one mode among many for handling the data of physics and chemistry, all of which are observationally equivalent. He likens these different modes to transcription from a polar to a rectangular system of coordinates. There is, Mach hints, no more reason to believe the atomic theory to be the uniquely true and natural representation of the world than there is for believing a specification of a point's location by polar coordinates to be the uniquely appropriate mode for characterizing its position.

Mach elaborates on this point at some length in his *Erkenntnis und Irrtum*. He there argues that a scientific theory must satisfy two conditions: it must explain the facts with precision, and it must be internally consistent. There are, in principle, many different theories or conceptions of the material world that will satisfy both conditions. Under such circumstances, preference for one mode of conceiving the world over another is entirely arbitrary. More to the point, he stresses that the availability of a potentially large set of empirically adequate and logically consistent conceptual systems rules out of court any easy slide from the empirical well-foundedness of a theory to its truth:

> Different concepts can express the facts in the observational domain with the *same* exactness. The facts must be distinguished from the intellectual images whose origins they have conditioned. The latter—the concepts—must be consistent with observation and moreover they must be logically consistent with one another. These two requirements are satisfiable in *numerous* ways.[33]

Another of Mach's important reservations about the inferential credentials of atomic/molecular theories concerns the kinds of "transempirical" inferences that a scientist is entitled to make. Mach's central argument here grows out of his interpretation of Newton's famous third rule of reasoning in philosophy. In brief, that rule seeks to specify the conditions under which we are entitled to make inferences from what we know of the macrostructure of the world to its microstructural properties. Oversimplifying, Newton's rule states that the only properties that we are entitled to ascribe to imperceptible bodies are those properties that we find to be shared to the same degree by *all* perceptible bodies. But what of atomic/molecular theories? Many of the properties attributed to atoms (for instance, their indivisibility or their non-deformability) are not properties that laboratory bodies exhibit. Indeed, as Mach puts it, "atoms are invested with properties that absolutely contradict the attributes hitherto observed in bodies."[34] It follows that atomic theories commit an elementary fallacy of inductive, theoretical inference that can only raise grave doubts about their inferential credentials.

It should be clear after this brief exegetical survey that the bulk of Mach's reservations about atomic/molecular theories arose—neither from his sensationalism nor from his physics—but rather from his conception of the role and significance of theory in the natural sciences.

Equally, it should be clear that Mach's stand on the atomic theories of his day was complex, characterized by a continuous spectrum of attitudes, depending upon whether he was dealing with atomism as "a working hypothesis," a "useful heuristic," a "proven physical theory," or as an established "research program." If this is true, it follows that any proper understanding of the atomic debates in general and of the Mach-Boltzmann controversy in particular must go behind and beneath the scientific and technical details of the dispute so as to unpack the very divergent views about the aim and structure of theory that those debates reveal and on which they depend. We must realize that here, as elsewhere, scientific controversies are firmly rooted in a wide range of logical, methodological, and philosophical differences of the first magnitude.

V. MACH AND NINETEENTH-CENTURY PHILOSOPHY OF SCIENCE

Although the aim of the paper thus far has been to locate Mach's specific attitude toward atomic and molecular theorizing within the context of his general philosophy of science, it is equally important to attempt to determine the relation between Mach's views on the nature of science and those of his scientific and philosophical predecessors and contemporaries. In part, this is important in order to realize just how much Mach's views were part of several mainstream traditions in nineteenth-century physics, chemistry, and methodology. But it is also important in order to understand why Mach's views received as sympathetic and widespread a hearing as they did. There seems much evidence to suggest that Mach's analysis of the methodological deficiencies of nineteenth-century atomistic theories struck a responsive chord in most of his contemporary readers, and that Mach's reticence about atomistic speculation, although couched in terms of his particular philosophy of science, was by no means either perverse or atypical of the views of his most reflective and able contemporaries.

A thorough study of the historical background to, and reception of, Mach's analysis of microphysical theories would require a full paper in itself. All I shall attempt here is to give a sampling of the views of some of Mach's predecessors and contemporaries in order to illustrate the extent to which Mach was far from being a voice in the wilderness. It is already well known that there were a number of prominent nineteenth-century figures who were less than enthusiastic about the atomic theory. Among these were Duhem, Comte, and Poincaré in France; Ostwald, Helm, and Avenarius in Germany; and Rankine, Stallo, and Brodie in England and America. To lump Mach together with these figures is, however, probably more misleading than helpful. For one thing, such a grouping tends to disguise some very significant differences among the "anti-atomists." Helm, Ostwald, Rankine, and Duhem were all "energeticists" whose primary objections to atomic/molecular theories were quite different from those of Mach—an elementary point that some historians have ignored at their peril. But more importantly, the grouping of Mach with the so-called anti-atomists leads one to ignore the degree to which Mach's general methodological worries about micro-theories were very broadly shared in the scientific community, shared alike by atomists and anti-atomists. Indeed, one of the unnoted ironies of the historical situation is that most of the atomists *accepted* Mach's methodological stand, and attempted to show—not that his methodology was wrongheaded—but rather that atomic theories could be legitimated *within* the framework of a generally Machist, generally positivistic philosophy of science. As a result, those who—like Einstein[35]—see the subsequent acceptance of the atomic/molecular hypothesis as a repudiation of Mach's general positivitism are being less than fair to the actual exigencies of the historical situation.

From the early years of the nineteenth century onward, there is a broad general consensus about the methodological problems of micro-theorizing that can be summarized in the following fashion: the atomic theory is nothing more than a possible (or, sometimes, plausible) hypothesis; whether we should retain that hypothesis depends on its fecundity in leading to the discovery of new empirical relations; even if it is retained, we should be most reticent about asserting the actual existence *in rerum natura* of atoms or any other non-verifiable entities.

Thus, Berzelius in his *Lehrbuch der Chemie* of 1827 argued that atomic theories were

mere methods of representation for the combining elements, through which we facilitate our understanding of the phenomena, but one does not thereby aim to explain the processes as they really occur in nature.[36]

411

A few years later, Auguste Comte articulated his theory of the "logical artifice"—specifically in connection with the atomic theory—which allows for the legitimate use of atomic and molecular theories, so long as one does not endow them with any objective reality. The organic chemist August Kekulé, though himself a frequent proponent of the atomistic hypothesis, conceded in 1867 that "I have no hesitation in saying that from a philosophical point of view, I do not believe in the actual existence of atoms."[37] Two years later, the chemist A. W. Williamson, though himself a proponent of the atomic theory, observed "that chemists of high authority refer publicly to the atomic theory as something which they would be glad to dispense with, and which they are rather ashamed of using."[38] The chemist E. J. Mills, in a passage that shows striking echoes of Mach, observes in 1871 that

> a phenomenon is explained when it is shown to be a part or instance of one or more known and more general phenomena. Isomerism is not, therefore, explained by assertions about indivisibles, which have neither been themselves discovered nor shown to have any analogy in the facts or course of nature.[39]

Mills continues:

> It would be a matter of the highest importance, one would imagine, especially on the part of experimental advocates (of atomism), to adduce, or at any rate to endeavor to adduce, an atom itself as the best proof of its own existence. Not only has this never been done, but no attempts have been made to do it; and it is probable that the most enthusiastic atomist would be the first to smile at such an effort, or ridicule the supposed discovery.[40]
>
> The atomic theory has no experimental basis, is untrue of nature generally, and consists in the main of a materialistic fallacy.[41]

Opinion among physicists was often even more scathing. In 1844, Michael Faraday pointed out that

> the word *atom,* which can never be used without involving much that is purely hypothetical, is often intended to be used to express a simple fact; but, good as the intention is, I have not yet found a mind that did habitually separate it from its accompanying temptations (i.e., the temptation to think of atoms as real).[42]

Another English natural philosopher, Colin Wright, put the argument against atomism differently. As he saw it, the atomic theory had been

> a mechanical conception suited, doubtlessly, to an age when an accurate knowledge of facts was only beginning to exist . . . it is unnecessary to express any facts, and incompetent (without *much* patching and blotching) to explain many generalizations . . . it is undesirable that the ideas and language of this hypothesis should occupy the prominent and fundamental part in chemical philosophy now attributed to them.[43]

Several partisans in the atomic debates, on both sides of the fence, noted that most of the critics of atomism were positivists and that many of their specific criticisms of the atomic theory had already been adumbrated by Comte and his philosophical and scientific disciples.[44]

Thus we can see that even among the most enthusiastic partisans of the atomic theory, there was a general acceptance that the methodological criteria by which Mach sought to evaluate such theories were sound and reasonable. Thus Ludwig Boltzmann and Max Planck both agreed with Mach that an economical representation of the facts is the central aim of science. Boltzmann even concedes Mach's point that the atomic theory has often been "a retarding influence and in some cases has served useless ballast."[45] Another prominent atomist, Adolphe Wurtz, is even more explicit about accepting Mach's yardsticks for theory evaluation. In his *La Théorie atomique*, he stresses that it is important not to confuse "facts and hypotheses." He goes on:

> We may retain the hypothesis (of atoms) as long as it permits us to interpret the facts faithfully; grouping them, relating them to each other and predicting new things, as long as, in a word, it will show itself fecund.[46]

Wurtz then goes on to show how, in his view, that chemical atomic theory has achieved just this. The important point here is that even among Mach's scientific opponents, there is still widespread acceptance of something very like the criteria for theory appraisal that Mach was espousing.

What such passages as these would suggest is that Mach's affinities with his contemporaries were considerably greater than the usual image of Mach as an eclectic crank would allow. In arguing for a fictional interpretation of atoms and molecules, in opposing the mechanistic program inherent in nineteenth-century atomism, in insisting that science is fundamentally descriptive, in demanding that micro-theories must be fertile at the level of observation and measurement, Mach was voicing not just his own but the anxieties of an entire generation of physicists and philosophers who were acutely concerned with the methodological credentials of the most widely utilized theories of the day.

I am very grateful to Wolfram Swoboda for several very useful discussions of Ernst Mach's early work and for drawing my attention to some of his important early writings.

1. Although Mach's scientific career began in the late 1850s, I am dating his opposition to the atomic theory from the mid-1860s. There is much evidence, unfortunately almost all of it ambiguous, that suggests that until the early 1860s Mach was a partisan of atomic and molecular hypotheses. My own belief is that Mach held serious reservations about atomism from the beginning of his scientific career, but that point requires much more elaboration than I can give it here. Useful discussions of Mach's early work can be found in S. Brush, "Mach and Atomism," *Synthese* 18 (1968): 192–215, and M. Faraday, "Speculation Touching Electric Conduction and the Nature of Matter," *Phil. Mag.*, Ser. 3, 24 (1844). Virtually all the claims of this particular paper concern Mach's views from about 1863 onward.

2. For examples of this approach to Mach, see J. Blackmore, *Ernst Mach* (Berkeley: University of California Press, 1972), pp. 321 ff., and F. Seaman, "Mach's Rejection of Atomism," *J.H.I.* 29 (1968): 389–93, both of which see Mach's 'phenomenalism' as the source of, and the motivation for, his rejection of atomism.

3. It should be pointed out that Mach had written a treatise on certain problems in psychophysics and the problem of perception in the early 1860s. However, given that Mach was very much under the influence of Fechner at this time, and given that Mach was to repudiate Fechner's views in the *Analysis of Sensations*, it is most unlikely that Mach in the 1860s adhered to anything like the sensationalism of that later work. (Unfortunately, the manuscript of his early treatise on psychophysics does not seem to be extant).

4. Mach stresses this point often himself. See, for instance, E. Mach, *Beiträge zur Analyse der Empfindungen* (Jena: Gustav Fischer, 1886); Eng. trans., *The Analysis of Sensations*, (Chicago: Open Court, 1914), p. 311. Again in E. Mach, *The Science of Mechanics,* 6th ed. (LaSalle, Ill.: Open Court, 1960), Mach argues that although atoms "cannot be perceived by the senses," that alone does not differentiate them from other objects since "all substances . . . are things of thought" (p. 589).

For yet another variant on this theme, see Mach's 1892 article in *The Monist*, where he argues that notions as divergent as "the law of refraction, caloric, electricity, lightwaves, molecules, atoms and energy *all and in the same way* must be regarded as mere helps or expedients to facilitate our view of things" (p. 202).

Even Mach's persistent opponent, Ludwig Boltzmann, points out that Mach is aware that a phenomenalistic epistemology cannot differentiate between atomistic conceptions and physical-thing conceptions, so far as their epistemic well-foundedness is concerned. See L. Boltzmann, *Populäre Schriften,* (Leipzig: J. A. Barth, 1905), p. 142.

5. I wish to make clear that, in denying that Mach's sensationalist epistemology had much to do with his stand on the cogency of atomic/molecular theorizing, I am *not* asserting a general claim about the complete independence of Mach's philosophy of science from his sensationalism. There

are numerous points of contact between the two which deserve careful exploration. Equally, I am not asserting that all of Mach's reasons for opposing atomic/molecular modes of explanation were independent of his sensationalism, for that claim, too, would be misleading. My line, rather, is that the bulk of Mach's stated reasons for opposing such theories are independent of his theory of perception and of epistemic 'elements'.

6. See E. Hiebert, "The Genesis of Mach's Early Views on Atomism," in R. Cohen and R. Seeger, eds., *Ernst Mach: Physicist and Philosopher*, (Dordrecht: Reidel, 1970), pp. 79–106.

7. Ibid., p. 95.

8. See especially section IV below.

9. See *Akademische Behörden, Personalstand und Ordnung der öffentlichen Vorlesungen an der k.k. Karl-Franzens-Universität zu Gratz*, Graz, 1863–66. Swoboda's study is entitled, "The Thought and Works of the Young Ernst Mach and the Antecedents to his Philosophy," Ph.D. dissertation, University of Pittsburgh, 1973.

10. E. Mach, *Erkenntnis und Irrtum*, (Leipzig: J. A. Barth, 1905), p. vii. I have used the 1917 edition.

11. Although Mach does not often explicitly identify himself as a 'positivist', his writings are strongly positivistic in tone and content, and he was regarded by many of his contemporaries as one of the leading exponents of positivism. Toward the end of his life, he did concede that he was a "positivist". See E. Mach, *Die Leitgedanken*, (Leipzig: J. A. Barth, 1919), p. 15.

12. Comte, for instance, often spoke of "l'introduction, strictement indispensable, des hypothèses en philosophie naturelle" (A. Comte, *Cours de philosophie positive*, 6 vols. [Paris: Bachelier, 1830–42], 2:434).

13. Thus, Comte writes: "Car, si d'un côté, toute théorie positive doit nécessairement être fondée sur les observations, il est également sensible, d'un autre côté, que, pour se livrer à l'observation, notre esprit a besoin d'une théorie quelconque" (*Cours de philosophie positive*, 1:8–9).

14. Mach, *The Science of Mechanics*, p. 161.

15. For a more detailed discussion of Comte's views, see L. Laudan, "Towards a reassessment of Comte's 'méthode positive'," *Phil. of Sci.* 38 (1971): 35–53.

16. E. Mach, *Popular Scientific Lectures*, 5th ed. (LaSalle, Ill.: Open Court, 1943), p. 248.

17. Here again, it is important to stress the chronology. In Mach's very early scientific writings, he is himself a reductionist, arguing for the reduction of all of physics to "applied mechanics" (E. Mach, *Compendium der Physik für Mediciner* [Vienna: Wilheim Braumuller, 1863], p. 55); for the reduction of physiology to "applied physics," ibid., p. 1; and, elsewhere, for the reduction of chemistry and psychology to mechanics. This is a view, and a program, that Mach repudiated during the mid-1860s and that he argued against for the rest of his life. We do not yet have any satisfactory account of this important shift in Mach's thought.

18. See especially J. Blackmore, *Ernst Mach;* J. Bradley, *Mach's Philosophy of Science* (London: Athlone, 1971); S. Brush, "Mach and Atomism"; G. Buchdahl, "Sources of Scepticism in Atomic Theory," *B.J.P.S.* 10 (1960): 120–34; E. Hiebert, "The Genesis of Mach's Early Views on Atomism"; M. J. Nye, *Molecular Reality* (New York: American Elsevier, 1972); and F. Seaman, "Mach's Rejection of Atomism."

19. For reasons that have never been clear to me, most of Mach's recent commentators have assumed that his doctrine that "science is description" precluded him from recognizing that there is any predictive element in science or that science can go beyond the known data. Harold Jeffreys, for instance, writes that "Mach missed the point that to describe an observation that has not been made yet is not the same thing as to describe one that has been made; consequently he missed the whole problem of induction" (*Scientific Inference* [Cambridge: At the University Press, 1957], p. 15). It is, I suspect, Jeffreys and others like him, such as Braithwaite (R. Braithwaite, *Scientific explanation*, [Cambridge: At the University Press, 1953], p. 348), who miss Mach's point. In stressing the view that the aim of science is description, Mach is contrasting it not with prediction but rather with

explanation (in the sense of identifying the underlying metaphysical causes of things). Mach stresses time and again the extent to which theories must anticipate new data, the degree to which every scientific hypothesis goes beyond a mere description of the known facts. One has only to glance at a work such as *Erkenntnis und Irrtum* to see the extent to which Mach did recognize several vital epistemic differences between descriptions of known facts and predictions of unknown ones and, correlatively, the extent to which he attempts to face up to the problem of induction in its various guises.

20. E. Mach, *History and Root of the Principle of the Conservation of Energy,* trans. P. Jourdain, (Chicago: Open Court, 1941), p. 57.

21. Ibid.

22. As he pointed out in *Wärmelehre:* "Der *heuristische* und *didaktische* Werth der Atomistik . . . soll keineswegs in Abrede werden" (E. Mach, *Die Principien der Wärmelehre,* [Leipzig: J. A. Barth, 1896], p. 430n).

23. Speaking of Boltzmann's use of the atomic theory, Mach notes that "Der Forscher darf nicht nur, sondern soll alle Mittel verwenden, welche ihm helfen können" (ibid.).

24. E. Mach, "Some Questions of Psycho-physics," *Monist* 1 (1890): 393ff., p. 396.

25. Mach explicitly points out in *Wärmelehre* that "it should be emphasized that an hypothesis can have great heuristic value as a working hypothesis, and at the same time be of very dubious epistemological value" (*Die Principien der Wärmelehre,* p. 430n).

26. E. Mach, "Some Questions of Psycho-physics."

27. The passage is probably worth quoting in full: "When a geometer wishes to understand the form of a curve, he first resolves it into small rectilinear elements. In doing this, however, he is fully aware that these elements are only provisional and arbitrary devices for comprehending in parts what he cannot comprehend as a whole. When the law of the curve is found he no longer thinks of the elements. Similarly, it would not become physical science to see in its self-created, changeable, economical tools, molecules and atoms, realities behind phenomena. . . . The atom must remain a tool for representing phenomena, like the functions of mathematics. Gradually, however, as the intellect, by contact with its subject matter, grows in discipline, physical science will give up its mosaic play with stones and will seek out the boundaries and forms of the bed in which the living stream of phenomena flows" (E. Mach, *Popular Scientific Lectures,* 5th ed. [LaSalle, Ill.: Open Court, 1943], pp. 206–7).

28. Mach, *The Science of Mechanics,* p. 596.

29. Ibid., p. 610.

30. Ibid.

31. Ibid.

32. E. Mach, *History and Root of the Principle of the Conservation of Energy,* p. 5.

33. Mach, *Erkenntnis und Irrtum,* p. 414.

34. Mach, *The Science of Mechanics,* p. 559.

35. See, for instance, the discussion quoted in S. Suvorov, "Einstein's Philosophical Views and Their Relation to His Physical Opinions," *Soviet Physics Uspekhi* 8 (1966): 578–609.

36. Quoted in G. Buchdahl, "Sources of Scepticism in Atomic Theory."

37. A. Kekulé, "On Some Points of Chemical Philosophy," *Laboratory* 1 (1867): 304.

38. A. Williamson, "On the Atomic Theory," *Jour. Chem. Soc.* 22 (1869): 328.

39. E. Mills, "On Statistical and Dynamical Ideas on the Atomic Theory," *Phil. Mag.,* 4th ser., 42 (1871): 112–29.

40. Ibid., p. 123.

41. Ibid., p. 129.

42. M. Faraday, "Speculation Touching Electric Conduction and the Nature of Matter," *Phil. Mag.,* Ser. 3, 24 (1844): 136.

43. C. Wright, *Chemical News* 24 (1874): 74–75.

44. As Brock has pointed out, in W. Brock, ed., *The Atomic Debates* (Leicester: U. P., 1967), pp. 145 ff., many of the most vocal members of the anti-atomist camp in the 1850s and 1860s were followers of Comte, including Berthelot, Wyrouboff, and Naquet.

45. Boltzmann, *Populäre Schriften,* p. 155.

46. C. A. Wurtz, *La théorie atomique,* (Paris: G. Baillière et cie., 1879), p. 2.

Alan Hausman

Chapter Fifteen

NON-EUCLIDEAN GEOMETRY AND
RELATIVE CONSISTENCY PROOFS

In this paper I shall explore the logical relationship between Euclidean and non-Euclidean geometry. The standard view of this relationship, as told by the positivists, has been accepted even by those, e.g., Strawson, who otherwise disavow positivistic claims.[1] The cornerstone of this view is an alleged inconsistency between Euclidean and non-Euclidean geometry. After laying out the standard story, I shall show how misleading, if not inaccurate, it is. It is, as we shall see, a very difficult matter to pinpoint the alleged inconsistency.

I. THE STANDARD VIEW

In much of the literature on the subject, the relation between Euclidean and non-Euclidean geometry is explained in terms reminiscent of a children's story. There are good guys (Lobachevsky, Riemann, Hilbert), bad guys (Kant), well-meaning dupes (Euclid), prophets without honor (Saccheri), and a Holy Quest (truth). Its difference from a fairy tale is that since it is most often told by mathematicians, or by hard-headed philosophers of science, it has, instead of mystery and ambiguity, the authority and finality of mathematics and logic. Freud has made us suspect the relationship between Little Red Riding Hood and the Wolf, but who could suspect the innocence and purity of a relative consistency proof?

Here in brief is the tale. Once upon a time Euclid told a beautiful story that he called Euclidean geometry. People who listened were very happy with its ending, and thought it mostly true, but they did not like a part of its beginning. There was a statement called the parallel postulate that annoyed them, and for hundreds of years readers tried to show that

Euclid's story could be even better if only that sentence could be left out. All of them thought it added nothing except doubt and confusion. After a very long time, when many had tried and failed to show the hated sentence redundant, a man named Saccheri tried a new tactic: he started the story like Euclid, but instead of telling the parallel postulate he told its opposite, and then tried to show that the rest of the story that came out was nonsense. Saccheri himself thought his whole story quite absurd and laughed a lot (though a bit uncomfortably, it is said). His audience, consisting mostly of a somber group of Russians and Germans, found the new story intriguing and not at all funny. They believed, or at least claimed they believed, that the new story made perfectly good sense. Lobachevsky (a Russian) and Riemann (a German) took over Saccheri's work and spun out new stories called hyperbolic and elliptical geometry. Most everyone was very happy except for a few doubters (there are always doubters), mostly followers of the arch-villain Kant. Kant had (before Lobachevsky and Riemann were even born) declared Euclid's story to be dogma; we had to believe it whether we liked it or not. There *could not* be another story, he said.

But people now said (safely, since Kant was dead) that Kant was just using Euclid for selfish ends.

The police were helpless. Finally, an inventive German quelled the doubts of these Kantian dunces once and for all. Since they liked Euclid so much, how did they like this: one could show that hyperbolic geometry was nonsense only if Euclidean geometry was nonsense too. Klein, which was the German's name, showed this by using what is called a relative consistency proof. The minions of Klein soon produced the same proof for elliptical geometry. They said that hyperbolic and elliptical geometry were extremely rich stories, and they could be interpreted to mean, or at least to be about, a lot of the same things Euclid talked about, even if they did not look that way. In fact (they said), the new geometries are not about what Euclid talked about, but you could look at this way anyway. After all (they said), *Moby Dick* is not just about a *whale*.

The populace danced in the streets. But on a few a problem dawned. Since *all* the stories made sense, which one was true? After all, we do not want a story that says the prince wakes the beautiful girl with a kiss, and another, just as good, that says his kiss puts her into a deeper coma. These are not *just* stories. Finally, the logical positivists, who claimed to dislike stories because they were mere expressions of emotion, settled everything. Euclid's story and Lobachevsky's and Riemann's were just yammerings, noises. One cannot really prefer one noise to another; but if one insists on choosing, then one must obey certain rules. As long as we obey the rules, we can let the stories mean pretty much what we want them to mean. Any of the stories then could be true. But we have to take a look to see which one *is* true. Looks, though, can be deceiving. One has to know how to look and how to obey the rules (the positivists were mostly Germans, you may be assured). The question of how one knows what to look *for*, if the stories are just yammerings, was raised by the last follower of Kant, and was considered a mere diversionary tactic. Herr Einstein now, the positivists said, he knew how to look and to obey the rules. By now people were losing interest. Everything was getting too complicated; some people confessed that though they enjoyed the hyperbolic and elliptical stories very much, they did not understand a word of them even with the damned rules. But evidently Einstein understood them: how could one argue with the bomb? And there things pretty much have stood.

II. THE STANDARD VIEW EXAMINED

In "Geometry and Empirical Science"[2] Hempel relates the saga I have related to you, though more formally, of course. To see what is wrong, we shall have to examine his argument in some detail.

He begins by posing a problem: Wherein lies the proverbial certainty of mathematics? His answer is that all theorems in a mathematical system are derived logically from its axioms, and the conditional statement whose antecedent is the conjunction of the axioms, and whose consequent is the theorem in question, is a logical truth. Thus the truth of a theorem is conditional upon the truth of the axioms. However, if one wishes to know whether the axioms are true, the problems get complicated. It can be shown, says Hempel, that mathematics does not unconditionally assert the truth of its axioms.

Take, for example, Euclidean geometry. Hempel weaves the familiar story: attempts to prove the parallel postulate all failed. Finally it was shown that the parallel postulate cannot be proved from the other axioms of geometry. Hempel's argument for this is rather cursory:

This was shown by proving that a perfectly self-consistent geometrical theory is obtained if the postulate of the parallels is replaced by the assumption that through any point P not on a given straight line l there exist at least two parallels to l. This postulate obviously contradicts the Euclidean postulate of the parallels, and if the latter were actually a consequence of the other postulates of Euclidean geometry, then the new set of postulates would clearly involve a contradiction, which can be shown not to be the case.[3]

Thus it seems that the independence of the parallel postulate in Euclidean geometry is shown by (1) assuming all postulates save the parallel postulate, which is replaced by a contradictory postulate; and (2) showing that this new set of postulates is consistent. For, if the new set is consistent, it cannot be the case that the postulates in common with Euclidean geometry, P1–P4, imply the parallel postulate, P5. If P1–P4 did imply P5, then the assumption of the contradictory of P5 would yield a contradiction. Since the new set consisting of P1–P4 and the contradictory of P5 is consistent, P1–P4 cannot imply P5, and P5 is independent of the other postulates of Euclidean geometry.

This all seems innocent enough, although we shall soon see it is not. At any rate, Hempel has now given a reason why mathematics may be said not to assert the truth of its postulates. For if there are different, consistent, and yet jointly inconsistent systems of geometry, there can be no question of mathematics accepting or preferring one over another with respect to their truth. That is, no *mathematical* reason can be given for a preference. We, of course, must assume with Hempel that mathematics is merely a deduction game, that mathematicians are deduction robots, and we must ignore the fact that many mathematicians *thought* that their axioms were true without the extralogical, empirical criteria that Hempel later sets out and favors.

Hempel goes on to give a second reason why it may be said that mathematics does not assert the truth of its axioms:

> . . . Geometry cannot be said to assert the truth of its postulates, since the latter are formulated in terms of concepts without any specific meaning; indeed, for this very reason, the postulates themselves do not make any specific assertion which could possibly be called true or false![4]

Now the two reasons we have been given for the claim that mathematics does not assert the truth of its postulates seem at first glance to be incompatible. One cannot hold both that P5 in Euclidean geometry contradicts a postulate in, say, hyperbolic geometry, and hold at the same time that neither has truth value. For what sense of contradiction would this be?

One might suggest that P5 and its contradictory be thought of as contradicting one another purely formally, in the sense of what we may call chicken-scratch contradiction. Two well-formed formulas, S and S′, contradict one another if one contains all and only the signs of the other in the same order, except that one is preceded by a sign called the negation sign. Leaving aside that we have two *different* systems of chicken-scratches to compare, this does not help Hempel. For if the sense of contradiction Hempel means is this purely formal one, nothing interesting has been said in presenting the first reason. The first reason would, on this view, be based on the assumption that the two geometrical systems are purely formal sets of uninterpreted marks, and hence become a corollary of the second. If one treats geometrical systems purely formally, then the question of truth of the postulates does not arise; the claim that one set of marks contradicts another would be of no interest with respect to the question of truth. That both could not be true is not a syntactic but a semantic claim.

It is not too difficult to see that treating geometry purely formally makes it impossible to state much of interest concerning the relation between Euclidean and non-Euclidean geometry. What, after all, is an uninterpreted geometry? There seem to be two ways to proceed to get one. As many of the positivists spoke, and many logicians still do speak, formal axiom systems are merely games in geometrical design. They have to be this way, or untoward empirical assumptions might sneak in and prejudice our deductions. Now sometimes the positivists, in speaking of the building of formal languages, tell a story that is clearly untrue. It is as if one simply begins in an ideal moment to write down sets of marks on paper, making up rules for forming strings of marks and getting new strings from old strings. How one would ever get any system that could be interpreted in this way, let alone a geometry—that is, to speak more precisely, a system that upon interpretation could be called a geometry—is an absolute mystery (the probabilities are about the same as those of the famous monkeys at the typewriter producing *Hamlet*). On the other hand, one could begin with statements in ordinary language that are said to be the axioms of a theory (e.g., Euclid's system), and step-by-step formalize them. One could first replace the predicate terms by marks, then fill in the logic of quantifiers, variables, and connectives. If we claim that the marks with which we replace the ordinary language predicates are constants, we may of course as logicians *ignore* their meanings, keeping in mind only that these marks have a different set of syntactic rules for their manipulation than other sorts of marks we call variables. It is then quite possible to treat a system formally that in fact has an interpretation; this is a formal system with signs that are treated syntactically as constants. Or we may do as Hempel suggests, namely, treat the predicate signs in the system language statements as free (or dummy) variables. Either way, we get a system that can be formally manipulated without regard to a semantic component, though one in fact has a semantic component.

Hempel, to repeat, speaks of formal geometries as containing variables:

> We see therefore that indeed no specific meaning has to be attached to the primitive terms of an axiomatized theory; and in a precise logical presentation of axiomatized geometry the primitive concepts are accordingly treated as so-called logical variables.[5]

This, I take it, implies that uninterpreted geometries are misnamed. What we have is a system of formulas written in terms of, say, first-order logic with free or dummy variables. These are axiom *schemata*. To call a given schema an axiom of geometry is inaccurate. Until we provide an interpretation, by adding constants to the language that have a fixed meaning and that are substitutable for the variables, we might just as well say we have a formal system of the relation between apples and pears. Such schemata may axiomatize many different sets of existing objects.

Now we clearly cannot get a contradiction between Euclidean and non-Euclidean geometry in this way. For if we have two uninterpreted sets of axiom schemata, we have no reason whatever to believe that every interpretation of one will contradict every interpretation of another. Thus if we have two axiom systems with the same axiom schemata P1–P4, but the first contains an axiom P5 while the second contains an axiom schema that is the negation of P5, we must assign the same constants to be substituted in the same way in both theories before one can claim the two are inconsistent.

All this is obvious enough. It immediately follows, though, that Hempel's first argument, to be of any interest, must be about interpreted rather than uninterpreted formal systems. This seems to return us to our original dilemma, that his two reasons for the claim that mathematics does not assert the truth of its axioms are incompatible. However, let us reformulate his argument in this way: if one treats a mathematical system such as Euclidean geometry as uninterpreted, then of course the question of truth does not arise. If one treats it as interpreted, then the question of truth must arise; but since we have consistent but mutually contradictory axiom systems for geometry, only one can be true. Which one is true cannot be decided mathematically.

We must, to grasp fully Hempel's claims interpreted in this way, be careful to separate the claim of the consistency of each set of axioms from that of their joint inconsistency. What complicates the issue is that relative consistency proofs for non-Euclidean geometry are often given in terms of Euclidean models. The following must be kept clear:

1. Relative consistency proofs are most clearly stated in terms of finding a model for a set of axiom schemata, which model is *assumed* consistent.[6] Whether we consider an uninterpreted system to be such a set of schemata, or whether we consider it a set of axioms containing constants whose meaning we can ignore, relative consistency is shown by finding a model for the set to be tested, with the assumption that the model is itself consistent.

2. That the model for hyperbolic geometry, historically, was given in terms of the interior of a circle (the Klein model), is merely accidental. It is of course possible to find models of hyperbolic or elliptical geometry in terms of non-geometrical systems, such as number theory.

3. That we can in fact give a model of hyperbolic geometry in terms of Klein's Euclidean figure, or a model of elliptical geometry in terms of the surface of a sphere, has given rise to the claim that if Euclidean geometry is consistent, then non-Euclidean geometry is also consistent. But here we must be careful not to blur this logical point with one over the content or *interpretation* of the axioms. All that has been shown by Klein's proof is that the axioms of hyperbolic geometry have the same *logical form* as some of the theorems of Euclidean geometry, so that if by the rules of deduction we could produce a contradiction in hyperbolic

geometry, we could produce one in Euclidean geometry. That this is so neither presupposes nor shows that hyperbolic geometry is inconsistent with Euclidean geometry.

4. Hempel's argument for the *independence* of the parallel postulate, P5, you will recall, goes in two steps:

a) Euclidean and hyperbolic geometry are mutually inconsistent.

b) Hyperbolic geometry is consistent.

It should be clear now that the sense of inconsistency needed *here* is only between two axiom schemata of the two systems. It is true that if there are such schemata, then no true interpretation of Euclidean geometry can also be a true interpretation of non-Euclidean geometry. But it is not necessary to assume that the intended interpretation of the primitive terms of the two systems is the same in order for either relative consistency proofs or independence proofs to be given.

To give a simple illustration, if we have a set of interpreted axioms about George Washington, P1–P5, and a set about Lincoln, P1′–P5′, such that P1–P4 have the same logical form as P1′–P4′, and the form of P5 is the negation of the form of P5′, we can, provided one can find a consistent model for both sets of axiom *schemata,* thereby prove the independence of P5. But it does not follow that P5, as about Washington, is the contradictory of P5′, as about Lincoln.

Let me summarize, then, the results we have obtained concerning Hempel's argument. Hempel wishes to maintain that mathematics does not assert the truth of its axioms. His second reason, that mathematical axioms are neither true nor false because they are uninterpreted strings of marks, we can accept, at least for the sake of argument. His first reason, that Euclidean and, say, hyperbolic geometry are each consistent, does not prove by itself that mathematics does not assert the truth of its axioms unless we add the premise that the two systems are inconsistent *as interpreted.* The only interesting sense of inconsistency, on which neither the relative consistency nor the independence proofs depend, is that both systems cannot have the same interpretation. But unless Hempel shows that both systems are intended to have the same interpretation and, even stronger, *could* have the same interpretation, he has not established his first argument. Euclidean and non-Euclidean geometry could both be consistent, could be contradictory in the sense of containing contradictory schemata, and yet on interpretation have all true axioms.

III. THE INTERPRETATION OF GEOMETRIES

Now surely, one can immediately object, hyperbolic, elliptical, and Euclidean geometry are at least intended to have the same interpretation. All three are intended to give us the laws of space. And it cannot be the case that what Euclidean geometry tells us about space is true at the same time that what elliptical or hyperbolic geometry tells us about space is true.

The notion of what geometry tells us about space, though, is certainly far from clear. Suppose one considers both Euclidean and hyperbolic geometry to be uninterpreted systems. It is easy to find—indeed, the relative consistency proofs tell us how to find—a geometrical interpretation that is perfectly consistent with Euclidean geometry. If we can find in the world straight lines and such, as Euclidean geometry describes them, we can also find straight lines as elliptical geometry describes *them*: we can find them on the surfaces of spheres.

It is clear, however, that non-Euclidean geometers, elliptical geometers, are not talking about the surfaces of spheres; at least, that is not what they take themselves to be talking about. Here is an example from *Introduction to Non-Euclidean Geometry,* by Harold E. Wolfe:

> It must not be inferred at all, however, that this type of Elliptic Geometry *is* spherical geometry. It merely happens that, on the curved surface known as a sphere, we find an exact representation of this kind of plane geometry, entity for entity, postulate for postulate, proposition for proposition. The reader will understand the significance of this relation better, if he is informed that there are curved surfaces in Hyperbolic Geometry and Elliptic Geometry upon which analogues of Euclidean Plane Geometry can be constructed.[7]

Wolfe adds, "In any attempt to visualize Elliptic Geometry, this resemblance to spherical geometry will be found quite helpful."

Obviously, Wolfe does not take non-Euclidean geometry to be an uninterpreted system (as Hempel has claimed geometers do or at least *should* do) but an interpreted one. For if it were uninterpreted, there would be no reason not to say that elliptical geometry is spherical geometry. But no mathematicians have wanted to say this. When it was discovered that elliptical geometry could be modeled on the surface of a sphere, spherical geometry was already well known. No mathematicians gave up their interest in non-Euclidean geometry because such a model was found. Thus mathematicians, historically at least, have believed that in some sense they knew what the axioms of non-Euclidean geometry *meant*.

Now it seems that we can easily show how Euclidean and non-Euclidean geometries are mutually inconsistent. Take elliptical geometry, and let us consider that it contains the primitive predicate 'straight line', as does Euclidean geometry. Consider: (A) Euclidean geometry attributes to straight lines properties that elliptical geometry denies that straight lines have.

For example, we can state Euclid's first postulate to be that two straight lines do not enclose a space. This is explicitly denied by elliptical geometry; it is a theorem of elliptical geometry that two straight lines always enclose a space. This seems contradictory enough. Given that we can now find a suitable candidate for straight lines in the actual world, it must be the case that two such lines either enclose a space or do not. If they do, elliptical geometry is correct; and if they do not, Euclidean (or hyperbolic) geometry is correct with respect to the first postulate.

But here we come upon one of the most puzzling aspects of our problem of the relation between different geometries. An axiomatized theory, according to the positivists, contains, besides the apparatus of logic, two sets of predicates: the undefined or primitive predicates, and defined predicates. All defined predicates are taken to be explicitly defined in terms of the primitives and hence eliminable. Now although Euclid takes 'straight line' to be a defined predicate, his definition, that a straight line is one which lies evenly with the points on itself, is taken by Heath and by most modern geometers to be unintelligible.[8] Hilbert, for example, takes 'straight line' as a primitive term. Our problem is this: geometry is supposed to be a complete theory of space. That is, a geometry is supposed to specify all the properties that space has *qua* space. But this means that all properties that straight lines have must be mentioned in the theory itself. And this seems to mean that in order to interpret a primitive, such as 'straight line', one must utilize at least some of the properties attributed to straight lines in the geometry under consideration. However, since 'straight line' is undefined, we are left

427

with the problem of which of the axioms to choose as somehow giving us the conditions that anything that is to be called a straight line must satisfy. If there is no way of deciding, or if no characterization of the meaning of 'straight line' can be given independent of the theory under consideration, then the claim that the two theories, Euclidean and, say, hyperbolic geometry, are inconsistent seems to make no sense.

To see the problem more clearly, consider an axiomatization of Euclidean geometry and one of hyperbolic geometry, each containing the same primitives, among them being 'straight line'. Assume the first four axioms to be the same, and the fifth axiom of Euclid to be the so-called Playfair axiom, namely, that through a given point off a given straight line, one and only one straight line can be passed parallel to the given straight line. The fifth postulate of the hyperbolic geometry says that through a given point not on a given straight line, more than one straight line can be passed parallel to the given straight line. We cannot maintain straight off (forgiving the pun) that 'straight line' *can be* interpreted the same way in both theories, even if it must be so interpreted if the theories are to contradict one another. I think it fair to say that all those who believe Euclidean and hyperbolic geometry mutually inconsistent do believe that such an interpretation can be given without difficulty.[9]

But how do we answer someone who maintains that part of what is meant by 'straight line' *is* what the fifth Euclidean postulate says? Notice that if this were true, if it were the case that the conditions laid down by the fifth postulate were somehow part of what it means to be a straight line, then no interpretation of the fifth axiom of hyperbolic geometry could be consistent. The claim that the fifth postulate of hyperbolic geometry makes would be inconsistent. True, the formal theory would not exhibit this contradiction; but anyone trying to apply the fifth postulate of hyperbolic geometry would find it impossible to do so.

Thus if we allow that the fifth postulate of Euclid gives what is sometimes called a (partial) implicit definition of 'straight line', then the claim that hyperbolic geometry contradicts Euclidean geometry really should be reformulated to read: If the fifth postulate of Euclid is true, then the fifth postulate of hyperbolic geometry is self-contradictory.

However, it only seems fair that if we allow P5 to be a partial implicit definition of 'straight line' in Euclidean geometry, we allow the same privilege to P5′ in hyperbolic geometry. If so, then P5 and P5′ certainly do not have to be taken as inconsistent. If I am told that it is axiomatic that banks are good places to store money, and also that they are poor places to store money, I might naturally conclude that 'bank' was ambiguous, meaning the great marble edifices in the first case, and great muddy edifices in the second. Thus the contradiction disappears.

Looked at in this way, one must wonder why it is that geometers ever believed that the new geometries they invented were inconsistent with Euclidean geometry. I think it fairly clear that it is indeed mathematicians, and not philosophers, who have long held the view of the implicit definitions of terms. One can only speculate that in the nineteenth century it was because geometers believed that the Euclidean postulates gave us the 'real' meaning of 'straight line' that they somehow, confusedly, thought of the new geometries as inconsistent with Euclidean geometry. But one cannot consistently hold that (1) Euclidean and non-Euclidean geometry are each consistent and at the same time hold that (2) Euclidean geometry gives us, implicitly, the meaning of the term 'straight line' for both theories. P5′ on such a view may not be a formal contradiction, but it would be a *semantic* contradiction.[10]

It may be that pre-positivist philosophers and mathematicians, when they spoke of the inconsistency of Euclidean and non-Euclidean geometries, meant something like the above. But I doubt that this is what either Hempel or Reichenbach meant to assert. Hempel and Reichenbach do not buy the notion of implicit definition. This is easy to see when one considers that both believe that which geometry is the geometry of our world must be decided somehow on empirical grounds.[11] Now if 'straight line' is implicitly defined by the axioms of the theory, we would not choose Euclidean geometry over non-Euclidean geometry on such empirical grounds. For if 'straight line' means in part an entity that satisfies P5, then we would not have to look at the world to decide whether or not hyperbolic geometry was true. The question to be answered by an empirical theory could then only be: are there Euclidean straight lines, and not, *given* that there are straight lines, are they Euclidean or hyperbolic.

Let us look at the same point from a different angle, to continue with geometrically appropriate language. Since Hempel holds that, say, 'straight line' is a primitive, he must hold that all the axioms in which the term appears, if they are not mere tautologies, are synthetic truths. Take, for example, Hilbert's first axiom of connection for Euclidean geometry, that two points always determine a straight line.[12] By 'determine' Hilbert does not mean 'define'; he means to assert that a certain relation obtains among two points and a straight line. Another way of putting his axiom, he says, is that two points *a* and *b* lie on a straight line A. Thus, by logical standards, it is only a matter of fact that a straight line stands in this relation to the points. Since all the axioms are stated in terms of such primitive predicates, both relational and non-relational, we must upon interpretation of the terms be able to specify what class of entities, or what properties and relations, these primitive predicates represent.

Hempel, Reichenbach, and Waismann speak of the interpretation of Euclidean and non-Euclidean geometry as if it were a rather straightforward procedure.[13] The key to their notion of interpretation is the *coordinative definition*. We can, to take their favorite example, coordinate the term 'straight line' in a given geometry to the path of a light ray, or to a taut string. Lurking in the background is the familiar positivist notion of *pointing;* we interpret a primitive by pointing to an object that has the property in question. Thus it seems that we can get around the objection raised earlier, that any interpretation of the terms of a geometrical theory must proceed by using terms of the theory, by using a non-linguistic method of interpreting the terms. There are, I think, two ways to take this notion of interpreting primitives.

1. Interpreting a term such as 'straight line' means specifying a set of entities that satisfy the predicate 'straight line'.

2. Interpreting a term such as 'straight line' means coordinating the term to a property of some set of entities, e. g., the property of being the path of a light ray.

Now the first notion is surely not what they meant to assert. (1) amounts to this: There is a class of entities A such that the determining characteristic of the class, say, being the path of a light ray, is coextensive with the characteristic of being a straight line (or, if you wish, the two predicates are coextensive). But in this case we have not been told the meaning of 'straight line', but rather what entities are as a matter of fact straight lines. This would not be an interpretation of the primitive at all. To judge whether or not the assignment of the class A was the correct one, we would have to know what a straight line is.

(2) is the notion that Hempel and Reichenbach adopt.[14] Hilbert's first axiom can be given the reading, for example, that given two points, they lie on one and only one path of a light ray. Notice that given this sense of interpretation, we can no longer ask, as we could with (1), whether the paths of light rays are really straight in a vacuum. To speak of a straight line is to speak of the path of a light ray. To ask whether our light-ray geometry is Euclidean or non-Euclidean is now to ask, for example, whether light rays (that is, their paths) have certain properties and stand in certain relations. Whether two points lie on one and only one light ray (given that we have of course also specified the meanings of 'lie on' and 'point') can be determined by experiment.

Have we used a geometrical theory to interpret 'straight line'? If (2) can be realized, the answer is *no*. Presumably we need no geometric theory to specify the path of a light ray or the path of a taut string. We could even verbally describe such entities without pointing. This seems obvious enough once we realize that in giving a physical interpretation we are dealing with physical entities that have other properties than those geometrical ones ascribed to them by a theory of geometry. We thus could, it seems, give 'straight line' a meaning in the sense of a coordinating definition, and we could give this term the same meaning in both a Euclidean and an elliptical or hyperbolic geometry. Then we could test to see which geometry is true. So, at last, we seem to have come to a genuine sense of inconsistency between Euclidean and non-Euclidean geometry; and by experiment we can decide which of the geometries is the right one.

But the joy is reserved only for those who have not read Hempel and Reichenbach. Reichenbach especially is at pains to point out that it is a mistake to think of finding the true geometry of our world.[15] Furthermore, he specifies in detail the logic of a certain claimed equivalence of a Euclidean geometry and a physics with a non-Euclidean geometry and a different physics. This we must explore. For if a given non-Euclidean geometry G' is inconsistent with a given Euclidean geometry G, we cannot conjoin any set of statements with either that will render the combined theories consistent.

First, Reichenbach points out that in order to *test* an axiom of a given geometry, it is not enough to specify the coordinating definitions for the primitive terms of the theory.[16] Suppose, for example, we find in testing Hilbert's first axiom that two different light rays, i.e., with different paths, have the same two points a and b lying on them. We now have a choice of what to do:

1. We could claim Hilbert's first axiom is false and one of elliptical geometry is true.

2. We could claim that one of the light rays is longer than the other.

3. We could claim that neither path is a straight line, that is, we must adopt another coordinating definition.

Consider (2). We must, to verify (2), have a way to measure the length of the two paths. And in order to do this, we must introduce another coordinative definition for the unit of measurement. Suppose then that we specify our unit of measurement, then find that the two paths are the same length. We are not therefore forced to (3). For if one wished to hold onto Euclidean geometry, we might claim that a force was distorting the measuring rods in one of the paths. Take (3). We might now, given that the paths of light rays do not act as Euclidean straight lines are supposed to act, choose another class of entities, say, taut strings. Even if no such class could be found, we could hold onto Euclidean geometry by claiming that universal forces distorted all paths, no matter what the kind of object. In such a theory 'straight line' would be a theoretical term such that no class of entities is its extension. The axioms of geometry would be like the law of inertia, true of ideal entities.

The point is that one does not succeed in interpreting the axioms of a geometrical theory unambiguously by simply giving a coordinating definition that aligns, say, 'straight line' with the path of a light ray. What Reichenbach is not clear about is whether or not, in specifying the metric and the physics necessary for deciding the truth of the axioms, we have thereby 'added' a new component of meaning to the primitive terms.

To make this clearer, consider again our case of the assignment of the predicate 'straight line' to be the path of a light ray. We interpret a Euclidean geometry this way, then find that we have options when the Euclidean axiom seems false. Finally, we end up with an elliptical geometry and the *definition* that there are no universal forces. Reichenbach claims that we can always find a Euclidean geometry and its corresponding physics that are equivalent to the elliptical geometry and its physics. But now, how shall we interpret the predicate 'straight line' in Euclidean geometry? In order to render the Euclidean theory true, it seems we cannot use the same interpretation of 'straight line' as we had in the non-Euclidean theory. For, if we do, the Euclidean first axiom is false. Furthermore, since the first axiom of Euclidean geometry and the first axiom of the chosen elliptical geometry are contradictory when we interpret its primitives the same way, we could not conjoin any statements to the Euclidean geometry to render it consistent with the elliptical geometry, let alone equivalent. Hence, it must be the case that to get the equivalence, 'straight line' must be interpreted in a different way. But how can this be done? Presumably, there will not be a class of objects that we can choose that do obey the Euclidean first axiom.[17] If there is no such class, how do we give a coordinating definition for 'straight line' in the Euclidean system? True, we have the term 'straight line' in the Euclidean theory, but if there is in fact no class of objects that satisfy the axioms, how can we possibly specify what the term 'straight line' means?

Reichenbach claims that instead of the elliptical geometry and its physics, we can choose Euclidean geometry and a physics in which we have introduced universal forces by definition. This means that, putting the matter with deliberate inaccuracy, we specify that Euclidean straight lines deviate from their true paths because of universal forces. But so far, we have no notion of what sort of entities these straight lines are. My conjecture at this point—and it is only a conjecture—is that what is being appealed to in such a move is some intuitive notion of the *concept* of a straight line, rather than some empirical notion of a straight line.[18]

Thus it appears that we cannot maintain the claim that Euclidean geometry and its physics are equivalent to elliptical geometry and its physics unless we claim that the coordinative definitions for the primitives of the geometries differ. But even worse, we must give up the claim that there is a coordinative definition for the Euclidean geometry term 'straight line'.

Before trying a final alternative in understanding the notion of an interpretation, I think it must be pointed out that so far we still have not fastened on to a coherent notion of the inconsistency of a Euclidean and a non-Euclidean geometry. For, as we have seen, Reichenbach maintains that given the same coordinative definitions for each of the two systems, we still cannot *test* the axioms to see which geometry is true without adding a physics to each. If we claim that specifying a physics is not specifying additional meaning to the primitives, then without the physics the interpreted axioms are still neither true nor false. If we cannot decide between Euclidean and non-Euclidean geometry on the basis of the initial coordinating definitions alone, then those 'interpreted' geometries are not verifiable as they stand. Hence they are not, at least on the view of Reichenbach and Hempel, either true or false.[19] We would be in a sort of limbo between chicken-scratch inconsistency and a full-blown interesting inconsistency. On the other hand, if we somehow add meaning to the primitives by the specification of a physics, it is no longer clear what sense it makes to speak of geometries as inconsistent with one another. 'Packages' of a geometry and a physics might be consistent or inconsistent with one another, but not geometries alone.

So we still have not found our inconsistency. But if it is the case that coordinating the path of a light ray to 'straight line' in the first axioms of Euclidean and elliptical geometry does not give us a genuine sense of contradiction, we are not forced, as I claimed before we were forced, to change the interpretation of 'straight line' in the Euclidean geometry in order to render it, together with a physics, equivalent to an elliptical geometry with *its* physics. What is needed is some clear statement, which I am unable to give, of the logic of the equivalence of Euclidean geometry with its physics and a non-Euclidean geometry and its physics. The + sign in the tired formula P + G = P' + G' must be interpreted in some way other than conjunction. When we look closely at the coordinating definitions given for 'straight line', for example, we often see that it is taken to mean the path of a light ray through a *homogeneous medium*. Thus it appears that the notion of physical forces may have appeared in the coordinating definition itself, so that, to put it a bit crudely, the P component of the formula is made part of the interpretation of the G component. Whether or not such a notion of interpretation will allow us to have our cake and eat it too, have at last the sense of

inconsistency we want and at the same time the equivalence between
$P + G$ and $P' + G'$, remains to be seen.

I am grateful to Professor Howard Kahane of Bernard Baruch College for many helpful suggestions.

1. P. F. Strawson, *The Bounds of Sense* (London: Methuen & Co., 1966), pp. 277 ff.

2. C. G. Hempel, "Geometry and Empirical Science," in *Readings in Philosophy of Science,* ed. Philip P. Wiener (New York: Charles Scribner's Sons, 1953), pp. 40–51.

3. Ibid., p. 44.

4. Ibid., p. 46.

5. Ibid.

6. Throughout this paper I use the terms 'model' and 'interpretation' differently. A model may be purely formal, e.g., the coordination of marks in one theory with marks in another. An interpretation is carried through by assigning constants to be substituted for the variables of axiom schemata.

7. Harold E. Wolfe, *Introduction to Non-Euclidean Geometry* (New York: Dryden Press, 1945), p. 178.

8. Sir Thomas L. Heath, *The Thirteen Books of Euclid's Elements* (New York: Dover Publications, 1956), 1:165 ff.

9. Wolfe, for example, claims that lines are every bit as straight in hyperbolic and elliptical geometry as they are in Euclidean geometry (p. 174 n.3).

10. For an excellent discussion of the *a priori* nature of Euclidean geometry, see S. F. Barker's *Philosophy of Mathematics* (Englewood Cliffs, N.J.: Prentice-Hall, 1964), pp. 50 ff. An interesting reply to Barker was given by F. Wilson in his "Barker on Geometry as *A Priori,*" *Philosophical Studies* 20 (1969): 49–53.

11. H. Reichenbach, *The Philosophy of Space and Time* (New York: Dover Publications, 1958), chap. 1; and Hempel, "Geometry and Empirical Science," p. 47.

12. D. Hilbert, *Foundations of Geometry* (La Salle, Ill.: Open Court Publishing Co., 1947), pp. 3–4.

13. This is not quite fair to Reichenbach, though it is to Hempel and Waismann. For the latter's view, see his *Introduction to Mathematical Thinking* (New York: Frederick Ungar Publishing Co., 1951), chap. 3.

14. Hempel, "Geometry and Empirical Science," pp. 47–48, and Reichenbach, *The Rise of Scientific Philosophy* (Berkeley: University of California Press, 1958), p. 130. However, it seems that Reichenbach vacillates here between (1) and (2).

15. *The Philosophy of Space and Time,* p. 33.

16. Ibid., pp. 10–36, and *The Rise of Scientific Philosophy,* pp. 129 ff.

17. Notice especially that we cannot use alternative (2) here, saying that some light rays do and some do not obey the axiom. For, so far 'straight line' means 'path of a light ray', and nothing else.

18. This of course brings us close once again to the notion of the *a priori* nature of Euclidean geometry. See Barker, *Philosophy of Mathematics*.

19. This point does not necessarily depend on some verificationist principle, however. The point is that the alleged interpreted axioms are in a sense ambiguous or, perhaps better, incomplete, in the same way that "The Boston Red Sox won the pennant" is incomplete.

Ronald Laymon

Chapter Sixteen

THE MICHELSON-MORLEY EXPERIMENT: DESCRIPTIVE
DEPENDENCE ON TO-BE-TESTED THEORIES

This analysis of the Michelson-Morley experiment, like many others, has been motivated by a philosophical problem. That problem, though, is not one that is usually associated with the experiment, for it does not deal with the concept of a crucial experiment, or with the concepts of "ad hocness" and proper scientific methodology.[1] It does not even deal with the so-called D-thesis in any of its many forms.[2] The problem that motivates this paper is in a sense logically prior to these problems, and is simply to answer the question, Do the descriptions that scientists give of their experiments ever depend on the very theories that are being tested? The interest of the Michelson-Morley experiment from the point of view of this question stems from the fact that the theories the experiment was used to test are the very theories of observation that were used to justify the given descriptions of the experiment and its result. The principal aim of this paper then will be to demonstrate this dependence, and to thereby give an affirmative answer to the question that is the motivation for this study. I shall at the same time show how, in this particular case, the experimental descriptions were a function of the to-be-tested theories. It will be possible to show these functional dependencies without having to appeal to dubious semantic principles,[3] or to stretched uses of pychological metaphors.[4] I shall also begin to show the relevance of this sort of functional dependence for theory testing and comparison, that is, to show the sense in which this analysis of the Michelson-Morley experiment is logically prior to most others.[5]

Before analyzing the experiment in these terms, however, I want to briefly (and because of that, unfairly) summarize some of the recent philosophical literature on the nature of experimental evidence so as to clarify the problems of this paper by placing them into a broader philosophical context.

It is, I think, generally agreed that the descriptions scientists give of their experimental evidence depend on their acceptance of *some* theories. This thesis has been frequently illustrated by examples, such as Galileo's telescopic discoveries and the science of optics;[6] the observations of positrons[7] and electrons[8] in cloud chambers and the theory of ionization; Dicke's purported refutation of the General Theory by the optical oblateness of the sun and assumptions concerning optical and gravitational size.[9]

Cases of the above sort are frequent in the philosophical literature, but it is considerably more difficult, if not impossible, to find discussions of cases where the very theory to be tested is used to justify the description of the experimental text of that theory. Perhaps there is this paucity because it is assumed that such a procedure is in some way "circular," and, hence, irrational. The procedure would *seem* to be particularly pernicious if it were used to justify the theory in question *vis-à-vis* some other theory. The exact sense, though, in which such a procedure is circular, and hence pernicious, is difficult to specify. That circularity, for example, is not exposed by the frequently used positivist schema[10] for theory testing and comparison, for on that account the envisioned procedure suffers not from circularity but from simple non-instantiation of the schema. Since my later focus will be on the description of the initial conditions of the Michelson-Morley experiment, I will show this non-instantiation only with respect to theoretically dependent initial conditions. This non-instantiation can be easily seen if the variables representing the initial conditions are super-scripted:

$$o_i^1 \ \& \ t_1 \rightarrow o_1$$
$$o_i^2 \ \& \ t_2 \rightarrow o_2$$

As usual, t_1 and t_2 are two alternative theories, o_1 and o_2 their respective predictions (or postdictions), and o_i^1 and o_i^2 the initial conditions. On most positivist accounts, though, it must be the case that o_i^1 and o_i^2 represent expressions that are either formally the same (i.e., are tokens of the same type), or are logically equivalent to one another.[11] This restriction guarantees the (logical) identity of the *descriptions* of the initial conditions of the experiment, and, hence, ensures it that there is agreement as to what those initial conditions are (assuming, of course, a common semantics). It is this agreement about the descriptions of the initial experiments, and via the common semantics, about the initial conditions themselves, that makes the testing procedure rational.[12] If, however, my account of the Michelson-Morley experiment is correct, then $o_i{}^1$ and $o_i{}^2$ are not the same, and one must look elsewhere for an account of the rationality of this particular testing situation.[13]

It may be objected that this consequence has been drawn too hastily, and that it does not in fact follow from what I purport to be the historical details of the Michelson-Morley experiment. If one identifies the o's (the observation sentences) of the positivist schema merely with what scientists have as a matter of fact explicitly counted as the initial conditions and results of their experiments (say, in their research reports), then it is true that the positivist test schema does not explain the rationality of the Michelson-Morley experiment. If, however, the o's are not restricted in this way, then it appears possible to utilize the positivist test schema to account for the rationality of the case, even if part of what Michelson and Morley explicitly counted as evidence (the initial conditions) was a function of the to-be-tested theories. To see the grounds of this possibility, first consider what it means on a positivist-like account to say that some statement of initial conditions is a function of a to-be-tested theory.[14] Such an account would hold that the to-be-tested theory had to be assumed as a premise in a reconstructible inference that has the initial conditions in question as a consequence. But since the to-be-tested theory is clearly not sufficient by itself (because no theory is) to generate this consequence, it must be the case that some reconstructible observation sentences appear as additional premises in the reconstructed inference. And here these reconstructed observation sentences must be the *same* even if different to-be-tested theories are used to derive what as a matter of fact have been explicitly counted as initial conditions. Again it is the sameness of the initial conditions that ensures rationality; only now this sameness is not a natural historical product, but a philosophical reconstruction. The testing schema exemplified by the Michelson-Morley experiment now becomes:

$$o_i \& t_1 \rightarrow o_i{}^1$$
$$o_i \& t_2 \rightarrow o_i{}^2$$

theoretical dependence of initial conditions

$$o_i{}^1 \& t_1 \rightarrow o_1$$
$$o_i{}^2 \& t_2 \rightarrow o_2$$

basic test schema

On this expanded positivist account the historical use of the Michelson-Morley experiment is neither circular nor irrational. While this is indeed a comforting conclusion, it has been purchased at the cost of the reconstructible observation sentence. Fortunately, I need not descend into a discussion of the virtues of the quickly decidable sentence[15] in order to show the philosophical relevance of my analysis of the Michelson-Morley experiment with respect to the positivist test schema. That analysis will show, first of all, that the historical case instantiates neither the simple positivist testing schema nor our modified schema. Secondly, it will show that the rationality of the case can be explained *without* having to resort to the type of reconstructed inference here envisioned. This is, of course, not to say that such a reconstruction is impossible; nor is it to deny the desirability of reducing all rational patterns of scientific inference to the simple model just discussed.

While examples of experiments utilizing to-be-tested theories for their description create difficulties for those who would utilize traditional positivist schema, it would seem that such examples would be received with open arms by those who reject the utility of such schema.[16] And in fact they are so received. For example, two of the principal supporting examples in Kuhn's *The Structure of Scientific Revolutions* are cases where an experimental situation is described in terms that are basic to each of a pair of competing theories, and where those terms are not the same.

Lavoisier . . . saw oxygen where Priestly had seen dephlogisticated air and where others had seen nothing at all. . . . After discovering oxygen Lavoisier worked in a different world.[17]

Contemplating a falling stone, Aristotle saw a change of state rather than a process. . . . Galileo, on the other hand . . . saw a pendulum, a body that almost succeeded in repeating the same motion over and over again ad

439

infinitum. . . . The immediate content of Galileo's experience with falling stones was not what Aristotle's had been.[18]

An often-made objection to this dramatic way of describing science is to note that historically more neutral descriptions were available, and were in fact used.[19] Thus Priestly and Lavoisier could both agree that they were observing the gas produced by heating mercury oxide, and Galileo and the Aristotelians could both agree that they were observing swinging stones. In fact, Kuhn in his less dramatic moments describes things exactly this way![20] Nevertheless, Kuhn and Feyerabend both go on to claim that even though the same terms are used in these cases, those terms have different meanings depending on one's theoretical point of view.[21] Therefore, because of this "incommensurability," the logic of the crucial experiment cannot be reconstructed according to the positivist schema.[22] Furthermore, it is presumably a consequence of this "incommensurability" that any circularity incurred by assuming a to-be-tested theory in order to describe a test of that theory is inconsequential. Since the meaning of the descriptive terms used to describe experiments is already a function of the theoretical context that surrounds their use, it can hardly matter much if that theoretical context is assumed in some *specific* way when those terms are actually used to describe some specific experiment.

Again, I need not descend into the philosophical depths in order to show the relevance of my analysis of the Michelson-Morley experiment for these problems. For that analysis will show that some of the ways in which competing theories are incommensurable *vis-à-vis* purportedly crucial experiments can be seen without having to make use of problematic semantic principles, and without having to rely on extended uses of psychological metaphors. This is, of course, not to say that these semantical and psychological elements are not important for the analysis of science. It is to say that some aspects of incommensurability can be explained in simpler and clearer ways. The crucial concept that is needed in order to display these aspects of incommensurability is that of an idealization in science, for it is the theoretical dependence of the particular idealizations used that is the missing parameter in explicating the sense in which competing theories are incommensurable *vis-à-vis* some experimental situation.

Given these concerns with idealization and theoretical dependence, the Michelson-Morley experiment would seem to be an ideal clarifying and test case because of the following three reasons (all of which will be further developed).

1. The experiment was, and is commonly, described and analyzed in a highly "idealized" way. Among other simplifications, the interferometer arms are described as being equal, the mirrors at the end of those arms as being at right angles, and, finally, the mirror at the intersection of the arms as being at a forty-five degree angle. These descriptions are given despite the existence of good reasons for believing them false.

2. This idealized description is incompatible with the very theory that it is supposed to help support. For if this description were correct, then given the special theory of relativitiy, the split beams of the experiment would be in phase and no interference pattern would be observable. (*A fortiori,* no fringe shift would be observable.) There was, however, an observable interference pattern (and shift) in the Michelson and all other versions of the experiment. Therefore, the special theory is false if these descriptions are correct. Nevertheless, the experiment is read as supporting the special theory.

3. The most accurate methods of length determination available to Michelson were those *optical* methods that he helped develop. And in fact an elementary optical test was implicitly appealed to as a justification of the description of the interferometer arms as equal. This justification, though, depends on the very optical theory that was supposedly being tested by the Michelson-Morley experiment.

The experiment is, therefore, an excellent test case from the point of view of the motivating philosophical problems of this paper, since it contains in problematic combination the elements of idealization and theoretical dependence of descriptions.

In order to simplify matters somewhat, I will focus the attention of this paper on the question of the equality of the interferometer arms.[23] To begin, none of the historical personages explicitly give the reasons why the arms of the interferometer were made either equal or approximately equal. Furthermore, no one, except Miller in his review article of 1933,[24] gave their reasons for taking the interferometer arms to be equal. That is, there is no explicitly given rationale for placing the demand that the arms be equal, nor, except for Miller, is there any explicitly given description of the methods used to determine that this demand had been met.

Given the extreme importance and, in the minds of some, notoriety of the experiment, it is perhaps surprising that the reasons we seek for were not explicitly given. Against this, it must be remembered, as Kuhn has emphasized, that scientific research reports are intended for a professional audience that has had a similar educational background; and because of this intended audience, the contents of these reports tend to be rather abbreviated and implicit. It is genuinely surprising, though, to find that none of the standard textbooks (Michelson's included) give these reasons.[25] Our immediate problem, then, is to retrieve these reasons from their historical context.

Fortunately, they can be reconstructed on the basis of the fairly explicit descriptions that were historically given of the *procedure* used to initially adjust and to operate the apparatus.[26] Since Miller's explicitly given reason for taking the arms to be equal can only be understood in terms of this procedure, my analysis begins with that procedure.

Basically, the adjustment of the apparatus consisted of three steps. First, ordinary laboratory rules were used to determine the ''approximate'' equality of light-path lengths. Second, ''monochromatic'' light was used to provide finer adjustment, that is, the various parts of the apparatus (though usually just one of the end mirrors) were adjusted until interference fringes appeared in the viewing telescope. Finally, ''white light'' was substituted for the monochromatic source, and the apparatus further adjusted until a central dark band appeared with colored fringes. This procedure was, as Miller notes, both difficult and time-consuming.

> When the apparatus was first assembled on Mount Wilson, the time required for the approximate adjustment of the distances between mirrors with the wood rods was about one hour, for the centering of the mirrors fifteen minutes, for finding the fringes with sodium light thirty minutes, and for finding the fringes with white light forty-five minutes, or two hours and a half for the entire operation. Upon another occasion, the fringes for sodium light were found with ten minutes of searching and the white-light fringes in thirty-five minutes more.[27]

The relevance of this adjusting procedure for the question of the arm lengths of the interferometer depends on the fact that the experiment was conducted, at least by the original cast of characters and their students, with "white" light.[28] The obvious question is, of course, why was white light used in preference to the apparently simpler homogeneous light. Simpler, that is, from the point of view of the theoretical analysis of homogeneous as opposed to inhomogeneous light. Furthermore, it is, as Miller notes, easier to experimentally produce fringes with homogeneous light. Again somewhat surprisingly, the reason for using white light appears in none of the early papers, and is not given until Miller's 1933 review article.

> White light fringes were chosen for the observations because they consist of a small group of fringes having a central, sharply defined black fringe which forms a permanent zero reference mark for all readings.[29]

This sort of central and unambiguous reference band was needed because of the following considerations. The experiment required (for reasons to be given below) the more or less continuous observation of the fringes *as they shifted*. Since the requisite photographic techniques had not yet been developed, the observations had to be made directly by eye (though aided by a small telescope). But, obviously, the human physiology is not capable of continuous observation, since after all one must blink, refocus one's eyes, clear one's throat, fight off pleasant daydreams and, most importantly, fatigue. In this connection it must be remembered that the fringe shifts were visually observed for hours at a time.[30] Now, if homogeneous light is used, the interference pattern produced consists of dark bands indistinguishable other than by position, alternating with equally indistinguishable light bands. This means, given the frailties of the human perceiver, that if homogeneous light were used, then it would be possible for the pattern to shift one or more band-widths without being noticed. White light with its unambiguous central reference band was used then in order to prevent these sorts of unnoticed shifts. More generally, white light makes both short- and long-term checks on the stability of the apparatus easier because, unlike homogeneous light, constant monitoring is not needed.[31] As usual, none of these considerations are explicitly noted by any of the historical or textbook figures, perhaps, because they are unproblematically implicit in the design and operation of the interferometer.

The connection between using white light and the claimed equality of the interferometer arms is, as Miller notes,

> for the production of the interference fringes in white light, the two light paths, which are at right angles to each other . . . must be exactly equal to the fraction of a wavelength.[32]

While Miller is the only one to explicitly note this implication, it seems unlikely, because of its obviousness, that he was the first to make use of that implication as a justification for the equality of interferometer arm lengths. Furthermore, Miller collaborated with Morley on some early interferometer experiments.[33]

It will be shown below that if one ignores the difficulties with the human observer, then the interferometer arms need not be equal in order to conduct the Michelson-Morley experiment. Assuming that Michelson and Morley were aware of this possibility, it follows that the reason the interferometer arms were made ''equal'' was to allow for the use of white light so that some of the difficulties with human observers could be overcome.

One possible objection here is to note an ambiguity in my claim to have reconstructed the reason why an equal-arm configuration was chosen for the interferometer.[34] That ambiguity is between giving the reason that *should* have been given (on the explicit assumption of certain experimental goals and methods), versus giving those reasons that were psychologically effective even if not explicitly given. On this first sense of reason my claim reduces to having shown how the aims of the experiment coupled with the existing experimental techniques logically required an equal-arm configuration for the interferometer. This logical ordering, however, of premises and conclusion need not represent the historical ordering of Michelson's thought since it is both possible and likely that Michelson, having set himself the problem of determining the ether drift, *began* with a consideration of equal-path devices, realized that the equal-arm interferometer could be used to measure that drift, and then realized that the device was experimentally feasible as well. There are two reasons that make this a likely temporal ordering.[35] First, most of the preexisting interference devices used paths that were at least approximately equal.[36] Secondly, it is natural to view small differences against a symmetrical or near symmetrical background. If this pattern accurately reflects Michelson's thinking, then the psychologically effective reason for the equality of arm lengths consisted in the fact that Michelson was educated in an environment where most of the interference devices were of equal-path configuration.[37] But whatever the historical facts of Michelson's invention may have been, the ultimate justification remains the same even if that justification was not given, and even if its need was not explicitly perceived.

Most importantly, the historical facts about Michelson's reasoning regarding his invention are irrelevant with respect to the logical problems involved in using the production of white light fringes *as a test* for the equality of the arm lengths of the interferometer. To begin, using this optical test means that the description of what are generally called the initial conditions of the experiment is a function of the to-be-tested optical theory. The situation gets further complicated by the realization that strictly speaking the optical test in question only implies that it is the so-called *optical* and not the *physical* paths that are equal to within a fraction of a wavelength. In this particular case the optical path can be identified with the path of the light ray with respect to the ether, while the physical path is with respect to, and as measured by, the arms of the interferometer.[38] This distinction between optical and physical paths complicates the optical test for arm equality in the following way. In terms of the Fresnel stationary ether theory, the theory initially assumed by Michelson, the optical and physical paths will coincide only if the apparatus is at rest in the ether. Therefore, the optical test will guarantee the equality of path lengths within the error range specified only if the apparatus is at rest with respect to the ether. But it was assumed by Michelson that if there is an ether, then it is highly unlikely that the apparatus is at rest in that ether. Assuming, then, that the apparatus is in motion with respect to the ether, the optical test must be modified so as to reflect this motion. But to so modify the optical test is to assume the truth of the theory that the interferometer is supposed to test. On the other hand, not to so modify the optical test is to assume the falsity of the to-be-tested theory. Hence, the initial conditions of the experiment cannot be described without assuming the truth or falsity of the to-be-tested theory. Furthermore, to determine the degree of modification required by the motion through the ether, it is necessary to know the velocity of the apparatus with respect to the ether. But it is that velocity that the interferometer was supposed to measure in the first place! Hence, it appears that the initial conditions of the experiment cannot be adequately described without first knowing what it is the experiment is supposed to determine.

The situation so far may be summarized by saying that the to-be-tested theory was used to "derive" the description of the initial conditions of the experiment, albeit in an apparently "circular" way.

Two paths would seem to have been open to our historical protagonists:

1. to show that the above discussed inconsistencies and circularities are only apparent; and

2. to show that the observational errors involved are inconsequential.

Unfortunately, *none* of the relevant personnel explicitly noted the above-mentioned difficulties that were involved in using the white light interference test for the equality of interferometer arm lengths. Therefore, the question of which of these two ways to overcome those difficulties did not arise in any public way. This apparent lack of sophistication is, I think, due to the fact that one of the consequences of the standard, though much simplified, accounts of the experiment is that the errors involved in not having the most accurate possible determinations of interferometer arm lengths, are in a general way inconsequential. Hence, there was no need to calculate in detail the various ways in which the observational errors might go wrong. It must be remembered though that the experiment was supposed to measure a second-order effect, roughly "one hundred-millionth part of the whole time of transmission," so that the relatively inconsequential nature of the interferometer arm length determinations is *prima facie* implausible. Given this implausibility, and also to set the stage for some later objections that *were* made by the historical personnel, the standard account of the experiment, first given in part by Michelson, will now be reviewed.

Michelson's account as given in his 1887 paper is relatively brief and, hence, can be reproduced in full.[39]

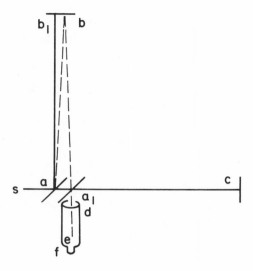

Let V = velocity of light.

   v = velocity of the earth in its orbit.

   D = distance ab or ac.

   T = time light occupies to pass from a to c.

   T_1 = time light occupies to return from c to a.

Then

$$T = \frac{D}{V - v} \, , \, T_1 = \frac{D}{V + v} \, .$$

The whole time of going and coming is

$$T + T_1 = 2D \frac{V}{V^2 - v^2} \, ,$$

and the distance traveled in this time is

$$2D \frac{V^2}{V^2 - v^2} = 2D(1 + v^2/V^2)$$

neglecting terms of the fourth order. The length of the other path is evidently $2D\sqrt{1 + v^2/V^2}$, or to the same degree of accuracy, $2D(1 + v^2/2V^2)$. The difference is, therefore, $D(v^2/V^2)$: If now the whole apparatus be turned through 90°, the difference will be in the opposite direction, hence, the displacement of the interference fringes should be $2D(v^2/V^2)$.

Note that Michelson by his use of the single variable D assumes in his analysis that the interferometer arms are equal.[40] Although this assumption simplifies the expression for the anticipated fringe shift, it does obscure the effect of the measurement errors associated with the lengths of the interferometer arms. The first person, so far as I have been able to determine, to modify this simple account by explicitly treating the interferometer arms as potentially unequal is Silberstein in his marvelous textbook of 1914.[41]

Letting $D_1 = ac$ and $D_2 = ab$, Silberstein notes that the time required for the horizontal path ac is

$$t_h = \frac{2D_1}{c(1 - \beta^2)} \tag{1}$$

where $\beta = v/c$, with $v = $ velocity with respect to the ether, and $c = $ the speed of light.

Similarly, the time required for the vertical path is

$$t_v = \frac{2D_2}{c(1 - \beta^2)^{1/2}} . \tag{2}$$

Therefore,

$$T = t_v - t_h = (2/c)\gamma(D_2 - \gamma D_1) \tag{3}$$

where

$$\gamma = \frac{1}{(1 - \beta^2)^{1/2}} .$$

The difference in optical paths D is therefore:

$$\Delta D = 2\gamma(D_2 - \gamma D_1) \tag{4}$$

That is, the difference in horizontal and vertical transit times is "very

448

nearly" (for small v) a function of the *difference* in lengths of the interferometer arms. Fortunately, though, the *change* in this difference of transit times as the apparatus is rotated is a function of the *sum* of the interferometer arms. That is, for a 90° rotation,

$$\Delta(\Delta T) = (\Delta T_0 - \Delta T_{90}) = (2/c)\gamma(\gamma - 1)(D_2 + D_1) \qquad (5)$$

$$\Delta(\Delta D) = 2\gamma(\gamma - 1)(D_2 + D_1) \qquad (6)$$

The dependence of $\Delta(\Delta T)$ on the sum of D_1 and D_2 means that the experiment does not require equality of interferometer arm lengths. Approximate equality is required, though, if white light is to be used.

Incidentally, *if* the interferometer arms were equal, then the interference pattern produced by the above 90° rotation would be *identical* to that produced before the rotation. This is because the two rays simply change the order of their arrival, keeping, however, the same degree of time separation. Hence, we see the "theoretical" reason for requiring that the interference pattern be kept under continuous observation as the apparatus is rotated. If the apparatus were viewed only upon 90° rotations, the anticipated fringe shifts would not be detectable. The concept of a theoretical reason in this context means simply a reason based upon the particular idealized treatment of the experiment that is used. A less-idealized reason for requiring continuous observation is based on the fact that there are disturbing causes that create fringe shifts of unpredictable intensity and periodicity. There must be continuous observation as the apparatus is rotated if the effects of these disturbing causes are to be analyzed away.

Returning to the question of the inconsequential nature of the differences in interferometer arm lengths, we note that because $\Delta(\Delta T)$ is a function of the *sum* of the interferometer arm lengths, this means that small errors (of the order of at most a couple of wavelengths) will be relatively *inconsequential* given that the total path lengths of D_1 and D_2 were about 40,000,000 wavelengths (of yellow light).[42] This functional dependence of $\Delta(\Delta T)$ on the *sum* of D_1 and D_2 would, therefore, seem to solve our problems about the determination of the initial conditions of the Michelson-Morley experiment. That is, the considerations we have adduced concerning the accuracy of the light test for arm equality are simply irrelevant when one considers the total packet of experimental errors.

449

It must be emphasized, though, that it was only what might be called an "accidental" fact about the particular functional relationships exhibited, and the particular values of the variables v and c, that made it possible to ignore the dependence of the description of the initial conditions on the to-be-tested theories. Had the facts been otherwise, the "logic" of this crucial experiment would have been considerably complicated. If, for example, $\Delta(\Delta T)$ had been a function of the *difference* of D_1 and D_2, then because of the limitations of the human observer and the non-existence of the necessary photographic techniques, it would not have been possible, as it was for Kennedy and Thorndike in their experiment, to maximize the difference of D_1 and D_2.[43] This means that the experiment would have required a double-description, one corresponding to relativity theory, and one corresponding to the stationary ether theory.

In order to better appreciate the philosophical consequences of these logical possibilities, the theoretical dependence of the length measurements of the interferometer arms will be developed in a more quantitative way. To begin, I want to consider whether the Michelson-Morley experiment could have been conducted by using a non-rotating interferometer and equation (4), which gives the optical path difference as a function of the relative velocity and (to within a small factor) of the difference in interferometer arm lengths.

$$\Delta D = 2\gamma(D_2 - \gamma D_1) \tag{4}$$

Of course, as a matter of historical fact neither this equation nor the Michelson "equivalent" was used to calculate the anticipated ethereal velocity. Still, it is worthwhile to consider why this apparent possibility was not acted upon since a non-rotating interferometer has several practical advantages when compared with a rotating one. In addition, if the description of the interferometer arms as equal is taken seriously, then the experimental results do seem to confirm the existence of an ethereal velocity. How, though, is equation (4) to be used to calculate the ethereal velocity? To answer this question it need only be noted that if the reference frame of the interferometer is moving with respect to the ether, then the fringes of the interferometer will vary from what they would be if the interferometer reference frame were at rest. Therefore, the interferometer apparently does not have to be rotated within the laboratory frame since the ethereal velocity can be calculated according to the following procedure. First, calculate on the basis of the inter-ferometer arm lengths where the fringes would be if the interferometer reference frame were at rest. Second, note the actual location of the fringes. Third, determine the difference between this calculation and the actual position.[44] Finally, calculate the ethereal velocity on the basis of that difference. This procedure besides its calculational simplicity has the practical advantage that it is much easier to stabilize the test condi-tions if the interferometer is not rotated. In this way one avoids all the variations due to stresses and strains on the apparatus due to rotation. And historically, not having to rotate the apparatus within the laboratory frame was one of the technical factors that made the Kennedy-Thorndike experiment possible. Why then was this procedure not followed by Michelson as well?

This question becomes even more pressing if Michelson's single variable D is used since if $D_1 = D_2 = D$, then according to the Fresnel theory there should be no interference at all if v equals zero. Since there was interference, however, the apparatus must therefore be moving with respect to the ether. A difficulty, though, arises if one attempts to translate this qualitative claim into a quantitative measure of that mo-tion. For equation (4) is based on the assumption that the apparatus is oriented with one of its arms perpendicular to the ethereal motion, while the other arm is parallel to that motion. But clearly no way has been given for determining that this orientation assumption has been met. This difficulty can be partially met once it is realized that equation (4) can be used to provide a *minimum* measure of the ethereal velocity even in the absence of the knowledge that the orientation assumption has been met. The equation can be used in this way because it gives the *maximum* effect to be expected from motion with respect to the ether. That is, all other orientations of the Michelson apparatus will give a smaller fringe displacement for the same over-all velocity with respect to the ether. Therefore, if fringes are observed, then those fringes can be used as the basis of a calculation of the smallest possible ethereal velocity consonant with that effect.

If the equality of D_1 and D_2 could be determined, then the existence of fringes would entail that ΔD was less than a wavelength, approximately 5×10^{-7} meters.[46] Let x represent this value. The ethereal velocity, v, then, could presumably be calculated on this basis. The difficulty with

451

this approach is that the only method for determining the equality of D_1 and D_2 with anything like the required accuracy also required the use of equation (4). If terms of order higher than the second are ignored, the functional dependencies exemplified in (4) are considerably simplified, and one obtains:

$$x \geqslant \Delta D = 2(D_2 - D_1) + \beta^2(D_2 - 2D_1) \tag{7}$$

The maximum difference in length between the interferometer arms is, therefore, $x/2$, and occurs when the ethereal velocity is equal to zero. These functional dependencies would appear to be exactly what are wanted for the purpose of fixing the initial conditions for the Michelson-Morley experiment, since a maximum error limit is set for the interferometer arm difference even if one is totally ignorant of the relative velocity of the apparatus with respect to the ether.

Unfortunately, the maximum set for v is not nearly so small, and is for an interferometer length of 10 meters, approximately 7×10^4 meters/ second. Even if one were to whittle D down by a factor of 10, the maximum ethereal velocity would still be 7×10^3 meters/second. Furthermore, it is obvious that equation (4) cannot be used to calculate v more accurately on the basis of its determination of the equality of the interferometer arms to within $\pm x$, because there is no way of dividing that part of the total effect D that is due to a simple difference in interferometer arm lengths and that part due to motion through the ether.

Fortunately, the Michelson apparatus could be rotated without undue ill effect so that the anticipated fringe *shift* could be tested, that is, the apparatus could ordinarily be rotated without causing the fringes to disappear. (This is not to say that rotation had no effect on the fringes.) If the construction materials available to Michelson had not been sufficiently immune to rotational and other environmental effects, then it would not have been possible to conduct the experiment.[47] In this case the lack of suitable construction materials would have made the descriptions of the initial conditions of the experiment overly dependent on the to-be-tested theory. It is the possibility of cases such as these that explains, in part, our intuitions about the "circularity" of any experimental procedure that utilizes a to-be-tested theory in the very test of that theory. It appears, though, that not all examples of such utilization suffer from these defects of circularity, since as the Michelson case shows, Michelson was able to use the to-be-tested Fresnel theory to justify his description of the interferometer arms as equal, and at the same time use that description as part of a successful test of that to-be-tested theory.

What was it, then, about the ability to rotate the apparatus that made both the theoretically dependent descriptions and the experiment itself possible? The most visible advantage gained by rotation is a doubling of the size of the anticipated effect, which on Michelson's account increases from $D\beta^2$ to $2D\beta^2$. Obviously, a mere doubling of effect will not help much by itself, since what is needed is some way of dividing the causes of the interference effects into those due to the ethereal motion and those due to simple differences in the lengths of the interferometer arms. Because Michelson's use of the single variable D obscures this division process, as well as the error behavior, equation (6) will be utilized instead of Michelson's.

$$\Delta(\Delta D) = 2\gamma(\gamma - 1)(D_1 + D_2) \tag{6}$$

Or ignoring terms of higher order,

$$\Delta(\Delta D) = \beta^2(D_1 + D_2) \tag{8}$$

Here, as if by magic, the functional dependence of the anticipated effect is now on the *sum* as opposed to the difference of the interferometer arm lengths.[48] On the assumption that D_1 and D_2 remain constant during rotation,[49] that part of the fringe production due to simple differences in the interferometer arm lengths drops out with the result that equation (8) measures only the algebraic sum, i.e., the *change* of effects due to motion. Therefore, because of this fortuitous behavior of the plus and minus signs, the necessary number of significant figures for the interferometer arm length measurements could be supplied by simple non-optical measuring methods that did not depend on the to-be-tested theory.

An additional benefit brought about by equation (8) is the periodicity of effect that it implies. It is this periodicity that enabled Michelson and others to separate out from the fringe data that part of the observed shift that was due to extraneous or disturbing causes. The assumption was made that these extraneously caused shifts would not have the periodicity of the shift due to the rotation of the interferometer within a moving ethereal frame.[50]

Being able to rotate the interferometer and, thus, to use equation (8), meant that the purpose of the experiment could be satisfied by the available measuring techniques. But what exactly is the purpose of the Michelson-Morley experiment? It is now time to more carefully distinguish between using the interferometer as a means of determining the ethereal velocity v, and using it as a means of testing the Fresnel ether theory. Michelson reported that the original purpose of the experiment was to measure the velocity of the earth with respect to the ether, that is, the interferometer was originally conceived as an ethereal speedometer and was only later seen as a test of the Fresnel ether theory. In order to use the interferometer as an ethereal speedometer, one need only solve for v in equation (6). Michelson, though, never made it to this stage. His own approach was to first *estimate* the magnitude of the anticipated effect as a test of the *feasibility* of the experiment. He made this estimate by using the earth's orbital velocity as the *minimum* ethereal velocity to be expected. On the basis of this value, Michelson notes:

> Considering the motion of the earth in its orbit only, this displacement should be
>
> $$2D(v^2/V^2) = 2D \times 10^{-8}$$
>
> The distance D was about eleven meters, or 2×10^7 wavelengths of yellow light; hence the displacement to be expected was 0.4 fringe.[51]

The anticipated effect, therefore, could be observed by the experimenter. The next question was, would this anticipated shift be swamped by shifts due to other causes? As Michelson notes, it very often was, especially in the first version of the experiment.

> In the first experiment, one of the principal difficulties encountered was that of revolving the apparatus without producing distortion; and another was its extreme sensitiveness to vibration. This was so great that it was impossible to see the interference-fringes except at brief intervals when working in the city, even at two o'clock in the morning.[52]

Fortunately, there were cases when shifts due to disturbing causes did not occur in a way where they could not be analyzed away. Unfortunately, these were also cases where the anticipated shift due to the ether velocity did not occur! That is, the fringe shifts observed were regular but not of the periodicity expected on the basis of the Fresnel theory. When these shifts, presumably due to "extraneous" causes, were subtracted from the data, the remaining shifts had neither the intensity nor the periodicity required by the Fresnel theory. Therefore, the situation appeared hopeless for the Fresnel ether theory since the estimate of 0.4 fringe represented the smallest possible shift given the minimum estimate of v as 30 kilometers per second. Furthermore, Michelson's subtractive technique of data analysis seemed to eliminate any hope for saving the theory by means of appeals to disturbing causes. Both of these claims were attacked historically, and some of these attacks will be discussed below. But before going on to these attacks and other matters, I want to summarize the situation so far with respect to the given descriptions of the lengths of the interferometer arms.

Michelson could get by with his description of the interferometer arms as equal because the white light optical test represented by equation (7) has as a consequence that the maximum difference between the arms is no more than a wavelength. On the other hand, that accuracy was insufficient to calculate the ethereal velocity via equation (7). Since, however, the apparatus could be rotated, that error became irrelevant because the fringe shift upon rotation is a function of the sum of the interferometer arm lengths. Therefore, it did not matter if Michelson continued to describe those arms as equal *even though* the null result of the experiment coupled with equation (7) has the result that the interferometer arms are *not* equal. It is, in part, the existence of this reason for holding that the arms are not equal that has motivated me to call Michelson's description of the interferometer arms as "equal" an idealization. (Other, and more important, reasons for calling this description an idealization are given below.)[53]

Now, as already suggested earlier, this happy (if somewhat muddled) state of affairs would not have obtained if some of the crucial facts and functional relationships had been otherwise. I want now to make good on an earlier promise and draw out some of the philosophical consequences of these logically possible alternative worlds.

We have already noted the importance of the availability of reasonably rigid materials and stable environments. The lack of such materials and environments would have meant that the theoretical dependency of the description, of the lengths of the interferometer arms, was such as to make the experiment impossible. The experiment would also have been, if not impossible, at least extremely difficult if some of the functional relations represented by equations (4) and (6) had been different. For example, consider the theory T_a which has as consequences:

$$\Delta D = v_1\beta^2 k\ (\Delta s) + \Delta s \tag{9}$$

$$\Delta(\Delta D) = (v_1 - v_2)\beta^2 k\ (\Delta s) \tag{10}$$

Here v_i is a simple function (e.g., $\sin\theta$) of the angular orientation of the interferometer with respect to the ethereal motion. Also, $\Delta s = (D_1 - D_2)$, and k is some constant.

So that this possibility is not considered just as a mere possibility, it is to be noted that the interferometer test for time dilation, as opposed to spatial contraction, is a function of Δs. This is why the differences in the interferometer arm lengths were maximized in the Kennedy-Thorndike experiment.[54] But, unlike the Kennedy-Thorndike case, assume that the photographic and other experimental techniques that made that experiment possible are not available to proponents of T_a as indeed they were not to Michelson. That is, assume that the interferometer experiment must be done with white light. Given these assumptions about available experimental technique, would it have been possible to test theory T_a using the Michelson interferometer?

To begin, it will be further assumed that the constant k is of suitable order of magnitude, such that ΔD calculated via (9) has the same order of magnitude as it would had it been calculated via (4). If this assumption is not made, then the envisioned experimental test might be impossible simply because the shift predicted by T_a is below the powers of observational discrimination. Is the experiment feasible then for suitable values of k?

Equation (9) is somewhat worse off than the historical (7) since (9) sets a maximum value only for Δs. Unfortunately, equation (10), unlike (6), still contains Δs, so that optical methods must still be used in order to obtain the necessary measurements of the interferometer arms. But this means that the error associated with that measuring technique is such as to swamp the predicted fringe shift, since the experimental error associated with ΔD (due to the error associated with Δs) is of at least the same magnitude as the anticipated effect.[55] Therefore, unlike the historical case, the Michelson interferometer could not be used to test theory T_a. This experimental impossibility means that the Michelson interferometer experiment could not be used to decide between T_a and the Fresnel ether theory, that is, the experiment could not be a crucial experiment. This impossibility of a crucial experiment is not because of any general semantic incommensurability between T_a and the Fresnel theory; nor is it due to a paradigm-induced blindness. Neither is it due to any radical disagreements as to experimental standards or techniques. Finally, it is not due to differences in auxiliary theories. It is due simply to the ways in which the descriptions of the initial conditions and their associated errors depend on the theories that are being tested.

This last claim needs to be refined somewhat since all of our arguments so far have been based on Michelson's simple ray analysis of the interferometer. That is, the *justification* for using the idealized description of the interferometer arms as equal had itself been based on a very idealized treatment of the operation of the experiment. Therefore, the given descriptions of the initial conditions of the Michelson-Morley experiment are a function not only of the to-be-tested theory, but also of a particular treatment or description of the operation of the experiment.

Historically, it was hoped by some that a more realistic treatment of the experiment would show that Michelson's results were really compatible with the Fresnel stationary ether theory. If this could be shown, then the Fitzgerald-Lorentz spatial contraction hypothesis would not be necessary. The following specific objections were raised against the Michelson ray analysis.[56]

1. The single-ray analysis given cannot explain the production of interference fringes since it deals only with the rays arriving at a single point.

2. Furthermore, the rays of the single-ray analysis do not even converge on a single point.

3. According to the Fresnel theory, the ordinary laws of reflection and refraction are violated in moving systems.

4. The apparatus because it is rotating during the experiment (both with respect to the laboratory and with respect to the sun) is constantly undergoing a (varying) acceleration.

457

There were two sorts of reaction to these difficulties. The one sort consisted of attempts to trace out in detail the total history of the respective wave fronts according to Huygens's principle. The other response was Lorentz's and consisted of the development of a *generalized* optical proof based on Fermat's principle and the calculus of variations.

All of the examples of the first sort shared, besides their common use of Huygens's principle, one important additional feature, namely, a mutual inconsistency. For example, some analyses claimed that a null result was to be expected given the stationary ether and no contraction. Others claimed a first-order effect of period π in addition to, or sometimes in place of, the second-order effect of period $\pi/2$. In fact, the only detailed Huygens's principle account consistent with Michelson's original simplified theory is Kennedy's relatively late 1935 account.

Lorentz's approach also had the advantage of reinforcing the result of the original simplified account. It had, though, the disadvantage of apparently not being well understood by those of Lorentz's contemporaries who were actively interested in the Michelson-Morley experiment. Lorentz promised during the 1927 "Summit Conference" on the experiment to further clarify the proof and to show its connection with the detailed Huygens's principle proofs. Regrettably his death, very soon after, made that impossible.

While a detailed analysis of these proofs would further clarify the nature of descriptive dependency on to-be-tested theories as well as be interesting in its own right, that detailed analysis need not be undertaken in order to appraise in general terms the relevance of these proofs for that dependency. And the relevance is as follows. Closely allied with the "idealized" description of the interferometer is a very simplified treatment of its "operation," where that simplified treatment is used to justify the particular idealization used in the description of the initial conditions of the experiment. Now just as the justification for the idealized description of the initial conditions was sought for in what might be called an idealized treatment of the experiment's operation, so, too, justification was sought for that idealized treatment in terms of a more developed analysis, but one that again is usually in many ways still idealized. There are, then, a hierarchy of justifications that one ascends only if pressed.

458

It is also to be noted that not only are the descriptions of initial conditions a function of these idealized treatments of the operation of experiments, but the specification of the *cause* of either actual or anticipated effects is also a function of these treatments. For example, on the elementary Michelson ray analysis the cause of the interference fringes is the difference in the optical path lengths, which, in turn, is caused by the difference in the interferometer arm lengths, as well as by the motion through the ether. On more realistic accounts the cause of the interference fringes is now the fact that the split beams of the experiment are no longer parallel when they converge on the viewing telescope. This lack of parallelness is caused by the inclination of the reflecting mirrors away from the ideal ninety degrees, as well as by the fact that the law of reflection is modified because of motion through the ethereal medium.

The basic descriptive and explanatory methodology seems to be to try to get by with as little as possible. If disaster strikes, remove the simplifications and idealizations as slowly as possible, i.e., piecemeal, until the disaster goes away. Unfortunately, for ether theorists, it appeared that proponents of the special theory of relativity could get by with very little, i.e., with very idealized and simplified descriptions and accounts of the Michelson-Morley experiment. Ether theorists, on the other hand, could not get by with so little, and hence were faced with the potentially (and, as it turned out, ultimately) thankless task of trying to fend off what was for them a disaster by the successive removal of simplifications and idealizations. Crucial experiments, it appears, rarely treat their theoretical suitors fairly. This sort of favoritism is serious since it may be impossible for proponents of some theories to remove enough simplifications and idealizations so as to satisfactorily explain the results of crucial experiments. This impossibility may be because of psychological limitations on man's inventiveness and endurance, or because of the absence of the necessary mathematical techniques, or because of the absence of the necessary data.

I have already noted that crucial experiments do not always parcel out measurement errors equally to theoretical competitors. To that unfairness must be added this latest example of favoritivism. Both are causes for two scientific theories to be incommensurable *vis-à-vis* some purportedly crucial experiment.

459

I would like to thank Professors Arnold Koslow, Roger Stuewer, Wesley Salmon, Kenneth Shaffner, Laurens Laudan, Peter Machamer, and Carl Nielson for their helpful comments on an earlier draft of this paper that was read at the Ohio State University. Roger Stuewer is especially to be thanked for his extensive and critical remarks.

1. I have in mind here primarily the work of Adolf Grünbaum, especially *Philosophical Problems of Space and Time* (New York: Alfred Knopf, 1963); "The Special Theory of Relativity as a Case Study of the Importance of the Philosophy of Science for the History of Science," *Delaware Seminar in the Philosophy of Science*, vol. 2 (New York: Interscience Publishers, 1963); "The Bearing of Philosophy on the History of Science," *Science* 118 (1964): 1406–12.

2. For varying statements of the D-thesis see: Phillip Quinn, "The Status of the D-thesis," *Philosopny of Science* 36 (1969): 381–99; Grünbaum, "Can We Ascertain the Falsity of a Scientific Hypothesis?", in *Observation and Theory in Science* (Baltimore: Johns Hopkins University Press, 1969), pp. 69–129.

3. For example, Paul Feyerabend's claim that the positivists' observation terms get their meaning from the theory they are attached to, in: "An Attempt at a Realistic Interpretation of Experience," *Proceedings of the Aristotelian Society* 58, n. s. (1958): 143–70; "Problems of Empiricism," in *Beyond the Edge of Certainty*, ed. Robert G. Colodny (Pittsburgh: University of Pittsburgh Press, 1965), pp. 145–260; "Reply to Criticism," in *Boston Studies in the Philosophy of Science*, ed. Robert S. Cohen and Marx W. Wartofsky (New York: Humanities Press, 1965), pp. 223–62.

4. The paradigmatic examples are, of course: Thomas S. Kuhn, *The Structure of Scientific Revolutions* (Chicago: University of Chicago Press, 1962); and Norwood Russell Hanson, *Patterns of Discovery* (Cambridge: At the University Press, 1965).

5. Grünbaum, for example, takes the historically given descriptions of the initial conditions and results of the Michelson-Morley experiment as unproblematic. He focuses his concern on the logical relations that exist between these descriptions and the competing theoretical explanations of them. The rationale of this focus is twofold, namely, to test the logical and methodological limits that can be placed on the addition of auxiliary hypotheses. My claim is that these neat logical relations disappear if one pays more attention to the problematic nature of the experimental descriptions. Grünbaum, though, does show in a footnote some concern for possible difficulties with these descriptions and suggests a way to alleviate them. I will show in note 53 that Grünbaum is mistaken here.

6. Feyerabend, "Problems of Empiricism, II," *The Nature and Function of Scientific Theories*, ed. Robert G. Colodny (Pittsburgh: University of Pittsburgh Press, 1970), pp. 275–353; Peter K. Machamer, "Feyerabend and Galileo: The Interaction of Theories, and the Reinterpretation of Experience," *Studies in the History and Philosophy of Science* 4 (1973): 1–46.

7. Hanson, *The Concept of the Positron* (Cambridge: At the University Press, 1963).

8. Marshall Spector, "Theory and Observation," *British Journal of the Philosophy of Science* 17 (1966): 17.

9. Grünbaum, "Can We Ascertain the Falsity of a Scientific Hypothesis?", pp. 74–85.

10. For example, these schemata are used in the following places: Quinn, "The Status of the D-thesis"; Grünbaum, "Can We Ascertain the Falsity of a Scientific Hypothesis?"; James W. Corman, "Craig's Theorem, Ramsey-Sentences, and Scientific Instrumentalism," *Synthese* 25 (1972): 82–128. The schemata are also an essential part for the Bayesians, e.g., Wesley C. Salmon, "Bayes's Theorem and the History of Science," in *Historical and Philosophical Perspectives of Science*, ed. Roger Stuewer (Minneapolis: University of Minnesota Press, 1970), pp. 68–86.

11. It might also seem reasonable to allow materially equivalent descriptions as well, assuming of course that these equivalences were reflected in the competing theories so that the deductive consequences remained either logically or materially equivalent to what they would have been in the absence of materially equivalent initial conditions.

12. On one reading, at least, Carnap's pragmatic criteria for observation sentences were an attempt

to guarantee rationality without having to make use of semantic or reference talk. Feyerabend attempted in "Problems of Empiricism" to develop this line but later gave it up.

13. A typical move is to try to identify a common reference even if the scientific proponents describe the presumably common object in radically different ways. The Priestly-Lavoisier oxygen-dephlogisticated-air case is a favorite here, as is the acids and bases case.

14. This account takes some of its inspiration from Grünbaum, "Can We Ascertain the Falsity of a Scientific Hypothesis?"

15. For some of the virtues and vices of the quickly decidable sentences see: Feyerabend, "An Attempt at a Realistic Interpretation of Experience" and "Problems of Empiricism"; Spector, "Theory and Observation."

16. For example, Kuhn, *The Structure of Scientific Revolutions,* pp. 145–47.

17. Ibid., p. 117.

18. Ibid.; this is an amalgam from pages 123, 118, 124.

19. This is exactly our modified positivist schema, only gotten in a less *a priori* way. For this latter approach see Mary Hesse, "Is There an Independent Observation Language," in *The Nature and Function of Scientific Theories,* pp. 35–78; Kenneth E. Schaffner, "Outlines of a Logic of Comparative Theory Evaluation with Special Attention to Pre-and-Post Relativistic Dynamics," in *Historical and Philosophical Perspectives of Science,* pp. 311–53.

20. Kuhn, *The Structure of Scientific Revolutions,* chap. 10.

21. Feyerabend, "An Attempt at a Realistic Interpretation of Experience" and "Problems of Empiricism"; Kuhn, *The Structure of Scientific Revolutions,* pp. 148–49.

22. For Feyerabend's attempt to reconstruct the logic of crucial experiments and incommensurable theories, see his "Reply to Criticism," pp. 230–34; "Consolations for the Specialist," in *Criticism and the Growth of Knowledge,* ed. Imre Lakatos and Alan Musgrave (Cambridge: At the University Press, 1970), p. 226; "Against Method," in *Analyses of Theories and Methods of Physics and Psychology,* ed. Michael Radner and Stephen Winohur (Minneapolis: University of Minnesota Press, 1970), pp. 88–90.

23. One of the reasons for choosing this focus is that the Michelson-Morley experiment is sometimes distinguished from the Kennedy-Thorndike experiment by contrasting the equality of Michelson interferometer arms with the inequality of those of the Kennedy-Thorndike apparatus. This distinction, though true in fact, does not capture the essence of the difference between the two experiments. See my "Michelson-Morley, Kennedy-Thorndike, and the Equality of the Interferometer Arms" (in preparation).

24. Dayton C. Miller, "The Ether-Drift Experiment and the Determination of the Absolute Motion of the Earth" (hereafter cited as "The Ether-Drift Experiment"), *Review of Modern Physics* 5 (1933): 204–42.

25. These include: L. Silberstein, *The Theory of Relativity* (London: Macmillan, 1914, 1st ed.; 1924, 2d ed.); A. A. Michelson, *Studies in Optics* (Chicago: University of Chicago Press, 1927); Georg Joos, *Theoretical Physics,* trans. I. M. Freeman (New York: Hafner Publishing Co., 1934); Peter Gabriel Bergmann, *Introduction to the Theory of Relativity* (Englewood Cliffs, N.J.: Prentice-Hall, 1942); S. Tolansky, *An Introduction to Interferometry* (London: Longmans, Green & Co., 1955); Wolfgang K. H. Panofsky and Melba Phillips, *Classical Electricity and Magnetism* (Reading, Mass.: Addison-Wesley, 1955); J. Aharoni, *The Special Theory of Relativity* (Oxford: Clarendon Press, 1965); Max Born, *Einstein's Theory of Relativity* (New York: Dover Publications, 1962); Max Born and Emil Wolf, *Principles of Optics,* 4th ed. (Oxford: Pergamon Press, 1970).

26. Albert A. Michelson, "The Relative Motion of the Earth and the Luminiferous Ether" (hereafter cited as "Relative Motion of the Earth"), *American Journal of Science* (3), 34 (1887): 339; Miller, "The Ether-Drift Experiment," p. 210.

27. Miller, "The Ether-Drift Experiment," p. 210.

28. An excellent bibliography of the interferometer tests for the ether drift is given in Loyd S. Swenson, Jr., *The Ethereal Aether* (Austin: University of Texas Press, 1972).

29. Miller, "The Ether-Drift Experiment," p. 214.

30. Ibid., p. 212.

31. Ibid., p. 215.

32. Ibid., p. 214.

33. Morley and Miller, "Report of an Experiment to Detect the FitzGerald-Lorentz Effect," *Proceedings of the American Academy of Arts and Sciences* 41 (1905): 321–28; "Final Report on Ether-Drift Experiments," *Science* 25 (1907): 525.

34. Roger Stuewer raised this objection against an earlier version of this paper.

35. For evidence that Michelson's interest in the problems of measuring the ether drift came before his study of optical interference, see R. S. Shankland, "Michelson-Morley Experiment," *American Journal of Physics* 32 (1964): 17.

36. For some examples see Michelson, *Studies in Optics*, pp. 15–28; Tolansky, *An Introduction to Interferometry*, pp. 43–117; Born and Wolf, *Principles of Optics*, pp. 260–322.

37. There is also, at least implicitly, a logical difference between Michelson's apparent reasoning and my reconstructed justification. On my justification the equal-arm configuration is necessary and sufficient for the satisfaction of the experimental goals given that white light is to be used. For my reconstructed Michelson only the sufficiency part of the above claim follows.

38. More accurately the optical path is equal to the light path with respect to the ether frame multiplied by the index of refraction of the medium and by the speed of light in vacuum.

39. Michelson, "Relative Motion of the Earth," pp. 335–36.

40. Actually, Michelson nowhere in "Relative Motion of the Earth" explicitly describes the interferometer arms as being equal. He tells us that "the lengths of the two paths were measured by a light wooden rod reaching diagonally from mirror to mirror, the distance being read from a small steel scale to tenths of millimeters. The difference in the lengths of the two paths was then annulled by moving the (adjustment) mirror." Michelson then describes the paths as being "approximately equal," and goes on to note the interference adjustments made with homogeneous and white light.

41. Silberstein, *The Theory of Relativity*, pp. 72–77.

42. Michelson, "Relative Motion of the Earth," p. 341; Miller, "The Ether-Drift Experiment," p. 209.

43. Roy J. Kennedy and Edward M. Thorndike, "Experimental Establishment of the Relativity of Time," *Physical Review* 42 (1932): 400–418.

44. One of the difficulties with this non-rotating version of the Michelson-Morley experiment is the determination of the difference between the actual fringes and those calculated on the basis of zero ether velocity. However, since I have so far avoided, in the interests of clarity and reasonable length, the difficulties of translating ΔT into fringes and $\Delta(\Delta T)$ into fringe shifts, I will not pursue these difficulties here.

45. If both of the interferometer arms are in any one of the planes defined by the etherial motion and $D_1 = D_2 = D$, and θ is the inclination of one of the interferometer arms to that etherial motion, then:

$$T = \frac{2D}{c^2 - v^2}\left[(c^2 - v^2 \sin^2\theta)^{\frac{1}{2}} - (c^2 - v^2 \cos^2\theta)^{\frac{1}{2}}\right].$$

46. I have simply used the wavelength of yellow light. This estimate can be refined either up or down depending on the particular details of the case. The particular value is more or less irrelevant to the points being made in this section of the paper. Furthermore, it cannot be refined on the basis of the simple ray account. More realistic treatments of the experiment are discussed below. Cf. Born and Wolf, *Principles of Optics*, pp. 300–302.

47. For some of the details on the rigidity of the interferometer see Miller, "The Ether-Drift Experiment," p. 215.

48. Although it is fairly easy to intuitively appreciate the sign behavior here, it is not easy to explain it in a way that is both clear and brief. This is the reason one never finds such an explanation even in the textbooks. This lack of an explanation is also supported by the maxim that if you are on the right track, the signs will cancel; if you are on the wrong track, it does not matter if you improperly cancel signs.

49. This is, of course, a very important assumption and one which cannot be justified, without circularity, by equations (4) or (7). Such a procedure would be nearly identical to the one already discussed using a non-rotating interferometer. More indirect and inductive arguments have to be and were used here. Unfortunately, an analysis of those arguments will have to wait until another time.

50. The most careful attempt to utilize this periodicity was by Miller. For an appraisal of Miller's work see R. S. Shankland et al., "New Analysis of the Interferometer Observations of Dayton C. Miller," *Reviews of Modern Physics* 27 (1955): 167–78.

51. Michelson, "Relative Motion of the Earth," pp. 340–41.

52. Ibid., p. 336.

53. As noted above, note 5, Grünbaum takes the historically given descriptions of the Michelson-Morley experiments as unproblematic. He does, however, in a footnote of his own raise a problem about the equality of the interferometer arms. (This footnote appears in both "Can We Ascertain the Falsity of a Scientific Hypothesis?", pp. 103–4, and *Philosophical Problems of Space and Time*, p. 394). Grünbaum notes:

> Since the aether-theoretically expected time difference in the second order terms is only of the order 10^{-15} second, allowance had to be made *in practice* for the absence of a corresponding accuracy in the measurement of the equality of the two arms. This is made feasible by the fact that, on the aether theory the effect of any discrepancy in the lengths of the two arms should *vary*, on account of the earth's motion, as the apparatus is rotated. For details see Bergmann, *Introduction to the Theory of Relativity*, pp. 24–26, and J. Aharoni, *The Special Theory of Relativity*, pp. 270–73. Indeed slightly unequal arms are needed to produce neat interference fringes. (My italics.)

Grünbaum has apparently overlooked the fact that $\Delta(\Delta T)$ is a function of the sum of the interferometer arm lengths. This oversight is especially surprising since the passage that Grünbaum refers to, in Bergmann and Aharoni, contains the proper expression for $\Delta(\Delta T)$. Furthermore, Bergmann and Aharoni (especially the latter) explicitly note that this was the method followed *in practice*. Also, neither Bergmann nor Aharoni shows how variations in the difference of the interferometer arm lengths would manifest themselves from frame to frame. For some of the difficulties involved in measuring the anticipated effects from frame to frame, see Miller, "The Ether-Drift Experiment," and Kennedy and Thorndike, "Experimental Establishment of the Relativity of Time."

54. These expressions are almost the reduced forms (i.e., ignoring higher order terms) of the basic equation for the Kennedy-Thorndike experiment:
$$n = v' \Delta s/c(1 - \beta^2)^{\frac{1}{2}}$$
The variable v', which here represents frequency as measured by a moving observer, has been changed to our trigonometric function. The wave number n has been changed to ΔT. Cf. Kennedy and Thorndike, "Experimental Establishment of the Relativity of Time," pp. 401–5. I have also assumed for the purpose of the philosophical point that Δs does not cancel out. One might imagine, for example, that the sign function was served by v_1 and v_2.

55. One can conceive of a possibility (perhaps not a real one) where the trigonometric functions v_i might conspire to make the experiment possible although with large error. This would happen if $\Delta(\Delta T) > \Delta T$. In this case T_a would have the advantage over the Fresnel theory (T_f) because T_a would not have to fit the facts as closely. On the other hand, any such victory over T_f would be correspondingly hollow. It would also be the case that the *methods* used to determine the initial conditions of the two experiments would be different since simple laboratory rules would be

463

sufficient for measuring D_1 and D_2 for T_f. Any use of optical methods by proponents of T_f would be purely gratuitous as it was for Michelson. Proponents of T_a, since it requires the difference $D_1 - D_2$, would have to use optical methods.

56. For examples and further references see the following: W. M. Hicks, "On the Michelson-Morley Experiment Relating to the Drift of the Aether," *Philosophical Magazine* (6), 3 (1902): 9–36; "Conference on the Michelson-Morley Experiment," *Astrophysical Journal* 68 (1928): 341–73; Roy J. Kennedy, "Simplified Theory of the Michelson-Morley Experiment," *Physical Review* 47 (1935): 965–68.

Kenneth F. Schaffner

Chapter Seventeen

SPACE AND TIME IN LORENTZ, POINCARÉ, AND EINSTEIN:
DIVERGENT APPROACHES TO THE DISCOVERY AND DEVELOPMENT
OF THE SPECIAL THEORY OF RELATIVITY

I. INTRODUCTION

In the first 1909 edition of his book *The Theory of Electrons,* the great Dutch physicist H. A. Lorentz made some comments on the relation between Einstein's then still recent special theory of relativity and his own older theory of the electrodynamics of moving bodies.[1] Though he found that Einstein's theory possessed a "fascinating boldness," and that it embodied a "remarkable reciprocity" between different frames of references, Lorentz still favored his own theory. Lorentz thought that Einstein had arrived at experimental results that were in essential agreement with those that he, Lorentz, had obtained, and that Einstein had "simply postulated" what he had "deduced, with some difficulty and not altogether satisfactorily, from the fundamental equations of the electromagnetic field."[2]

By 1915 Lorentz had come to see deeper into the value of Einstein's achievement and into the conceptual differences between their approaches. In a long footnote appearing in the second edition of *The Theory of Electrons,* Lorentz wrote:

> If I had to write the last chapter now, I should certainly have given a more prominent place to Einstein's theory of relativity (§ 189) by which the theory of electromagnetic phenomena in moving systems gains a simplicity that I had not been able to attain. The chief cause of my failure was my clinging to the idea that the variable t only can be considered as the true time and that my local time t′ must be regarded as no more than an auxiliary mathematical quantity. In Einstein's theory, on the contrary, t′ plays the same part as t; if we want to describe phenomena in terms of x′, y′, z′, t′ we must work with these variables exactly as we could do with x, y, z, t.[3]

Lorentz's assessment of the central contribution of Einstein's approach is supported by diverse historical testimony. The revolutionary approach to the concept of time, and the simplicity that such a *kinematical* approach introduced into the study of the electrodynamics of moving bodies, is stressed by a number of the early writers on relativity theory. H. Minkowski in his important 1908 essay "Fundamental Equations for Electromagnetic Processes in Moving Bodies," after attributing the discovery of relativity to Lorentz, added that "Einstein has brought out the point very clearly that this (relativity) postulate is not an artificial hypothesis but is rather a new way of comprehending the time-concept which is forced on us by observation of natural phenomena."[4] A famous letter of recommendation for Einstein supporting him for membership in the Prussian Academy of Sciences and written in 1913 by Max Planck and three other distinguished scientists, Nernst, Rubens, and Warburg, discussed Einstein's contributions to relativity theory and asserted that "[Einstein's] new interpretation of the time concept has had sweeping repercussions on the whole of physics, especially mechanics and even epistemology."[5]

W. Pauli, in his superb 1921 monograph on relativity, similarly wrote concerning Einstein's contribution and its relation to Lorentz's (and Poincaré's modifications of Lorentz's theory):

> It was Einstein, finally, who in a way completed the basic formulation of this new discipline. His paper of 1905 was submitted at almost the same time as Poincaré's and had been written without previous knowledge of Lorentz's paper of 1904. It includes not only all the essential results contained in the other two papers, but shows an entirely novel, and much more profound, understanding of the whole problem. . . .
>
> . . . Lorentz and Poincaré had taken Maxwell's equations as the basis of their considerations. On the other hand, it is absolutely essential to insist that such a fundamental theorem as the covariance law should be derivable from the simplest possible basic assumptions. The credit for having succeeded in doing just this goes to Einstein. He showed that only the following single axiom in electrodynamics need be assumed: *The velocity of light is independent of the motion of the light source.* . . .
>
> . . . Still further, Einstein showed in particular that the distinction between "local" and "true" times disappears with a more profound formulation of the concept of time. Lorentz's local time is shown to be simply the time in the moving system k'. . . . It is also of great value that Einstein rendered the theory independent of any special assumptions about the constitution of matter.[6]

466

Essentially what Einstein did in this area of inquiry, then, was to demonstrate that the various perplexing problems were due to employing classical notions of *kinematics* that had outlived their usefulness. A conceptual reanalysis of the time concept and a realization that a theory of the optics and electrodynamics of moving bodies needed to be constructed without a dependency on the "constructive" theories of matter and electrodynamics provided the radically new viewpoint of Einstein's approach.

In spite of Lorentz's clear but tantalizingly brief statement of the conceptual difficulties he had experienced with Einstein's theory, and in spite of the stress on the significance of Einstein's radically novel approach to the time concept by his contemporaries and near contemporaries, there has been no in-depth historical study of the Lorentz and Poincaré analyses of time and space in electrodynamics and their comparative relation to Einstein's account.[7] It is one of the purposes of this paper to provide such a conceptual contrast. I shall also in my discussion of Einstein dwell at length on the problem of the genesis of the special theory of relativity in Einstein's mind. This facet of the paper will allow us to look at the interactions between theory and experiment in the discovery and development of the theory, and also to suggest certain reasons why Einstein pursued an independent path from Lorentz and Poincaré. I believe this account will indicate the relevance of what I shall term transempirical factors in both the discovery and justification of the theory and, furthermore, that it will suggest that there is more parity between the process of discovery and that of justification than is normally recognized in this case. I also believe that a study of the genesis of the special theory of relativity will underscore the conceptual differences between Einstein's and the ether theorists' approach to the electrodynamics of moving bodies.

II. LORENTZ'S ELECTRON THEORY: LORENTZ'S "LOCAL TIME," LARMOR'S "TIME DILATION," AND POINCARÉ'S "RELATIVITY"

A. Maxwell's Theory and Lorentz's Modifications

A brief overview of some aspects of Maxwell's theory of electricity and magnetism, which contained within it a theory of optics, will best provide the proper background for examining Lorentz's ideas. This is so even though Lorentz's first explorations into electromagnetic theory were pursued from the perspective of Helmholtz's eclectic electromagnetic theory of 1870, which contained Maxwell's theory as a limiting case.[8] By the time that Lorentz developed the theory of electrons in 1892, he had come to accept Maxwell's theory, essentially in the simplified 1890 Hertz-Heaviside form, as the proper basis for theoretical advance.[9]

In his own theory Lorentz accepted the Maxwellian concepts of an electric force, which Lorentz treated as equivalent to a dielectric displacement, and a magnetic force. The displacement and forces were associated with largely unknown modifications of the electromagnetic-optical ether. Disturbances of this ether were propagated as transverse waves with the speed of light, c, and the electric and magnetic intensities were known to vibrate transverse to the motion of the wave front and perpendicular to each other.

Maxwell's equations for the charge-free ether in their simplified form can be written as:

$$\text{div } d = 0 \tag{1}$$

$$\text{div } h = 0 \tag{2}$$

$$\text{curl } h = \frac{1}{c} \frac{\partial d}{\partial t} \tag{3}$$

$$\text{curl } d = -\frac{1}{c} \frac{\partial h}{\partial t} \tag{4}$$

These equations are sufficient to explain many electromagnetic and optical results. To apply them to matter and to obtain boundary conditions, they must be supplemented with additional terms associated with the dielectric and magnetic permeabilities of the media. One of the virtues of Lorentz's 1892 theory was to be able to show that these "macroscopic" media quantities could be derived from the equations of the free ether supplemented with four additional hypotheses: (1) There are very small charged particles, known at first as "ions" and later (beginning about 1900) as electrons. These particles are the origin of the dielectric displacement and in their interior equation (1) above becomes modified to div $\mathbf{d} = \rho$, where ρ is a measure of the charge density. It is

468

only, with these particles that the electromagnetic ether has any causal connection. (2) Lorentz also postulated that the motion of these particles constituted a current, of magnitude $\rho\mathbf{v}$ where $\mathbf{v}$ is the "absolute" velocity (in the ether) of the moving electrons. This results in an additional term in equation (3) above. (3) Lorentz also accepted the Young-Fresnel hypothesis concerning the relation between ether and matter, namely that the ether permeates and can move freely through all matter (including electrons) and vice versa. (4) Finally Lorentz provided an equation relating the mechanical force per unit charge the electrons experience as a result of the electrical and magnetic forces affecting them: $\mathbf{F} = \mathbf{d} + \frac{1}{c}\,(\mathbf{v} \times \mathbf{h})$. These modifications of Maxwell's theory yield the *fundamental equations of the Lorentz electron theory:*

$$\operatorname{div} \mathbf{d} = \rho \tag{5}$$

$$\operatorname{div} \mathbf{h} = 0 \tag{6}$$

$$\operatorname{curl} \mathbf{h} = \frac{1}{c}\frac{\partial \mathbf{d}}{\partial t} + \rho\,\mathbf{v} \tag{7}$$

$$\operatorname{curl} \mathbf{d} = -\frac{\partial \mathbf{h}}{\partial t} \tag{8}$$

$$\mathbf{F} = \mathbf{d} + \frac{1}{c}\,(\mathbf{v} \times \mathbf{h}) \tag{9}$$

From 1892 to 1909 Lorentz utilized these equations to account for a growing variety of experimental results including dispersion, conduction in metals, and the Zeeman effect, for which he shared the Nobel Prize in Physics with Zeeman in 1902. Most important for our purposes, on the basis of these equations, Lorentz developed a theory of the electrodynamics of moving bodies. This theory was successively modified from 1892 to 1904, and in 1905 was taken over by Poincaré, associated with a principle of relativity, and corrected slightly so as to be in agreement with such a principle. Lorentz became aware of Einstein's distinctive and bold approach in his 1909 book, *The Theory of Electrons,* from which I provided some quotes above. Let us examine Lorentz's theory of the electrodynamics of moving bodies and in particular the FitzGerald-Lorentz contraction and the Lorentz conception of "local time".

469

B. The Electrodynamics of Moving Bodies in Lorentz's Theory

As mentioned above, Lorentz developed his account of what occurs to electromagnetic phenomena in bodies moving through the ether over a period of 14–19 years. It is not possible in these pages to provide a detailed account of the many modifications in Lorentz's theory over this period—this has been done elsewhere by me and others.[10] What I wish to do is to focus on only the major facets of the Lorentz theory as they contrast in interesting epistemological and ontological ways with Einstein's approach. We will begin with Lorentz on length.

Lorentz first proposed the famous contraction hypothesis in late 1892. He did so acting independently of G. F. FitzGerald, who had suggested the hypothesis in an unknown letter to *Science* in 1889.[11] To both FitzGerald and Lorentz the contraction was thought to be plausible because it would follow if forces holding the molecules in bodies were electrical in nature. Such a contraction is, to use Reichenbach's terminology invented for a different context, caused by a "universal force" dependent on the absolute velocity of a body through the ether.[12] (This force affects all bodies equally.) The magnitude of the conjectured contraction was exactly enough to account for the puzzling null result of the Michelson-Morley interferometer experiment of 1887. Though the contraction was legitimized by associating it with the result of modified intermolecular forces, Lorentz still thought of it as somewhat *ad hoc* when he first proposed it. He could see no way out of it, however, and by 1909 had come to think of the null result of the Michelson-Morley experiment as *proving* the contraction. In Lorentz's words:

> The [contraction] hypothesis certainly looks rather startling at first sight, but we can scarcely escape from it, so long as we persist in regarding the ether as immovable. We may, I think, even go so far as to say that, on this assumption, Michelson's experiment proves the changes of dimensions in question, and that the conclusion is no less legitimate than the inferences concerning the dilatation by heat or the changes of the refractive index that have been drawn in many other cases from the observed positions of interference bands.[13]

The contraction hypothesis appears in Lorentz's very influential 1895 monograph, *Versuch einer Theorie der Elektrischen und Optischen Erscheinungen in Bewegten Körpern* (hereafter referred to as the *Versuch*).[14] It is also introduced in a formally more direct way (though the

470

physical cause of the contraction is not mentioned as being different) in Lorentz's 1899 and 1904 papers on the electrodynamics of bodies in motion. In these papers the contraction hypothesis appears as part of the *transformations* for quantities in the ether rest system to the system in motion.

In 1892 Lorentz had been able to derive the important Fresnel partial dragging coefficient, $1 - 1/n^2$, for transparent bodies in motion through the ether from his theory on the basis of considerations about the velocity of light waves emitted from the oscillating electrons in the moving bodies. In 1895 in the *Versuch*, however, Lorentz proposed a very novel means of handling aberration problems to the first order of quantities of v/c. In chapter 5 of that work he introduced the notion of a "local time" (*Ortzeit*) that, when employed together with Galilean transformations for space and rather standard first order transformations for electromagnetic quantities, yielded *approximate* covariance of the fundamental Lorentz equations (to the first order of v/c). This covariance is introduced in the form of Lorentz's *theorem of corresponding states*. In Lorentz's own (but translated) words:

> If, for a system of bodies at rest, a state of things is known where
>
> $$D_x, D_y, D_z, E_x, E_y, E_z, H_x, H_y, H_z$$
>
> are certain functions of x, y, z, and t, then in this same system, provided it moves with a velocity v, a state of things can exist in which:
>
> $$D_x', D_y', D_z', E_x', E_y', E_z', H_x', H_y', H_z'$$
>
> are the same functions of x', y', z', and t'.[15]

(Here Lorentz understands D to represent the Lorentz force, E to represent the electric force, and H the magnetic force.) The new time parameter is given by the transformation

$$t' = t - vx'/c^2 \qquad (10)$$

for a system K' moving in the positive x direction with an absolute velocity v. Unknown to Lorentz, this new time transformation had been proposed earlier in 1887 by W. Voigt in a paper on the Doppler effect in light, to obtain first-order invariance of the equation for light propagation in two reference systems.[16] Neither Voigt nor Lorentz commented on the significance of introducing a new time transformation. In the succeeding years Lorentz modified the formal expression for local time, in 1899 proposing the transformation

$$t' = t - vx'/c^2 - v^2 \tag{11}$$

for first-order effects and in a later part of the same paper

$$t' = \frac{1}{\varepsilon} \cdot \frac{1}{\sqrt{1-v^2/c^2}} (t - vx'/c^2 - v^2) \tag{12}$$

for second-order effects (involving measurements of v^2/c^2).[17] Lorentz had no means at this time (1899) of determining the unknown coefficient ε, but by 1904 he had, on dynamical grounds, seen that $\varepsilon = 1$, and that the local time could best be written

$$t' = \frac{t - vx/c^2}{\sqrt{1-v^2/c^2}} \tag{13}$$

which is formally equivalent to the later Einstein transformation for time.[18]

It will be useful to analyze the concept of local time in some detail. Lorentz seems to treat it as a mathematical change of variable in its early forms. Even in the 1895 form (10), however, it entails that clocks at different points in the moving system will be out of synchronization by a factor vx/c^2. In 1901 Poincaré commented on this notion of local time by noting that the effect is very small and that, more specifically, if two clocks were placed one kilometer apart on the earth, which is moving through the ether, say, at 10 km/sec, that their difference (or the second clock's difference with the real time) would be only $^1/_9 \times 10^9$ seconds.[19] It is important to note that in the 1895 form local time does *not* involve a change in the scale, i.e., *there is no time dilation* in this form, only what may be viewed as a failure in the synchronization of clocks placed at different points in a moving system. (I shall return to this point below in a discussion of Poincaré's analysis.) The first scientist to propose correct *second-order* transformations for electromagnetic quantities and to discern that there would entail a time dilation in moving systems was Joseph Larmor in 1897.[20]

472

C. *Larmor's Second-Order Theory of 1897 and 1900*

Larmor, largely independently of Lorentz, had arrived at an "electron" theory of electrodynamics in 1893–94 by working from a tradition that is distinctively British and Dublin-based. Beginning initially from Lord Kelvin's conception of a vortex atom in the ether, he developed a theory of electrons as singularities in an ether that was a slight modification of the optical ether of MacCullagh as electromagnetically interpreted by FitzGerald in 1878–80.[21] In 1897 Larmor proposed essentially correct second-order transformations for a theory of the electrodynamics of moving bodies, and in 1900 in his book *Aether and Matter* indicated explicitly that such second-order transformation involved a time scale dilation of $1/\sqrt{1-(v^2/c^2)}$ or $1 + (v^2/2c^2)$, dropping terms higher than the second order. More specifically Larmor noted:

> As a simple illustration . . . consider the group formed of a pair of electrons of opposite signs describing steady circular orbits round each other in a position of rest. . . . When this pair is moving through the aether with velocity v . . . the period will be changed only in the second order ratio: $1 + \frac{1}{2}\, v^2/c^2$.[22]

D. *Lorentz's 1904 and 1909 Advances and Poincaré's Modifications*

Lorentz of 1904 and 1909. By 1904 additional experimental and methodological objections had forced Lorentz to modify his transformations but not his fundamental equations. I mentioned above that the local time in 1904 is formally equivalent to the well-known relativistic transformation. Similarly, Lorentz's space transformation, which in 1899 had possessed an undeterminate coefficient ε, was then put into the now-well-known relativistic form $x' = x - vt/\sqrt{1 - v^2/c^2}$. There are a number of other transformations that were necessarily introduced in the 1904 paper, but rather than cite them all, it is more important to realize the *intent* of the transformations: they, plus the fundamental equations, are intended to yield a new, more powerful form of the theorem of corresponding states, i.e., covariance, which will hold for *many (but not necessarily all)* electromagnetic phenomena to *all* orders of v/c. Lorentz had accepted Poincaré's criticism made in 1900,[23] and again in 1901 in

473

his textbook *Electricity and Optics* (see note 19 below) that additional and more precise experiments yielding null effects of the ether wind always seemed to require additional hypotheses from Lorentz's point of view to account for them. This continual adding of hypotheses lent Lorentz's theory an unwieldy *ad hoc* appearance, and Poincaré felt that a *principle of relativity* asserting null results for *all* experiments to *all* orders of v/c was almost certainly involved in all natural processes. In Poincaré's 1900 address, which Lorentz cited in his 1904 paper (and which was also reprinted in *Science and Hypothesis*), Poincaré asked that there be no new hypotheses proposed: "Hypotheses" he said, are what we lack the least." Rather, he added:

> The same explanation must be found for the two cases [first and second order null results] and everything tends to show that this explanation would serve equally well for the terms of the higher order, and that the mutual destruction of these terms will be rigorous and absolute.[24]

Several experiments, principally the Rayleigh-Brace and Trouton-Noble experiments, had in fact yielded new null second-order results by 1904;[25] and Poincaré's comment notwithstanding, Lorentz felt he had to introduce a number of additional hypotheses to remedy his theory. Lorentz postulated that (1) the contraction held for the individual electron and not just at the macroscopic level. He thus removed the "justification" he had earlier provided for the contraction hypothesis in terms of the intermolecular forces. The contractile electron would also require non-electromagnetic forces to account for the electron's stability and for the contraction, a fact that was not lost to proponents of alternative electron theories such as Abraham, who criticized Lorentz on this point. Lorentz also postulated that (2) all the mass of the electron is electromagnetic in origin, i.e., is not inertial but due to the self-interaction of the charge with its own field. In addition, Lorentz found it necessary to assert that (3) *all* forces "are influenced by a translation in quite the same way as the electric forces in an electrostatic system." (I shall refer to this hypothesis as the *strong form* of the Lorentz force transformation law.) It then followed from (3) that "the proper relation between the forces and accelerations will exist in the two cases [i.e., will conform to Newton's second law for bodies in both absolute reference frame K_0 and in the moving frame K], if we suppose *that the masses of all particles are influenced by a translation to the same degree as the electromagnetic masses of the electrons''* (Lorentz's italics).[26] This means that *all* mass is a function of its velocity through the aether.

It also followed from (3), though Lorentz did not make it explicit until 1909, that such a force transformation, i.e., the strong force transformation law, would provide an explanation in a way,[27] of time dilation in moving systems. In 1909 Lorentz wrote:

> With the clocks of K the case is the same as with his measuring rod. If we suppose the forces in the clockwork to be liable to the changes determined by . . . [hypothesis (3) above], the motion of two equal clocks, one in K_0 and the other in K will be such that the effective coordinates of the moving parts are, in both systems, the same functions of the effective [or local] time. Consequently, if the hand of the clock in K_0 returns to its initial position after an interval of time θ, the hand of the clock in K will do so after an increment equal to θ of the effective time t'. Therefore a clock in the system K will indicate the progress of the effective time, and without his knowing anything about it, K's clocks will go $[\sqrt{1 - (v/c)^2}]$ . . . times slower than those of K_0.[28]

We see then that the two aspects of Lorentz's theory that are most often viewed as crucially anticipating Einstein's theory, namely, the contraction hypothesis and the new time transformation, are part of a complex electrodynamical theory of moving bodies: the space and time deviations from classical rods' and clocks' behaviors are deviations explained by special forces that are a consequence of motion through the ether.

Poincaré. I have already referred to some of Poincaré's critical comments and elucidations of Lorentz's theory above. Poincaré was deeply interested in the problem of electrodynamics and especially of the electrodynamics of moving bodies, and in his 1899 lectures at the Sorbonne, published as part of *Electricity and Optics* in 1901, he weighed the relative merits of Hertz's, Lorentz's, and Larmor's approaches to this sector of physics. It was his conclusion that although each of the theories was in some way unsatisfactory, Lorentz's provided the best explanation of the experimental results in this area. Poincaré was thus primed to seize on the further-developed theory of Lorentz that appeared in 1904 and was discussed briefly above. Poincaré modified the theory in papers published in 1905 and 1906, bringing the theory into

475

accord with his own belief, cited above, that the principle of relativity held rigorously and not, as Lorentz still thought in 1904, only approximately.

I cannot speak in detail of the modifications that Poincaré made in Lorentz's theory. This has been treated of elsewhere recently by other scholars, the most detailed account being by Arthur Miller.[29] Suffice it to point out that Poincaré found it necessary to modify Lorentz's velocity addition expression to make it constant with his other transformations, and also to alter the Lorentz charge density transformation to bring it into accord with a strict interpretation of the principle of relativity. These two modifications yielded complete covariance for both charge-free ether systems (Lorentz had this result already) *and* for moving systems with charges.

Poincaré also anticipated Einstein, in a way, by conceiving of the local time as involving a reanalysis of our conception of simultaneity. In his 1904 address to the International Congress of Arts and Sciences at Saint Louis, Missouri, Poincaré had discussed Lorentz's theory and the need to rethink the basis of mechanics, because of the implications of Lorentz's theory, Abraham's work with electromagnetic mass, and the experiments by Kaufmann on beta rays, which tended to support the idea that mass was a function of velocity. It is of interest to our inquiries that in his account of Lorentz, Poincaré noted that Lorentz's "most ingenious idea was that of local time."[30]

Earlier, in 1898, Poincaré had published a short essay on "The Measure of Time" in the *Revue de Métaphysique et de Morale*. In this essay he had pointed out that "*we have not a direct intuition of the equality of two intervals of time*" (Poincaré's italics). Time, Poincaré indicated, was involved with a definition of simultaneity. In the body of the 1898 essay he was concerned with both the problem of reduction of the qualitative psychological time to a precise quantitative time, as well as with what he viewed as the more important question: "Can we reduce to one and the same measure of time facts which transpire in different worlds."[31] In practice, Poincaré suggested in addressing himself to the second question, astronomers make use of light signals moving at a *presumed* constant velocity. Poincaré wrote:

[The astronomer] has begun by *supposing* that light has a constant velocity, and in particular that its velocity is the same in all directions. That is a

476

postulate without which no measurement of this velocity could be attempted. This postulate could never be verified directly by experiment; it might be contradicted by it if the results of different measurements were not concordant.

And again:

It is difficult to separate the qualitative problem of simultaneity from the quantitative problem of the measurement of time; no matter whether a chronometer is used, or whether account must be taken of a velocity of transmission, as that of light, because such a velocity could not be measured without *measuring* a time.

To conclude: We have not a direct intuition of simultaneity, nor of the equality of two durations.

Poincaré was thus sensitized to problems of defining common time measurements and his comments in his Saint Louis address, to return from our excursion to the Poincaré of 1898, are most perceptive and worth quoting *in extenso*. On Lorentz's local time in his 1904 speech Poincaré wrote:

Imagine two observers who wish to adjust their timepieces by optical signals; they exchange signals, but as they know that the transmission of light is not instantaneous, they are careful to cross them. When station B perceives the signal from station A, its clock should not mark the same hour as that of station A at the moment of sending the signal, but this hour augmented by a constant representing the duration of the transmission. Suppose, for example, that station A sends its signal when its clock marks the hour O, and that station B perceives it when its clock marks the hour t. The clocks are adjusted if the slowness equal to t represents the duration of the transmission, and to verify it, station B sends in its turn a signal when its clock marks O; then station A should perceive it when its clock marks t. The timepieces are then adjusted.

And in fact they mark the same hour at the same physical instant, but on the one condition, that the two stations are fixed. Otherwise the duration of the transmission will not be the same in the two senses, since the station A, for example, moves forward to meet the optical perturbation emanating from B, whereas the station B flees before the perturbation emanating from A. The watches adjusted in that way will not mark, therefore, the true time; they will mark what may be called the *local time*, so that one of them will gain on the other. It matters little, since we have no means of perceiving it. All the phenomena which happen at A, for example, will be late, but all will be equally so, and the observer will not perceive it, since his watch is slow; so, as the principle of relativity would have it, he will have no means of knowing whether he is at rest or in absolute motion.[32]

This analysis accounts for *one* aspect of the local time transformation, i.e., non-synchronicity; it does *not* however yield the time dilation effect. The latter is not explicitly mentioned by Poincaré but *could* have been accounted for by appealing to the modifications of the forces in the clockwork mechanisms as pointed out later by Lorentz in 1909 (see above, p. 475). It thus might be relevant to mention that Poincaré cites in the Saint Louis address the need to accept the strong form of the Lorentz force transformation expression as well as the contraction hypothesis in order to complete the Lorentz theory.

That the new time transformation can be interpreted up to a point in the context of an absolute theory and using Lorentzian concepts of special forces and contraction can be more clearly seen in an interpretation that Lorentz provided in 1909.[33] Lorentz pointed out that if we synchronize two clocks at two points P and Q that are moving through the ether with a constant velocity v, and if we do so by a light-signal synchronization process, the clocks will indicate the local time corresponding to their positions, and will embody the Lorentz time transformation equation of 1904. This is an important point, and it is worth sketching the proof so as to reveal how the new dynamics and contractions affect temporal measurements.

Let us first assume that clock P lies at the origin of the K' (or moving) system of coordinates and that the clock at Q is at x'. Since the clock at P is moving through the ether at velocity v, its time t' will be reduced because the forces in its spring pendulum, or whatever mechanism, are reduced, and it will continually read behind the *true* time:

$$t' = \sqrt{1-v^2/c^2}\ \mathrm{t} \tag{14}$$

Now let an observer at P send a light signal to Q at the moment $t' = 0$. $\bar{Q}$ is at x' which is in *true distance* units $x'\ \sqrt{1-v^2/c^2}$ to the right of $\bar{P}$. Since both $\bar{P}$ and $\bar{Q}$ are in motion, the *true relative velocity* of the light signal sent to Q is $c - v$. It therefore takes true time:

$$\Delta t = x'\ \sqrt{\frac{1 - v^2/c^2}{c - v}} \tag{15}$$

for the signal to move from $\overline{P}$ to $\overline{Q}$. (This is simply the true distance divided by the true relative velocity.) In this time the clock at Q has advanced an apparent time of x'/c. At any other *true* time t, the clock at Q will read

$$t' = \frac{x'}{c} + \sqrt{1 - v^2/c^2} \left(t - x' \sqrt{\frac{1 - v^2/c^2}{c - v}} \right) \qquad (16)$$

which gives $\overline{Q}$'s reading *correcting* for the true elapsed time, Δt, for the light signal to move from P to Q, and also including the time dilation factor on $\overline{Q}$'s readings. But this expression, since $x' = x - vt / \sqrt{1 - v^2/c^2}$, yields on simple algebraic simplification:

$$t' = \frac{t - vx/c^2}{\sqrt{1 - v^2/c^2}} \qquad (13)$$

which is the Lorentz (and formally equivalent to the Einstein) time transformation. This proves, as Lorentz pointed out in 1909, that "when the clocks in a moving system are adjusted by means of optical signals, each of them will indicate the local time corresponding to its position."[34]

I can summarize this long discussion of space and time in Lorentz and Poincaré, and in Larmor to some extent as well, in the following points:

1. Lorentz was initially unaware of the implications of his concept of "local time". The idea seems to have been a convenient but essentially inconsequential deviation from classical concepts of time occasioned by motion through the ether. We have published evidence that as late as the years 1910–12 Lorentz still thought that an absolute simultaneity could be defined and a clear distinction made between "local" and "universal" time, though at this point Lorentz had run out of *experimental* suggestions for introducing absolute simultaneity.[35]

2. Poincaré was apparently the first scientist to realize that "local time" could be explicitly viewed as due in part to a failure of classical synchronization. This was explained by him as due to motion through the ether, however, and it is clear from his papers

of 1905 and 1906 that he understood "relativity" as a set of compensations produced by motion through the ether. Thus there is, for Poincaré, a true reference frame in which electrons, for example, are true spheres, only we can never know it.

3. Larmor seems to have been the first person to realize that if "local time" were extended to second-order quantities, this would yield a change in the scale of time for moving systems, i.e., a time dilation.

4. Lorentz, only later and after learning of Einstein's papers, realized that his strong force transformation law of 1904 could be utilized to account for the time dilation aspect of the local time transformation. This is questionable as an *explanation,* however, since it *assumes* that forces (and lengths too by 1904) are mysteriously altered by motion through the ether by factors involving the expression $\sqrt{1 - v^2/c^2}$.

5. Lorentz, Larmor, and Poincaré all believed that there was one privileged frame of reference, the ether, and that alterations in space and time as embodied in the new kinematical transformations were in fact alterations in lengths of rods and rates of clock mechanisms due to special forces called into play by motion through the ether. Further, Lorentz and Larmor, at least, do not seem to have sensed the *reciprocal* nature of the transformations with which they worked. Clearly Lorentz did not realize this until after he heard of Einstein's account, and then he attempted to explain it away by ascribing it to "fictitious" readings an observer in motion would obtain because his rods and clocks were contracted and running on "local time."

III. EINSTEIN'S SPECIAL THEORY OF RELATIVITY

Thus far I have been examining the concepts of length and time in pre-Einsteinian theories of the electrodynamics of moving bodies. In so doing I have introduced some distortion into the historical record in order to provide analytical and philosophical clarity. Lorentz, Poincaré, and Larmor as indicated above were concerned with new length and time transformations, but these were *derivative* of their more fundamental

480

interests in formulating what we may, following Einstein, refer to as *constructive theories* of the electrodynamics of moving bodies; namely, theories based on true ontologies and employing the fundamental equations describing those ontologies of, e.g., atoms, electrons, and (at least in Einstein's case) light quanta. The new space and time transformations were clearly unwillingly forced on these pre-Einsteinian thinkers by the received theories and new experiments with which they worked. It was Einstein who, with the characteristic move of genius that is able to view a problem from a radically new perspective, proposed an alternative approach to these problems that *relocated the problem* in the classical conceptions of space and, especially, of time. In the introduction to this paper I discussed the opinion that many of Einstein's fellow contributors to relativity had of his achievement. Let us recall those comments about his striking breakthrough and analyze what the conditions and meaning of his achievement might be.

A. The Psychological, Physical, and Epistemological Roots of Einstein's Approach

Prior to examining the actual structure of the 1905 theory, it will be useful to cite some of Einstein's own remarks about the background that led him to special relativity. Later I shall discuss in some detail what the more proximate reasons were and relate these to the 1905 work in some detail.

I think it would be instructive to consider the background of Einstein's thought via three types of influence: (1) the psychological cast of Einstein's mind; (2) his awareness of the experimental and theoretical situation in physics in the early twentieth century—this will overlap with my later discussion on proximate reasons for the genesis of relativity; and (3) the epistemological views that influenced and motivated Einstein in the years prior to 1905.

1. Psychological background. Einstein tells us in his "Autobiographical Notes" that he suffered a crisis of religious faith at age twelve.[36] This was due to his realizing that the Bible and elementary or popular science were inconsistent, and it resulted in a very skeptical and critical attitude toward *all* received views in his later years. This was attested to by a number of Einstein's associates in the scientific world.

481

Among them, Poincaré, who wrote a letter of recommendation for Einstein in November 1911 for a position at the Zurich Polytechnic, commented:

> Einstein is one of the most original minds that I have known; in spite of his youth he has already achieved a most honorable position among the best scientists of our time. What we can admire above all in him is the facility with which he adapts to novel conceptions and is able to draw all consequences from them. He does not remain attached to classical principles, and, faced with a problem in physics, is quick to envisage all possibilities.[37]

2. Physical background. Here I will be brief and treat what I believe to be the major points in a summary fashion.

a) Einstein was clearly troubled by a *dualism* between the mechanical and the electromagnetic pictures of the world. He was acquainted with Maxwell's theory and the 1892 and 1895 works of Lorentz. The electromagnetic theory suggested to Einstein that "electromagnetic fields" could be substances or independently existing entities. This idea was very likely a necessary prelude to being able to eliminate the *ether* from a theory of the electrodynamics of moving bodies, since heretofore fields were thought to be mere *conditions* of the substance ether. Einstein wrote in his "Autobiographical Notes":

> The electrodynamics of Faraday and Maxwell and its confirmation by Hertz showed that there are electromagnetic phenomena which by their very nature are detached from every ponderable matter—namely the waves in empty space which consist of electromagnetic "fields."

Einstein further noted that after zealous but fruitless attempts to interpret Maxwell's equations mechanically, physicists "got used to operating with these fields as independent substances without finding it necessary to give . . . an account of their mechanical nature."[38]

Einstein was also aware of the possibility of explaining material of inertial mass via electromagnetic mass, such that "only field-energy would be left and the particle would be merely an area of special density of field-energy."[39] Such an electromagnetic reduction had not been carried out, however, and Einstein saw no way to formulate in a "non-arbitrary" manner the necessarily non-linear field equations that would yield the electron as a solution. It does seem that Einstein did

482

spend some effort in trying to solve this problem, but it did not directly lead to any success.

b) Einstein indicates in his "Autobiographical Notes" that a crisis in physics set in with Planck's discovery of the quantum nature of energy.[40] He further points out that Planck's argument

> presupposes implicitly that energy can be absorbed and emitted by the individual resonator only in "quanta" of magnitude hν, i.e., that the energy of a mechanical structure capable of oscillation as well as the energy of the radiation can be transferred only in such quanta—in contradiction to the laws of mechanics *and electrodynamics* (my italics).[41]

Einstein had worked in thermodynamics and statistical mechanics in the early years of this century and with characteristic boldness did not hesitate to apply the methods he had developed in that area of inquiry to electromagnetic radiation.[42] With Planck's discovery in his mind, he was able to show that the photoelectric effect could be accounted for by imputing a quantum, or particulate, aspect to electromagnetic radiation. He also became aware that certain types of radiation pressure on small moveable mirrors could only be explained by light quanta. These discoveries raised extremely serious problems for the electromagnetic theories of Maxwell and Lorentz, and caused Einstein much anguish as he attempted to work out a "constructive" theory that would take light quanta and the problems associated with the electrodynamics of moving bodies into account. Eventually he solved the latter problem by moving to a different type of theory and putting aside the quantum problem. To appreciate the notion of different types of theories and also to outline the background for Einstein's willingness to attack the time concept, it is necessary to develop some of Einstein's epistemological background.

3. Epistemological background. As I shall again point out below in the context of Einstein's discussion of the principle of special relativity, Einstein wrote in his "Autobiographical Notes" that the "critical reasoning" required for the reanalysis of the concept of time "was decisively furthered . . . especially by the reading of David Hume and Ernst Mach's philosophical writings." Einstein was one of the most philosophically sensitive scientists of recent history, and his insights into the nature of scientific theorizing and scientific discovery are well worth attention, especially for this inquiry as they are often associated with the genesis of the special theory of relativity.

It is generally accepted that in his early years and perhaps into the nineteen-teens Einstein was much affected by Mach's philosophy of science and especially by his reading of *The Science of Mechanics,* which he obtained in 1897. In the "Autobiographical Notes" Einstein mentions that "this book exercised a profound influence on me [and] in my younger years Mach's epistemological position also influenced me greatly."[43] Gerald Holton has recently examined the Einstein Archives at Princeton and has obtained letters that corroborate this influence of Mach, but that also reveal the growing philosophical separation between Mach and Einstein that occurred after 1905.[44]

In his own writings Einstein seems unsure himself what he specifically obtained from Mach, and suggests that some of the effect might well be "unconscious." In a letter to M. Besso written in 1948 he said: "The extent to which . . . [Mach's *Mechanics* and *Wärmlehre*] influenced my own work is, to say the truth, not clear to me. As far as I am conscious of it, the immediate influence of Hume on me was greater."[45] In his "Autobiographical Notes" Einstein was somewhat more specific about what aspect of Hume's thought he found important: "Hume saw clearly that certain concepts, as for example that of causality, cannot be deduced from the material of experience by logical methods."[46]

Though Mach's influence may well have been unconscious—it is difficult to determine this because there is very little that bears on this point written by Einstein in his early years—there is a methodological approach to the reanalysis of fundamental scientific concepts that one finds even in the early editions of Mach's *Science of Mechanics,* which could not have failed to make some impression on Einstein, and which may well have served as an unconscious methodological guide for Einstein in his 1905 reanalysis of simultaneity.[47]

In his section on a "Synoptical Critique of Newtonian Enunciations" Mach criticized Newton's mode of presenting his definitions and laws of mechanics and suggested that even if the problems he had raised earlier in the book regarding time and space were momentarily disregarded, "it is possible to *replace Newtonian enunciations by much more simple, methodologically better arranged, and more satisfactory propositions*" (my italics).[48] What Mach then provided was a series of *definitions* that were sharply distinguished from *experimental propositions.* These sentences, Mach added, "satisfy the requirements of simplicity and parsimony which, on economico-scientific grounds, must be exacted of them. They are, moreover, obvious and clear; for no doubt can exist with respect to any one of them either concerning its meaning or its source; and *we always know whether it asserts an experience or an arbitrary convention.''* [49]

Einstein would have found a similar approach taken in Mach's *Principien der Wärmlehre,* which he also read. Several quotations from the *Wärmlehre* will indicate the methodology that seems to have influenced Einstein in his analysis of the concept of time. In his chapter ''Critique of the Concept of Temperature'' Mach stressed the ''conventional'' and ''arbitrarily definitional'' aspects of the concept and then added:

> It is remarkable how long a period elapsed before it definitively dawned upon inquirers that the designation of *thermal states* by *numbers* resposed on a *convention.* Thermal states exist in nature, but the concept of temperature exists only by virtue of our arbitrary *definition,* which could very well have taken another form. Yet until very recently inquirers in this field appear more or less unconsciously to have sought after a *natural* measure of temperature, a real temperature, a sort of Platonic Idea of temperature. [50]

Importantly, Mach also drew attention to parallels between the concept of temperature and the Newtonian kinematical concepts:

> Newton's conceptions of ''absolute time,'' ''absolute space,'' etc., which I have discussed in another place [in the *Science of Mechanics*], originated in quite a similar manner. In our conceptions of time the *sensation of duration* plays the same part with regard to the various measures of time as the sensation of heat played in the instance just adduced. The situation is similar with respect to our conceptions of space. [51]

The awareness that scientific theorizing involves important definitional or conventional elements would have been underscored in Einstein's mind by his reading of Poincaré's *Science and Hypothesis,* which we have evidence he became acquainted with, probably in 1903, as part of a discussion group. [52] Though we have no direct acknowledgement from Einstein of the influence of Poincaré on the methodological and epistemological background of his reasoning, it is interesting nonetheless to note that in the preface to *Science and Hypothesis* Poincaré wrote:

> We shall see that there are several kinds of hypotheses; that some are

verifiable, and when once confirmed by experiment become truths of great fertility; that others may be useful to us in fixing our ideas; and finally that others are hypotheses only in appearance, and reduce to definitions or conventions in disguise.[53]

Poincaré further added:

> In mechanics we shall be led to analogous conclusions, and we shall see that the principles of this science, although more directly based on experience, still share the conventional character of the geometrical postulates.[54]

It seems then that there were likely sources of conventionalist epistemology on Einstein in the years prior to his articulation of a new time concept, which, as we shall see below, involves a conventionalist assumption in a most crucial way.

Another important epistemological facet of Einstein's background, and which he cites in connection with the genesis of the special theory of relativity, is the distinction between "constructive" and formal or "principle" theories in physics. Einstein asserts, and I shall quote him extensively below on this point in a more appropriate context, that it was only by proposing a "universal formal principle" that he was able to resolve the problem of the electrodynamics of moving bodies. In 1919 in a note that he wrote for the *Times* of London, Einstein amplified his views on these two approaches to physical theorizing. Einstein wrote:

> We can distinguish various kinds of theories in physics. Most of them are constructive. They attempt to build up a picture of the more complex phenomena out of the materials of a relatively simple formal scheme from which they start out. Thus the kinetic theory of gases seeks to reduce mechanical, thermal and diffusional processes to movements of molecules—i.e., to build them up out of the hypothesis of molecular motion. When we say that we have succeded in understanding a group of natural processes, we invariably mean that a constructive theory has been found which covers the processes in question.
>
> Along with this most important class of theories there exists a second, which I will call "principle-theories". These employ the analytic, not the synthetic, method. The elements which form their basis and starting-point are not hypothetically constructed but empirically discovered ones, general characteristics of natural processes, principles that give rise to mathematically formulated criteria which the separate processes or the theoretical representations of them have to satisfy. Thus the science of thermodynamics

seeks by analytical means to deduce necessary connections, which separate events have to satisfy, from the universally experienced fact that perpetual motion is impossible.

The advantages of the constructive theory are completeness, adaptability and clearness, those of the principle theory are logical perfection and security of the foundations.

The theory of relativity belongs to the latter class.[55]

Both "principle" and "constructive" types of theories are criticizable from two points of view. Einstein, in his "Autobiographical Notes," proposed that empirical adequacy was most important in judging a theory: "the theory must not contradict empirical facts."[56] There is, however, *another* desideratum for adequate theories or, as Einstein put it, another point of view from which one can judge a theory. He wrote:

> The second point of view is not concerned with the relation to the material of observation but with the premises of the theory itself, with what may briefly but vaguely be characterized as the "naturalness" or "logical simplicity" of the premises (of the basic concepts and of the relations between these which are taken as a basis.) This point of view, an exact formulation of which meets with great difficulties, has played an important role in the selection and evaluation of theories since time immemorial.[57]

Einstein continued, pointing out the difficulty of specifying precisely in what this notion of simplicity consisted:

> The problem here is not simply one of a kind of enumeration of the logically independent premises (if anything like this were at all unequivocally possible), but that of a kind of reciprocal weighing of incommensurable qualities.[58]

Einstein added that what one was judging from this second point of view might be termed the "inner perfection" of a theory, and that though there were difficulties with the notions of simplicity and inner perfection, Einstein was neither a pessimist nor an obscurantist concerning further elucidation and application of such criteria. He further wrote:

> I believe . . . that a sharper formulation [of such notions] would be possible. In any case it turns out that among the "augurs" there usually is agreement in judging the "inner perfection" of the theories and even more so concerning the "degree' of "external confirmation."[59]

We do not know whether such notions concerning simplicity and "naturalness" were necessarily involved in Einstein's genesis of the special theory of relativity, but there is some evidence that they were. Einstein gives no indication that his respect for simplicity came late in his career and certainly he would have been sensitized to the importance of simple theories due to his being influenced by Mach's "economical" view of science. Further, as we shall see below, Einstein has said that he did find certain aspects of Lorentz's theory "*ad hoc*" and "artifical" and distrusted it on that account.[60]

We see then that there were some interesting philosophical or epistemological issues that affected Einstein in the years prior to his discovery of relativity theory. These were, to briefly summarize: (1) the critical and analytical approach of Hume and Mach to fundamental and generally accepted concepts; (2) the mode of reconstructing fundamental concepts by ascertaining the empirical and the definitional or conventional components of such concepts, which Einstein also found in Hume, Mach, and Poincaré; (3) the distinction between different types of theory construction, namely, between theories of principle and constructive theories; and (4) the significance of simplicity and "naturalness" as important constraints on the acceptability of a scientific theory. Let us now turn to the details of the impact of all the background factors on Einstein's genesis of the special theory of relativity.

B. Einstein's Kinematical Approach to the Electrodynamics of Moving Bodies

In this section I wish to examine some of the details of both the genesis of the special theory in Einstein's own mind and the form in which he presented the theory in his 1905 fundamental paper in the *Annalen der Physik*.[61] I believe that what the history of the genesis of the theory discloses is interestingly relevant to the logical analysis of the theory since it will clarify both the relation of the theory to experiment and the trans-empirical constraints, as well as highlight the differences between the special theory of relativity and Lorentz's theory.

In his "Autobiographical Notes" Einstein tells us, in an often-quoted passage:

Reflections of this type [and here Einstein is referring to problems the

quantum hypothesis had occasioned for Maxwell's and Lorentz's theories]
made it clear to me as long ago as shortly after 1900, i.e., shortly after
Planck's trailblazing work, that neither mechanics nor thermodynamics
could (except in limiting cases) claim exact validity. By and by I despaired of
the possibility of discovering the true laws by means of constructive efforts
based on known facts. The longer and the more despairingly I tried, the more
I came to the conviction that only the discovery of a universal formal
principle could lead us to assured results. The example I saw before me was
thermodynamics. The general principle was there given in the theorem: the
laws of nature are such that it is impossible to construct a *perpetuum mobile*
(of the first and second kind). How, then, could such a universal principle be
found? After ten years of reflection such a principle resulted from a paradox
upon which I had already hit at the age of sixteen: If I pursue a beam of light
with the velocity c (velocity of light in a vacuum), I should observe such a
beam of light as a spatially oscillatory electromagnetic field at rest. However, there seems to be no such thing, whether on the basis of experience or
according to Maxwell's equations. From the very beginning it appeared to
me intuitively clear that, judged from the standpoint of such an observer,
everything would have to happen according to the same laws as for an
observer who, relative to the earth, was at rest. For how, otherwise, should
the first observer know, i.e., be able to determine, that he is in a state of fast
uniform motion?

One sees that in this paradox the germ of the special relativity theory is
already contained. Today everyone knows, of course, that all attempts to
clarify this paradox satisfactorily were condemned to failure as long as the
axiom of the absolute character of time, viz., of simultaneity, unrecog-
nizedly was anchored in the unconscious. Clearly to recognize this axiom
and its arbitrary character really implies already the solution of the problem.
The type of critical reasoning which was required for the discovery of this
central point was decisively furthered, in my case, especially by the reading
of David Hume's and Ernst Mach's philosophical writings.[62]

This passage is unfortunately somewhat obscure and it has, as a
result, given rise to a number of perplexities and misinterpretations.
Since it does not mention specific experiments such as the Michelson-
Morley experiment, and since it uses the phrase "from the very beginning it appeared to me intuitively clear," it has lent credence to anti-
empiricist reconstructions of the genesis of relativity.[63] Let us recall
what theoretical background Einstein had been exposed to and also what
other accounts based on Einstein's own recollections tell us about this
situation.

In 1916 Max Wertheimer had "hours and hours" of conversations
with Einstein about the genesis of the special theory of relativity, which

he recounts in a chapter in his later published book *Productive Thinking*.[64] In the same year Einstein completed a popular work on relativity *Über die spezielle und die allgemeine Relativitätstheorie*, in the preface of which Einstein stated that "the author has spared himself no pains in his endeavor to present the main ideas in the simplest and most intelligible form, and on the whole, in the sequence and connection in which they actually originated."[65] These accounts together with various other statements that Einstein made, and together with the evidence provided by the 1905 paper itself, allow for the following type of reconstruction.

The light beam experiment that Einstein performed in his own mind at age 16 served to raise a *problem* but *not* to result in any *solution* at this point. Maxwell's equations *themselves* do not prohibit relative velocities of light less than or greater than c; in fact, taken together with the hypothesis of a stationary ether and Galilean transformations, they *entail* such velocities.[66] However, there is a principle of relativity for mechanical experiments performed in empty space that does prohibit experimental determinations of absolute rest; this is simply the proposition that Newton's laws of motion are invariant with respect to the Galilean transformations. Wertheimer questioned Einstein closely on this point:

> When I asked him whether, during this [early] period, he had already had some idea of the constancy of light velocity, independent of the movement of the reference system, Einstein answered decidedly: "No, it was just curiosity. That the velocity of light could differ depending on the movement of the observer was somehow characterized by doubt. Later developments increased that doubt."[67]

We may conjecture as to what those "later developments" might have been. First, but not necessarily most importantly, it would seem that Lorentz's theorem of corresponding states for first-order experiments associated with aberration would have corroborated Einstein's doubts. This theorem entails the apparent constancy of the velocity of light c to *first*-order experiments of precision v/c. There are a number of experiments that, seen from the perspective of the Lorentz theory, indicate that the motion of the earth, and of matter in general, has no effect on light propagation or on the form of Maxwell's equations. To corroborate this influence a letter from Einstein to Shankland written in 1952 can be cited.[68] There Einstein noted that he "was also guided [to relativity] by the result of the Fizeau experiment and the phenomenon of aberration.'' Secondly, and perhaps more important, it is clear that Einstein was very forcefully struck by the electromagnetic induction experiments which indicated that only relative motion had any empirical effect on electromagnetic interactions. The received theories of Einstein's time, both Hertz's theory and, more importantly, Lorentz's theory, treated the situation of a magnet in motion and a wire at rest (in the ether) differently from a wire in motion and a magnet at rest.[69] Einstein was sufficiently impressed with the importance of this to begin his 1905 paper by citing such asymmetries that were not "inherent in the phenomena." In the same letter to Shankland quoted above, Einstein asserted: "What led me directly to the Special Theory of Relativity was the conviction that the electromotive force induced in a body in motion was nothing else but an electric field." Finally, it is clear that Einstein *was* aware of the null result of the Michelson-Morley experiment. The general position that I believe can be extracted from the protracted discussion in the literature on this topic[70] is best put by saying that Einstein by the time he became aware of the experiment had already anticipated its null result. This is not particularly puzzling, however, as Poincaré too would have anticipated null results for the Trouton-Noble and Rayleigh-Brace experiments performed in the years 1902–4 on the basis of his own empirical generalization of a principle of relativity.

Einstein's thinking that led to the reanalysis of simultaneity, then, was licensed by experimental considerations as enlightened by physical theory, and as further constrained by a desire for simplicity and a minimization of *ad hoc* or "artificial" hypotheses. How such considerations might have led to the special theory of relativity and in particular the new analysis of the time concept will be taken up in the following pages.

Wertheimer suggests that the Michelson-Morley experiment was no surprise to Einstein but that nevertheless it was "very important and decisive." This comment on the face of it does not appear to accord with a number of other comments Einstein has made concerning the "indirect" and "negligible" effect that the Michelson-Morley experiment had on the genesis of relativity.[71] This apparent contradiction admits, I think, of a resolution that also clarifies the long quotation from the "Autobiographical Notes" above concerning the import of the light beam *gedankenexperiment* performed by Einstein at age sixteen.

We may hypothesize that on the basis of first-order ether drift experiments and the electromagnetic induction analysis that Einstein had become convinced prior to learning of the Michelson experiment that light had no property of absolute rest. The Michelson experiment confirmed this, but in fact such a belief is stronger than anything that the Michelson experiment licenses, since the null result of the experiment can be explained by a contraction hypothesis. Such a route, however, would have been barred to Einstein by his reluctance to accept *ad hoc* hypotheses, and it may well be that this is why he did not move in the direction of developing a theory that would have employed both the contraction hypothesis and the second-order local time transformation. Such a "doubly amended" ether theory[72] would have accounted for the fact that light *apparently* had no property of absolute rest, but it would be inconsistent with a principle of relativity interpreted so as to deny the existence of any special reference frame. That the principle of relativity, extraordinarily well confirmed for mechanical phenomena, was deeply entrenched in Einstein's mind is attested to by both Wertheimer's account and by Einstein's popular work of 1916. Thus in 1905 Einstein was confronted by the dilemma of interpreting how a light beam could have the same velocity for a reference system K' as it did for a system K in motion with respect to K'. Let us call this hypothesis of constant light velocity the CLV hypothesis.

The CLV hypothesis is, as mentioned above, confirmed by the Michelson or Michelson-Morley experiment, though it goes beyond it. The experiment, however, might offer a means of *concretizing* the relations between the CLV hypothesis and the assumptions of classical electromagnetic theory, as well as offering some empirical control over the speculations concerning light velocity in moving reference frames.

In the Einstein literature there seems to be only one source that explicitly touches on the crucial stage of Einstein's reasoning when he conjectured that the classical conceptions of time and simultaneity might require reanalysis. This source is Wertheimer's account.[73] Wertheimer couches his discussion of Einstein's reasoning in terms of an analysis of the Michelson experiment, a fact that is both interesting and troublesome in the light of Einstein's disavowal of the importance of the experiment. However, if we can assume that Einstein used the theoretical picture of the Michelson apparatus in connection with analyzing the

CLV hypothesis, as essentially equivalent to his light beam *gedankenexperiment,* we reach a reasonable scenario that brings both the story told in the "Autobiographical Notes" and the following Wertheimer account into accord. Wertheimer recounted Einstein's reasoning as follows (I have italicized certain passages in order to highlight the relation of the analysis to the CLV hypothesis):

> Einstein said to himself: "Except for that [null] result, the whole situation in the Michelson experiment seems absolutely clear; all the factors involved and their interplay seem clear. But *are* they really clear? Do I really understand the structure of the whole situation, especially in relation to the crucial result?" During this time he was often depressed, sometimes in despair, but driven by the strongest vectors.
>
> In his passionate desire to understand or, better, to see whether the situation was really clear to him, he faced the essentials in the Michelson experiment again and again, *especially the central point: the measurement of the speed of light under conditions of movement of the whole set in the crucial direction.*
>
> This simply would not become clear. He felt a gap somewhere without being able to clarify it, or even to formulate it. He felt that the trouble went deeper than the contradiction between Michelson's actual and the expected result.
>
> He felt that a certain region in the structure of the whole situation was in reality not as clear to him as it should be, although it had hitherto been accepted without question by everyone, including himself. His proceeding was somewhat as follows: There is a time measurement while the crucial movement is taking place. "Do I see clearly," he asked himself, "the relation, the inner connection between the two, between the measurement of time and that of movement? Is it clear to me how the measurement of time works in such a situation?" *And for him this was not a problem with regard to the Michelson experiment only, but a problem in which more basic principles were at stake.*
>
> It occurred to Einstein that time measurement involves simultaneity. What of simultaneity in such a movement as this? To begin with, what of simultaneity of events in different places?
>
> He said to himself: "If two events occur in one place, I understand clearly what simultaneity means. For example, I see these two balls hit the identical goal at the same time. But . . . am I really clear about what simultaneity means when it refers to events in two different places? What does it mean to say that this event occurred in my room at the same time as another event in some distant place? Surely I can use the concept of simultaneity for different places in the same way as for one and the same place—but can I? Is it as clear to me in the former as it is in the latter case? . . . It is not!"[74]

493

The above account is oversimplified, since it does not mention the problems with electromagnetic theory occasioned by Planck's quantum discovery, does not indicate the need to formulate a principle type of theory, and in addition does not cite the Fizeau experiment or the electromagnetic induction analysis. There is, however, no reason to demand that it do so as the account is concerned with a very specific point in Einstein's reasoning and accords with our other information that can be understood as playing an earlier and background role at this point.

The scenario is conjectural but has several interesting features. First, it does give us a plausible reading of the available documents,[75] fills in some gaps, and resolves some paradoxes. Secondly, it suggests that there are some interesting parallels between the discovery of the special theory of relativity and the way in which Einstein presented the theory in its kinematical aspect in his 1905 paper. To show that this is so, and also to provide a somewhat detailed analysis of Einstein's revolutionary approach to the time concept, let us look at the arguments as presented in the 1905 paper of Einstein.[76]

Einstein indicated that to examine the concept of motion one could give Euclidean or Cartesian coordinate system readings "as a function of the time." This presumed that the time concept was clear, but Einstein suggested that any *definition* of time presupposed some means of judging events *simultaneous*. Such judgments were unproblematic when they referred to events contiguous with the observer, but *distant simultaneity*, Einstein added, was more difficult to ascertain. He proposed, rhetorically, that one *might* station an observer at the origin of a system of coordinates and impute times to distant events by checking light signals emitted from those points coincident with the events' occurrences. Such a procedure, however, would not be satisfactory since, Einstein asserted, "it is not independent of the standpoint of the observer with the clock, as we know from experience." Einstein could be referring here to the need to know the *distances* to the events, but it is more likely he is citing the need to know the velocity of the observer with respect to classical electromagnetic theory, since this would affect the elapsed time the observer would calculate. (Recall if he were moving *toward* the source the velocity would be, classically, $c + v$.)

"A much more practical determination," Einstein contended, would involve both definitional and empirical components. Taking two distant points A and B, we "establish *by definition* that the time required by light to travel from A to B equals the time it requires to travel from B to A'' (Einstein's italics). This stipulation defines two clocks to be in synchrony, letting t_A be the time the light leaves A, t_B the reading of B's clock on receipt of the signal, and $t_{A'}$ the time the signal, reflected from B, returns to A, if:

$$t_B - t_A = t_{A'} - t_B \tag{17}$$

A definitional component is clearly required, as Einstein pointed out in his 1916 popular work cited above, since the velocity of light in a *one way* direction cannot be *empirically* ascertained prior to a determination of distant simultaneity.

The empirical component in the definition is that the round-trip velocity of light

$$c = 2AB/t_{A'-tA}, \tag{18}$$

which can be determined from a Fizeau type of light velocity experiment at point A.

This definition allows symmetry and transitivity of synchronization, but as has been forcefully argued by Reichenbach and Grünbaum, alternative non-standard *definitions* of simultaneity could be given that would not be empirically falsifiable, though they would result in considerable loss of *descriptive simplicity*.[77]

(It should also be added here that the need for a definition, and the permissibility or unfalsifiable character of such a definition, is dependent on the world being an Einsteinian world. As Grünbaum has clearly argued, if clocks were *not* affected by their motion or if there *were* signals with a greater velocity than c, such a definition would not be permissible. In addition, I myself believe that the permissibility of the definition depends on the Einstein principle of the independence of the velocity of light from its source holding. For if a Ritz ballistic hypothesis were true in nature, absolute measurements of simultaneity would be definable.)

Having implicitly questioned the received view of time by providing a rational reconstruction of simultaneity, Einstein then went on in his paper to formulate a new kinematics on the basis of two simple principles:

1. The laws by which the states of physical systems undergo change are not affected, whether these changes of state be referred to one or the other of two systems of coordinates in uniform translatory motion.

2. Any ray of light moves in the "stationary" system of co-ordinates with the determined velocity c, whether the ray be emitted by a stationary or moving body. Hence velocity = light path/time interval where time interval is to be taken in the sense of the definition [given above].

These two principles are then immediately applied to *two* reference frames and used to demonstrate that (1) classical kinematical measurements of a moving length in a rest system involve simultaneity determinations of end points of that length and (2) employing the definition of synchronicity, observers moving at uniform motion with respect to one another would synchronize their clocks differently, and therefore, Einstein asserted, "we cannot attach any *absolute* signification to the concept of simultaneity."

A more-detailed analysis of the two systems in relative motion and the connections between the time and length measurements made in their respective systems follows in Einstein's paper. This analysis is a *gedankenexperiment* involving clocks and rods and light signals, and together with a hypothesis of the homogeneity of space and time that insures linear transformations, the "Lorentz" transformation equations for space and time are *derived*.

Einstein's derivation of these connections, which are formally equivalent to the Lorentz transformations, employs in a part of it a system moving with velocity $-v$ and develops the inverse transformations that differ only in the sign of the velocity, providing a hint of Einstein's reciprocity interpretation of the transformations. In the following two sections Einstein explicitly brought out the *reciprocal* contraction effect that followed from his understanding of the transformations, and also discussed time dilation and the addition of velocities from the perspective of the new kinematics. In closing his kinematical section, he noted:

We have now deduced the requisite laws of the theory of kinematics corresponding to our two principles, and we proceed to show their *application* to electrodynamics [my italics].

Einstein now had at his disposal a theory that asserted the electrodynamical *equivalence* of all inertial frames of reference and provided a set of transformations for space and time which showed that the values of t and t', x and x', are ontologically equivalent. These equivalences eliminated the ether as a substance in any mechanical sense, as it had been denied even the property of simple position. The transformations, moreover, which had been obtained without the necessity of employing any "constructive" theory of electrodynamics, depended only on the hypothesis of the independence of the velocity of light from its source. The theory thus had both *security of foundation* and *simplicity,* especially when contrasted with Lorentz's theory, which was built on a number of hypothesized compensation effects and a series of non-experimentally corroborated distinctions between true and "effective" or "local" quantities. The new understanding of time had been legitimitized by a searching conceptual analysis of the notion of time and of the conditions of length and time measurements, an analysis that had uncovered certain ambiguities and gaps in the classical conceptions.

C. Einstein's New Kinematics and Its Application to Electrodynamics

The electrodynamical part of Einstein's paper shows the power of his orientation and represents the reason why many physicists found the Einstein approach most useful. The transformation for the electromagnetic parameters E, the electric field, and H, the magnetic field, were obtained from a direct application of the principle of relativity—by writing Maxwell's equations in exactly the same form for the "moving" system as for the "stationary" system except that the quantities were primed. The new space and time transformations were employed for the primed kinematical parameters, and transformations were derived for E' and H' in terms of E and H and a *relative* velocity term. The complete reciprocity and relativity of his approach allowed Einstein to resolve the paradox of electromagnetic induction with which he had begun his paper. He pointed out that the Lorentz force represented an "old manner of expression," and that a transformation of the electric or magnetic field, using his just obtained transformation, to "a system of co-ordinates at rest relatively to the electric charge" or magnets, was sufficent to treat any such problem. As a result, Einstein noted:

It is clear that the *asymmetry* mentioned in the introduction as arising when we consider the currents produced by the relative motion of a magnet and a conductor, now disappears. Moreover, questions as to the "seat" of electrodynamic electromotive forces (unipolar machines) now have no point.

A derivation of a new, relativistic, form of Doppler's principle followed, as did a general expression for the aberration effect due to motion of source and observer, which gave the customary expression v/c for the aberration coefficient as a limiting case. Einstein further showed that the problem of energy and radiation pressure of light waves could be easily computed by his theory, and in general that:

> All problems in the optics of moving bodies can be solved by the method here employed. What is essential is, that the electric and magnetic force of the light which is influenced by a moving body, be transformed into a system of co-ordinates at rest relatively to the body. By this means all problems in the optics of moving bodies will be reduced to a series of problems in the optics of stationary bodies.

The principle of relativity and the kinematical transformations were also applied to the Lorentz theory of 1895, yielding a new and relativistically correct charge density transformation. This permitted complete covariance of the fundamental Lorentz equations, (5)–(8) above. (Lorentz had only obtained covariance for his charge-free equations, though Poincaré, independently of Einstein, was able to formulate correct charge density (and velocity addition) expressions.)[78]

The electromotive force transformations are finally applied by Einstein to a *slowly* accelerated electron—i.e., an electron that is presumed to accelerate without radiating energy. The use of Newton's second law of motion in two systems yields the conclusion that, on the customary definition of force,[79] the mass of the electron was a function of its relative velocity. The definition of force employed by Einstein implied that the inertial reactions for accelerations in the same direction as the relative velocity (longitudinal direction) were different from accelerations at right angles to this direction (transverse direction). The expressions for the mass values were:

$$m \text{ longitudinal} = \frac{m}{\sqrt{(1 - (v/c)^2)^3}} \tag{19}$$

$$m \text{ transverse} = \frac{m}{1 - (v/c)^2} \tag{20}$$

These expressions are identical to Lorentz's formulae for mass changes as given in his 1904 paper, only for Lorentz v represents the "absolute" velocity. Einstein did not have to know of that paper, however, since familiarity with the electrodynamical literature would have acquainted him with these modes of locution.[80] Three properties of motion of such slowly accelerated electrons that might be tested by experiment closed the electrodynamical part of Einstein's paper.

In a sequel[81] to his fundamental paper—published later in 1905—Einstein pointed out that the theory of his earlier 1905 paper could be easily applied to a system of a body giving off plane waves of light, and that if such a *"body gives off the energy L in the form of radiation, its mass diminishes by L/c²."* Einstein added:

> The fact that the energy withdrawn from the body becomes energy of radiation evidently makes no difference, so that we are led to the more general conclusion that the mass of a body is a measure of its energy-content; . . . It is not impossible that with bodies whose energy-content is variable to a high degree (e.g., with radium salts) the theory may be successfully put to the test.

Scientists working with the electrodynamics of moving bodies in the years 1905–10 were struck with the coherence of the special theory of relativity, its simplicity, the directness with which it could be applied, and, especially, its focus on the concept of time and kinematics. Planck immediately saw its significance and gave a seminar on Einstein's theory in late 1905. Sommerfield was an early convert and wrote to Lorentz, probably in late 1906, and commented on the similarity of Einstein's results to Lorentz's, but on the different epistemological point of view that Einstein took.[82] Max Born writes that though he had been "quite familiar with the relativistic idea" in the Lorentz sense, when he came on Einstein's papers in 1907, "Einstein's reasoning was a revelation to me."[83] Born added:

> For me—and many others—the exciting feature of . . . [Einstein's] paper was not so much its simplicity and completeness, but the audacity of challenging Isaac Newton's established philosophy, the traditional concepts of space and time.[84]

Perhaps most important of those accepting Einstein's theory was Einstein's former teacher, H. Minkowski. Minkowski's work was very important both for completing the application of the theory to electrodynamics and for developing a very general and powerful mathematical method for applying the theory to mechanics.[85] Minkowski's "geometrical" account of the relativistic kinematics,[86] with his picture of an imaginary time coordinate, also made the implications of the special theory easier to visualize and brought out the revolutionary kinematical aspects of the theory more clearly. Minkowski seems to have excited a larger mass of physicists than Einstein's presentation had done, and it is also clear that Minkowski's contributions were most important for Einstein's later research on the general theory of relativity.[87]

IV. SUMMARY AND CONCLUSION

I would like to bring this paper to a close by reemphasizing three epistemological points developed earlier. The first concerns the important role of earlier theories in the genesis of a new theory; the second concerns *conceptual* differences between successive theories; the third point focuses on the significance of transempirical factors in both the genesis and the justification of a new theory. I believe that because all of these points have been overlooked there have been many misinterpretations of the history of the theory of special relativity.

Experimental considerations, and especially the Michelson-Morley experiment, have often been taken as crucial generators of the special theory of relativity. It is to the merit of Polanyi and, especially, of Holton,[88] to have effectively criticized such simplistic accounts. Nonetheless, I think that there has been an overreaction on the part of the critics of empiricism, and that a more balanced and historically accurate picture is obtained if we consider the effect of experiments as interpreted by earlier *theories*.

As argued above, experimental considerations such as induction experiments and aberration effects of both the first (Fizeau's experiment) and second order (Michelson-Morley) played an important role in Einstein's genesis of the special theory of relativity, but only via their association with Maxwell's and Lorentz's *theories*. Einstein implicitly cites Lorentz's theorem of corresponding states in the opening paragraphs of his fundamental paper on relativity, remarking that the induction experiments and null result aberration experiments "suggest . . . as has already been shown to the first order of small quantities, the same laws of electrodynamics and otpics will be valid for all frames of reference."[89] Theories possess more empirical content than do the reports of experiments, and exercise greater constraints on possible innovative attempts to explain experiments, even in those cases in which an experiment appears to falsify such a theory. Inattention to the theoretical background on the part of historians of science can make the birth of a novel theory appear inexplicable, and as the result of a "poetic" intuition, rather than the consequence of rational deliberation.[90]

One central theme of this paper was the import of Einstein's conceptual reanalysis of the traditional notion of time. Other appraisals of the central contribution of the special theory of relativity take a different point of view. Positivist philosophers such as Mach's disciple Petzold,[91] Reichenbach,[92] and Bridgman[93] have taken Einstein's special theory of relativity as representing the triumph of empiricism over metaphysics. From such a perspective Lorentz's theory involves an ether metaphysics with no physical content, which if purged of its obfuscations, would point to Einstein's special theory of relativity. For example, Reichenbach has written:

> The physicist who wanted to understand the Michelson experiment had to commit himself to a philosophy for which the meaning of a statement is reduced to its verifiability, that is, he had to adopt the verifiability theory of meaning if he wanted to escape a maze of ambiguous questions and gratutitous complications. It is this positivist, or let me rather say, empiricist commitment which determines the philosophical position of Einstein.[94]

Petzold believed that "Lorentz's theory is, at its conceptual center, pure metaphysics, nothing else than Schelling's or Hegel's *Naturphilosophie*."[95] Bridgman thought that Einstein had provided a method of approaching physical concepts that, if strictly followed, would immunize physics against further revolutions. Revolutions for Bridgman were consequences of realizing that a central concept, such as time, as it had been understood involved *a priori,* non-empirical components that were illegitimate.

Such a view of Einstein's accomplishment indicates both a weak understanding of the historical predecessors of Einstein's theory, and a lack of sensitivity to *conceptual* innovation. The ether in Lorentz's and Poincaré's theories of electrodynamics played important causal roles,

501

and until 1910 Lorentz thought there were experiments that were likely to detect absolute motion.[96] Einstein's most significant contribution, his reanalysis of the time concept, revealed very clearly a new *interpretation* of Lorentz's "local" time. This new interpretation of time did not rest on any new experiments, and it, in contrast to Bridgman's understanding, contained an important *a priori* component in the definition of synchronicity. The point I wish to stress is that a radical conceptual innovation, not necessarily involving clear empirical or metaphysical amelioration, can be most significant and stimulating to a scientific community. This is especially true if the conceptual innovation is associated with an increase in coherence and simplicity in a subject area, which brings me to my third and last point.

I believe that the above account of the evolution of the time concept from Lorentz to Einstein demonstrates the importance of transempirical criteria of theory evaluation. I indicated that there is strong historical evidence that indicates that Einstein could not accept the Lorentz approach to the electrodynamics of moving bodies because he viewed Lorentz's theory as lacking simplicity and logical perfection, and that Einstein was also influenced by the still unconfirmed hypothesis of light quanta from working at the "constructive" level. The Einstein theory, with its kinematical approach and its bold reanalysis of the time concept, seized the imagination of eminent physicists soon after its publication, in large measure because of its systematic simplicity and the paucity of its premises. I have analyzed some of the senses of simplicity elsewhere,[97] and here only wish to emphasize the role that transempirical constraints such as simplicity and inter-theoretic accord have played in both the genesis and the acceptance of the special theory of relativity.

Grateful acknowledgment is made to the National Science Foundation for support of research.

1. A. Lorentz, *The Theory of Electrons*, 1st ed. (New York: Columbia University Press, 1909); 2d ed. (New York: Dover Publications, 1915); pagination same except in notes.

2. Lorentz, *The Theory of Electrons*, p. 230.

3. Ibid., 2d ed., p. 321.

4. H. Minkowski, "Die Grundgleichungen für die elektromagnetischen Vorgänge in bewegten Körpern," in *Göttingen Nachrichten* (1908), p. 53.

5. Letter is in the Einstein-Sammlung der Eidgenossische Technische Hochschule Bibliothek in Zürich, and is dated 12 June 1913.

6. W. Pauli, *The Theory of Relativity* (Oxford: Pergamon Press, 1958) (originally published in German in 1921).

7. A number of studies of Einstein and his background have appeared in recent years, and are cited below in notes 10, 21, 29, 42, 58, and 60. None of these are specifically addressed to the comparative analysis of pre- and post Einsteinian theories of time, though they all touch on the problem to some extent.

8. See Hirosige's analysis for the Helmholtzian background of Lorentz's ideas in T. Hirosige, "Origins of Lorentz' Theory of Electrons and the Concept of the Electromagnetic Field," *Historical Studies in the Physical Sciences* 1 (1969): 151.

9. See Hirosige, "Origins of Lorentz' Theory of Electrons," and also R. McCormmach, "H. A. Lorentz and the Electromagnetic View of Nature," *Isis* 61 (1970): 459.

10. See K. Schaffner, "The Lorentz Electron Theory and Relativity," *Am. J. Phys.* 37 (1969): 498–513, and also S. Goldberg, "The Lorentz Theory of Electrons and Einstein's Theory of Relativity," *Am. J. Phys.* 37 (1969): 982–94. See also references in notes 8 and 9 above.

11. S. G. Brush, "Note on the History of the Fitzgerald-Lorentz Contraction," *Isis* 58 (1967): 230–32.

12. See H. Reichenbach, *The Philosophy of Space and Time* (New York: Dover Publications, 1957).

13. Lorentz, *The Theory of Electrons,* p. 196.

14. H. A. Lorentz, *Versuch einer Theorie der elektrischen und optischen Erscheinungen in Bewegten Körpern* (Leiden: Brill, 1895).

15. Ibid., p. 85. The "theorem of corresponding states" is put into a simpler form in the first-order theory section of Lorentz's 1899 paper (see note 17 below), in which all electromagnetic parameters are transformed.

16. W. Voigt, "Über das Dopplerische Prinzip," in *Göttingen Nachrichten* (1887), p. 41. See also Lorentz, *The Theory of Electrons,* p. 198 n. 1.

17. H. A. Lorentz, "Simplified Theory of Electrical and Optical Phenomena in Moving Systems," *Proc. Roy. Acad. Amsterdam* 1 (1899): 427. Reprinted in K. Schaffner, *Ninteenth-Century Aether Theories* (Oxford: Pergamon Press, 1972).

18. H. A. Lorentz, "Electromagnetic Phenomena in a System Moving with Any Velocity Less Than Light," *Proc. Roy. Acad. Amsterdam* 6 (1904): 809–30. Partially reprinted in A. Einstein et al., *The Principle of Relativity,* trans. W. Perrett and G. B. Jeffery (New York: Dover Publications, n.d.).

19. H. Poincaré, *Electricité et optique* (Paris: Gauthier-Villars, 1901).

20. J. Larmor, "A Dynamical Theory of the Electric and Luminiferous Medium" *Phil. Trans. Roy. Soc.* 190 (1897): 205–300. See also my discussion in *Nineteenth-Century Aether Theories,* pp. 91–98, 110–12.

21. See *Nineteenth-Century Aether Theories,* pp. 59–68, 84–91.

22. J. Larmor, *Aether and Matter* (Cambridge: At the University Press, 1900).

23. H. Poincaré, *Congrès de physique de 1900* (Paris), 1 (1900): 22. See also Poincaré's *Science and Hypothesis* (New York: Dover Publications, 1952), chap. 10, for a translated version of that address.

24. *Science and Hypothesis,* p. 172.

25. See Schaffner (note 10 above) for a description of these experiments.

26. Lorentz, "Simplified Theory." My analysis of the additional hypotheses and their interconnection is somewhat different from G. Holton's reading in "On the Origins of the Special Theory of Relativity," *Am. J. Phys.* 28 (1960); 627–36, and from McCormmach's analysis in "H. A. Lorentz and the Electromagnetic View of Nature."

27. "In a way," because the factor $\sqrt{1-(v/c)^2}$ is *postulated* for the force transformation law without good physical reasons for doing so.

28. Lorentz, *The Theory of Electrons,* p. 224. (Letters have been altered for consistency with other parts of this paper.)

29. A. Miller, "Study of Henri Poincaré's 'Sur la dynamique de l'électron,'" *Archive for History of the Exact Sciences* 10 (1973): 207–328. See also G. Holton, "On the Thematic Analysis of Science: The Case of Poincaré and Relativity," in *Mélanges Alexandre Koyre* (Paris: Hermann, 1964), p. 257; C. Scribner, "Henri Poincaré and Einstein's Theory of Relativity," *Am. J. Phys.* 32 (1964): 672–78; S. Goldberg, "Henri Poincaré and Einstein's Theory of Relativity," *Am. J. Phys.* 35 (1967): 934–44, and "Poincaré's Silence and Einstein's Relativity: The Role of Theory and Experiment in Poincaré's Physics," *Brit. J. Hist. Sci.* 5 (1970): 73–84.

30. H. Poincaré, *The Value of Science* (New York: Dover Publications, 1958), has a translation of Poincaré's Saint Louis address as chapters 7–9.

31. "The Measure of Time" was reprinted in *The Value of Science,* chap. 2.

32. Ibid., chaps. 6–7.

33. The derivation that follows is patterned after Lorentz, *The Theory of Electrons,* p. 225.

34. Ibid., p. 226.

35. H. A. Lorentz, "The Principle of Relativity for Uniform Translations," in vol. 3 of his *Lectures on Theoretical Physics* (London: Macmillan, 1931), lectures originally given in 1910–12 at the University of Leiden.

36. A. Einstein, "Autobiographical Notes," in P. A. Schilpp, ed., *Albert Einstein: Philosopher-Scientist* (New York: Tudor Publishing, 1959), p. 5. Hereafter the "Autobiographical Notes" will be cited as AN.

37. Letter is quoted in Carl Seelig's important biography *Albert Einstein: A Documentary Biography* (London: Staples, 1956), pp. 228–29. (I have modified the translation from the French original in slight ways.)

38. AN, pp. 25, 27.

39. Ibid., p. 37. See M. Jammer, *Concepts of Mass* (Cambridge, Mass.: Harvard University Press, 1961), chaps. 11–12, for background on this notion; see also McCormmach's searching treatment of the electromagnetic world-view in note 9 above.

40. AN, p. 37.

41. Ibid., p. 45.

42. See M. J. Klein's insightful "Thermodynamics in Einstein's Thought," *Science* 157 (1967): 509–16.

43. AN, p. 21.

44. G. Holton, "Mach, Einstein, and the Search for Relativity," *Daedalus* 97 (1968): 636–73.

45. Ibid.

46. AN, p. 13.

47. E. Mach, *Science of Mechanics,* 2d ed. (LaSalle, Ill.: Open Court, 1893). This is a translation of the German edition of 1883.

48. Ibid., p. 243.

49. Ibid., p. 244.

50. E. Mach, *Principien der Wärmelehre*. Quotations are from the translation by T. J. McCormack published in *Open Court* 17 (1903): 157.

51. Ibid., p. 157.

52. Seelig, *Albert Einstein,* pp. 57, 61.

53. Poincaré, *Science and Hypothesis,* p. xxii.

54. Ibid., p. xxvi.

55. The note in the *Times* is reprinted in a collection of Einstein's essays, *The World As I See It* (New York: Covici-Friede, 1934), p. 73.

56. AN, p. 21.

57. Ibid., p. 23.

58. Ibid.

59. Ibid., pp. 23–25.

60. For evidence that Einstein viewed the contraction hypothesis as *ad hoc* and artificial, see the letter from Einstein to R. Shankland published in Shankland, "The Michelson-Morley Experiment," *Am. J. Phys.* 32 (1964): 23–34, and also Max Wertheimer, *Productive Thinking,* enlarged ed. (New York: Harper & Bros., 1959), chap. 10.

61. A. Einstein, "Zur Elektrodynamik bewegter Körper," *Ann. d. Phys.* 17 (1905): 891–921 (translated in Einstein et al., *The Principle of Relativity*).

62. AN, pp. 51–53.

63. For the most famous of such anti-empiricist analyses see M. Polanyi, *Personal Knowledge* (Chicago: University of Chicago Press, 1958), pp. 9–15. See also A. Grünbaum, "The Genesis of the Special Theory of Relativity," in H. Feigl and G. Maxwell, eds., *Current Issues in the Philosophy of Science* (New York: Holt, Rinehart & Winston, 1961), pp. 43–53. G. Holton seems to have some sympathy with anti-empiricist analyses of the genesis of special relativity in his fine article "Einstein, Michelson, and the 'Crucial experiment'," *Isis* 60 (1960); 133–97. For a discussion of the relative merits of Grünbaum and Holton see G. Gutting, "Einstein's Discovery of Special Relativity," *Phil. Sci.* 39 (1972): 51.

64. See Wertheimer, *Productive Thinking,* chap. 10.

65. A. Einstein, *Relativity—The Special and General Theory* (New York: Crown Publishers, 1961), p. v.

66. This point is clearly made by Grünbaum, "The Genesis of the Special Theory of Relativity."

67. Wertheimer, *Productive Thinking,* p. 215.

68. See Shankland, "The Michelson-Morley Experiment"; see also Holton's translation in "Einstein, Michelson, and the 'Crucial Experiment.'"

69. See M. von Laue's analysis of this problem from the perspective of Hert's theory in his *Das Relativitätsprinzip* (Braunschweig: Viewig, 1911), p. 26. See also the same problem dealt with from Lorentz's perspective in R. McCormmach, "Einstein, Lorentz, and the Electron Theory," *Historical Studies in the Physical Sciences* 2 (1970): 56.

70. See the references cited in note 63 above.

71. See Shankland, "The Michelson-Morley Experiment," Polanyi, *Personal Knowledge,* Holton, "Einstein, Michelson, and the 'Crucial Experiment,'" and Gutting, "Einstein's Discovery of Special Relativity."

72. The expression is Grünbaum's, used in his article "The Bearing of Philosophy on the History of Science: Philosophical Mastery of the Special Theory of Relativity Is Required for Unraveling Its History," *Science* 143 (1964): 1406.

73. Gutting, "Einstein's Discovery of Special Relativity," uses Wertheimer's account but does not attempt to reconcile the paradox of Wertheimer's heavy reliance on the Michelson-Morley experiment with his, Gutting's, acceptance of Holton's position that the Michelson experiment was "subsidiary in Einstein's discovery of STR."

74. Wertheimer, *Productive Thinking,* pp. 218–19.

75. Since the above scenario of Einstein's steps to the special theory of relativity was written in May 1973, additional historical documentation has become available in the form of further reports on conversations with Einstein by R. S. Shankland, "Conversations with Albert Einstein. II," *Am. J. Phys.* 41 (1973): 895. It is Shankland's position that the Michelson-Morley experiment was important for Einstein. Shankland believes that Wertheimer's account is accurate, and writes:

As clearly reported by Max Wertheimer, who in 1916 discussed with Einstein the development of his ideas in special relativity in great detail, it is evident that the importance of the Michelson-Morley experiment for Einstein was that it gave positive confirmation to his belief that the speed of light is invariant in all inertial frames, independent of the motion of source, apparatus, or observer. Such invariance in c was necessary for his interpretation of Maxwell's equations and for his derivation of the Lorentz transformation, as well as for his conviction that the "local time," first introduced by Lorentz, is indeed the only true time for the description of physical phenomena. Professor Einstein's statement to me that "at last it came to me the [absolute] time was suspect," and that the new absolute for physics must be the speed of light in vacuum, rather than space and time.

. . . .

This writer finds the account of Wertheimer entirely consistent with his own notes and recollections of Einstein's attitude in 1952 toward the Michelson-Morley experiment.

The significance of the Michelson-Morley experiment and the complete veracity of Wertheimer's account is questioned in a forthcoming paper by A. Miller, "Max Wertheimer and Albert Einstein: A Gestalt Psychologist's View of the Genesis of Special Relativity Theory." I do not believe that Miller's analyses vitiate in any way the conjectural account of Einstein's discovery process developed in the text above, but cannot in these pages present detailed arguments.

76. References and quotations are from Einstein, "Zur Elektrodynamik bewegter Körper," as translated in *The Principle of Relativity*.

77. For Reichenbach's position see *The Philosophy of Space and Time;* Grünbaum's views are set out in his *Philosophical Problems of Space and Time* (New York: Alfred A. Knopf, 1963).

78. H. Poincaré, "Sur la dynamique de l'électron," *Rendiconti del Circolo Matematico di Palermo* 21 (1906): 129–76.

79. M. Planck, in "Das Prinzip der Relativität und die Grundgleichungen der Mechanik," *Verh. dtsch. phys. Ges.* 4 (1906): 136–41, showed that an alternative definition of force based on the Lorentz force expression is the only natural definition, inasmuch as only this definition allows one to obtain the force as the time derivative of a momentum. Such a definition eliminates the need to distinguish a longitudinal from a transverse mass, all mass governed by the equation:

$$m = \frac{m_2}{\sqrt{1 - (v/c)^2}}$$

80. M. Abraham used this terminology in his 1903 article published in a journal that we know Einstein tended to read. See M. Abraham, "Prinzipien der Dynamik des Elektrons," *Ann. d. Phys.* 10 (1903): 105–79.

81. A. Einstein, "Ist die Trägheit eines Körpers von seinen Energieinhalt abhängig?", *Ann. d. Phys.* 18 (1905): 639–41.

82. Letter, A. Sommerfeld to H. A. Lorentz, 12 December 1906, Lorentz Collection, Algemeen Rijksarchief, The Hague, Holland.

83. M. Born, *Physics in My Generation* (New York: Springer-Verlag, 1969), p. 103.

84. Ibid., p. 105.

85. See Minkowski, "Die Grundgleichungen für die elektromagnetischen Vorgänge in bewegten Körpern." Minkowski's influence is discussed by T. Hirosige, "Theory of Relativity and the Ether," *Jap. Stud. Hist. Sci.* 7 (1968): 37–53.

86. Minkowski's stimulating geometrical interpretation of special relativity is to be found in the translation of his 1908 lecture in *The Principle of Relativity,* p. 75.

87. See Hirosige, "Theory of Relativity and the Ether," and McCormmach, "Einstein, Lorentz, and the Electron Theory."

88. See note 63 above for Polanyi's and Holton's arguments.

89. Einstein, "Zur Elektrodynamik bewegter Körper," p. 37.

90. I would not want to totally deny the existence of "free speculation" in the genesis of new hypotheses and theories, but only wish to emphasize that there are very strong constraints channeling such speculations.

91. J. Petzold "Die Relativitätstheorie der Physik," *Zeit, posit. Phil.* 2 (1914): 10.

92. H. Reichenbach, in Schilpp, ed., *Albert Einstein: Philosopher-Scientist.*

93. P. W. Bridgman, *The Logic of Modern Physics* (New York: Macmillan, 1928).

94. Reichenbach, pp. 290–91.

95. Quoted in Holton, "Einstein, Michelson, and the 'Crucial Experiment.'"

96. See Pauli, *The Theory of Relativity,* p. 1 n. 3. Pauli also notes, pp. 4–5, citing Einstein (*Ann. Phys. Lpz.* 38 [1912]: 1059), that "it might seem that the postulate of relativity is immediately obvious once the concept of an aether has been abandoned. Closer reflection shows however that this is not so."

97. K. Schaffner, "Outlines of a Logic of Comparative Theory Evaluation with Special Attention to Pre- and Post-Relativistic Electrodynamics," in R. Stuewer, ed., *Minnesota Studies in the Philosophy of Science* 5 (Minneapolis: University of Minnesota Press, 1970): 311.

Wesley C. Salmon

Chapter Eighteen

CLOCKS AND SIMULTANEITY IN SPECIAL RELATIVITY
OR, WHICH TWIN HAS THE TIMEX?

In October 1971 J. C. Hafele[1] placed several cesium beam clocks aboard ordinary commercial around-the-world jet flights. One flight circumnavigated the earth traveling in an eastward direction; the other in the opposite direction. These clocks were compared with similar clocks that remained at the U.S. Naval Observatory during the trip (see figure 1). Relative to the Naval Observatory clocks, the clocks that traveled eastward lost 59 ± 10 nanoseconds (1 nsec $= 10^{-9}$ sec), while those traveling westward gained 273 ± 7 nanoseconds. These results are in excellent agreement with the theoretical predictions of standard relativity theory; according to Hafele and Keating, they "provide an unambiguous empirical resolution of the famous clock 'paradox'[2] with macroscopic clocks."[3]

This outcome is not in the least disturbing; it is if anything reassuring, especially as it bears upon the special theory of relativity. The time dilation effect had long since been dramatically confirmed by muon decay phenomena. The special theory is supported by the fact that the Stanford Linear Accelerator is two miles long instead of a bit under one inch.[4] Indeed, special relativity is a standard item in the physicist's tool kit, and it is tested every day in physical laboratories throughout the world. It is nice to see that macroscopic clocks exhibit the same retardation effects as do microscopic clocks, but it would have been a great surprise if they had not.[5] Thus, we can say with even greater confidence than before that the twin who takes a very fast trip on a rocket ship will return younger than the twin who remains at home. There is nothing contradictory or unintelligible about this fact; at worst it is in the category of "strange but true."

Hafele's traveling clocks were, of course, transported through gravi-

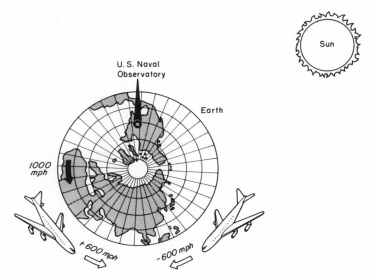

Fig. 1.   Clock paradox

tational fields. Their behavior, consequently, depends in part upon general relativistic considerations, though special relativistic effects were also involved. In this paper it is the special theory that is my object of concern, and it is the special theory that has been charged with

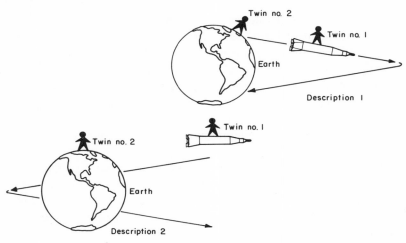

Fig. 2.   Twin paradox

509

outright inconsistency in connection with the traveling twin. For, it has been argued, if the first twin travels from A to B and back again, he will be younger than the second twin who remained at A. But, the argument continues, since there is no such thing as absolute rest, we could just as well say that the first twin remained at rest while the second twin (whom we previously characterized as remaining at A) has traveled away from the first twin and then returned to rejoin him. On this way of looking at the situation, however, the second twin should be younger than the first (see figure 2)! There is nothing paradoxical in saying that one twin is younger than the other (twin paradox), but it is blatantly inconsistent to say that each is younger than the other when they meet after the trip (clock paradox).[6]

As has often been pointed out, this method of attempting to generate a paradox is defective for a very elementary reason. Such formulas as the time dilation equation apply only to clocks situated in inertial frames and not subject to accelerations. In order for the twins to move apart and then back together again, however, at least one of them has to experience accelerations.[7] Suppose that the second twin (who remains at A) experiences no accelerations. Then his brother has to accelerate to an appreciable fraction of the speed of light; moreover, when he reaches B he has to reverse his direction in order to travel back to A. This involves further acceleration. Which twin suffers the accelerations can be established empirically in an unambiguous way: crashing an automobile into a concrete wall at 200 miles per hour[8] is negligible compared with instantaneous reversal of direction when one is traveling at, say, 9/10 the speed of light. In the immortal words of John Cameron Swayze, the watch of the first twin really has to "take a licking and keep on ticking."[9] Only one of the twins need undergo what is known colloquially as a "bad trip." When accelerations are taken into account, any apparent symmetry in the situation vanishes, and the general theory can be invoked to ascertain unambiguously which of the two twins will be the younger.

While I do not deny the correctness of the general relativistic treatment of the clock paradox, I do agree with Adolf Grünbaum and others that it is somehow not completely satisfying intellectually.[10] It is, after all, the time dilation of special relativity that is suspected of spawning an inconsistency.

Contradictions are not best treated by invoking a more complex theory, for it is a general and fundamental principle of logic that contradictions in a set of premises can never be eradicated by *adding* new premises; the only way to get rid of a contradiction is by *removing* some premises of the original set. If the special theory does contain an inconsistency, we had better locate it, rather than covering it up with an augmented theory. Of course, it may be replied that the clock paradox holds no difficulties for special relativity because it cannot even be formulated in terms to which the restricted theory is applicable, and so the question of inconsistency cannot even arise. This answer does not seem fully adequate, however, for it is possible to formulate a version of the clock paradox that does not involve any accelerations. This version has been attributed to Lord Halsbury.[11]

Assume we have three clocks, C, C', C''. C is situated at rest at point A in its inertial frame K. K' and K'' are the frames in which C' and C'', respectively, are at rest. C' travels from left to right at constant velocity v relative to frame K. At the moment C and C' meet, the two clocks are synchronized with each other by setting both to read zero. C' continues moving uniformly to the right; when it reaches a point B in K, it meets a clock C'' traveling in the opposite direction. At the moment of meeting, C' and C'' have the same reading, i.e., they are locally synchronous at that point. C' continues moving to the right (and on out of the picture); C'' moves inertially toward A where C has been situated for the whole time, and it passes C with a relative velocity $-v$. At the moment of meeting, the readings of C and C'' are compared (see figure 3). In this formulation we have dispensed with the twins, for there is no way to bring them back together again without subjecting them to accelerations, but we have made the relevant time comparisons. To be sure, we have not brought clock C' back to clock C, but we have brought the time reading of C' back via a third clock C'' that was synchronized (locally) with C' as the two of them met at B. Local synchronization is taken as absolute and unproblematic; when two clocks are located at (approximately) the same place, we can make a direct empirical comparison of their readings. Since no accelerations are involved and all clocks are moving inertially, we can use the special theory to calculate the results. No considerations from general relativity are involved.

Letting events E_1, E_2, and E_3 represent, respectively, the meeting of C with C' at A, the meeting of C' with C'' at B, and the meeting of C'' with C at A, we attempt to ascertain the times of these events according to the

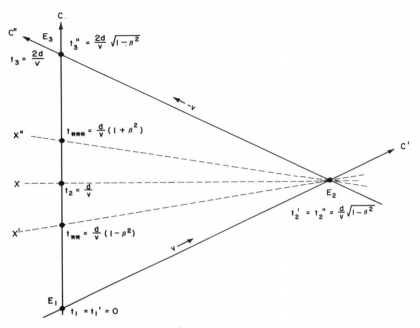

Fig. 3.   Clock paradox as seen from K (frame of clock C)

proper times of the various clocks. We let t stand for the time of clock C, t' the time of clock C', and t'' the time of the clock C'', and we let d be the distance between A and B with respect to frame K. As usual, we let $\beta = v/c$ (or $-v/c$ indifferently since this quantity is always squared). We have set the clocks C and C' to read $t_1 = t'_1 = 0$ at E_1. With respect to frame K, C' travels a distance d at a velocity v, arriving at B (where it meets C'') at $t_2 = d/v$. This is the time of event E_2 according to C. Applying the standard time dilation formula, we find that C' reads $t'_2 = (d/v)\sqrt{1 - \beta^2}$; this is the time of E_2 according to C'. We have stipulated, moreover, that C'' be set to agree with C' at E_2; hence, $t''_2 = t'_2$. Since, obviously, with respect to frame K, it takes C'' just as long to get from B to A as it took C' to travel from A to B, C'' will read $t''_3 = (2d/v)\sqrt{1 - \beta^2}$ at E_3, while C will show $2d/v$ as the combined time for the trips of C' and C'' over distance d at speed v. Hence, C'' is retarded relative to C when they meet. This outcome seems unambiguous when the analysis is carried out from the standpoint of frame K.

In spite of these straightforward results, one might still harbor a

512

suspicion that the clock paradox has not been fully resolved; for even though we are using three clocks instead of two, the special theory of relativity says unequivocally that when clocks are in uniform motion with respect to one another *each* is retarded with respect to all the others. Until we have shown that this general fact is compatible with our analysis, we have not completely handled the clock paradox.

In figure 3, the broken lines X, X', X" represent the spatial axes in the frames K, K', and K", respectively; that is, they represent the simultaneity relations in each of the three inertial frames. Thus, in frame K, the clock reading of C which is simultaneous with E_2 is $t_2 = d/v$. With respect to frame K, clock C' runs more slowly than C, for C' reads $t'_2 = (d/v)\sqrt{1 - \beta^2}$ when C reads $t_2 = d/v$. C" also runs slower than C with respect to frame K for reasons of obvious symmetry. When we ask, from the standpoint of frame K', what clock-reading on C is simultaneous with event E_2, the answer is different. According to K', $t^{**} = (d/v)(1 - \beta^2)$ is the reading on C when C' meets C", and that is less than $t'_2 = (d/v)\sqrt{1 - \beta^2}$; consequently, with respect to K', C runs slower than C'. Similarly, the clock-reading on C that is simultaneous with E_2 with respect to K" is $t^{***} = (d/v)(1 + \beta^2)$. Again, from the standpoint of K", C is retarded with respect to C". The key to the whole problem lies in the discrepancy between the simultaneity relations of K' and K"; from the combined standpoints of K' during the interval from E_1 to E_2, and of K" during the interval from E_2 to E_3, the clock times of C between t^{**} and t^{***} simply drop out of the picture. The moments in that interval are not simultaneous (with respect to the simultaneity relations of K') with any part of the trip of C' from A to B, and they are not simultaneous (with respect to the simultaneity relations of K") with any part of the trip of C" from B to A.

In order to be quite sure that no paradox can be generated on the basis of Lord Halsbury's formulation, let us look at the whole situation from the standpoint of the frame K" in which C" is at rest (see figure 4). From this viewpoint, there are two clocks, C and C', approaching C" from the left at different velocities v and V. Using the composition of velocities formula, we find that $V = 2v/(1 + \beta^2)$. Before they arrive, the faster-moving clock C' catches up with C (this is event E_1) and moves on to meet C" before C gets there. The meeting of C' with C" is E_2. Somewhat later, C arrives; this is event E_3. The lack of symmetry between the situation as seen from the standpoint of C and as seen from that of C"

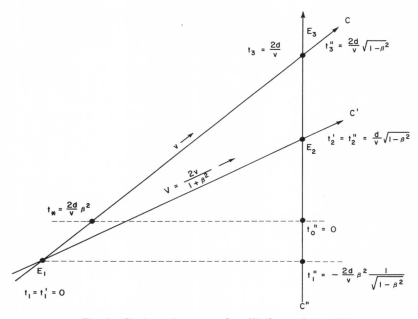

$$t_3 = \frac{2d}{v} \qquad t_3'' = \frac{2d}{v}\sqrt{1-\beta^2}$$

$$t_2' = t_2'' = \frac{d}{v}\sqrt{1-\beta^2}$$

$$V = \frac{2v}{1+\beta^2}$$

$$t_* = \frac{2d}{v}\beta^2$$

$$t_0'' = 0$$

$$t_1'' = -\frac{2d}{v}\beta^2 \frac{1}{\sqrt{1-\beta^2}}$$

$$t_1 = t_1' = 0$$

Fig. 4. Clock paradox as seen from K″ (frame of clock C″)

becomes apparent by comparing figures 3 and 4; they are in no sense reflections of each other.

The crucial point in analyzing the situation from the standpoint of K″ is recognition that E_1 (when C and C′ register $t_1 = t' = 0$) is *not* simultaneous with the reading t′ $= 0$ on C″ according to the simultaneity relations of K″. That this must be the case is obvious from inspection of figure 3, which shows the simultaneity relations of each of the three frames. Rather, from the standpoint of K″, the reading $t* = (2d/v)\beta^2$ is simultaneous with t″ $= 0$. In K″, the time t''_1 of E_1 has the negative value $(-2d/v)\beta^2(1/\sqrt{1-\beta^2})$.[12] According to the simultaneity relations of K″, when C reads zero, C″ has a negative reading; when C″ reads zero, C has a positive reading. C is thus ahead of C″ on both of these occasions, and although C continues to run slower than C″ until they meet, C is still ahead of C″ at E_3 because C″ does not succeed in catching up with C before their meeting. From the standpoint of clock C″, it turns out, we get the same result as we got in analyzing the situation from the standpoint of C. On the occasion of their meeting (event E_3) clock C″ is retarded with respect to clock C. Precisely the same outcome can be calculated from the standpoint of K'.[13]

The foregoing resolution of the clock paradox obviously depends heavily upon appeal to the relativity of simultaneity. Indeed, the concept of simultaneity constitutes the key to the entire special theory of relativity. The well-known length contraction and time dilation effects rest directly upon the relativity of simultaneity. As a boy, Einstein remarks in his "Autobiographical Notes," he had *wondered* about the role of the speed of light as a constant in Maxwell's electromagnetic theory, but it was only some years later that he saw how to *handle* the problem through a radical reanalysis of the concept of simultaneity.[14] The crucial character of Einstein's treatment of simultaneity has, of course, been widely recognized, but the fact that the revolutionary new analysis was a two-stage affair has not always been clearly noted. Einstein, however, treated the two stages separately, and he explicitly acknowledged the distinction between them.

Einstein's first discussion of simultaneity in the 1905 paper relates to the problem of establishing simultaneity relations within a single frame of reference. This is the problem of synchronizing clocks that are at rest with respect to one another in any inertial reference frame. This problem must be treated before moving on to consideration of the relativistic effects that arise from relative motion between two or more inertial frames, and Einstein discusses it in section 1 of his original paper. His resolution of this problem is deceptively simple. Given two clocks situated at widely separated points A and B in an inertial frame, "we established *by definition* that the 'time' required by light to travel from A to B equals the 'time' it requires to travel from B to A."[15] That Einstein meant to emphasize the definitional character of such synchrony is evidenced by the fact that the italics in the quoted passages are his, and by the fact that the title of section 1 is "Definition of Simultaneity."

If a light signal sent from A is reflected to B so that it returns to A, a clock at A can be used to ascertain the total time required by light for the round trip from A to B and back again. Letting t_A be the time at which the light signal is sent from A, and letting t'_A be the time at which the light signal returns to A, Einstein goes on to say, "In agreement with *experience* we *further* assume the quantity

515

$$\frac{2AB}{t'_A - t_A} = c$$

to be a universal constant—the velocity of light in empty space'' (my italics).[16]

Einstein thus enunciates two distinct principles regarding the speed of light in the first section of his 1905 paper. The second of these principles is an *empirical* hypothesis stating that the *average* speed of light on a *round trip* in a vacuum is a constant c. This principle, Einstein claims, is supported by experience; if true, it describes a *fact* of nature. The first principle equates the speed of light on each of the two legs of a round trip in a vacuum. Unlike the second principle, it is a *definition*. There can be no question of its truth of falsity, and under no circumstances can it properly be construed as articulating any fact of nature. The two principles can be contrasted and summarized as follows:

1. One-way light principle (a convention): On any round trip *in vacuo* the speed of light in the outbound direction is equal to its speed on the return trip.

2. Two-way light principle (a factual hypothesis): On any round trip *in vacuo* the average speed of light for the entire trip is equal to a constant c.

In conjunction, of course, these two principles imply that the speed of light on any one-way trip *in vacuo* is equal to the constant c. In most discussions of relativity it is the combination of the two principles that goes under some such heading as ''the principle of constancy of light velocity,'' and it is this combined principle that we want to use in most contexts. Einstein, himself, uses the combined principle in subsequent portions of the 1905 paper. When we are trying to be careful about logical foundations, however, as was Einstein at the beginning of his famous paper, it is crucially important to distinguish the conventional from the factual components of the combined principle.

In order to underscore the distinction between the two light principles, let us consider a classical method for measuring the speed of light. This method is due to Armand Fizeau, who devised a cog-wheel arrangement to make the measurement (see figure 5). A wheel with gaps between its teeth is made to rotate in front of a light source, which is sending a beam

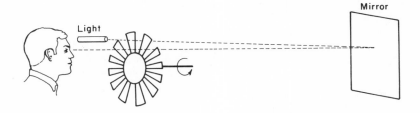

At a suitable speed of rotation, the wheel will move the distance between two adjacent gaps in the time required for a light pulse to travel from the wheel to the mirror and back.

Fig. 5.   Fizeau's method for determining speed of light

toward a mirror. When a tooth is in the light path no light gets through, but when a gap is there a light pulse passes through to the mirror, where it is then reflected back. Upon its return the reflected light pulse may encounter a tooth or a gap; if a gap is present it passes through once more, but if a tooth is present it will be blocked. By varying the angular velocity of the wheel and observing those velocities that allow the reflected pulse to go through, Fizeau was able to measure the speed of light. His method was similar in principle to that used later by Michelson; the measurement is patently the determination of a round-trip speed. It is worth noting that the Michelson-Morley experiment, which is often cited as evidence for the constancy of the speed of light, also compares average round-trip speeds over paths that are perpendicular to one another (see figure 6). These are the kinds of experimental evidence that support the two-way light principle, but clearly, since they are methods of ascertaining and comparing round-trip speeds, they have no bearing whatever on the one-way light principle.

Einstein could not reasonably have offered the one-way light principle as a definitional convention unless he held that the equality or inequality of the speed of light in the two opposite directions is *not* a matter of fact. Such an attitude is hard to accept. If we believe that the round-trip speed is *not* a matter of convention, it is difficult to see how the one-way velocity can fail to be a matter of fact. We all realize, of course, that *every* scientific statement rests in some fashion upon a bunch of semantic conventions. Once the standard semantic conventions have been adopted, however, the average round-trip speed of light

517

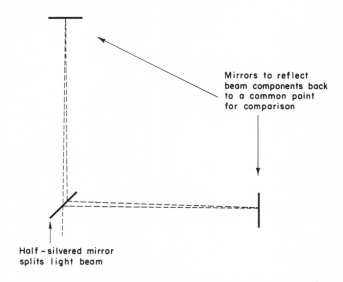

Mirrors to reflect
beam components back
to a common point
for comparison

Half - silvered mirror
splits light beam

Fig. 6. Michelson-Morley experiment (schematic diagram)

is (as we have just seen) a matter of empirical fact, amenable to experimental ascertainment; the one-way speed, in contrast, must still remain a matter of convention after the *standard* semantic conventions have been adopted if Einstein's *definition* of simultaneity is to make sense. Thus, we must consider with some care whether it is possible to measure the one-way velocity of light.

To measure the one-way velocity, we need, in principle, a path along which the light is to travel with a clock located at each end. One of these clocks measures the time of departure of a light signal, and the other measures its arrival at the other end of the path. Such measurements can have a significant bearing upon the determination of a one-way velocity only if they tell the same time—i.e., only if they are synchronized with each other. Otherwise, the difference between the readings of the two clocks cannot represent a genuine time interval used in the one-way trip. But since the synchronization of spatially separated clocks is precisely the question at issue, we must consider further how that can be accomplished. Two methods suggest themselves.

First, we might try synchronizing the two clocks by sending messages back and forth. A person situated at A could send a radio message to someone located at B saying that his clock at A now reads (say) 12:00.

518

The person at B who receives the message knows, however, that it took some time for the message to travel from A to B; hence, when he receives the message he knows that the clock at A no longer reads 12:00. As a matter of fact, radio waves travel at the same speed as light rays, so to know how long after 12:00 the message arrived would be tantamount to knowing the one-way speed of light.

This would pose no serious problem within Newtonian mechanics. If we wish to use two spatially separated clocks to measure the one-way speed of light, we should simply use much faster signals to synchronize them. It would be analogous to ascertaining the one-way speed of sound. We can tell how fast sound waves travel in one direction by noting, for example, how long after the lightning flash one hears the associated thunder (assuming, of course, that we know how far away the lightning bolt struck). The velocity of light is so much greater than the velocity of sound that it is harmless to regard the light signal as being transmitted instantaneously (i.e., at infinite velocity). Similarity, in ascertaining the one-way speed of light, the time required for transmission of a signal that travels fast enough could simply be ignored. The clocks could be synchronized with super-light signals and then used to measure the one-way speed of light.

Newtonian physics imposes no upper limit on the speed at which messages can be transmitted, for material particles can be accelerated to arbitrarily large velocities. You can simply write your message on a piece of paper, enclose it in a suitable container, and then impose sufficient force for enough time to accelerate it to the desired velocity. Whether, according to Newtonian theory, gravitational influences are propagated infinitely rapidly does not matter; it is sufficient that material particles can travel at arbitrarily large finite velocities. Classical physics thus contains, in principle, the resources to measure the one-way speed of light to any desired degree of accuracy.

The situation in the special theory of relativity is fundamentally different. The speed of light is a limiting velocity. Even if we are not yet in a position to assign a numerical value to the one-way speed of light, we can say that it constitutes a speed limit for the transmission of signals. No signal sent from A at the same time as a light signal will reach B earlier than the light signal (sent directly *in vacuo*) does. For this reason Reichenbach called light a "first signal". The speed of light is an upper limit for the propagation of any causal process, according to the special

519

theory, and only causal processes can be used to transmit messages (and to synchronize clocks).[17] We need not worry about "tachyons" at present, for, although they are the subject of interesting speculation, all attempts so far to discover them empirically have failed.[18] I shall return below to the question of the bearing it would have on special relativity if the existence of tachyons were established.

Reichenbach summarized the problem of synchronizing spatially distant clocks by means of signals in the following succinct terms:

> To determine the simultaneity of distant events we need to know a velocity, and to measure a velocity we require knowledge of the simultaneity of distant events. The occurrence of this circularity proves that simultaneity is not a matter of knowledge, but of a coordinative *definition,* since the logical circle shows that a knowledge of simultaneity is impossible in principle.[19]

Reichenbach is evidently attempting to reconstruct the rationale for Einstein's introduction of a stipulation regarding the one-way speed of light as a basis for a *definition* of simultaneity. We may begin by sending a light signal from A to B and, by reflection, back to A, and we can measure the time for the round trip on our clock at A. We may then announce the conventional decision that half of that time was taken by the light in going from A to B, and the remaining half was consumed in the return trip. The total round-trip time can be communicated to our colleague at B, and we can tell him of our *decision* to regard half of that as the time taken for the signal to go from A to B. The next time we send him a message telling him what the clock at A reads, he can, without difficulty, synchronize the clock at B with ours. This is known as "standard signal synchrony".

Reichenbach introduced a useful notation for discussing synchronization of clocks by means of light signals.[20] If t_1 is the time at which a light signal departs from A traveling toward B, and t_3 is the time at which that signal, reflected at B, returns to A, then the time at which the signal arrived at B can be represented as

$$t_2 = t_1 + \varepsilon (t_3 - t_1).$$

In order to secure the fact that the signal arrived at B after it was sent from A and before it returned to A, he imposed the condition

$$0 < \varepsilon < 1.$$

Einstein's definition of simultaneity results when

$$\varepsilon = 1/2.$$

Reichenbach insists that this choice of ε results in far simpler description of the physical world than could be achieved with other choices, but the choice is motivated by convenience. It is not demanded by the ''fact'' that light travels at the same speed in the two directions of a round-trip, for *there is no such fact*. Other choices of ε would result in equivalent descriptions of the same physical facts. If someone were to make an alternative choice, Reichenbach claims, there would be no experimental way of proving him wrong by virtue of his commitment to the consequence that light travels at different speeds in the two opposite directions. This is the ''cash value'' of the claim that the one-way light principle is a convention.

Fizeau's method of measuring the speed of light is, as mentioned above, applicable only to the round-trip speed, but another historically important method of ascertaining the speed of light seems to give the one-way velocity we have been looking for. This method was involved when Olaf Römer accounted for periodic discrepancies in the eclipsing of the moons of Jupiter on the basis of the finitude of the velocity of light. When Earth and Jupiter are in opposition, light from Jupiter has to

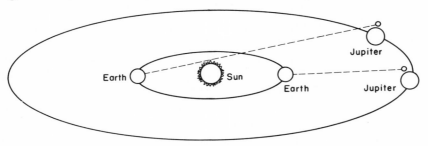

Because of the large size of Jupiter's orbit, Jupiter moves through only a small part of it during half of an Earth year.

Fig. 7.   Römer's method for ascertaining light velocity

521

travel a path that is longer (by the diameter of the orbit of Earth) to reach us than it does when the two planets are in conjunction (see figure 7). Knowing the diameter of Earth's orbit and using the apparent delay in the eclipsing of Jupiter's moon, Römer was able to make a rough determination of the speed of light on a *one-way* trip across the earth's orbit.

Römer's measurement of the one-way speed of light requires consideration of a second method (other than light signals) of attempting to synchronize clocks located at A and B. We begin with two clocks located at A and synchronized locally with one another. One of these clocks is then transported to B, where it is used subsequently to indicate the time. The most obvious difficulty with this method lies in the prediction by special relativity of the fact that transported clocks are affected in a way that depends upon the path and speed of transport. The relevant part of this consequence of the special theory can be checked without making any assumptions or stipulations about distant simultaneity. We can simply observe that the clock that was previously synchronized locally with the clock at A and then transported to B will, if returned to A, no longer be synchronized with the clock that remained at A. It would, consequently, be an entirely unwarranted assumption to suppose that the clock transported from A to B retains its synchrony with the clock at A.

It is easily seen that Römer's method for measuring the one-way velocity of light makes use, in effect, of a clock that is transported from one end of the one-way path to the other. In the half-year (approximately) between the approximate conjunction and the approximate opposition of Earth and Jupiter, the clock we use to establish the time of apparent eclipse of a moon of Jupiter travels with the Earth from one end to the other of a diameter of its orbit. Unless the clock that travels with the earth is assumed to remain in synchrony with two *hypothetical* clocks that remain fixed (relative to the frame of the solar system) at the two ends of that diameter, the apparent discrepancy in the times of eclipsing of Jupiter's moons would not constitute a measure of the one-way velocity of light across Earth's orbit.[21] Hence, to construe Römer's method as a way of measuring the one-way velocity of light requires assumptions about clock synchrony that we are not yet entitled to make.

522

The standard Lorentz transformation equations of special relativity imply that the retardation R to which a clock traveling a distance d at velocity v is subject is given by

$$R = \frac{d}{v}(1 - \sqrt{1 - \beta^2}).$$

Although the expression on the right has v in its denominator, it is quite easy to show (using l'Hospital's rule) that R goes to zero as v approaches zero. Thus, the retardation of a clock we wish to send any distance d can be made as small as we like by choosing a sufficiently slow mode of transport. It has been noted, for example, that the orbital velocity of the Earth, which is of the order of 20 mi/sec, is a very tiny fraction of the speed of light, and consequently, Römer's Earth-bound clock would suffer negligible retardation in its semiannual trip from one side of the orbit to the other.

There is, however, a basic logical problem that must somehow be handled. The term v in the foregoing equation designates a one-way velocity, but until we have clocks synchronized with one another at the two ends of the path, no meaning can be assigned to *any* one-way velocity. It is not just the one-way speed of light that is in question. P. W. Bridgman, one of the authors who recognized the potentiality of slowly transported clocks for the establishment of synchrony relations, dealt with this problem by using the "self-measured velocity" of the clock.[22] Let d be the distance between A and B as measured in the frame in which the clocks at A and B are at rest. If we send a clock C' from A to B, it will show a certain time t'_1 at its departure from A and another time t'_2 upon arrival at B. The self-measured velocity at which C' traveled from A to B is $d/(t'_2 - t'_1)$. This is a sort of bastard velocity, for it involves a spatial interval as measured in the rest frame of the clocks at A and B divided by a time interval as measured in the rest frame of C'. Nevertheless, as Bridgman shows, this self-measured "velocity" can be used to define convergence to the limiting transport velocity of zero. In other words, we can substitute the self-measured velocity for v in the retardation formula, and we can then allow convergence of v to zero as a criterion of convergence of R to zero. It can be shown that the "slow clock transport synchrony" that is established in this way coincides

523

exactly with Einstein's standard signal synchrony. Bridgman did not view this fact as altering the conventional status of distant simultaneity; rather, he merely regarded slow transport as providing an alternative operational *definition* of synchrony.

Brian Ellis and Peter Bowman, in a widely discussed paper, took the opposite point of view. They argued that the facts of slow clock transport render the concept of distant simultaneity non-conventional (except in a trivial sense).[23] In order to handle the problem of defining the transport velocity of the moving clock, they introduce an "intervening velocity" that is quite different from Bridgman's self-measured velocity, but it does the same job. The Ellis-Bowman slow clock transport synchrony coincides exactly with Bridgman's, and both coincide with Einstein's standard signal synchrony. Ellis and Bowman, in contrast to Bridgman, regard it as a physical *fact* that clocks synchronized by slow clock transport show the same time. They conclude that spatially separated clocks that have been synchronized by slow clock transport can be used to determine objectively whether light travels at equal speed on the two legs of a round trip. If the special theory of relativity is correct, then the speed of light is the same in the two directions, for that is the theory from which the coincidence of Einstein's standard signal synchrony (which stipulates the equality of the speeds in the two directions) with the Ellis-Bowman slow clock transport synchrony is deduced. They further maintain that Römer's method of ascertaining the one-way speed of light is free from any non-trivial conventional elements, and that consequently it provides an objective, factual determination of the one-way speed of light.

The Ellis-Bowman paper elicited a response consisting of papers by Grünbaum, van Fraassen, and me, in which we tried to show that the method of slow clock transport, while entirely suitable for the purpose of establishing relations of distant simultaneity, is just as infected with non-trivial conventions as is Einstein's standard signal synchrony.[24] There is, I believe, no need to rehearse these arguments here. In my contribution to the aforementioned groups of papers,[25] which has come to be known informally as the "Pittsburgh Panel," I enunciate a criterion for the non-triviality of conventions[26] and, by applying it to synchrony by slow clock transport, show that this method involves conventions that are as non-trivial as those involved in standard signal synchrony. I attempt to show, moreover, that this conventional element in clock transport synchrony does not depend upon the fact that synchrony can be destroyed by relative motion of clocks. Even if, contrary to fact, the clock that is transported from A to B and back to A were always in agreement with the clock that remained at A, clock transport synchrony would still involve the same kind of conventionality. Consequently, the fact that the retardation can be made arbitrarily small by transporting clocks slowly enough has no bearing upon the conventionality of distant simultaneity. The crucial basis for the conventionality of simultaneity in special relativity is not the time dilation phenomenon, but rather, the limiting character of the speed of light. This feature of special relativity constitutes one of its most fundamental departures from classical mechanics. If distant simultaneity is conventional, it is so because of a pervasive fact about the physical world: namely, that light is a first signal.

There have been many ingenious proposals for measurement of the one-way velocity of light. Reichenbach discusses several in his classical work, *The Philosophy of Space and Time,*[27] and others are still being advanced.[28] As one example, consider the familiar "light clock" used to show how the constancy of the round-trip speed of light implies time dilation (see figure 8). This device consists of two mirrors mounted a fixed distance apart, perpendicular to a support joining their midpoints, with a photon of light that is reflected back and forth between them. A unit of time can be defined as the interval required by the photon to make the round trip from one mirror to the other and back again. If we have

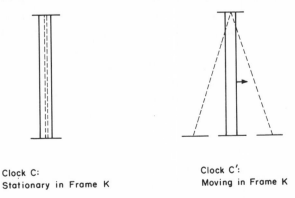

Clock C:
Stationary in Frame K

Clock C':
Moving in Frame K

Fig. 8.   Light clock

525

such a clock in our frame K, and compare it with another such clock located in a frame K' that is in motion with respect to K, its unit interval (as seen from our frame) must be longer than ours because the path of its photon is longer than the path in our clock, and both photons travel at the same average round-trip speed with respect to any inertial frame of reference. Hence, the time of the moving clock is dilated.

It has been suggested that we can establish the equality of the two one-way speeds by examining the path of the photon in the moving clock C' from the standpoint of our frame of reference. If angle 1 (the angle of incidence) is equal to angle 2 (the angle of reflection), then the photon is traveling at the same speed before and after the reflection from the upper mirror. This argument is inconclusive, however, for it contains a tacit synchrony assumption. In order for the equality of angles 1 and 2 to entail the equality of the speeds in the two directions, it is necessary that the support bar of clock C' be oriented perpendicular to the line of motion of C' in reference frame K. This in turn requires that points A and B in our frame coincide with points A' and B' in the moving frame *at the same time*. Thus, the use of this method to ascertain empirically the equality of the two one-way speeds is vitiated by the need to have already established the simultaneity of events occurring at A and B, the endpoints of the one-way path.

If the foregoing argument about the behavior of the light in the moving clock were adequate to establish the equality of the two one-way speeds, then a similar argument based upon the well-known phenomenon of aberration of starlight would be equally adequate. Suppose we have a star that lies in a line perpendicular to the plane of the earth's orbit (the ecliptic) at a given point in the earth's orbit, and assume that it is distant enough to render negligible any discrepancy from perpendicularity at any point of the orbit. Assume, that is, that the star would appear in a line normal to the plane of the earth's orbit by an observer at rest in the frame of the fixed stars at any point in that orbit. Such a star would, of course, have no measurable parallax. The phenomenon of aberration of starlight consists in the fact that to view such a star from a telescope moving with Earth relative to the fixed stars, the telescope must be deflected slightly from the perpendicular orientation in order to compensate for the finite time interval required by the light to traverse the length of the telescope (see figure 9). Knowing the length of the telescope, and knowing the speed of the earth relative to the fixed stars, we could

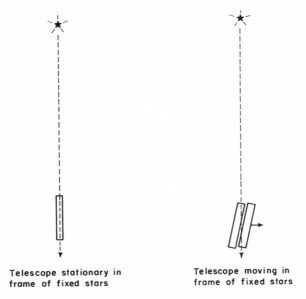

Telescope stationary in
frame of fixed stars

Telescope moving in
frame of fixed stars

Fig. 9.   Aberration of starlight

calculate the one-way speed of light as it passes through the telescope, if we could ascertain what orientation of the moving telescope is absolutely parallel to the path of the light ray in the rest frame of the fixed stars. But establishing parallelism or perpendicularity of a moving telescope (or light clock) begs the question of simultaneity of events located at opposite ends of the one-way path, and consequently fails to provide a basis for empirical determination of the one-way velocity of light. It is worth noting that the difficulty in using the light clock or the aberration of starlight as a means of ascertaining the one-way speed of light is identical to that involved in the use of such "unreal sequences" (discussed by Reichenbach) as the motion of the point of intersection of two rulers in motion with respect to one another.[29] It is also interesting to note that, according to Shankland, Einstein had been studying experiments on aberration of starlight very intensively at the time just before he composed the famous 1905 paper on special relativity.[30]

I do not wish to assert dogmatically that no method can be devised for genuinely circumventing the need for a non-trivial convention such as Einstein employed regarding the one-way speed of light, for that would amount to a claim of ability to foresee the results of all future experiments. At present, I do not know of any such method. It seems to me, therefore, that to the best of our present knowledge, the equality or inequality of the speed of light in the two opposite directions of a round trip must be regarded, not as a matter of fact, but rather, as a matter of convention. Our choice of a value of ε seems, therefore, to characterize the language we choose from our description of the world; it does not seem to characterize any non-linguistic feature of the physical world itself.

Given the difficulty of the problems associated with Einstein's definition of simultaneity, one might wonder why a great many textbook introductions to special relativity make no mention of the need for any such definition. Such presentations frequently mention the constancy of the speed of light, but they fail to make any reference to the distinction between the one-way light principle and the two-way light principle. Nothing is said about methods of measuring the one-way velocity of light. It is hard to believe that authors of such books want to maintain that Fizeau's experiment or the Michelson-Morley experiment establishes the constancy of the one-way speed of light when manifestly they do no such thing. There is, I believe, a different explanation.

If one is anxious to deal with the interesting parts of the special theory, it is tempting to get busy and derive the Lorentz transformation equations for frames that are in uniform motion with respect to one another. A frequent approach is to introduce, casually and without fanfare, what is known as the "reciprocity condition", namely, if frame K' is moving at velocity v with respect to frame K, then frame K is moving at velocity $-v$ with respect to frame K'. This assumption seems so natural that it goes by almost unnoticed; nevertheless, it turns out that the one-way light principle can be deduced from it.[31] Thus, in the context of special relativity, the reciprocity condition is tantamount to Einstein's stipulation concerning the one-way speed of light. Once the reciprocity condition is introduced, the Lorentz equations can be derived, but they are derived on the basis of the assumption that the speed of light is equal in the two legs of the round trip. This stipulation can be expressed in Reichenbach's notation by choosing 1/2 for the value of ε. Hence, the Lorentz equations in their standard form embody the convention, $\varepsilon = 1/2$. The convention has been smuggled in quite surreptitiously via the

528

reciprocity condition. The reciprocity condition is, consequently, a good deal more significant than it may appear at first blush.

The thesis that the value of ε may be freely chosen within wide limits is known as the *conventionality of simultaneity*. Once the normal choice $\varepsilon = \frac{1}{2}$ has been made, and we adopt Einstein's standard signal synchrony in all inertial frames, it is easy to show that spatially separated events that are simultaneous with respect to one inertial frame are not simultaneous with respect to any other inertial frame that is in motion relative to the first frame. This result is known as the *relativity of simultaneity*. The conventionality and relativity of simultaneity constitute the above-mentioned two stages in Einstein's revolutionary analysis of simultaneity.

The relativity of simultaneity rests upon the conventionality of simultaneity in the following sense. Two events e_1 and e_2 that are simultaneous in frame K on the basis of standard signal synchrony will not be simultaneous in K' (where K' is in motion relative to K) *according to standard signal synchrony*. By a judicious selection of a value other than $\frac{1}{2}$ for ε in frame K', however, it is possible to set up nonstandard synchrony relations that will render e_1 and e_2 simultaneous in K' as well as in K. The conventionality of simultaneity makes it possible to erase the relativity of simultaneity.[32] The relativity of simultaneity thus embodies the standard convention, $\varepsilon = \frac{1}{2}$.

Similar remarks apply to length contraction and time dilation. Since the length of a moving object is given by its simultaneity projection in the stationary frame, a change in the synchrony relations entails a change in the length contraction relations. Indeed, it can be shown that, by judicious selection of a nonstandard value of ε in frame K, the length of an object in motion in frame K can turn out to be equal to its own rest length. Furthermore, if we have a clock C' moving from A to B in frame K, which is locally synchronized with the clock at A when it departs, it will be retarded with respect to a clock at rest at B upon its arrival there, *if the clock at B is in standard signal synchrony* with the clock at A. Again, however, by judicious choice of a value of ε other than $\frac{1}{2}$, it is possible to synchronize the clocks at rest at A and B in K in such a way as to yield the result that C' is locally synchronous with the clock at rest at B when it arrives there. The length contraction and time dilation relations thus involve the standard choice of ε, and within limits, they can be erased by adoption of suitable nonstandard synchrony relations.

529

It might be tempting, at this point, to wonder whether the special theory of relativity has any distinctive content at all, or whether, rather, it involves nothing more than a definition of simultaneity which departs from that used in classical physics. The answer to this query is already available: as we saw at the outset, special relativity has certain testable consequences that differ from the consequences of classical physics (e.g., the observed retardation in Lord Halsbury's three clock setup; the muon decay phenomena; the length of the Stanford linear accelerator). Special relativity in its standard form has factual content, but it also has results that express conventional choices. To sort out these two kinds of elements can be a tricky job, but it is one that has to be done if we are to have a clear conception of the logical foundations of special relativity.

Both Einstein and Reichenbach maintain that the choice of ε is a matter of convention, and that one could, without contradicting any empirical fact, make a different choice. The price of a nonstandard choice is simply a great deal of added complexity in the statement of the theory, without any change of content. They then proceed to invoke the usual convention and set ε equal to ½. From there they go on to derive the Lorentz transformations. The equations they derive are, therefore, infused with this conventional element.

In a most interesting and constructive response to the thesis of Ellis and Bowman, John Winnie has taken a different tack.[33] Instead of choosing a particular value for ε and deriving the Lorentz equations in their customary form, he deliberately forgoes any choice and proceeds to derive the equations of the special theory with ε present as a free variable. His rationale is quite straightforward. If the value of ε is a matter of convention, then it ought to be possible to state the "factual core" of special relativity without introducing the convention at all. To carry out this task he derives what he calls the "ε-Lorentz transformations." The burden he thus assumes is twofold: first, he must show how these transformation equations can be derived from premises that do not contain hidden one-way velocity assumptions, and second, he must show that these equations yield the observational results for which the special theory must be held accountable.

Winnie adopts three postulates for special relativity, all of which are free from assumptions about one-way velocities.[34] They are:

1. *The round-trip light principle.* The average round-trip speed of any light-signal propagated (*in vacuo*) in a closed path is equal to a constant c in all inertial frames of reference.

This is simply our two-way light principle, generalized to hold for any closed path.

2. *Principle of equal passage times.* Let K and K′ be two inertial frames in relative motion, and let A and A′ be arbitrary points on the x-axes of K and K′ respectively. Let Δt be the time-interval in K of the passage of a rod at rest in K′ of rest length s past the point A in K, and let $\Delta t'$ be the time-interval in K′ of the passage of a rod at rest in K of rest length s past the point A′ in K′. Then $\Delta t = \Delta t'$.

This principle is *not* equivalent to the reciprocity condition, for it does not imply that the velocity of K with respect to K′ is equal to minus the velocity of K′ with respect to K. Nor is it tantamount to the condition of reciprocity of relative lengths, for it does not imply that the lengths of the two rods whose rest lengths are both s in their respective frames have equal lengths as measured in the systems with respect to which they are in motion.[35] The principle of equal passage times entails only that the ratio of moving length to relative velocity is equal for the two rods. It is clear that this principle is free from one-way velocity assumptions, for only one clock in each inertial frame is involved. This principle is thus independent of all considerations relating to the synchronization of spatially separated clocks in any inertial system.

For reasons of expository convenience Winnie adopts a linearity principle, which says that any motion that is uniform straight-line motion in one inertial frame must likewise be uniform straight-line motion in any other inertial frame. This principle says, in effect, that Newton's first law holds equally in all inertial frames. But, inasmuch as the linearity condition is not obviously free from one-way velocity assumptions, Winnie remarks that it could be replaced by

3. *Principle of proportional passage times.* The passage time in K (or K′) of a rod of rest-length L_0 is directly proportional to the rest length L_0.[36]

This principle, in conjunction with the other two, is sufficient to derive the linearity principle, and it is evidently free from one-way velocity assumptions. From these three principles Winnie derives his ε-Lorentz transformations.[37]

The main value in Winnie's approach lies in its ability to distinguish those consequences of the special theory of relativity that depend upon a particular choice of ε from those that are valid regardless of the choice. Thus, for example, we cannot uniquely assign a velocity to a muon traveling in a straight line relative to our frame of reference nor a time of travel without making a definite choice of ε. By applying the ε-Lorentz transformations, we can, however, uniquely determine the place at which the average muon will decay if we know the average lifetime of a muon at rest in our frame of reference.[38] Observable facts that occur every day in particle accelerators can be derived from the factual core of the special theory without invoking any conventional choice of ε; consequently, the observations of such results are irrelevant to the question of the one-way speed of light, for they are compatible with any decision we choose to make.

For another example, consider our earlier discussion of the clock paradox in terms of Lord Halsbury's three clock setup. In that context, we noted that the readings of adjacent clocks could be compared directly. Thus, we could ascertain the agreement of C with C′ when they met (event E_1) and the agreement of C′ with C″ when they met (event E_2), and we could determine empirically whether C″ was or was not retarded relative to C when they met (event E_3). The predicted retardation is an observable consequence that must occur, regardless of any conventions or stipulations, if special relativity is a true theory. This is a time dilation effect, and Winnie shows that its occurrence (and amount) is derivable from his ε-Lorentz transformations without introducing any one-way velocity assumptions.[39] This fact is another consequence of the "factual core" of special relativity.

When the standard Lorentz transformations (which embody the convention $\varepsilon = \frac{1}{2}$) are used, one can derive statements about clock readings that are not directly verifiable by observation. For example, referring back once more to our discussion of the clock paradox, we can ask what reading appears on clock C′ when (from the standpoint of frame K) C reads d/v. We can derive the answer $(d/v) \sqrt{1-\beta^2}$, but we cannot check the answer directly because the two clocks are not adjacent. If we had a clock at B that had previously been synchronized with C, we could compare its reading with that of C′, but the result obviously depends upon the manner in which the clock at B was synchronized with the clock at A. Winnie's ε-Lorentz transformations do not provide unique answers to questions of this sort; the answers turn out to be functions of ε. This illustrates the manner in which some questions have convention-laden answers, while others have convention-free answers that can be tested by direct observation.

A similar distinction can be made concerning length contraction phenomena. Using the standard Lorentz transformations, we can compute the length of a moving rod; but as we have already seen, this answer depends upon our choice of ε. A different choice of ε will yield a different length of the moving rod. This sort of contraction is, therefore, convention-laden. But there is no way to check such contraction phenomena empirically apart from performing a simultaneity projection of the moving rod, and that obviously involves commitment to some particular synchrony.

Einstein, however, proposed a "twin rod" experiment that gives rise to an observational result that is independent of any synchrony or one-way velocity assumptions. Suppose we have two rods of equal rest length moving in opposite directions in our frame of reference. Let the left-hand ends of the rods be A and A′, respectively, and the right-hand ends B and B′, respectively. As the rods pass one another, there will be a point A* in our frame at which the two left-hand ends coincide, and another point B* at which the right-hand ends coincide. We make no assumption about the one-way speeds at which the two rods are moving, and we make no commitment to whether the coincidence of the left-hand ends at A* is simultaneous with the coincidence of the right-hand ends at B*. We merely mark these two points in our frame when each coincidence occurs, and we measure the distance between them at our leisure. Using the ε-Lorentz transformations, it is possible to show that the distance A*B* $<$ AB $=$ A′B′ . This is an observable contraction effect whose occurrence follows from the factual core of the special theory; consequently, its occurrence does not have any bearing upon the question of the one-way speed of light.[40]

Although Winnie does not show exhaustively that all observational consequences of special relativity are derivable by means of the ε-Lorentz transformations of the convention-free formulation, he does deal with enough of the typical problems to establish a strong presumption that this is true. It seems to me that the burden of proof now falls upon anyone who wishes to maintain the factual character of the one-way speed of light to produce an observable consequence of the theory that does not follow from the ε-Lorentz transformations. And it is strongly to be recommended that any alleged experimental demonstration of the equality of the two one-way speeds be analyzed in terms of the ε-Lorentz transformations, in order to guard against the unwitting introduction of the convention $\varepsilon = \frac{1}{2}$ via the standard Lorentz transformations. For example, it follows from the ε-Lorentz transformations that standard signal synchrony must coincide with slow clock transport synchrony. From this it follows that Römer's method does not constitute an independent method for ascertaining the one-way speed of light within the special theory. It shows that, whatever value we assign to ε, slow clock transport synchrony must agree with standard signal synchrony. Römer's method does not constitute a measurement of the value of ε; instead, it constitutes a test of the factual content of special relativity. If an experimental determination of the one-way speed of light by Römer's method were to establish it to be other than c, this would not be an experimental proof that $\varepsilon \neq \frac{1}{2}$; it would, instead, be an experimental disproof of the special theory of relativity. For Römer's method involves adoption of slow clock transport synchrony, and the ε-Lorentz transformations entail that this must coincide with $\varepsilon = \frac{1}{2}$.[41]

According to Newtonian mechanics, as I mentioned above, material particles can be accelerated to arbitrarily high velocities, well beyond the speed of light. If Newtonian mechanics were true, such particles could, in principle, be used to synchronize clocks and to establish relations of absolute simultaneity. Suppose a light signal is sent from A toward B at time t_1 according to a clock C at rest at A; upon reaching B it is reflected back, the time of its return being t_3 according to the same clock C (see figure 10). If we assume that the speed of light is the same in both directions, we set the time of arrival of the light signal at B equal to $t_2 = t_1 + (\frac{1}{2})(t_3 - t_1)$. In the context of Newtonian mechanics, this is a physical hypothesis that can be verified to any desired degree of accuracy through the use of super-light signals. At a time t^*_1, which is later than t_1 but earlier than t_2, a super-light signal could be sent from A that would arrive at B at the same time as the light signal sent at t_1. In

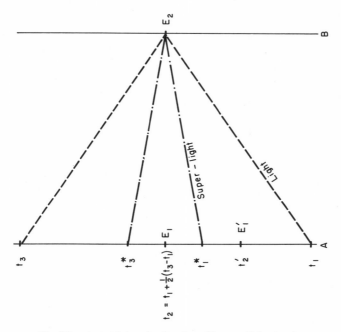

Fig. 10. Ascertaining simultaneity with super-light signals

addition, let the super-light particle be reflected back to A upon arrival at B—such "reflection" might consist simply in the triggering of an emission of another particle similar to the one that traveled from A to B. In any case, the super-light signal that returns from B to A arrives at A at t^*_3, a time (on clock C) that is between t_2 and t_3. In view of these facts, we could now say that the time of the arrival of the light signal and the super-light signal must lie somewhere between t^*_1 and t^*_3. This interval is smaller than that between t_1 and t_3; moreover, it can be made arbitrarily narrow by the use of arbitrarily fast super-light signals. We are not making any assumption about the relative one-way velocities of the fastest super-light signal that is employed; all we are relying upon is the assertion that the super-light signal arrived at B sometime between its emission from A and its return to A. If we were to say that our light signal arrived at B at a time t'_2, which is earlier than t^*_1, then we would be involved in asserting that our super-light signal (which arrived at B at the same time as the light signal) arrived at B before it was sent from A. This would contradict the causal features of our assumption that the

535

super-light signal was sent *from* A *to* B, and that it is capable of transmitting information from A to B. Similar difficulties would obviously arise if we said that the light signal arrived at B after $t*_3$.

Suppose that there are two events, E_1 and E_2, that occur at A and B respectively; let each of these events be the radioactive decay of an unstable nucleus. E_1 occurs at the clock time t_2, while E_2 occurs at the moment at which the light signal and the super-light signal reach B. Suppose, moreover, that no matter how fast a super-light signal we employ, it must be sent from A before E_1 occurs if it is to reach B by the time E_2 occurs. Analogously, suppose that no matter how fast a super-light signal we use, it must be sent from B before E_2 if it is to reach A by the time E_1 occurs. E_1 is thus singled out as the unique event at A that is simultaneous with E_2. If, for example, an event, E'_1, that occurred at A before E_1 were said to be simultaneous with E_2, we could refute the claim by pointing out that the event E'_1 is earlier than the sending of a super-light signal that arrives at B just as E_2 is occurring. Hence E'_1 must be earlier than E_2 and cannot be simultaneous with it. By parallel reasoning, we can say that no event later than E_1 can be regarded as simultaneous with E_2.

The foregoing situation has been described from the standpoint of a particular frame of reference, as must be done if we want to assign space-time coordinates to the events we are discussing. But there are certain features of the situation that hold true in any reference frame we might choose—these are the *invariants*.[42] They are objective physical facts that do not depend upon our way of describing them. Regardless of the reference frame we choose, event E'_1 is causally connected with the emission of a super-light signal from A at $t*_1$ (e.g., by a causally connected series of time readings on the clock C), and this super-light signal constitutes a causal connection between its own departure from A and its arrival at B at the moment E_2 is occurring. Hence, E'_1 and E_2 cannot be considered simultaneous *in any reference frame whatever,* for the causal connection between them is an invariant that must be acknowledged to hold regardless of reference frame. If these same events, E'_1 and E_2, were observed from a different reference frame that is in motion relative to our frame, and if these two events were pronounced simultaneous on account of the time of their occurrence as registered on the clocks of that reference frame, we would simply have to say that something had gone wrong with the clocks in that frame and that they are no longer synchronized.

It is equally an invariant fact that the events E_1 and E_2 cannot be causally connected with one another; this is true regardless of reference frame, and, consequently, these two events must be considered simultaneous in every reference frame. Our friends in the moving reference frame must reset their clocks accordingly. We see, therefore, that in the Newtonian world of arbitrarily fast signals and causal processes, both the conventionality and the relativity of simultaneity are untenable. Arbitrarily fast signals yield absolute simultaneity of the strongest sort; the presence of the relativity of simultaneity in special relativity hinges crucially upon the existence of a finite upper speed limit on the propagation of causal processes and signals.

Although it has been recognized from the beginning that special relativity precludes the acceleration of particles from velocities less than that of light to velocities greater than c, it has also been observed that its equations (e.g., for composition of velocities) do not rule out the possibility of particles that *always* travel at super-light velocities. Such particles could not, of course, be decelerated to speeds lower than c. In recent years there has been considerable speculation about the existence

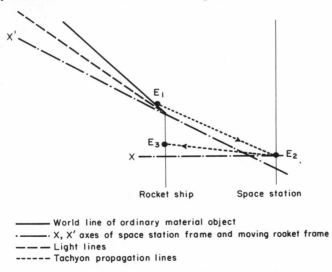

Fig. 11.   Causal anomaly with tachyon signal

of such particles, called "tachyons", and some serious effort to detect them experimentally.[43] Although all such efforts to date seem to have failed (as mentioned above), it is still interesting to consider what the implications for special relativity would be if tachyons were to be discovered. We can say, at the very least, I believe, that their existence would have severe repercussions for the concept of causality.

Suppose we have a rocket ship that stands for some time at rest relative to a space station, but at a substantial distance away from it, and then it rapidly accelerates to a speed that is a very large fraction of c (see figure 11). Immediately after achieving the desired velocity the rocket ship sends a tachyon message back to the space station at a velocity much greater than light. E_1 is the sending of the message and E_2 its arrival at the space station. Immediately upon receipt of this message the space station sends a reply by tachyons that reaches the rocket ship sometime before it began its acceleration. In fact, the message from the space station to the rocket ship might trigger a detonator that destroys the rocket ship. Let E_3 be this explosion. Thus, the rocket ship is destroyed before it sends the message that initiated the process leading to its destruction. It seems to me that we are faced with the following dilemma: either tachyons cannot be used to send messages (and if not, why not?) or they can be used to establish absolute synchrony and absolute simultaneity, thereby eliminating the relativity of simultaneity, which is so fundamental to the entire special theory.[44]

In special relativity, although the spatial distance between two distinct events as well as their temporal separation are relative to a frame of reference, the space-time interval between them is an invariant. This interval, whose square is equal to the square of the temporal separation minus the square of the spatial separation, is a constant. It depends in no way upon the choice of the reference frame in which the spatial and temporal separations are measured. If the square of the space-time interval between two events is positive, the interval is said to be timelike (see figure 12). The interval between E_0 and E_1 is timelike; this means that it is physically possible to have a material reference frame in which both E_0 and E_1 occur at the same place, though at different times. In any other material reference frame there is some time later than the occurrence of E_0 at which a light signal can be sent from the place at which E_0 occurred, and which will arrive at the place at which E_1 occurs at precisely the time of its occurrence. When the square of the interval

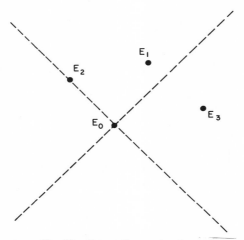

Fig. 12.   Types of space-time interval

between two events is zero, we say that they have a lightlike interval. The interval between E_0 and E_2 is lightlike. This means that it is physically possible for a light signal that originates in the event E_0 to travel directly to the space-time point at which E_2 occurs. If the square of the interval is negative the interval is said to be spacelike. E_0 and E_3 have a spacelike separation. This means that (according to standard signal synchrony or any other permissible synchrony definition) there is a reference frame in which E_0 and E_3 are simultaneous. It means, moreover, that if one wanted to send a light signal from the place at which E_0 occurs (with respect to any possible material reference frame) which will arrive at the space-time point at which E_3 occurs, it is necessary to send it at a time earlier than the occurrence of E_0.[45]

This is one natural way of interpreting the sign of the square of the interval. A zero interval squared means that two events are directly connectable via a first signal; a positive interval squared indicates that direct connectability via a first signal lies in the future; a negative interval squared indicates that connectability via a first signal lies in the past. If, however, light were not a first signal, there would hardly seem to be any physical justification for according light such a significance in our space-time schemas. If there were another first signal, faster than light but still propagated at a finite velocity, it could, of course, replace light in the foregoing discussions. But if, as in Newtonian theory or in

539

tachyon theory, there were no first signal with a finite velocity, it would seem that the physical foundation for the distinction between time-like and space-like separations would vanish. Thus, I should think, a convincing experimental demonstration of the existence of tachyons would demand severe revisions in the foundations of the special theory of relativity, perhaps even to the extent of eliminating the conventionality and relativity of simultaneity altogether.

Although Reichenbach was clearly aware of the distinction between conventionality of simultaneity with respect to a single inertial reference frame and the relativity of simultaneity between two or more reference frames in motion with respect to one another, he was not particularly scrupulous in his use of such terms as "convention", "definition", and "relativity". The equivocal use of such terms does create a danger of confusion, and the more careful usage of subsequent authors (e.g., Grünbaum, Winnie) helps to lessen the risk. It also suggests that we look carefully at the concepts of conventionality of simultaneity and relativity of simultaneity to ascertain the relations between them. We have seen that in Newtonian mechanics simultaneity is neither (non-trivially) conventional nor relative, and we have been led to suspect that the same situation would obtain if there were, in fact, such things as tachyons. We have also seen that the Einstein-Reichenbach interpretation of special relativity renders simultaneity both conventional and relative. We have seen, moreover, that the conventionality of simultaneity can be exploited to eliminate some types of relativity; indeed, it can easily be used to eliminate the relativity of simultaneity entirely. By a conventional decision we could designate some particular inertial system the privileged frame of reference, and by another conventional decision we could choose standard signal synchrony *in that frame* as determining the relations of simultaneity that obtain *nonrelatively* among all events in the entire universe. Such conventional choices would be contrary to the spirit of the special theory of relativity in some sense; in addition, they would prove inconvenient and inefficient. I cannot see, however, that they would lead to any refutable claims (if special relativity is true). On Reichenbach's construal of the special relativity, such a set of conventions would lead simply to another description to the physical world that is factually equivalent to the special theory in its standard formulation.

These considerations lead us naturally to wonder about the fourth possibility, namely, a simultaneity relation that is non-conventional but still relative. This is a possibility that, I believe, should be considered seriously. It is the result to which one is led by accepting the Ellis-Bowman thesis of non-conventionality of simultaneity by slow clock transport. Quite clearly, the simultaneity relation established by slow clock transport is relative to a particular inertial reference frame, for slow clock transport synchrony coincides with standard signal synchrony. It is well-known that standard signal synchrony leads to relativity of simultaneity; two events that are simultaneous with respect to one frame of reference will be non-simultaneous in a different reference frame. The fact that simultaneity becomes relative to a reference frame does not, however, entail that the relation is non-trivially conventional. It means, merely, that a relation that had been considered binary in Newtonian mechanics is ternary in special relativity. Ternary relations are just as capable of being objective as are binary relations. We have maintained, for instance, that one event can stand in the relation of *earlier than* to another event objectively and invariantly when a suitable sort of causal relation connects them. Similarly, under easily specifiable causal relations, we are entitled to say objectively and invariantly that one event is *between* two others in time. No non-trivial conventions need to be invoked for either type of statement.

Similarly, one might argue, it is possible for the statement "Event E_1 is simultaneous with event E_2 with respect to reference frame K" to represent an objective state of affairs that involves no use of non-trivial conventions. The fact that the statement "Event E_1 is not simultaneous with event E_2 with respect to reference frame K'" may be true is obviously irrelevant to the question of a non-trivial conventional component. One could, for instance, argue that simultaneity relative to a given frame is an objective fact of nature by maintaining that it is an objective fact that the one-way speed of light is c in every reference frame. To maintain that the one-way speed of light is a constant in every reference frame (an invariant of the theory), one needs to provide an objective method for ascertaining the one-way speed (without invoking non-trivial conventions such as are contained in Einstein's standard signal synchrony). Ellis and Bowman believe they have provided such a method based on slow clock transport synchrony, and that consequently, Römer's method of ascertaining the one-way speed of light does exactly the desired job. Thus, they claim, the one-way speed of light is an invariant fact of nature, and this implies that the ternary relation of simultaneity relative to an inertial frame also represents an invariant factual relation.

I do not believe that Ellis and Bowman provided a satisfactory argument to establish the relativity and non-conventionality of the relation of simultaneity. The most that has been shown, and perhaps the most that can be shown, is that it is an objective and invariant fact that standard signal synchrony coincides with slow clock transport synchrony (and that both of these coincide with my clock transport synchrony, which does not require slow transport).[46] The coincidence among these various types of synchrony entails neither the conventionality nor the non-conventionality of simultaneity. Although I still believe that the relation of distant simultaneity in a single frame of reference is conventional, I am less sure than I once was of the impossibility of rendering it non-conventional without at the same time undermining its relativity to a given reference frame. Perhaps someone will provide an objective method for ascertaining the one-way speed of light.

Although I am convinced that the thesis of conventionality of simultaneity is correct, my main purpose has not been to argue for that view. Rather, I have tried to exhibit the importance of the problem, and to investigate the extent of its ramifications. I will be satisfied if I have made a convincing case for the indispensability of considering the question if we hope for a thorough understanding of the logical foundations of the special theory of relativity. Almost everyone would agree that all physical theories have elements of factual content and elements of conventional decision. The basic philosophical task is, I believe, to try to discover which aspects of a physical theory describe the physical world, and which merely reflect the notations we have chosen for the purposes of that description. A further task, which is also of primary philosophical importance, is to try to ascertain the grounds and status of the various conventions we discover. Some conventions turn out to be strikingly non-trivial; the conventional choice of a value for ε (if it is, indeed, a convention) is perhaps the most outstanding example.[47]

1. Then of Washington University, but more recently—ironically—with the Caterpillar Tractor Company of Peoria, Illinois.

2. I would prefer to call this the "twin paradox", reserving the term "clock paradox" for the problem to be discussed below under that name.

3. J. C. Hafele and Richard E. Keating, "Around-the-World Atomic Clocks: Predicted Relativistic Time Gains and Observed Relativistic Time Gains," *Science* 177 (1972): 166–70.

4. See Edwin F. Taylor and John Archibald Wheeler, *Spacetime Physics* (San Francisco: W. H. Freeman & Co., 1963), pp. 16–17.

5. The cesium beam clock is an "atomic clock," but it is nevertheless a macroscopic object. The fact that it has atomic parts does not distinguish it from other macroscopic objects.

6. W. P. Montague, "The Einstein Theory and a Possible Alternative," *Philosophical Review* 23 (1924): 156ff.; A. O. Lovejoy, "The Paradox of the Time-Retarding Journey" and "The Time-Retarding Journey: A Reply," *Philosophical Review* 40 (1931): 48, 152, 549.

7. Of course, both of them might experience accelerations, but this case only introduces further complications without adding anything of interest to the problem.

8. At the time of writing: yesterday, Art Pollard was killed at the Indianapolis Speedway when his car hit a concrete wall.

9. Historical note: John Cameron Swayze is a television announcer who narrated many commercials in which Timex watches were subjected to severe accelerations (e.g., being attached to the ski tip of an Olympic ski jumper during a jump) without impairment of function. At the conclusion of each such demonstration Swayze would triumphantly proclaim of Timex watches that they can "take a licking and keep on ticking." Some years earlier, when home permanents were being introduced to the public via television, the manufacturers of Toni home permanents had a series of commercials in which one twin had a permanent by a professional hairdresser while the other had a Toni home permanent. The motto, "Which twin has the Toni," was designed to suggest indistinguishability.

10. Adolf Grünbaum, "The Clock Paradox in the Special Theory of Relativity," *Philosophy of Science* 12 (1954): pp. 249–53.

11. By H. Bondi, "The Space Traveller's Youth," *Discovery,* Dec. 1957, pp. 505–10. This formulation is also employed by Grünbaum, "The Clock Paradox."

12. These clock readings can be computed as follows: In the interval t'_0 to t'_3 clock C″ registers an elapsed time of $(2d/v)\sqrt{1-\beta^2}$. Compared with C″ the time of C is dilated by a factor $\sqrt{1-\beta^2}$. Hence

$$t_3 - t^* = \frac{2d}{v}(1 - \beta^2)$$

$$t_3 = \frac{2d}{v}$$

$$t^* = \frac{2d}{v}\beta^2$$

Furthermore, in the interval t_1 to t_3 C registers an elapsed time of $2d/v$. Since the time of C is dilated by the factor $\sqrt{1-\beta^2}$ with respect to K″ the interval

$$t''_3 - t''_1 = \frac{2d}{v} \cdot \frac{1}{\sqrt{1-\beta^2}}$$

$$t''_3 = \frac{2d}{v}\sqrt{1-\beta^2}$$

$$t''_1 = \frac{2d}{v}\sqrt{1-\beta^2} - \frac{2d}{v} \cdot \sqrt{\frac{1}{1-\beta^2}}$$

$$= \frac{2d}{v}\frac{(1-\beta^2)}{\sqrt{1-\beta^2}} - \frac{2d}{v}\frac{1}{\sqrt{1-\beta^2}}$$

$$= -\frac{2d}{v} \beta^2 \; \frac{1}{\sqrt{1 - \beta^2}}$$

13. This calculation is carried out in Grünbaum, "The Clock Paradox," but the crucial calculation, from the frame of C″, is surprisingly omitted. Hence, the analysis just presented is merely the completion of the work done earlier by Grünbaum.

14. In P. A. Schilpp, ed., *Albert Einstein: Philosopher-Scientist* (Evanston, Ill.: Library of Living Philosophers, 1949).

15. "On the Electrodynamics of Moving Bodies," in Einstein et al., *The Principle of Relativity,* trans. W. Perrett and G. B. Jeffrey (New York: Dover Publications, n.d.), p. 40.

16. Ibid.

17. Certain pseudo-causal processes can travel faster than light, but they are useless for transmission of information. See Hans Reichenbach, *The Philosophy of Space and Time* (New York: Dover Publications, 1957) §23, "Unreal Sequences."

18. Michael N. Kreisler, "Are There Faster-than-Light Particles?" *American Scientist* 61 (1973): 201–8.

19. Reichenbach, *The Philosophy of Space and Time,* pp. 126f. Reichenbach talks about simultaneity rather than synchrony in this passage, but this is an inconsequential matter of terminology, for clocks are synchronized if they show the same readings simultaneously.

20. Ibid., p. 127.

21. Hans Reichenbach, "Planetenuhr und Einsteinsche Gleichzeitigkeit," *Zeitschrift für Physik* 33 (1924): 628–34, provides a detailed discussion of Römer's measurement.

22. P. W. Bridgman, *A Sophisticate's Primer of Relativity* (Middletown, Conn.: Wesleyan University Press, 1962), pp. 64–67.

23. Brian Ellis and Peter Bowman, "Conventionality in Distant Simultaneity," *Philosophy of Science* 34 (1967): 116–36.

24. Adolf Grünbaum et al., "A Panel Discussion of Slow Clock Transport in the Special and General Theories of Relativity," *Philosophy of Science* 36 (1969): 1–81. The contribution by Janis deals with general relativity.

25. "The Conventionality of Simultaneity," ibid.

26. Ibid., p. 61.

27. Reichenbach, *The Philosophy of Space and Time,* §20, "Attempts to Determine Absolute Simultaneity." The original German edition, *Philosophie der Raum-Zeit-Lehre,* was published in 1927.

28. One such proposal was made by H. Weinberger and M. Mossel "Theory for a Unidirectional Interferometric Test of Special Relativity," *American Journal of Physics* 39 (1971), but it was shown invalid by G. E. Stedman, "A Unidirectional Test of Special Relativity?", *American Journal of Physics* 40 (1972).

29. See Reichenbach, *The Philosophy of Space and Time,* p. 147f., for discussion of this example.

30. R. S. Shankland, "Conversations with Albert Einstein," *American Journal of Physics* 31 (1963).

31. This is shown by Ellis and Bowman, "Conventionality in Distant Simultaneity," pp. 123f., as well as by John Winnie in the paper I am about to discuss.

32. See Adolf Grünbaum, *Philosophical Problems of Space and Time* (New York: Alfred A. Knopf, 1963), pp. 359–68.

33. "Special Relativity without One-Way Velocity Assumptions," *Philosophy of Science* 37 (1970): 81–99, 223–38.

34. Ibid., pp. 229–31.

35. This principle, like the reciprocity condition, is sufficient for the derivation of $\varepsilon = 1/2$. See ibid., p. 230.

36. Ibid., p. 231 n.

37. Ibid., pp. 231–37.

38. Ibid., pp. 91–93.

39. Ibid., pp. 97–98.

40. Ibid. See also John Winnie, "The Twin-Rod Thought Experiment," *American Journal of Physics* 40 (1972): 1091–94.

41. Winnie, "Special Relativity without One-Way Velocity Assumptions," pp. 223–38.

42. See Max Born, "Physical Reality," *Philosophical Quarterly* 3 (1953): 139–49, for an excellent discussion of the philosophical significance of invariants in physical theories. Reprinted in Max Born, *Physcis in My Generation* (London and New York: Pergamon Press, 1956).

43. Gerald Feinberg, "Possibility of Faster-than-Light Particles," *Physical Review* 159 (1967): 1089.

44. This example was provided by Professor Roger Newton, Department of Physics, Indiana University, in a public lecture.

45. These concepts are explained in a clear and elementary manner in Taylor and Wheeler, *Spacetime Physics*.

46. See Salmon, "The Conventionality of Simultaneity," sec. 1.

47. See ibid. for discussion of the nature of this nontriviality.

David Furley is professor of classics at Princeton University. He is the editor of *Pronesis* and translator of *Aristotle's "On the Cosmos"* and author of *Two Studies in the Greek Atomists*.

Edward Grant is professor of history and philosophy of science at Indiana University. He is the author of *Physical Science in the Middle Ages* and editor and translator of *Nicole Oresme and the Kinematics of Circular Motion*.

Thomas Hankins is associate professor of history at the University of Washington and the author of *Introduction to d'Alembert's "Traité de Dynamique"* and *Jean d'Alembert, Science, and the Enlightenment*.

Erwin Hiebert is professor of the history of science at Harvard. Among his published works are *The Impact of Atomic Energy, Historical Roots of the Principle of Conservation of Energy,* and *The Conception of Thermodynamics in the Scientific Thought of Mach and Planck*.

Arnold Koslow is professor of philosophy, The Graduate Center, City University of New York. He is editor of *The Changeless Order* and author of articles on Newton, Leibniz, and other scientists and philosophers.

Norman Kretzmann is professor and chairman of the Department of Philosophy at Cornell University. Among his published works are *Elements of Formal Logic; William Ockham: Predestination, God's Foreknowledge, and Future Contingents;* and *William of Sherwood's Treatise on Syncategorematic Words*.

Laurens Laudan is professor of history and the philosophy of science at the University of Pittsburgh. He is the general editor of the Cass Library of Science Classics and was coeditor of *Studies in History and Philosophy of Science*.

James Edward McGuire is professor and director of graduate studies in the Department of History and Philosophy of Science at the University of Pittsburgh. He is consulting editor for *Studies in History and Philosophy of Science* and author of many articles and reviews for scholarly journals and anthologies.

Peter Machamer is associate professor of philosophy at the Ohio State University. He has written articles and presented many papers on a wide variety of topics in the history and philosophy of science.

G. E. L. Owen is Laurence Professor of Ancient Philosophy at Cambridge University. He is the editor, with I. Düring, of *Aristotle and Plato in the Mid-Fourth Century* and the author of numerous articles in journals and collections of essays.

Wesley Salmon is professor of philosophy at the University of Arizona. He is the author of *Logic, Foundations of Scientific Inference, Zeno's Paradoxes,* and *Statistical Explanation and Statistical Relevance.*

Kenneth Schaffner is professor and chairman of the Department of History and Philosophy of Science at the University of Pittsburgh. He is the author of *Nineteenth-Century Aether Theories* and numerous articles in scholarly journals, anthologies, and the Encyclopedia Americana.

Robert Turnbull is professor and chairman of the Department of Philosophy at the Ohio State University. He is the author of many articles on philosophy and metaphysics, and on ancient and medieval topics.

Margaret Wilson is associate professor of philosophy at Princeton University. She is the editor of *The Essential Descartes,* coeditor of *Philosophy: An Introduction,* and the author of several articles in scholarly journals and books.

SUBJECT INDEX

INDEX OF NAMES